Recent Studies of Rodent Ultrasonic Vocalizations and Their Use in Experimental Models

Recent Studies of Rodent Ultrasonic Vocalizations and Their Use in Experimental Models

Editors

Stefan M. Brudzynski
Jeffrey Burgdorf

MDPI • Basel • Beijing • Wuhan • Barcelona • Belgrade • Manchester • Tokyo • Cluj • Tianjin

Editors

Stefan M. Brudzynski
Psychology
Brock University
St. Catharines
Canada

Jeffrey Burgdorf
Falk Center for Molecular
Therapeutics
Northwestern University
Evanston IL
United States

Editorial Office
MDPI
St. Alban-Anlage 66
4052 Basel, Switzerland

This is a reprint of articles from the Special Issue published online in the open access journal *Brain Sciences* (ISSN 2076-3425) (available at: www.mdpi.com/journal/brainsci/special_issues/Ultrasonic_Vocalizations).

For citation purposes, cite each article independently as indicated on the article page online and as indicated below:

LastName, A.A.; LastName, B.B.; LastName, C.C. Article Title. *Journal Name* **Year**, *Volume Number*, Page Range.

ISBN 978-3-0365-3285-1 (Hbk)
ISBN 978-3-0365-3284-4 (PDF)

Contents

About the Editors

Stefan M. Brudzynski

Stefan M. Brudzynski, Ph.D., D.Sc. is a neurophysiologist and neuroscientist, presently Professor Emeritus in the Department of Psychology at Brock University, Canada. Professor Brudzynski, a former Director of the Centre of Neuroscience and Professor of Psychology and Biology, is the recipient of the prestigious Outstanding Achievement Award from the International Behavioral Neuroscience Society for his contribution to the field of Behavioral Neuroscience. He is the editor of two handbooks on mammalian vocalization (2010) and ultrasonic vocalization (2018) published by Elsevier. His main research interest is in animal behavior; particularly, the brain control of ultrasonic vocalization and emotional communication in rodents.

Jeffrey Burgdorf

Jeffrey S. Burgdorf, Ph.D. is a neuroscientist, presently the Senior Director of research at Gate Neurosciences Inc., and is an adjunct Research Associate Professor at Northwestern University, U.S.A. He studies the brain mechanisms that underlie positive and negative emotions using rodent ultrasonic vocalizations and has been able to translate these findings into the development of novel therapeutics that are currently in clinical trials for psychiatric disorders.

Preface to "Recent Studies of Rodent Ultrasonic Vocalizations and Their Use in Experimental Models"

This book contains publications included in the Special Issue of *Brain Sciences* entitled "Recent Studies of Rodent Ultrasonic Vocalizations and Their Use in Experimental Models", which appeared in print in 2021. The aim of this Special Issue was to review the most recent studies of rat ultrasonic vocalizations and to demonstrate their growing importance in many rodent models of neuropsychiatric and neurodevelopmental disorders and diseases. The Special Issue has not been planned as an exhausting review of publications on this topic, but rather as a presentation of selected studies from a vast array of research topics, in which rat vocalizations offer a reliable measure of animals'emotional arousal and changes in their physiological and pathological conditions. Thus, the goal of this book is to present exemplary studies from a diverse line of research, which will further demonstrate the importance of vocalizations as key behavioral measures and inspire researchers to explore other areas of behavioral neuroscience and neuropsychiatric modelling studies.

Emissions of ultrasonic vocalizations by rodents serve as innate intraspecies communication and a physiological regulator of behavior. Ultrasonic calls, however, may also be useful for researchers and play critical role in behavioral studies as indicator of animal emotional arousal, emotional valence, motivation, behavioral strategies and choices, as well as indicator of pathological changes in the animal brain, particularly in the limbic system, which influences vocal expression.

The book consists of different types of publications, as extensive review articles, short reviews, experimental papers, and methodological presentations that are based on exhaustive literature. The contemporary field of studies of rat ultrasonic vocalizations consists of several hundreds of published papers and is steadily growing. We believe that this book will provide a useful update of the newest results, trends, and lines of research in this field and will assist in planning further studies.

The editors would like to express thanks to the MDPI Publisher, and particularly to Ms. Seal Li, Section Managing Editor, for providing extensive help, guidance, and encouragement in preparing this publication.

Stefan M. Brudzynski, Jeffrey Burgdorf
Editors

brain sciences

MDPI

Editorial

Recent Studies of Rat Ultrasonic Vocalizations—Editorial

Stefan M. Brudzynski [1,*] and Jeffrey S. Burgdorf [2]

1 Department of Psychology, Brock University, St. Catharines, ON L2S 3A1, Canada
2 Falk Center for Molecular Therapeutics, McCormick School of Engineering and Applied Sciences, Northwestern University, Evanston, IL 60201, USA; j-burgdorf@northwestern.edu
* Correspondence: sbrudzynski@brocku.ca

check for updates

Citation: Brudzynski, S.M.; Burgdorf, J.S. Recent Studies of Rat Ultrasonic Vocalizations—Editorial. *Brain Sci.* **2021**, *11*, 1390. https://doi.org/10.3390/brainsci11111390

Received: 12 October 2021
Accepted: 16 October 2021
Published: 23 October 2021

Since the realization that human emotional experiences and behavior evolved from mammalian ancestors and are evolutionary continuations of animal emotional behavior [1], there has been no clear subdiscipline focusing on the phenomenon of emotion in animals. Although many common limbic structures were identified that were responsible for control of emotional processes in animals and humans [2,3], progress was impeded by a widespread criticism that there is no linguistic means in animals to communicate their emotional states, and thus, researchers cannot know animals' feelings. However, after extensive studies in the field of behavioral neuroscience for more than 50 years using numerous techniques, such as permanent or reversible brain lesions, electrical or chemical brain stimulation, and behavioral techniques (e.g., place preference, self-stimulation, switch-off behavior, approach and avoidance tests, etc.), the discipline of affective neuroscience was justified and finally defined in 1990s [4–6].

The cornerstone of affective neuroscience studies relates to brain mechanisms of emission of animal vocalizations because they convey animal emotional arousal. Animal calls are honest advertisement of their affective and motivated states and were used as measures of these states. The most studies involved rodent ultrasonic vocalizations because they express affective arousal better than many other experimental animals and are convenient for pharmacological studies. This was demonstrated in the first report of pharmacological induction of the emission of rat 22 kHz ultrasonic vocalizations by direct intracerebral injection of carbachol, a drug mimicking the action of acetylcholine [7]. The field of studies of ultrasonic vocalizations started rapidly growing and led to the identification and validation of two types of rat ultrasonic calls that signal two different emotional states with opposite valences. Emission of 22 kHz calls signals aversive arousal and a negative state, and emission of 50 kHz calls signals appetitive arousal and a positive state [8].

Pharmacological and behavioral studies of vocalizations enabled researchers to identify the brain systems responsible for aversive/punishing and appetitive/rewarding states. The valence-specific calls provide an invaluable insight into the rodent emotional processes and expression of their basic internal states. Although the emission of animal vocalizations presents a phylogenetically older and simpler communication system than human speech, it reliably serves animals as an overt announcement of their emotional arousal, and its valence is highly adaptive. Vocal expression of emotional arousal in animals is a direct homolog to expression of emotional states by human vocalizations such as crying or laughing [9,10].

Interest in emission of rodent ultrasonic vocalizations and their biological functions has markedly increased in recent years, particularly their use in animal models of neuropsychiatric diseases and disorders. Alteration in emission of ultrasonic vocalizations appeared to be a sensitive index of many pathologies. The present Special Issue offers a collection of recent studies on rats using recording and analysis of ultrasonic vocalizations as a guide to animals' emotional arousal and emotional states. This is particularly useful in modelling numerous human neurodevelopmental and neuropsychiatric disorders and diseases.

The publications in the Special Issue are not representing predetermined topics but rather serve as examples of many approaches demonstrating different studies of ultrasonic vocalizations and different ways of modelling psychopathological states from infancy to adulthood.

The opening article of the Special Issue thoroughly reviews the socio-physiological functions of ultrasonic communication in rats, as well as the brain mechanisms of emotional arousal, and discusses how emission of these calls should be interpreted. Examples of these interpretations are presented in the following articles.

Studies of the emission of distress vocalizations in infant rats caused by maternal separation demonstrated their sensitivity to levels of vasopressin having anxiogenic effect. Application of angiotensin antagonists decreased emission of pup distress ultrasonic calls (D. Zelena laboratory). Pharmacological activation of the maternal immune system also had an effect on the emission of ultrasonic vocalizations of the offspring and impaired socio-communicative functioning of adolescent rats (A. Nikiforuk laboratory). Consumption of alcohol by pregnant rats has been known to have detrimental effects on pups' development, which was reflected in their ultrasonic distress calls. However, lactational exposure of a mother to alcohol appeared to have more devastating effect on pups and their distress calls than gestational exposure to alcohol (M.A. Shahrier laboratory).

Emission of ultrasonic vocalizations is coordinated with respiration rate, and call emission and respiration rate collectively affect brain oscillatory activities (A-M. Mouly laboratory). Sex differences in emission of ultrasonic vocalizations were also reviewed and showed differences not only in structures of the rat larynx but also in the acoustic parameters of the vocalizations between sexes (M.R. Ciucci laboratory).

In a couple of articles, emissions of adult ultrasonic vocalizations were studied in stressful situations as observation fear learning, which is associated with social transmission of the demonstrator's emotional state and induction of an empathy-like or anxiety state in the observer rat. Although fear learning was more pronounced in rats with higher trait anxiety, it appeared in this study that communication by ultrasonic vocalizations was not critical for observational fear learning (M. Fendt laboratory). However, if a rat (observer) was witnessing other rats receiving foot shock (demonstrator), the emission of ultrasonic calls was dependent on an earlier experience of these rats with the foot shock and on the presence of a warning sound before each foot shock. Particularly the warned rats, in addition to typical, long 22 kHz vocalizations, emitted short 22 kHz calls (less than 100 ms in duration), and in warned pairs with a naïve observer, 22 kHz calls were intermixed with the emission of 50 kHz vocalizations, which suggested vocal social buffering (A. Hamed laboratory). Earlier experience with foot shocks also had an influence on the subsequent responses of rats to playback of 50 kHz or 22 kHz vocalizations. Initial foot shock experience caused that rats vocalized more often with longer duration calls and a higher sound frequency than the control animals; these rats also showed a lower heart rate, higher locomotor activity in response to playback of 50 kHz vocalizations, and decreased their activity following playback of 22 kHz calls (R. Filipkowski laboratory).

Ultrasonic vocalization responses in tickling tests (heterospecific play of juvenile rats with human hand) showed that rats could be divided into two groups having high or low positive affectivity. It was shown that dopamine in the nucleus accumbens regulated responses to stress in different way in these groups. Stress increased the release of dopamine in rats with high positive affectivity but decreased it in the low positive affectivity rats (J. Harro laboratory). In another study, rats were subjected to a tickling procedure and then tested for novel object recognition and for the alternation behavior in a Y-maze. Rats emitted 50 kHz vocalizations but not 22 kHz calls during these tests; however, responses to novel objects were impaired if the animals consistently emitted 22 kHz in the initial tickling tests (N. Simola laboratory).

Emission of ultrasonic vocalizations recorded in rats awaiting play partner or food reward showed that emission of trill type calls of 50 kHz was increased by play reward but not by food reward (S.M. Pellis and D.R. Euston laboratories). In another experiment, using chemogenetic reversible inhibition of the nucleus accumbens neurons suppressed the emission of trill 50 kHz calls without changes in emission of the flat (unmodulated) 50 kHz vocalizations or other vocalizations (S.V. Mahler laboratory).

Publications contained within this Special Issue are also further demonstrating the use of the emission of ultrasonic vocalizations in modelling many neuropsychiatric dysfunctions and diseases, such as Parkinson's disease, affective disorders, autism, addiction, developmental abnormalities, and many other pathologies. This is demonstrated in a 6-hydroxydopmine model of Parkinson's disease (M.R. Ciucci laboratory); autistic spectrum disorder, modelled by viral-like activation of the maternal immune system during pregnancy, which may affect the neurodevelopment of offspring (A. Nikiforuk laboratory); in models of posttraumatic stress disorder (R. Filipkowski laboratory); and in studies of the mechanisms of drug addiction, particularly with self-administration of the powerful opioid drug fentanyl (M.O. West laboratory). Other approaches may target many neuropsychiatric disorders together, as studied using the cross-disorder risk gene CACNA1C, which is strongly associated with multiple neuropsychiatric dysfunctions. Using Cacna1c haploinsufficiency in rats, robust deficits in socio-affective communication through 22 kHz and 50 kHz vocalizations and associated alterations in social behavior were found (M. Wöhr laboratory).

Analysis of recorded vocalizations is time consuming and subjected to numerous errors given the researcher's experience. A web-based automated ultrasonic vocalization scoring tool, called Acoustilytix, is described. This system implements a machine learning methodology in the detection of ultrasonic calls and their classification, which is not related to the environment or the scorer's experience. It uses automated learning principles without the need for an expert to be present (Ch.L. Duvauchelle laboratory).

This Special Issue is a useful update on the current research directions using recordings of ultrasonic vocalizations and on the different methodological approaches.

Author Contributions: This is an original article jointly written by both co-authors. All authors have read and agreed to the published version of the manuscript.

Funding: There was no external funding support for writing this article.

Conflicts of Interest: The authors declare no conflict of interest.

References

1. Darwin, C. *The Expression of the Emotions in Man and Animals*, 3rd ed.; with commentaries by Ekman, P.; Oxford University Press: New York, NY, USA, 1872/1998.
2. Papez, J.W. A proposed mechanism of emotion. *J. Neuropsychiatry Clin. Neurosci.* **1937**, *7*, 103–112. [CrossRef]
3. Maclean, P.D. Psychiatric implications of physiological studies on frontotemporal portion of limbic system (visceral brain). *Electroencephalogr. Clin. Neurophysiol.* **1952**, *4*, 407–418. [CrossRef]
4. Panksepp, J. Affective neuroscience: A conceptual framework for the neurological study of emotions. In *International Reviews of Emotion Research*; Strongman, K., Ed.; Wiley: Chichester, UK, 1991; pp. 59–99.
5. Panksepp, J. A critical role for "affective neuroscience" in resolving what is basic about basic emotions. *Psychol. Rev.* **1992**, *99*, 554–560. [CrossRef] [PubMed]
6. Davis, K.L.; Montag, C. Selected principles of Pankseppian affective neuroscience. *Front. Neurosci.* **2019**, *12*, 1025. [CrossRef] [PubMed]
7. Brudzynski, S.M.; Bihari, F. Ultrasonic vocalization in rats produced by cholinergic stimulation of the brain. *Neurosci. Lett.* **1990**, *109*, 222–226. [CrossRef]
8. Brudzynski, S.M. Ethotransmission: Communication of emotional states through ultrasonic vocalization in rats. *Curr. Opin. Neurobiol.* **2013**, *23*, 310–317. [CrossRef] [PubMed]
9. Panksepp, J.; Burgdorf, J. "Laughing" rats and the evolutionary antecedents of human joy? *Physiol. Behav.* **2003**, *79*, 533–547. [CrossRef]
10. Brudzynski, S.M. Emission of 22 kHz vocalizations in rats as an evolutionary equivalent of human crying: Relationship to depression. *Behav. Brain Res.* **2019**, *363*, 1–12. [CrossRef] [PubMed]

brain sciences

MDPI

Review

Biological Functions of Rat Ultrasonic Vocalizations, Arousal Mechanisms, and Call Initiation

Stefan M. Brudzynski

Department of Psychology, Brock University, St. Catharines, ON L2S 3A1, Canada; sbrudzynski@brocku.ca

Abstract: This review summarizes all reported and suspected functions of ultrasonic vocalizations in infant and adult rats. The review leads to the conclusion that all types of ultrasonic vocalizations subserving all functions are vocal expressions of emotional arousal initiated by the activity of the reticular core of the brainstem. The emotional arousal is dichotomic in nature and is initiated by two opposite-in-function ascending reticular systems that are separate from the cognitive reticular activating system. The mesolimbic cholinergic system initiates the aversive state of anxiety with concomitant emission of 22 kHz calls, while the mesolimbic dopaminergic system initiates the appetitive state of hedonia with concomitant emission of 50 kHz vocalizations. These two mutually exclusive arousal systems prepare the animal for two different behavioral outcomes. The transition from broadband infant isolation calls to the well-structured adult types of vocalizations is explained, and the social importance of adult rat vocal communication is emphasized. The association of 22 kHz and 50 kHz vocalizations with aversive and appetitive states, respectively, was utilized in numerous quantitatively measured preclinical models of physiological, psychological, neurological, neuropsychiatric, and neurodevelopmental investigations. The present review should help in understanding and the interpretation of these models in biomedical research.

Keywords: evolution of vocalization; ultrasonic vocalization; 22 kHz calls; 50 kHz calls; infant isolation calls; emotional arousal; mesolimbic dopaminergic system; mesolimbic cholinergic system; anxiety; hedonia; rat

check for updates

Citation: Brudzynski, S.M. Biological Functions of Rat Ultrasonic Vocalizations, Arousal Mechanisms, and Call Initiation. *Brain Sci.* **2021**, *11*, 605. https://doi.org/10.3390/brainsci11050605

Academic Editor: David Conversi

Received: 11 April 2021
Accepted: 5 May 2021
Published: 9 May 2021

1. Introduction

Production of vocalization is one of the best means of communication in most terrestrial vertebrates, even though many physical conditions and environmental objects influence and impede sound transmission. Vocal communication is not dependent on daylight and visibility or on the proximity of organisms, does not leave permanent traces, and in most situations, is not critically influenced by air currents, humidity, or temperature. It is, thus, not surprising that the emission of vocalization for intraspecies communication is one of the oldest features present in vertebrates and tetrapods, ranging from lung fish to humans [1–3]. The neuronal mechanisms for the regulation of fundamental features of vocalization, such as call duration and sound frequency, are located in the deep hindbrain, bordering the spinal cord [1,4]. They are conserved in the vertebrate evolution and could be demonstrated in species ranging from toadfishes, such as midshipmen fish [5], to mammals, such as rats [6]. Sound frequency and duration of vocalizations are regulated by separate hindbrain nuclei [5], which allow for the generation of a large number of combinations of sound parameters, and thus, the generation of different signals with different information content that still use the same acoustic mode of communication.

The old phylogenetic history of vocalizations suggests that the vocal form of communication is highly adaptive and has been biologically important for animal behavior for hundreds of millions of years (approx. 400 million years for tetrapods) [7]. In this review, the emphasis will be on rat vocalization, which is the most extensively studied in rodents. In addition, some other mammalian species will be mentioned because only mammals have developed ventral myelinated vagal innervation, which originates from the nucleus ambiguous, innervates the larynx, and is critical for the generation of vocalization and the regulation of developed social interactions by the system termed social engagement system [8].

Studies of behavioral situations with the emission of ultrasonic vocalizations as indexes of emotional states have been extensively used as models of different neurodevelopmental, neurological, and psychiatric dysfunctions and diseases [9]. The magnitude and type of emitted ultrasonic calls were used in these models as a measure of the relevant effects. Therefore, there is a need for better understanding of the origin, nature, and role of rats' ultrasonic vocalizations, their initiation mechanisms, their interpretation, and their equivalence to human vocal emissions.

Rat ultrasonic calls cannot be compared with human speech because speech is only a human function, and rodents do not have the necessary neural mechanisms and developed cognitive brain to have this way of communication. However, rat ultrasonic calls may be compared with human vocalizations. This is a valid comparison because human non-verbal vocalizations, such as crying, laughing, grunting, groaning, moaning, or shrieking, do not have lexical content, are generated by subcortical limbic mechanisms, and are evolutionary counterparts of other mammalian vocalizations (for classification of human verbal and non-verbal vocalizations as well as pathological vocalizations, see [10]). Although animal vocalizations have many subtypes and may convey referential information or be situation-specific (e.g., in species of prairie dogs [11–13]), they do not represent language in a human sense and do not have grammatical structure, sentences, words, syllables, or even fully translatable meaning. Naming mammalian vocalizations as syllables, particularly those emitted in series, is a misnomer and mistake that is still repeated in the literature. Animal calls (and human vocalizations) remain simple signals, even though they may have some specific situational content or may be emitted repeatedly or in a combination of calls. Vocalizations represent an evolutionarily older system of communication than human language with different neural regulation and different semiosis ("meaning"), and these two types of vocal communication should not be confused. Moreover, animal vocalizations were suggested to be interpreted as means of influencing the behavior of other individuals in a general way, rather than signals sending specific (e.g., lexical type) information to conspecifics as we know it from human language [14].

This review will attempt to facilitate understanding of functions of rat vocalizations, i.e., answering the question of why rats emit their calls. Different functions of vocalizations reflect our understanding of situations favoring vocal communication, and they do not mean that animals have many "understandings" or many intentional scenarios of call emission. Moreover, the emission of ultrasonic calls may serve more than one function at the same time. Thus, the classification of the functions of emitted vocalizations is used as a heuristic tool for the classification of the behavioral roles that vocalizations play in phylogenetic and ontogenetic history. Such a classification of documented or suspected functions of rat ultrasonic vocalizations has not yet been fully accomplished [15–18], and it will aid the ultimate goal of this review, which is to cumulate evidence supporting the hypothesis that all types of rat vocalizations, serving all biological functions, are driven by emotional arousal. Neural mechanisms that initiate emotional arousal, positive or negative, are, therefore, common in fulfilling any of these functions.

2. Evolution and Functions of Rat Vocalizations

2.1. *Functions of Vocalization Originating from Mother-Infant Interactions*

2.1.1. Self-Preservation Function

The emission of vocalization evolved to serve many functions, but its primary and most important role was associated with saving the individual's life and protecting the species since birth as we can observe today in altricial rat pups. The emission of infant vocalizations serve a self-preservation function, which is regulated by an innate, ancient emotional survival mechanism ascribed to the basic functions of the mammalian brainstem limbic system [19–21]. The communication of mothers with their offspring with vocalizations is regarded as the phylogenetically oldest form of vocal communication in rats and in all mammals [22]. The emission of calls by rat infants in the ultrasonic range was probably caused by increased air pressure in the respiratory system with constricted vocal folds that evolved in response to the cold [23,24], and this range of sound frequencies appears to have been highly adaptive. It may be speculated that, initially, the crying of infants and juveniles emerged and it was paralleled by the vocal responses of the mothers [22] since mostly females were primary caregivers to the offspring due to nursing and additional critical care (grooming, licking, nest attendance, etc.), without which rat infants could not survive.

In broader sense, maternal behavior was termed epimeletic behavior (from Greek *epimeleteon*—caregiving), which also includes paternal and biparental behavior [25]. On the other hand, pup or young vocalizations directed to their parents were termed et-epimeletic behavior (from Greek, *aeteo*—to beg, + epimeletic), and this label also includes signals other than vocalizations [25]. These terms were introduced at the beginning of the 20th century to make studies on behavior unbiased, devoid of colloquialisms, and easier to compare across species. While maternal vocalizations directed to pups were not carefully studied, etepimeletic vocalizations of pups focused the extensive attention of researchers.

2.1.2. Locating Function

The emission of juvenile broadband vocalizations contains a primary locating function. The infant's calls inform its mother as to where the pup is and help her to retrieve it back to her if it fell out of the nest, or to move it within the nest when the mother's body compresses too strongly on the infant's body [26]. The infant calls, termed sometimes as separation or isolation calls, or distress calls [27,28], provide mothers with critical information about the location of the vocalizing pup. Mothers orient toward the calling pup and approach it (phonotaxis) [29,30]. This maternal behavior is regulated not only by vocalizations but also by olfactory cues. In playback experiments, it was shown that mothers showed enhanced orientation toward the source of the infant's calls if a silent pup was placed under a speaker [31]. Ultrasonic calls provide critical directional information for mothers, while pup odors determine the urgency and speed at which mothers begin searching [32,33].

Initial calls of pups after birth are not well exercised, and infants learn to emit the separation calls in such a way that maximizes the locating of the calling infant. In successive days after birth, the isolation calls gradually become longer and more complex, and the sound frequency within each call has a fluctuating character across many frequencies in a fashion similar to an ambulance siren [34], and they may reach significantly deep fluctuations. This pattern of emitting vocalizations could appear by the natural vocal selection that was demonstrated in infant rats [35]. Although there are some differences in the number of emitted calls, peak frequency, and frequency modulation among male and female pups of the main laboratory strains of Wistars, Long–Evans, and Sprague Dawley rats [36,37], the principle of locating the calling pup remains the same.

Pups of 14 days of age and older, however, will markedly reduce their emission of these calls when an adult male rat is nearby [38,39]. If a pup was isolated from the nest for a short time and retrieved, the mother will spend more time with it and pay more attention to that pup as compared to pups that have always been in the nest [40]. Maternal care is conserved in evolution and it can be demonstrated from rats to humans [41]. It has been shown that maternal care and proper vocal communication with the offspring contributes to development of the infants' social brain and increases the offspring's survival and their future reproductive success [42].

The importance of communication by vocalizations in rat infants may be further demonstrated by experiments showing that pups with less maternal help than usual were more anxious and emitted more infantile ultrasonic vocalizations than controls when they were separated from their mother and litter [43]. Additionally, experimental daily 30 min maltreatment of pups caused increased emission of their isolation calls as compared to pups receiving expected maternal care, and this procedure caused detectable epigenetic changes in the development of the brains of the maltreated rats [44]. Moreover, rats selected for higher emotionality traits emitted more ultrasonic calls when isolated as compared to pups with lower emotionality [45]. With the prolonged separation of pups from their mother, the acoustic parameters of their vocalization changed, and the pups emitted a larger proportion of high sound frequencies than the controls [46,47].

Pups pay constant attention to their mother and to her proximity. It was demonstrated that the interaction of pups with their mother just before the pup's isolation further increased the pups' vocalizations to a subsequent isolation (named maternal potentiation). If male rat contributed to caring for infants, pups would also show paternal potentiation [48,49]. It was concluded that infant separation vocalizations express infantile anxiety, and these vocalizations could be pharmacologically decreased by numerous anxiolytics [50,51].

2.1.3. Protective Function of Maternal Care

The expression of infantile anxiety with the emission of calls also represents a protective function, which secures continued maternal help and safeguarding and builds a bond between the mother and her infants. Although many cues contribute to the development of the mother–infant bond, a repeatable emission of calls is one of the important signals. As studied in mice, infants at Postnatal Day 17 and 21 were able to recognize their own mother in a two-choice tests and preferred their own mother to a foreign mother [52]. In addition, based on the emitted calls, mouse mothers located their own pups faster than a stranger pup [52]. These bonds are the precursor of adult social bonds and are established by the release of oxytocin both in mice and rats [53]. The development of this bond is important at the infantile age. The creation of new stable bonds in adult rat social groups seems to be difficult or not possible, as it was studied in pairs of adult female rats [54].

The same protective function applies to the vocalization of human infants; it was suggested that excessive human infant crying may express anxiety of being abandoned, and crying prevents the withdrawal of parental help and secures continued care [55]. Loss of contact with the mother and the nest environment seems to be the primeval aversive emotional state expressed vocally [56]. Ensuring the continuation of mother–infant contact is one of the evolutionarily oldest and fundamental functions of emitting vocalizations.

2.2. Functions of Vocalization in Non-Agonistic Adult Social Interactions

2.2.1. Phatic Communication Function

The mother–infant relationship developed another related function of vocalization, termed phatic communication, that is mostly characteristic of the rats' adult life and seems to be appetitive. The term "phatic" was coined initially in anthropology as "bonding by language" [57], i.e., by emission of words (in humans) or vocalizations (in animals) that serve to create and maintain social bonds and closeness. The phatic communication in animals aims at maintaining connection between individuals, a reassuring proximal presence, and maintaining the cohesiveness of social groups in gregarious species [15]. The category of calls for establishing and maintaining contacts between adult members of the social group has been known for a long time. These vocalizations were later termed contact calls in classical ethological studies and were demonstrated in numerous species [58]. It has been even suggested that human humming may be regarded as the human equivalent of contact calls in social animals [59].

In rats, short-duration vocalizations, classified as flat 50 kHz calls, are used as contact calls and are emitted toward familiar conspecifics, even if these conspecifics are not present nearby [60–62]. Rats will particularly emit these calls when they detect fresh olfactory traces of other rats, and the more scent traces they detect, the more calls they emit, usually of a frequency-modulated type [63]. In the case of possible contact with many individuals, these calls may also play another, affiliative function (see below). It was also reported that rats may emit contact vocalizations in dyadic interactions or when being alone in a cage without detectable traces of other rats but shortly after separation from other companions [60,61].

Phatic communication has a character of the social announcement of presence, acceptance, and reassurance, and is not expected to be associated with approach. Phatic vocalizations may initiate similar reply calls from other rats but without further behavioral consequences. Such a mutual, infrequent calling is also evidence of social tolerance and potential social support, which would be associated with mutually positive emotional states. These calls may also have potential anxiolytic properties.

2.2.2. Affiliative Function

Related to phatic communication is the emission of ultrasonic calls that may play an affiliative function. This function includes assuring non-agonistic, close interactions, a signaling approach, promoting direct contacts among individuals, causing grouping, and huddling. Unlike the phatic role of vocalizations, the affiliative function is associated with the approach and even direct contact among rats [64,65]. There is some recent evidence suggesting that 50 kHz calls are emitted during grooming [66] so they may contribute to anti-stress and to close-contact social behaviors driven by rat-positive emotional arousal. In a recent study summarizing results from selective breeding, devocalization experiments, and playback studies, a general, broad-term conclusion was reached that 50 kHz ultrasonic calls "serve as situation-dependent socio-affective signals with important communicative functions" [67].

The affiliative function of calls may be observed in many situations. Rats spend most of the daytime in underground burrows where vocal communication with ultrasonic calls is needed and particularly effective [68]. However, subterranean social mole-rats evolved communication with low-frequency vocalizations of 1.6–6.3 kHz [69]. The rat emission of affiliative calls has a calming effect on approaching conspecifics and prevents unexpected, aggressive attack. During a non-aggressive, 'friendly' approach [18], particularly in dark tunnels, the approaching rat will emit 50 kHz calls, usually in a characteristic short series of three frequency-modulated vocalizations emitted in rapid succession, presumably announcing its movement toward the other rat(s) (unpublished observations).

On the part of receivers, there is ample evidence that, when rats hear abundant 50 kHz calls, they approach the emitter or the source of the calls (e.g., a loudspeaker [65,70,71]. This effect is dependent on previous social experience [64], call-specific (no approach to 22 kHz calls), stronger in juveniles and females [67,70], and species-specific, thus other rodents, for instance, bank voles, do not show an approach response to the rat's 50 kHz calls [72]. The perception of 50 kHz vocalizations and approach to them are clearly appetitive and rewarding responses to such an extent that rats can learn to self-administer the 50 kHz calls [73].

The extreme form of affiliative rat behavior is huddling. Huddling behavior, which is prevalent among infants, will also continue, to some extent, in adulthood [74,75]. While in infancy, one of the predominant roles of huddling is group thermoregulation [76], although other non-thermal stimuli are also important [77,78], adult huddling occurs less frequently and is associated with potential external danger or unfavored conditions, such as bright light or other dangers [79,80]. In these situations, rats have a tendency to crawl under other conspecifics, but this is not a blind behavior because males do not crawl under females but only under other males [81]. There are no studies recording the emission of vocalizations during this behavior.

2.2.3. Passive Defensive Function

Ultrasonic vocalizations of infants, promoting huddling and nestling behavior, and calls serving affiliative function in juveniles and adults, that cause approaching and staying in close social groups have been jointly regarded as a form of "passive" defensive behavior of social groups and associations [82]. This basic defensive function is a form of primary defense, i.e., behavior without any detected presence of danger or predator (as distinguished from secondary defense, which appears when danger or a predator is present). This defensive behavior is a preventive form of behavior and is driven by an ancient limbic mechanism, aiming at securing potential social support and protection while being among other conspecifics. It is dubbed "strength and safety in numbers", and it is present in all social mammals (and many other animal groups, e.g., fish and birds), including humans [83]. This defensive tendency must be a very old evolutionary development regulated by the oldest limbic mechanism.

2.2.4. Socio-Coordinating Function

In general terms, phatic, affiliative, and related defensive functions of the emission of ultrasonic vocalizations are associated with a social, regulatory function not only at the juvenile level but in later social life, particularly in larger groups. This function was termed socio-coordinating function and particularly flat 50 kHz calls are involved [43]. Specific 50 kHz ultrasonic vocalizations are not one-to-one related to individual movements but are associated with specific patterns of motor behaviors and are suggested to coordinate moment-to-moment during social interactions among rats [84]. Experiments with the newly developed *Cacna1c* haploinsufficiency rat model demonstrated the importance of the socio-coordinating function, particularly during rough-and-tumble play behavior and during female interactions [85,86]. The deletion of *Cacna1c* in rats reduced the number of emitted 50 kHz calls, reduced social approach behavior during the playback of 50 kHz vocalizations, and revealed general deficits in communication and coordination during social behavior [85]. In other mammals, the coordinating role of calls is important in the initiation of movement of animal groups, which was well documented for white-faced capuchin monkeys [87,88].

2.2.5. Social Buffering Function

It has been observed that the repeated tickling of rats (heterospecific play with a human hand) that was associated with the emission of vocalizations (mostly 50 kHz vocalizations) had a buffering effect on anxiety caused by handling or by the intraperitoneal injection of saline in these rats [89,90]. Although tactile stimulation during play and other cues are responsible for the buffering effect, it has been noticed that vocalization itself can also have a buffering function in other mammalian species [91]. It is conceivable that the emission of 50 kHz calls during grooming [66] may, jointly with tactile stimuli, have a buffering effect and may reduce stress and anxiety by the release of oxytocin. It was recently shown that juvenile and young rats that received repeated tactile stimulation with a human hand increased the emission of 50 kHz calls and showed the activation of oxytocin neurons in the hypothalamic paraventricular nucleus [92,93].

The abovementioned huddling behavior in rats also has an acute emotional buffering effect [80]. It is not known, however, what call types that rats emit in this situation, if any. However, the results with self-administration of 50 kHz calls by adult rats without any tactile stimulation may support the notion that the calls themselves could have a social buffering effect [73].

2.2.6. Investigative Function

Rats are known to have well-developed responses to novelty, although not all individuals are high responders [94]. High responders to novelty have increased locomotor activity in the new environment and an enhanced level of dopaminergic activity in the nucleus accumbens, and the novel stimuli are rewarding for them [94,95]. In such novel situations, rat will emit vocalizations that play a positive, investigative role and are associated with rewarding novelty-seeking behavior and positive expectation. However, this response is dependent on the rat's preliminary or pre-existing assessment of the environment or novel object. If the new environment seems to be rewarding, the rat will emit vocalizations, mostly 50 kHz calls. If, however, the new environment seems to be dangerous, the rat will not vocalize. This animal's initial bias in its evaluation of the new environment or novel stimuli has been shown experimentally. Rats were initially trained to respond differently to acoustic stimuli as being positive or negative stimuli. Then, upon hearing the playback of 50 kHz calls or 22 kHz calls, the rats were presented with a new, ambiguous cue that was neither positive nor negative. The rats responded positively to the ambiguous cue when they heard 50 kHz calls, but responded to the same cue negatively when they heard 22 kHz calls [96].

In a recent experiment in a semi-novel environment, i.e., in a cage that had holes in the walls and was familiar to the rats but the lighting was changed so that the illumination of the cage had a novel element, rat ultrasonic vocalizations were recorded. Any time the experimental animal nose-poked the hole, the light was switched off for 5 s. The yoked control group was unable to switch the light off, but the on–off lights were controlled by the experimental group of animals [97]. It was found that rats that could control lighting performed more nose-pokes and emitted more vocalizations than the yoked rats. Long, alarm 22 kHz calls were not emitted; however, the experimental rats had significantly more nose-pokes with the emission of 50 kHz calls (with predominance of flat calls) and more nose-pokes with the emission of short 22 kHz calls (approx. 10–20 ms in duration) as compared to the control yoked group. Although the behavioral situation was not a typical novel environment, there was an element of novelty and novel exploration, and it was associated with calling [97]. It was concluded that 50 kHz calls and short 22 kHz calls could be associated with the investigative function and novelty seeking. The role of ultrasonic calls emitted by rats in novelty situation needs, however, more studies.

The role of call emission, particularly 50 kHz vocalizations, in novel situations could be interpreted as signaling a rewarding novelty-seeking behavior but it is situation dependent. This conclusion is supported by studies on mice tested in novel or familiar environments. Mice emitted significantly more high-frequency ultrasonic calls and of longer duration in dimly lit novel environments than in the brightly illuminated novel environment. In the bright-light conditions (aversive stimulus) of the novel environment, more calls were emitted with lower sound frequencies [98].

2.3. *Functions of Vocalization in Social Play and Teamwork*

2.3.1. Ludic Function

The emission of vocalizations during social interactions within the nest evolved to perform another fundamental role during play behavior in juveniles and play-like behavior in adult rats [99]. The role of vocalizations aiding play is called ludic function or ludic behavior (from Latin *ludere*—to play), which is characteristic of immature animals [100]. The emission of 50 kHz calls is abundant during natural rough-and-tumble play (play fighting) in rats and occurs in anticipation of and during such play in juvenile rats [73,101,102]. The emission of 50 kHz vocalizations is associated with specific components of play behavior, and the calls function as play signals and signals maintaining playful mood and activity [99,103,104].

In general, the number of emitted 50 kHz calls was used as a quantitative index of the animal state both in juveniles and adults [15,105]. The number of emitted 50 kHz calls is not only the quantitative index of the magnitude of emotional arousal but the low levels of emission of these calls in infants and juveniles may inform about abnormal phenomena and may be indicative of potential prenatal damages to the limbic system as it was recently shown for prenatal exposure to valproic acid [106].

The emission of 50 kHz play calls can be also induced by a tickling procedure (heterospecific play with humans) that mimics natural rough-and-tumble play, and it has rewarding properties for rats [107–109]. The tickling procedure should follow two main events of rough-and-tumble play—dorsal contacts and pins separated by a short break—and should be repeated daily for the full effect [90]. The repeated tickling procedure can select groups of rats that will emit a particularly large number of 50 kHz vocalizations, signaling the appetitive value of the play [110]. Rats that refuse to play will emit low numbers of 50 kHz vocalizations, and even some 22 kHz calls, at the beginning of the play. Restrain stress applied to rats before tickling sessions significantly decreased the number of tickling-induced 50 kHz calls afterward [111].

Although adults rarely play, the appetitive value of light tactile stimulation, which only partially resembles rough-and-tumble play, still retained rewarding play value in young adults, and tactile stimulation induced emission of 50 kHz calls in habituated animals [60]. Interestingly, other forms of tactile stimulation by human hand in different parts of the rat's body, which had low resemblance to natural rough-and-tumble play, or were entirely unnatural to rats (e.g., holding rats in a vertical position and touching their flanks), also induced the emission of 50 kHz calls, although at a lower rate than during natural play [60,112].

Tickling stimulation of adolescent and young adult rats is appetitive and has rewarding value, as it was demonstrated by the tickle-induced release of dopamine in the nucleus accumbens [113]. On the other hand, very light touching of the rats' skin that failed to induce calling was not associated with the release of dopamine 113]. Therefore, tickling is not a purely tactile phenomenon but a procedure inducing a positive emotional arousal. A recent study has confirmed that the number of emitted 50 kHz vocalizations induced by tickling is proportional to the magnitude of positive emotional arousal and is, indeed, a good quantitative measure of this arousal as compared to other measures [114].

Another recent study reported that rats can play with humans in a "hide-and-seek" game. Rats not only learned how to play "hide" versus "seek" but they were emitting 50 kHz vocalizations during play with predominance of flat and frequency-modulated 50 kHz calls (approx. 72% of all emitted calls with 32% of frequency modulated calls) [115]. Since the authors were using a brief, abdominal tickling procedure any time the rat found a person, or was found by a person, a question arises as to whether rats really were playing "hide-and-seek" game in the human sense or had just learned some rules to merely try to get the tickle-like experience. Somewhat similar behavior was observed during the daily tickling procedure when the rats were escaping from the human hand that was trying to pin them and were approaching and chasing the human hand before dorsal contact [90,108].

The emission of 50 kHz calls during juvenile play fighting, which occurs more frequently in males, facilitates and maintains play behavior [99] that is important for male-typical brain development [116]. The emission of 50 kHz vocalizations contributes to maintaining play, regulates play, and indirectly serves as preparation for the young organisms to develop aptitude for general sociability (gregariousness). Play develops motor skills, even strengthens the skeletal apparatus, develops exploration skills, establishes social ranks and dominance, and prepares juveniles for aggressive and sexual behaviors with always present vocal communication [100,117]. As emphasized by Berlyne (1960) [118], who first proposed the term ludic behavior, play consists of a multitude of functional components, such as perceptual, cognitive, and motor activities, and emotional arousal [118]. Play is a vigorous and highly emotional positive behavior with a crucial role of vocalizations.

2.3.2. Conative Function

The emission of 50 kHz vocalizations during ludic behavior also has another function, a conative function. This is an intentional action of an animal to catch the attention of one or more of the social group members and eventually influence their behavior in a general, non-specific way, or to mobilize them to common action/play. The emission of vocalizations subserving this function in young individuals may play just a general activation role but may change, e.g., into invitation to play. This function is well known in mammals, particularly observed in the vocalizations of domestic animals living with humans, and in vocalizations of attention-seeking human infants, where they are often interpreted as excessive crying for the "manipulation of parents" [55,119,120]. In rats, conative function of calling may be also well illustrated in infants as maternal/paternal potentiation. The vocalization of the isolated pup is significantly increased when the pup has been in contact with its mother immediately before isolation [48]. In this situation, calling cannot be directly associated with a lack of food or other stimuli, but aims at rapid maternal/paternal attention in a general way.

2.3.3. Cooperative Function

Along with the development of the social life in rats, a new function of vocalization appeared that is termed cooperative function. An interesting experiment demonstrated this behavior in rats [121]. To receive a sucrose reward, pairs of familiar (to each other) rats were trained to simultaneously nose poke the holes to receive the reward. Cooperative behavior gradually increased over 44 days of training along with an increased emission of 50 kHz vocalizations. When the pair of rats was separated by a partition, blocking acoustic signals, the cooperative success deteriorated but reappeared again when the rats were separated only by a wire mesh partition and could hear each other. Thus, direct physical contact between the rats was not needed but the rats needed to communicate by ultrasonic calls to achieve cooperative success [121]. The emission of some ultrasonic calls was also suggested as a form of cooperation during common play actions in juvenile rats [122] and in sexual interactions [123].

2.4. Functions of Vocalization in Intraspecies Agonistic Interactions

2.4.1. Agonistic Function

Adult rats use vocalizations for the regulation of their social life, e.g., for the establishment of dominance hierarchies, during aggressive encounters (mostly with intruders), during sexual behavior, feeding, defending territory, and in other situations requiring significant emotional arousal [75]. In all these situations, the emission of ultrasonic calls plays an important communicative role. Even an overall, rough analysis of all emitted vocalizations in these situations shows significant changes in many acoustic parameters of calls, suggesting that rats emit, at least, some calls specific for a given situation and behavior [124].

Aggressive behavior focused the most attention in research and emissions of 50 kHz and 22 kHz type of vocalizations were recorded during the aggressive attacks and defensive actions of the rats [125,126]. Aggressive/defensive behavior is associated with the highest emotional and autonomic arousal because it may be associated with significant body damage. The general label "aggressive behavior" includes many behavioral patterns, such as threatening, aggressive sideway posturing, directed attack, wrestling, boxing, kicking, punching, jumping, submissive posturing, chasing, and flight, so both offensive and defensive elements. Each of these elements is associated with a potentially different type and combination of emitted vocalizations, which may further differ among different rat strains [127]; however, detailed studies about the role of particular types of calls in these components of behaviors have not been systematically conducted.

The 50 kHz vocalizations are emitted mostly by the attacker and the 22 kHz calls by the defeated rat [127,128]. In the intruder–resident interactions, the intruder emitted mostly 50 kHz vocalizations that were changed to 22 kHz calls after its defeat [128,129]. Audible squeals are frequently emitted during fighting and intermingled with ultrasonic calls. Study of vocalizations emitted in the resident/intruder situation with a wire mesh preventing physical contact confirmed that mostly intruders emitted ultrasonic calls [130]. Interestingly, the ultrasonic calls emitted by the intruder were decreased by systemic morphine, but audible vocalizations were not sensitive to morphine, suggesting that their function is different from ultrasonic calls [130].

The emission of the ultrasonic vocalizations in these offensive/defensive situations plays an agonistic function (from Greek *agonisticos*—combative). This term was first suggested by Scott and Fredericson in 1951, mostly in relation to the complex behavior of rats and mice, and it comprises offensive and defensive groups of behaviors, including withdrawal, avoidance, and escape [25,131,132]. In addition to that, agonistic function includes elements of territorial behavior and defense against intruders.

Rats live in large groups that need space for nesting, hiding, and foraging, and will defend this space against rats from neighboring groups. It was postulated that one of the important functions of the evolution of adult vocalizations was spacing among neighboring animal groups. Vocalizations that can be received from a distance serve this purpose very well [133]. The territories rats defend are rather small and there is no good evidence that rats are defending large and defined boundaries around their living burrows [132]. However, agonistic behavior and relevant calling was shown to play broader functions in controlling population density, group stability, and partner choice [75]. Ultrasonic vocalizations used during territorial defense were not studied in rats, but in a study on mice, several types of calls were demonstrated, which were important in territorial defense [134].

2.4.2. Appeasement Function

Many researchers have observed that the emission of long 22 kHz vocalizations during agonistic encounters by the defeated rat may have an appeasement function, which is widely observed in animal behavior [135]. The emission of these calls would decrease or inhibit further attacks of the aggressive or dominant rat [126,127]. This effect probably does not occur immediately and requires some repeated calling, and the opposite might not be true, i.e., the lack of emission of appeasement calls will not necessarily increase attacks of the aggressor, as it was observed in studies with devocalized rats in which deprivation of ultrasonic signals failed to increase aggressive behavior of the attackers [136]. In this experiment, however, rats could not communicate by any type of calls. Appeasement 22 kHz calls may also be emitted to prevent attack in establishing a dominant–submissive relationship. It was observed that a face-to-face encounter with a dominant rat immediately induced the emission of 22 kHz vocalizations in the submissive rat (usually smaller in size) [137].

It was also suggested that during play behavior (both in juvenile and adult rats), the emission of 50 kHz calls may not only be a play signal but also an appeasement signal that de-escalates agonistic behavior during play and prevents aggressive outcomes, which can happen particularly in playing rats that are unfamiliar to each other [99,138].

Finally, it should be mentioned that audible squealing (sonic threat calls) in young rats that are emitted after bites as pain signals, have been also suggested in the past to have an appeasement effect [58] (p. 126).

2.5. Functions of Vocalization in Reproductive Behavior

2.5.1. Mating Function

The emission of ultrasonic calls was well studied in the mating and reproductive behavior of rodents. This is a complex and partially ritualized behavior, so the emission of many types of calls were observed. These vocalizations play a mating function, i.e., they contribute to the regulation of the selection of partners, soliciting sexual contact, and initiating copulation. Fifty kilohertz vocalizations are emitted mostly during solicitation and mounting activity, while 22 kHz are emitted by males during the postejaculatory refractory period [73,123]. Male rats emit 50 kHz before successful mating, which may facilitate female responsiveness because females were less responsive when paired with a devocalized male [123,139]. Females also emit ultrasonic calls before copulation that were suggested to play a regulatory role in mating [140,141]. The emission of 50 kHz calls was dependent on female sex hormones because ovariectomized females exhibited few, if any, of the vocalizations [142] and their responses were graded depending on the hormonal condition [143].

Some recent studies, however, could not fully confirm the behavioral role of 50 kHz vocalization during mating [144,145]. Many factors could cause a lack of this response, such as rat strain, too-frequent repetition of tests, rat experience, or stress. Ultrasonic communication during mating is highly dependent on the gonadal status of both partners. For instance, it was found that females produced more 50 kHz ultrasonic vocalizations to intact males than to castrated males but produced similar numbers of calls to both relevant groups of females [146]. Using a playback paradigm, the role of ultrasonic calls in mating was studied in another laboratory, and it was concluded that female rats displayed high levels of social approach behavior in response to the playback of male, 50 kHz ultrasonic vocalizations and did not respond to amplitude-matched white noise [147]. The emission of 50 kHz calls plays an important role in establishing social proximity [147], and it may be concluded that is important in the regulation of mating behavior.

2.5.2. Social Detachment Function

Male postejaculatory 22 kHz vocalizations are particularly long calls [148]. However, unlike other types of 22 kHz vocalizations, they may have some limited frequency modulation, particularly in the medial and terminal fragments of the call. These calls were postulated to represent a different emotional state of the rat than during the emission of flat-type alarm 22 kHz vocalizations in other situations [149]. The emission of shorter 22 kHz calls than the postejaculatory calls may appear during mating and these calls are associated with unsuccessful intromissions or failed mountings [148], so aversive situations to males.

The postejaculatory 22 kHz vocalizations represent a state of behavioral inhibition with prolonged immobility, a withdrawn or socially depressed state, and an absolute refractory period with "desist-contact" function ([149–152]. During the postejaculatory calling state, rat males do not copulate [153]. It was suggested that the postejaculatory calls have function of keeping the females away [123,150]. It has been observed that experienced females leave the male during the emission of these calls [154]. Contrary to that, the prolonged presence of females together with males increased the duration of the emission of postejaculatory 22 kHz calls [152,155]. This notion that females would avoid male during emission of postejaculatory 22 kHz calls was, however, not always detected [156]. Many factors may influence this behavior and sexual experience is one of them. Nevertheless, the 22 kHz ultrasonic vocalizations in this situation may have a social detachment function or social disaffiliation function, i.e., a role opposite to the affiliative function. This function would be equivalent to the appeasement function in agonistic encounters, discussed above.

Acoustic analysis of the postejaculatory of 22 kHz vocalizations emitted by males during sexual contacts, or intended contacts, revealed that the calls are heterogenous and consist of long flat 22 kHz calls (20–35 kHz range) during the postejaculatory period, and anther class of 22 kHz calls. The other class of calls has higher sound frequency (23–45 kHz range) and some modulated frequency components that were observed during encounters of males with a female in a cage with a physical, perforated barrier, where animals could not have physical contact [157]. This last precopulatory category of calls would be compatible with a negative state of frustration caused by the presence of an inaccessible female, and not as a social detachment function [157] (for more details, see Sections 2.8 and 3.5).

2.6. *Functions of Vocalization in Ingestive Behavior*

2.6.1. Alimentary Function

The phenomenon of the social transmission of information about food has been studied in rats for some time [158–160]. The results indicated that the transfer occurs via olfactory cues, and the observers rely on smelling the breath of the demonstrator rat, who has had direct contact with the food [161,162]. Since rats are very vocal in social interactions, the question as to whether rats can convey food preferences via ultrasonic vocalizations was still open and was studied first in female rats. The results suggested that ultrasonic vocalizations do not play role in this communication because information conveyed by the demonstrator rat had no significant influence on the food choices of the observers when the rats were devocalized [163].

This experiment, however, could not fully explain the mechanism of the vocal transmission of feeding information in rats, and the possibility of the vocal transmission of food preferences was recently raised again [164]. Some clues may come from studies performed on female mice, showing that the observer mouse emits ultrasonic vocalizations toward the demonstrator mouse that has been recently fed, but these vocalizations are dependent on the motivational state of the observer. Non-deprived animals emitted more calls toward demonstrators that were fed on palatable food, while food-deprived animals vocalized more to mice that were fed on any food regardless of its palatability [165]. These calls facilitated the proximity of the mice; however, the exact motivation for the emission of these calls and their communicative value need further studies.

These and other experiments justify distinguishing the category of alimentary calls in rats. In an earlier study, the structure of rat vocalizations emitted by pairs of rats (and recorded in pairs) was studied and categorized, and then the categories were assigned to specific behaviors [166]. Three clusters of ultrasonic calls were identified, roughly referring to frequency-modulated 50 kHz calls, flat 50 kHz calls, and 22 kHz calls. It is of interest that the middle cluster, which was equivalent to the flat type, 50 kHz vocalization (with frequency range between 35 and 55 kHz), contained calls that were emitted mostly during feeding and their emission was consistent [166]. The communicative role of flat versus frequency-modulated 50 kHz vocalizations is dissimilar. For example, during experiments with the self-administered playback of 50 kHz vocalizations, rats reliably self-administered frequency-modulated 50 kHz calls with trills but not the flat 50 kHz calls [73].

2.6.2. Food Provisioning Function

In another recent study, pairs of rats were tested in a mutual food-provisioning task [167]. Firstly, it was found that receiver rats emitted 50 kHz ultrasonic vocalizations toward their donor partners, and the donors provided food to the receivers by pulling a tray with a treat toward the partner rat without a reward for themselves. This food delivery was done in a proportional way to the receivers' communication [167]. These results justify distinguishing a category of ultrasonic calls that have a food-provisioning function. Further research, however, is needed to better understand the type and behavioral role of these calls. It is not clear whether rats were only expressing the need and/or requesting food, or advertising sources of food to other rats.

It seems, however, that the 50 kHz vocalizations in this situation have positive signaling value, i.e., they would be associated with appetitive expectation of food, approach, and eating behavior. On the other hand, it has been shown in the past that alarming 22 kHz vocalizations had an opposite effect on eating behavior. When 22 kHz alarm calls signaled the proximity of a predator (cat), eating behavior was inhibited for up to 2 h [168]. When the alarm subsided, rats emerged from their burrows and resumed eating, but eating bouts were shortened and frequently interrupted by careful observing of the environment [169]. Thus, alarm calls discourage eating, so they have the opposite function to the food provisioning one.

2.7. *Functions of Vocalization in Defense against External Threat*

2.7.1. Predator Alarming Function

The most known and well-studied function of rat ultrasonic vocalizations is the alarming function. It evolved as one of the fundamental antipredator behaviors [168–171]. The alarm calls are long-duration 22 kHz vocalizations with relatively constant, i.e., unmodulated, sound frequency and are emitted for a prolonged time, call after call, after detection of the predator and for about 30 min after the predator has left [170,171]. The alarm calls are directed to the members of the entire social group (audience effect) and related to the approaching danger. However, the social effect may not be present, i.e., a rat may emit alarm calls when it is isolated from the group (particularly in the laboratory) or when it may not know where other conspecifics are [172]. Alarm calls are emitted from the place of a relative safety (not in the immediate reach of the predator when the fear response appears) and are not directed to the predator [173]. The alarm vocalizations usually cause a freezing response of the recipient rats, or their escape to the burrows. This effect was reproduced in an experimental situation in which rats that were chased by a fast-moving object (as a potential predator) showed an escape response with the emission of 22 kHz vocalizations and freezing episodes [174].

It has been documented that rats are also highly afraid of predator odors and consistently respond to them with defensive behavior [175]. The odors originate from predators' skin and fur, urine, feces, and anal gland secretions [176]. Rats respond to odors of many predators (e.g., cat, fox, or lion) but the alarm response to the cat's odor is the strongest [177,178]. When rats were placed in a protective tube within the cage of a predator, they reliably emitted 22 kHz alarm calls to the odor of a cat but emitted only a few calls to that of a snake, and no calls to the odor of a ferret or a control, clean cage [179]. Interestingly, rats did not raise an alarm to the odor of ferrets, which are large carnivores and pose a danger to rats. Ferrets, related to polecats and weasels, however, have been domesticated for a very long time, probably for 2000 years [180]. They live with humans and this could cause some changes in their bodily odors, and it may explain why rats did not recognized the ferret's odor as a threat in this study.

Some analyses of alarming vocalizations of many species led initially to the suggestion that alarm calls evolved to be communicated to predators [181], but other observations have not supported this view. Although cats and other large land predators can hear ultrasonic calls of rats, it was postulated that the ultrasonic alarm calls in rats evolved to protect them from birds of prey (a couple of hundred of species of them) that are the most dangerous predators to rodents [182]. This protection against predators could evolve by adaptation or by exaptation, i.e., by use of naturally preexisting ultrasonic sounds produced by narrow airways for disguised communication. Birds of prey cannot hear ultrasonic calls and their usual audibility is between 1 and 4 kHz (with the exception of the tawny owl, hearing up to 20 kHz) [183]. Thus, communication in the ultrasonic range is adaptive and protects rats. When cornered by a cat, rats have the capacity to defend themselves and often do that successfully as it was indirectly confirmed by a recent publication providing evidence that feral cats were ineffective in hunting for urban rats [184]. Rats have very sharp incisors, and their bites leave deep and not-well-healing wounds infected by bacteria carried by rats [185]. However, rats are defenseless against fast moving birds of prey, which can reach a velocity of 52–70 m/s in extreme situations, as measured for the falcon [186], or developed adaptation for a silent flight as that one of owls [187].

Behavioral analysis of the emission of 22 kHz alarm calls and audible squeals that were emitted in dangerous confrontations with predators or large mammals led to the conclusion that ultrasonic 22 kHz calls are directed to other rats and are associated with an audience effect, while audible squeals are emitted as warning calls directly to predators and other large animals (including humans) and do not require the presence of other conspecifics [173].

The alarming function of 22 kHz ultrasonic calls does not serve individual protection but is a form of social anti-predator defense, i.e., emission of these calls warns the entire social group. This social behavior was regarded as a higher order of defense [18,188]. Hearing the alarm, rats will respond (usually by escape and hiding) regardless of whether the individual colony members have detected or not detected the presence of the predator [171]. Once initiated (usually by alpha male), the alarm is maintained and emitted repeatedly by everyone in the colony for a prolonged time as studied in the visible burrow system [171]. At that time, animals reduce their activity or freeze as it was studied in many independent experiments with the replay of 22 kHz calls [178,189,190].

2.7.2. Alarming-Warning Function

Alarming 22 kHz calls are emitted not only in response to predators, but also when rats encounter any other direct danger. Thus, the alarm calls serve as a general danger signal, so they possess an alarming–warning function to other rats. The alarm calls may be emitted in contact with an unfamiliar human [191,192], in response to a sudden noise (startling acoustic stimulus) [193,194], an unpredictable tactile stimulus with a hissing sound (air-puff) [188,195], or an electric foot shock or tail shock [196–201]. Thus, the semiotic value of the 22 kHz calls in these situations is not related to the predator but is sent as a signal of a general but real and present danger, even though the animal might not fully recognize the nature of the danger (e.g., air-puff) and the danger may not necessarily relate to other rats. These alarming–warning calls are like predator alarm calls and are often interpreted as alarm calls but they differ from them by the circumstances. The alarming–warning category of calls was usually studied in single rats in laboratory cage conditions.

2.7.3. Security Function

It has been also suggested that the emission of 22 kHz vocalizations may serve as a signal of potential danger that is not actually perceived at a given time by the rat. This emission would be initiated by a special motivational system, termed security motivation system, which was well studied in rats [202–204]. It was argued that this system evolved to cope with unpredictable environmental risks, uncertainty, and potential, but not directly observed or detected, dangers [205]. In this situation, the emission of 22 kHz calls would have a security function. Such alarm 22 kHz vocalizations would serve as precautionary signaling of a potential danger, i.e., as an apprehension signal before any danger appeared [205]. The triggering events could be cues originating from the similarity of an environmental situation to the past aversive events, or the lack of stimuli that the animal would expect to detect (e.g., disappearance of a nearby predator), or some weak or new stimuli unknown to the animal.

The emission of calls from security motivation has not been described; however, there are some fragmentary observations suggesting such motivation. We observed such an emission in the laboratory, when a rat sitting quietly and silently in a cage suddenly started emitting 22 kHz calls without any provocation or other stimuli detectable to humans. The category of anticipatory calls has already been demonstrated for vocalizations expressing a positive emotional state associated with drugs of abuse [206]. Security motivation signaling requires further systematic studies.

All the functions described in the subsections above are associated with significant endocrine and autonomic changes, such as release of ACTH, changes in blood pressure, heart rate, body temperature, and respiration rate and confirm the stressful and emotional nature of situations associated with the emissions of ultrasonic vocalizations [207–210]. Although initial recordings of heart rate during the playback of ultrasonic calls did not detect changes in heart rate [211], more detailed and frequent sampling of heart rate detected changes caused by the playback. It is interesting that receivers of the ultrasonic calls develop relevant emotional arousal with autonomic changes [212]. Playback of 22 kHz vocalizations decreased heart rate in the receiver rats, while the playback of 50 kHz calls increased the rats' heart rate. These effects were stronger in singly housed rats as compared to pair-housed rats [212].

2.8. Functions of Vocalization in Expressing Internal Discomfort and Frustration

Frustration Expression Function

It should be also mentioned that the emission of 22 kHz vocalizations may express an anhedonic state, originating from other situations than external predatory or other dangers. The best example is a cycle of positive (euphoria) and negative (dysphoria) affective events observed in organisms addicted to drugs of abuse. It has been observed that during the withdrawal phase from a drug (e.g., from cocaine, heroin, amphetamine, opiates, and ethanol), rats will emit large numbers of 22 kHz vocalizations for many hours after discontinuation of the drug [213–217]. The emission of 22 kHz vocalizations signaled a negative affective state (dysphoria, anhedonia, or frustration), and the calls appeared right at the time when drug levels in the rat body started decreasing, even between binges of self-administration and later during the withdrawal state [217]. This negative emotional state was signaled by 22 kHz calls and initiated by deprivation of the expected drug delivery. This shift from a positive to negative emotional state was suggested to present a salient motivational factor for seeking more drugs, which is well known from the behavior of human drug addicts [218]. As it is known from studies on human patients, the withdrawal state is a powerful psychopathological state and even in former addicts with extinguished drug-seeking behavior, the state can be reversed and cause strong craving relapse when subjects are exposed to environmental situations previously paired with drug-taking situations [219].

In rats, when the addictive drug is not available, the emission of 22 kHz calls signals dysphoric frustration, so the calls may have a frustration expression function. A similar situation with the emission of 22 kHz vocalizations as expression of frustration and frustration-induced anxiety was also reported in rats. This type of call was observed during sexual contacts between male and female rats when the physical contact between the animals was prevented by three physical barriers with not-aligned holes that allowed for olfactory, visual, and auditory contact but not physical, tactile contact. Male rats emitted long 22 kHz vocalizations but with altered frequency structures (for acoustic details, see Section 2.5.2, above) as compared to postejaculatory 22 kHz calls. These calls were associated with the exploration of holes in the barriers and were compatible with irritation and frustration [157]. These frustration calls were also observed in rat sexual encounters during unsuccessful mounting attempts or failed intromissions [148].

3. Vocalization as Expression of Emotional Arousal

3.1. All ultrasonic Vocalizations Are Emotional Expressions

3.1.1. Characteristics of Vocal Expression of Emotional Arousal

The major functions played by rat ultrasonic vocalizations are summarized in Table 1. The circumstances causing animals to emit vocalizations lead to several conclusions. The emission of ultrasonic calls for all the functions has an emotional nature, so the calls represent the expression of emotional arousal with motivation to influence the situation, which instigated this arousal. Although the particular behavioral circumstances differ, these states have common features typical for emotional response, such as increased arousal, prolonged and focused attention, increased muscular tension and/or motor activity, emission of vocalizations, increased activity of the autonomic and endocrine systems, and certain persistence of the response [220,221]. It could be argued that vocal expression of emotion for its own sake does not exist. Animals express their emotional states vocally in specific situations only as means of changing or modifying the social and biological circumstances that induced these states. For this reason, the function of expressing emotion as an independent category was not distinguished in this review.

The understanding that rodent vocalization is produced by arousal was first clearly stated in 1974 [222]. The notion that the emission of 22 kHz ultrasonic calls in rats specifically expresses emotional arousal is also old and was postulated over 30 years ago [223,224]. In later studies, the vocal expression of emotional states in rats was confirmed for infant calls and for adults emitting 22 kHz or 50 kHz calls by many laboratories [23,30,43,214,221,225–230].

In adult rats, ultrasonic vocalizations express two different basic emotional states: an aversive state (displeasure) or appetitive state (pleasure). Each of the functions listed in Table 1 may be assigned to one or the other state (or both) and is labeled in the table as a positive or negative state. Vocal signaling in infants is interpreted as distress and the expression of an early anxiety state [23]. Early infant vocal signaling is a reflexive and automatic process because pups do not hear calls until Postnatal Day 12 [231].

Vocalizations evolved as a social adaptive strategy and are directed to other members of the social group [205]. Size of the social group, its organization and complexity will have influence on the vocal repertoire (for review, see [205]). The production of ultrasonic vocalizations is a complex process from the brain control point of view. Complicated sound production by the larynx in rodents (as in all mammals) is simultaneously coupled with the control of respiration and heart rate [232–234]. Calling is energetically costly, particularly for prolonged vocalizations, as it was directly documented in frogs continuously vocalizing for 2–3 h [235]. Thus, prolonged vocalizations are emitted only as a necessary activity initiated by growing emotional arousal.

Table 1. Summary of biological and social functions of rat ultrasonic vocalizations.

Function of Vocalization	Type of Calls	Deduced Valence	Selected References
	Mother–infant communication		
Self-preservation function	Isolation calls	Negative	[22,24]
Locating function	Isolation calls	Negative	[30–34,36,48]
Protective function	Isolation calls	Negative	[53,56]
	Non-agonistic adult social interactions		
Phatic communication function	50 kHz calls	Positive	[60,61,63]
Affiliative function	50 kHz calls	Positive	[63–65,67,80]
Passive defensive function	22 kHz or 50 kHz	Negative or positive	[82]
Socio-coordinating function	Flat 50 kHz	Positive	[43,84]
Social buffering function	50 kHz calls	Positive	[66,80,81]
Investigative function	50 kHz calls and Short 22 kHz	Positive or negative	[100,102,103]
	Social play and teamwork		
Ludic function	FM 50 kHz calls	Positive	[99,103,104,107–109]
Conative function	50 kHz calls	Positive	[48,49]
Cooperative function	50 kHz calls	Positive	[121–123]
	Intraspecies agonistic interactions		
Agonistic function	22 kHz calls	Negative	[75,126–128,130]
Appeasement function	Long 22 kHz calls	Negative	[99,126,127,136,137]
	Reproductive behavior		
Mating function	50 kHz or 22 kHz	Positive or Negative	[73,123,139–141,147]
Social detachment function	Long 22 kHz calls	Negative	[123,148–155]
	Alimentary behavior		
Alimentary function	Flat 50 kHz calls	Positive	[164,166]
Food provisioning function	Flat 50 kHz calls	Positive	[167]
	Defense against external threat		
Alarming function	Long 22 kHz calls	Negative	[168–171,177,188]
Warning function	Long 22 kHz calls	Negative	[191–201]
Security function	Long 22 kHz calls	Negative	[205]
	Expression of internal discomfort and frustration		
Frustration expression function	Short 22 kHz calls	Negative	[148,157,213–217]

3.1.2. Initiation of Emotional Arousal by the Brain

Emotional arousal is a powerful and extensive central process that changes the state of the entire brain [236] and as a result, emotional arousal leads to functional changes in the entire body, from autonomic adjustments to changes in the motor and sensory systems [237,238] and changes in animal behavior. Some manifestations of emotional arousal and the emerging emotional state might be marginal [239]; others, such as the emission of vocalizations, are powerful and carry significant semiotic value to conspecifics. The semiotic content of calls does not serve as sending specific (lexical) information, but it is always an emotional instrument of influence on other conspecifics to control their behaviors [222,240]; also, it is a behavioral plea for change, even if the change is impossible. From these reasons, it is not possible to directly translate the semiotic content of rats' emotive vocalizations to human lexical language, an idea that was formulated for the first time by McLean [241].

Emotional arousal is triggered by innate brainstem limbic mechanisms in response to incoming environmental stimuli and cues (complex stimuli) or lack of thereof, although the exact mechanism of this initiation is not fully known. These phylogenetically old mechanisms are located in the medial brainstem reticular core [242,243], and more precisely, in the oldest part of the idiodendritic core with neurons having overlapping dendritic fields. The reticular core of the brainstem reaches up to the diencephalon, the hypothalamus, and the septum [21]. This extensive system remained relatively unchanged in the process of evolution and deals with arriving afferent signals of heterogenous origin [244]. The core is part of the larger structure, reticular formation, stretching from the spinal cord to septum, although it is lacking precise neuroanatomical delineation [21]. The reticular core evolved for broadly understood sensorimotor integration and control of behavior [245].

The most general function of the reticular core was described as an activity leading to adaptive stability of the organism [21]. The generation of emotional arousal in relevant situations serves this function. Nauta understood the adaptive stability as an analog of homeostatic mechanisms. While classical homeostasis is concerned with stability of the internal environment of the organism, adaptive stability pertains to the stability of the relationships between the organism and the external environment [21]. The emission of vocalization is one of the fundamental tools in interacting with this environment (mostly social environment). In recent decades of studies on emotional states, the attention has been diverted from the brainstem to numerous other structures, including the neocortex; however, recently, the critical importance of the brainstem in the initial generation of emotional arousal and emotional state has been again acknowledged [246].

3.2. *Dichotomy of Emotional Arousal*

3.2.1. Limitations of Infantile Vocalizations as Relics of Paleomammalian Communication

Very young rat pups express only primeval aversive states associated with a basic self-preservation function. This aversive arousal is based on early parasympathetic regulation and most probably evolved before the evolution of the sympathetic control [8,247]. The pups' brain is immature, and growth of the myelinated innervation of the larynx is not yet completed [248]. The developing myelinated recurrent laryngeal nerves reach the larynx by Postnatal Day 15, and the formation of neuromuscular junctions in the larynx is not finished sooner than Postnatal Day 19 [249]. Only after this innervation emerges can intrinsic laryngeal muscles fully develop [250]. This happens just about the time when pups stop emitting juvenile isolation calls and begin the transition to adult forms of vocalization (about the Postnatal Days 21–23) (unpublished observations and [251]). This stage also coincides with the development of homoiothermy [252].

Thus, without full laryngeal innervation, rat infants are not capable of emitting adult-type vocalizations and initially rely on inborn mechanisms of poorly regulated, heterogenous broadband calling. Since the myelinated ventral vagal complex that innervates the larynx evolved as the last component of the autonomic nervous system and is responsible for the generation of adult ultrasonic calls [8], one may speculate that infantile isolation calls may be similar to primitive vocal communication at the paleomammalian stage, i.e., mammalian ancestors' stage of evolution. The term paleomammalian brain was coined by McLean [241] and this evolutionary ancestral brain was identified with the basic limbic system. At this earliest stage, only negative arousal was signaled; hence, the infantile calls have only negative valence.

3.2.2. Dichotomy of Adult Emotional Arousal Systems and Emotional Signaling

The adult rat ultrasonic vocalizations fall into two categories of different emotional valences and are labeled as 22 kHz and 50 kHz calls. These two categories of calls (with some limited variation of frequencies within each category) differ by 2–10-fold in all acoustic parameters [236] so they are easily discriminated by rats. Although many acoustic features of rat ultrasonic vocalizations may play a role in this discrimination, the sound frequency band proved to be the most informative and critical for this discrimination [253].

The mean sound frequency of any vocalization and any valence was approximately three times more likely to serve for the proper discrimination of calls than frequency modulation within the call, and 6.5 times more likely to discriminate a call than that based on its duration [253]. This call discrimination is biologically important because the 22 kHz and 50 kHz vocalizations signal two different emotional states that should be recognized by rats.

The aversive 22 kHz vocalizations are initiated by the ascending mesolimbic cholinergic system, while the appetitive 50 kHz vocalizations are initiated by the ascending mesolimbic dopaminergic system [254]. Unlike the cognitive arousal system, these two emotional arousal systems are targeting predominantly subcortical, limbic regions (Figure 1). This dichotomy in mesolimbic innervation evolved as an extension of the dichotomy in the autonomic nervous system that forms sympathetic and parasympathetic divisions. Two parallel, ascending mesolimbic emotional arousal systems have different and antagonistic functions, which prepare the animal for two different and behaviorally opposite outcomes, i.e., for danger in an aversive situation (negative state), and for affiliation and hedonia in an appetitive situation (positive state). Thus, the valence of the emotional arousal is mostly predetermined by the dominating activity of the type of the ascending mesolimbic system that initiates it.

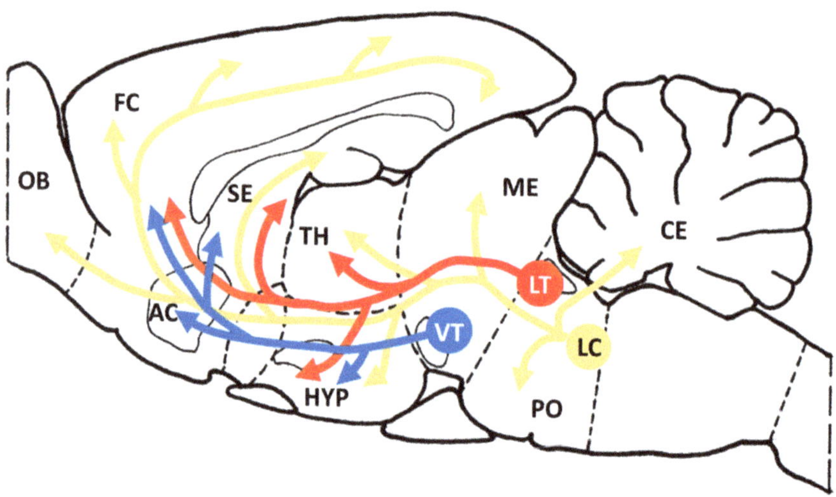

Figure 1. The diagram presents a rough outline and relationship between the ascending cognitive arousal system and two emotional arousal systems in the rat brain. The cognitive arousal system (yellow arrows) originates from the locus coeruleus (LC), releases norepinephrine and targets most of the brain but particularly the neocortex. The mesolimbic aversive emotional arousal system (red arrows) originates from the laterodorsal tegmental nucleus (LT) and targets extensive limbic regions through hypothalamus (HYP) to lateral septum (SE) and releases acetylcholine. The mesolimbic appetitive emotional arousal system (blue arrows) originates from the ventral tegmental area (VT) and targets predominantly the nucleus accumbens (AC) and neighboring regions and releases dopamine. Both mesolimbic arousal systems are most probably also reaching the frontal cortex (FC). The diagram shows only the essential parts of these two emotional arousal systems, which represent relevant functional fragments of all cholinergic and dopaminergic neurons in the brain. It is clear at the first glance that the ascending emotional arousal systems are targeting predominantly subcortical limbic regions. Other abbreviations: CE—cerebellum, ME—mesencephalon, OB—olfactory bulb, PO—pons, TH—thalamus.

3.2.3. Aversive and Appetitive Arousals Are Antagonistic Processes

The existence of two emotional arousal systems leads to the conclusion that aversive and appetitive arousals are mutually exclusive. The aversive and appetitive behaviors are controlled by different mechanisms, are based on different neurotransmitters, and form separate processes that cannot guide animal behavior at the same time. This dichotomy seems to be a general rule not only in vertebrates but also in invertebrates as it was recently shown for crabs [255]; otherwise, it would be maladaptive.

In rats, pharmacological experiments in which the arousal states were induced by intracerebral injections of cholinergic or dopaminergic agents provided evidence that the aversive state and appetitive state are antagonistic processes. The results showed that pharmacological initiation of an aversive state signaled by the emission of 22 kHz calls was significantly attenuated by a subsequent direct pharmacological initiation of the opposite, appetitive state in the brains of the same animals [256,257]. Not only the aversive cholinergic state may be inhibited by the activity of the dopaminergic system but there is evidence for the oppositive inhibition. In anesthetized rats, the identified dopamine neurons in the ventral tegmental area were all inhibited by an aversive stimulus [258]. In behavioral tests with the measurement of rat locomotor activity, a similar result was obtained. An amphetamine-induced increase in locomotor activity was antagonized by intracerebral application of carbachol into the anterior preoptic–hypothalamic region (part of the terminal fields of the ascending mesolimbic cholinergic system) [259]. This antagonism between opposite emotional states provides researchers with an additional way of assessing changes in the emotional valence. Thus, a rapid decrease in the emission of 50 kHz calls or a rapid decrease in the emission of 22 kHz calls may be interpreted as an aversive or appetitive shift, respectively.

Although the initiation of a positive emotional state (dopamine) functionally antagonizes the initiation of a negative emotional state (acetylcholine) in rats, these two systems do not work in a mirror-image way. Pharmacological antagonism of the mesolimbic dopaminergic system did not automatically increase the emission of 22 kHz vocalizations. On the other hand, cholinergic overstimulation of the aversive system with abundant emission of 22 kHz vocalizations caused a delayed rebound effect in the form of the spontaneous generation of 50 kHz calls in a proportional way to the intensity of the initial aversive response [260]. Moreover, the rebound emission of 50 kHz vocalizations was entirely blocked by haloperidol, proving that the emission of 50 kHz from whatever reason is generated by dopamine [260]. The question arises as to how cholinergic stimulation can initiate a delayed rebound with an underlying dopaminergic mechanism.

The rebound could be explained by the activity of a branch of the ascending cholinergic system from the laterodorsal tegmental nucleus to the ventral tegmental area [261]. These cholinergic fibers terminate on dopaminergic neurons of the mesolimbic (mesoaccumbens) dopaminergic system and have excitatory effects [262]. Cholinergic activation of the ventral tegmental dopamine neurons was shown to occur by cholinergic M5 type of muscarinic cholinergic receptors and caused the release of dopamine in the nucleus accumbens, particularly in a delayed phase of the prolonged release of dopamine [263]. This mechanism could explain the appearance of the 50 kHz rebound phenomenon. The exact role of the cholinergic input to the ventral tegmental area is not yet clear, but this is a different sub-system than that one for the initiation of the negative emotional arousal. In the aversive arousal, D1, D2, and D3 dopaminergic receptors are involved [62,264] while the cholinergic input to the tegmental dopaminergic neurons utilizes D5 dopamine receptors with a different pharmacological characteristic. Prolonged activity of the cholinergic neurons of the laterodorsal tegmental nucleus, as that one induced by long-lasting action of cholinergic agents, seems to initiate "a break" by activating the dopaminergic system, which gradually takes over.

On the other hand, the opposite situation may happen with the dopaminergic system. Prolonged stimulation of the dopamine neurons in the ventral tegmental area may decrease their activity and result in aversive arousal. Recent results have shown that the loop between the nucleus accumbens and the ventral tegmental area may be involved in inhibiting the activity of ventral tegmental dopaminergic neurons depending on the duration of stimulation. In the most recent study, brief optogenetic stimulation of the accumbens medium spiny neurons increased ventral tegmental neuronal activity and increased rewarding responses while prolonged stimulation of these neurons induced aversion and decreased rewarding effects [265]. A functional relationship between these two mesolimbic systems and the mechanism of the initiation of emotional arousal are complex and need further studies.

3.2.4. Emotional Arousal versus Cognitive Arousal

Emotional arousal is a separate process from cognitive arousal that is carried out by the classical reticular activating system innervating entire neocortex by noradrenergic axons [266] (see Figure 1, yellow arrows). These two arousal modes (emotional and cognitive) are functionally coupled together and can directly interact with each other, at least in the brainstem [267]. Emotional and cognitive arousal work in concert but target different structures (limbic structures and only limited frontal neocortical regions versus vast areas of neocortex).

Since the predominantly noradrenergic cognitive arousal maintains the awake state and vigilance [268], it is expected that this system needs to be active to allow emotional arousal to perform its function. This was demonstrated in a pharmacological experiment. During amphetamine-induced emotional arousal with the emission of vocalizations, pharmacologic antagonism of selected subtypes of receptors of the noradrenergic system significantly decreased the emission of 50 kHz calls or selectively decreased some subtypes of 50 kHz calls, such as trill calls, the most characteristic components of emotional expression [269]. In another study with the emission of 50 kHz calls by male rats in response to a female (initially present but removed for recordings), noradrenergic agonists led to an increase in the intensity and duration of ultrasonic calls while antagonists reduced the call rate, intensity, and bandwidth of 50 kHz calls [270].

There is not much research on this topic that is published but it seems that the role of emotional arousal (positive or negative) is to enhance neocortical information processing for emotionally important stimuli (salient stimuli) and, at the same time, decrease the processing of stimuli that are not biologically important at that time [271]. This process most likely occurs right in the brainstem by the interaction of the ascending arousal systems. In electrophysiological studies, it was observed some time ago that the ascending noradrenergic system exerts a tonic influence on the neocortex to maintain the waking state; however, the ascending cholinergic system provides additional input in a phasic manner in response to novel, unfamiliar, or threatening stimuli (the emotional component) [272,273].

3.3. *Pharmacology of the Systems for the Initiation of Emotional Arousal*

There are several diffuse ascending systems that originate from the brainstem that are involved in the generation and/or modulation of arousal, and the concomitant general animal state and functioning of the whole brain. All these systems have extensive ascending axon pathways reaching most of the brain, although the density of innervation varies among structures. Each of these systems utilizes a single main neurotransmitter that is massively released during activity mostly by numerous varicosities, suggesting a volume transmission in vast areas of the brain [274]. The following major systems have been identified as arising from the brainstem and associated with changes in brain functions and arousal: (1) the noradrenergic system arising from the locus coeruleus [275]; (2) the ventral dopaminergic system arising from the ventral tegmental area and substantia nigra [276,277]; (3) the brainstem cholinergic system arising predominantly from the laterodorsal tegmental nucleus and pedunculopontine nucleus of pontomesencephalic reticular formation [278]; (4) the serotonergic system arising from raphe nuclei [279]; (5) the histaminergic system arising from basal hypothalamus, mostly tuberomammillary nucleus of the posterior hypothalamus [280]; and (6) the orexinergic system arising from neurons in the lateral and posterior hypothalamus [281,282]. The volume of literature published on these ascending systems is particularly large, so detailed discussion of these systems and their projections is beyond the scope of this review. Although all these ascending systems are, directly or indirectly, involved in emotional mechanisms, there are only two basic systems that are critical in the initiation of emotional arousal with the emission of vocalization.

As demonstrated in the previous sections, emotional arousal in the rat's overt behavior is signaled by the emission of ultrasonic vocalizations. The following question arose: which of the six ascending systems mentioned above, when stimulated, can quickly and efficiently induce species-specific vocalizations and other behavioral manifestations of emotional arousal? It appeared, in rats, that the direct cholinergic stimulation of vast areas of the medial diencephalic and forebrain structures, up to the lateral septum, induced abundant aversive 22 kHz vocalization with other signs of a negative emotional state (such as decrease in activity, freezing, crouching, signs of anxiety, etc.) [283–287]. The emission of 22 kHz calls was dose-dependent and antagonized by atropine, suggesting muscarinic mechanism. This aversive system is marked with red arrows in Figure 1. On the other hand, direct dopaminergic stimulation of the nucleus accumbens and adjoining regions uniformly induced abundant emission of 50 kHz vocalizations with increased locomotor activity [105,288–291]. The response was dose-dependent and antagonized by raclopride, suggesting, at least some, dopamine D2 receptor involvement. This appetitive system is marked with blue arrows in Figure 1.

The emission of 50 kHz calls could also be induced from the hypothalamic–preoptic regions by intracerebral glutamate, but this emission was dependent on dopaminergic neurotransmission and was antagonized by haloperidol [289]. Additionally, emission of 22 kHz calls could be released by direct glutamate stimulation of the laterodorsal tegmental nucleus, and this emission was antagonized by atropine [285]. Although glutamate can initiate 22 kHz or 50 kHz vocalizations, their generation and emission remain dependent on the dopaminergic system for 50 kHz calls or on the cholinergic system for 22 kHz calls.

For comparison, numerous pharmacological–behavioral studies were unable to unconditionally induce emotional states with continuous emission of ultrasonic calls after direct intracerebral application of neurotransmitters utilized by any of the other ascending brainstem systems. Intracerebral application of norepinephrine [269,270], nicotine [292,293], serotonin [294,295], or application of orexin [296,297] appeared ineffective in inducing emotional arousal with the emission of ultrasonic calls. All the mentioned neuroactive agents, however, had a modulatory effect on the ongoing emissions of ultrasonic calls that were induced naturally or pharmacologically. It was, therefore, concluded that only the ascending mesolimbic cholinergic and dopaminergic systems have the capacity of initiating emotional arousal that leads to overt behavioral manifestations with the repeated emission of ultrasonic vocalizations.

It may be further concluded that the magnitude of the emotional arousal is proportional to the amount of released neurotransmitter—acetylcholine for the aversive state, or dopamine for the appetitive state—because emissions of pharmacologically-induced vocalizations were proportional to the doses of injected agents that initiated the arousal [105,283]. Transmitters of all other extensive ascending systems have only a modulatory influence on the two basic systems (dopaminergic and cholinergic).

Considering the anatomy of mammalian brains, it may be postulated that these two parallel and behaviorally opposite emotional arousal systems are homologous systems in the brains of all mammalian species and are universally responsible for two basic emotional states: positive (appetitive) or negative (aversive). As for species other than rats, so far, only the aversive arousal state with a consistent, growling vocalization was thoroughly studied in cats, and the results were similar to those for the rat species, with a homolog ascending cholinergic mesolimbic system, comparable terminal fields, comparable pharmacology of aversive vocalization, and comparable emotional valence (for a full review of studies on cats and comparison of the results with those on rats, see [284]).

3.4. *Transition of Infant Isolation Calls to Adult Calls*

3.4.1. Development of Rat Auditory Cortex

In addition to self-preservation and protective functions, infant calling also serves to develop the mother–infant bond [298] The initial development of pup vocalizations is a highly autonomous process that is not much influenced by external stimuli [299,300]. The question arises of how the infant responds to the mother's calls and how its vocalizations could rapidly change from automatic infantile calls to "meaningful" and behaviorally relevant signals within several days of development. A partial answer to this question may be provided by the mechanisms of brain development itself, and particularly, capabilities of the pups' auditory cortex.

The cortical auditory representation of ultrasounds contained in ultrasonic vocalizations is particularly well developed in the primary auditory area (A1) of the rat cortex [301]. Close to 40% of the primary auditory cortical (A1) responses represents an octave-wide band for critical sound frequencies used in ultrasonic vocalization (32–64 kHz) (i.e., all 50 kHz calls and some 22 kHz calls), while the responses to other sound bands that are below 32 kHz form only 20% of the A1. The group of frequencies for 22 kHz calls is somewhere at the border of these two cortical regions. The 32–64 kHz frequency bin occupies more surface area of the auditory cortex than any other single bin, from 1 to 32 kHz [301]. The adult rat auditory cortex has a clear overrepresentation of neurons responding to sounds characteristic for ultrasonic calls.

The overrepresentation of ultrasounds in the rat auditory cortex, however, needs early life acoustic experience and rapidly develops from the third week of postnatal life [301]. Each day of postnatal life makes a big difference in the development of cortical representation and, for example, the difference between postnatal Day 20 and 21 makes a highly significant increase in the cortical representation for sounds of about 60 kHz [301]. This is a developmental process, but it is based on exposure to ultrasounds and their perception. Hearing loss caused by ear ligation significantly prevented the developmental increase in the percentage of the A1 auditory region for sounds of 32–64 kHz [301]. This experiment may illustrate how early in a rat infant's life the acoustic system develops and most probably makes already early associations between vocalizations and some behavioral situations.

Another question was raised of how the rat brain can distinguish among biologically important sounds (calls) and other unimportant environmental sounds. Numerous studies have shown that sounds that are overrepresented in the acoustic cortex of rats are those that are in the ethological range and are frequently repeated within the critical period of the development. This repetition-dependent cortical plasticity generates the overrepresentation, i.e., more cortical neurons are tuned to these sounds [302].

3.4.2. Mechanisms of Transition from Infantile to Adult Vocalizations

Repeated exposure to natural vocalizations has further influence on the developing cortex, promoting categorical acoustic perception. Categorical perception depends on the development of additional neurons responding selectively to complex sounds of entire vocalizations and fewer neurons responding to individual sound frequencies within the calls [303]. This mechanism facilitates recognition of species-specific vocalization types from an early age. Even if the yet undeveloped brain cannot "understand" the semiotic content of the vocalization, the statistical property of incoming sensory signals (i.e., vocalizations repeated most frequently that are likely biologically relevant) will preferentially create their categorical representation and then recognition [302].

The parallel development of the brain, and particularly the limbic system, is needed to develop control of behavioral responses and enable utilizing the categorical information formed in the auditory cortex and its association with behavioral situations. When the limbic system matures, rats begin to emit a repertoire of species-typical adult ultrasonic vocalizations and they abandon the juvenile isolation calls. Since the pup isolation calls represent aversive vocalizations, the natural extension of these calls (with negative valence) after weaning are mature, constant-frequency 22 kHz calls. Maintaining constant frequency within the call requires some regulatory skills that young pups do not have. These skills of keeping the frequency flat develop gradually from Postnatal Day 7. Maturation of oligodendrocytes and the beginning of the intensive myelination process in the brain occurs from Postnatal Day 7 [304]. At the same time, the duration of calls gradually increases from Day 7.

We studied in our laboratory the development of only flat calls selected from the repertoire of infantile and later juvenile vocalizations over the first month of life (Figure 2).

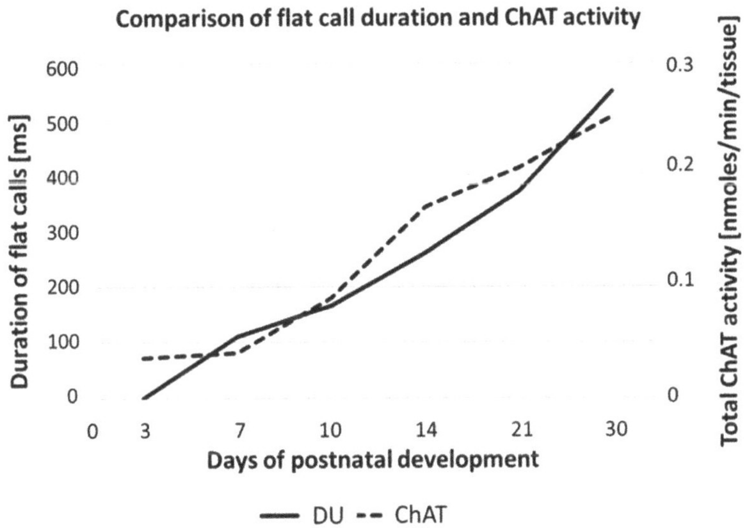

Figure 2. Comparison of developmental changes of duration of selected flat call (solid line) that were emitted by 3–30-day old Long–Evans rat infants with the developmental increase in the activity of choline acetyltransferase (ChAT) (dashed line) in the laterodorsal tegmental nucleus of Sprague Dawley rat pup brains. The pups' flat calls that had peak sound frequency between 20 and 35 kHz and call duration $\geq$ 100 ms were selected from all emitted vocalizations in response to an air puff. There were almost no such vocalizations below the postnatal age 7, and then, 3 to 5 such flat calls were collected per day of development. Data for each point were usually collected from different pups because of the low number of emitted flat calls. The calls were measured by QML S-200 bat detector and the duration of calls were measured sonographically. The bioacoustic data are a fragment of an unpublished study that was partially reported as an abstract [305]. The measurement of ChAT activity was taken from the study by Ninomiya et al. (2001) [306]. Because the data have variable n-vales and were collected from different rat strains, the error bars were omitted. The graph shows parallel trajectories of developmental changes. Abbreviations: DU—duration of flat calls, ChAT—activity of choline acetyltransferase in the laterodorsal tegmental nucleus.

During the first 17 postnatal days, such calls are very rare and short [34] and they gradually appear in older rats. The constant-frequency 22 kHz calls of juvenile and adult rats are initiated by the activity of the ascending cholinergic system from the laterodorsal tegmental nucleus. This ascending cholinergic system develops poorly during the first week of postnatal development, and then rapidly accelerates over the next 7 days (Postnatal Days 7–14) and continues until weaning, which is paralleled by the increase in the laterodorsal tegmental nucleus volume [306,307]. The capability of prolonging the infantile flat calls and reaching adult 22 kHz calls is paralleled by the increase in the activity of the choline acetyltransferase, the enzyme synthetizing acetylcholine, in the laterodorsal tegmental nucleus (Figure 2) and by maturation of the respiratory system. Vocalizations that fulfilled the criterion for 22 kHz alarm calls appeared not sooner than at postnatal Day 16, although they were still relatively short.

The relationship between the growing cholinergic innervation and cholinergic initiation of early flat calls appears late, probably close to weaning. During the second week of postnatal development, the cholinergic innervation is not yet finished. Thus, the systemic application of cholinergic muscarinic agonist, oxotremorine, between Postnatal Days 10 and 17 did not potentiate the pups' vocalizations but instead, inhibited them and it was a central effect [308]. In another study, adult rats, juveniles, and infants were subjected to standard foot shock. The rats showed the emission of different classes of ultrasonic calls to the same aversive stimulus (foot shock) [309]. While adults emitted typical 22 kHz vocalizations, juveniles emitted similar 30 kHz calls, but the infants responded with many calls grouped in two classes of calls (1) with an average main frequency of 40 kHz calls and a 300 ms call duration, and (2) an average frequency of 66 kHz calls of about 20 ms duration. Thus, the development of the brain was not equally prepared at a younger age for species-typical adult signaling. The cholinergic functions are fully developed not sooner than between Postnatal Days 20 and 25. Pilocarpine, a cholinomimetic drug, decreased amphetamine-induced psychomotor activation in 20–25 day-old rats but not in younger rats [310].

3.5. *Interpretation of Rat 22 kHz Vocalizations*

3.5.1. Emission of 22 kHz Calls as Expression of Anxiety

Emission of long 22 kHz vocalizations by adult rats have been unequivocally associated with aversive situations (see Table 1 and Sections 2.4 and 2.7, above). Regardless of the behavioral situation and function, the common denominator of these emissions is emotional arousal, reflecting a state of anxiety (not fear) [18,211,311–315].

It might be beneficial to provide a brief explanation of the difference between anxiety and fear that is frequently confused in publications [316]. Each of these states has different neurochemical setting and different behavioral outcome. In brief, anxiety is defined as a lasting negative state to an unknown and/or unpredictable threat, whereas fear is an acute response to a known and perceived external threat. The difference between the state of anxiety and fear is explained in the best way by a "predatory imminence continuum" that is defined by the physical (spatial and temporal) distance from rats to the approaching predator [317]. When a predator is approaching from a certain far distance (relative safety, but the predator's behavior cannot be predicted), or its exact location is unknown, a state of anxiety appears and it may last for a prolonged time. When the predator is too close and ready to strike and the rat is without possibility of escaping from it, the fear response is initiated and it is a short-lasting response, forcing the rat to immediate action. During the anxiety state, rats vocalize intensively with alarm 22 kHz calls. On the other hand, the fear response is either silent or audible squeals are emitted directly to the predator as a warning, and the rat is ready for "fight or flight" [173].

It has been suggested that the emission of rat 22 kHz calls represents the evolutionary vocal homolog of human crying and that 22 kHz calls and human crying both express anxiety and anhedonia [315]. The emission of these aversive vocalizations is stereotypic for the species, repetitive, and innate (for both rats and humans), so the organisms do not need to learn how to emit the crying calls. A comparative study of ultrasonic vocalizations among the main strains of laboratory male rats confirmed that species-typical, adult 22 kHz ultrasonic calls were comparable among Wistar, Long–Evans, and Sprague Dawley strains with only minor acoustic differences [36]. Results from female rats from these strains were also comparable, although many females, particularly of the Wistar strain, did not emit 22 kHz calls during the fear conditioning paradigm [318], so their sensitivity to aversive situations may be different than males.

3.5.2. Emission of 22 kHz Calls in Depression and Pain

The emission of 22 kHz calls and crying vocalizations in other mammals do not directly signal depression, although they are often the secondary, comorbid result of a depressive mood. The anxiety-driven emission of calls is an outward response directed to other members of the species, while depression is a withdrawn, inward response with different characteristics and without social signaling. Thus, the emission of 22 kHz vocalizations signals anxiety and should not be regarded as a direct index of depression in rats [315].

The emission of 22 kHz, however, was used as an indirect measure of a mixed affective state after social defeat in rats with some elements of depression, but the predominance of anxiety was signaled by these calls [319].

It should also be also emphasized that ultrasonic 22 kHz vocalizations do not directly signal pain itself [320–322]. Although the emission of 22 kHz vocalizations was increased during chronic pain (chronic polyarthritis or repeated electrical stimuli) as compared to healthy rats and these calls were suggested to serve as evaluation of analgesic drugs [323,324], 22 kHz vocalizations express an affective component (anxiety) of ongoing or repeated painful experiences, not pain itself, and these calls were sensitive to morphine [321,325–328]. Pain stimuli can even inhibit ultrasonic calling, which led in the past to a very confusing interpretation [328]. In a recent study, it was shown that the emotional response to acute pain (single injection of formalin that, however, caused long lasting pain), with the emission of vocalization presented by the demonstrator rat, showed contagion to cage mates but not to non-cage mates, or to cage mates separated by a visual barrier [322]. Thus, the familiarity among rats and visual contact both contribute to emotional contagion conveyed by vocal expression of anxiety caused by lasting painful experiences.

3.5.3. Emission of 22 kHz Vocalizations Requires Some Learning Experience

Although rats emit and recognize 22 kHz innately, some associative learning is needed to link these calls with aversive stimuli and situations [329]. The initial association happens most probably in the infancy stage of life (see Section 3.4.2, above). It has been shown that association of danger (foot shock) with the playback of 22 kHz vocalizations produced defensive responses that were better encoded and consolidated in memory than responses associated with any other ultrasonic call type or signal; these responses were resistant to extinction and were retained in memory for a longer time than other responses [329]. This associative learning depends on the perception of calls of other conspecifics but not the emitters' own calls [330]. Despite the need for this associative learning, it was concluded that rats are predisposed (primed) to learn defensive behavior in response to alarm calls, even without learning [329].

The emission of 22 kHz alarm vocalizations is the principal alarming signal in rats, which was demonstrated by the observation of the behavior of pairs of naïve or fear-experienced rats. A naïve or fear-experienced receiver rat was observed in contact with another demonstrator rat that was fear-conditioned to foot shock. The receiver repeated 22 kHz alarm vocalizations of the demonstrator and showed a freezing response but only when the receiver was experienced with the foot shock (although not conditioned to it). Naïve rats did not repeat the alarming calls of the demonstrator. In addition to that, rats with a damaged auditory system failed to repeat the calls of the demonstrator rat, even if they were fear-experienced [331]. Thus, the emission of 22 kHz ultrasonic calls is the main vehicle for the social transmission of anxiety; however, learning is needed for the proper recognition of the danger signaled by 22 kHz alarm calls [331].

Perception and recognition of the aversive value of 22 kHz alarming calls produced by adult rats significantly enhanced the acoustic startle response (an index of the anxiety-type emotional response) of adult receiver rats but garnered weak response from these rats if the emissions of alarming 22 kHz calls originated from young rats [332]. This result may further imply that the structure or pattern of emissions of 22 kHz calls by experienced rats contain some additional signaling features that are recognized by the recipients and can initiate anxiety in them.

Recognizing the aversive 22 kHz calls and learning the association between these calls and behavioral situations is a critical process that occurs from a very early stage of life at the infancy level and is continued over the life span. This process is significantly aided by acoustic cortex plasticity, recognizing and responding to whole categories of vocalizations [302].

3.5.4. Expression of Internal State of Anhedonia by 22 kHz Calls

In addition to what was described, the emission of 22 kHz vocalizations may express an anhedonic internal state caused by events other than external danger or a predator. The dysphoric state during withdrawal from drugs of abuse is accompanied by the abundant emission of 22 kHz calls (for details, see Section 2.8, above).

The general features of 22 kHz calls emitted by rats during withdrawal dysphoria are compatible with anxiety driven by the affective distress and frustration associated with drug withdrawal. The same type of "inconsolable crying" or "high-pitched crying" was observed in human pediatric patients during the withdrawal phase from their addictive behavior as one of the most common symptoms [333,334].

In rats, many of the withdrawal-induced 22 kHz calls were reported as short 22 kHz calls of 10–500 ms in duration [217,335], while most long 22 kHz vocalizations are 300–3000 ms in duration [188]. Short 22 kHz calls that are less common were initially reported in rats as calls of 20–300 ms in duration [336] but their behavioral role has not been defined. Based on the observations that very long 22 kHz calls were emitted in the predator situation or in response to an air-puff, while the short calls were observed during drug withdrawal, it may be suggested that long calls are emitted in the face of external danger while short calls are characteristic of an internal dysphoric state, irritation, and displeasure without a direct, external threat [217].

The emission of 22 kHz calls during withdrawal when the drug of abuse is not available, or during frustration caused by lack of availability and access to a receptive female, being separated from the male by a partition [157], may be interpreted as signals sent to other conspecifics, even if they might not be available or cannot help. These aversive situations and the resulting behavior have been explained as frustration-induced anxiety [337], and the anxiety is signaled to conspecifics. Such behavior of irritation, frustration, stress, and resulting anxiety might be associated with the activation of additional and supplementary brain mechanisms supporting emotional arousal, i.e., augmenting the emotional arousal when the goals cannot be reached. This conclusion is supported by human studies with a concurrent, functional magnetic resonance imaging recording, in which individuals were subjected to experimentally induced frustration [338]. The results showed increased activity of structures directly involved in performing the frustrating task (sensorimotor activation) and activity of structures involved in acute stress, such as the striatum, cingulate cortex, insula, and middle frontal gyrus. Thus, the brain activity during the frustrating situation increased its activity to possibly find a solution [338] while still remaining in a state of anxiety expressed vocally.

3.6. Interpretation of Rat 50 kHz Vocalizations

3.6.1. Emission of 50 kHz Vocalizations as Expression of Hedonia

The emission of 50 kHz vocalizations has been observed predominantly in appetitive behavioral situations (see Table 1 and Sections 2.2, 2.3, 2.5 and 2.6, above). It has been suggested that the emission of rat 50 kHz calls represents an evolutionary counterpart of human laughter [108,339]. This homology was particularly appropriate for comparing joyful childhood laughter during active play with the emission of 50 kHz calls during juvenile rats' rough-and-tumble play [73,101,104]. The emission of 50 kHz vocalizations expresses a positive or hedonic emotional state that may be termed hedonia (a state of pleasure, from Greek *hedone*—pleasure), a pleasurable (joyful) state within physiological limits. It should be distinguished from the obsolete and unclear meaning of this word as a pathological "abnormal cheerfulness" in human psychiatric patients [340], which was earlier called delusional amenomania [341]. In a physiological sense, hedonia is signaled by rats in most appetitive states by the emission of frequency-modulated 50 kHz calls, and particularly trill calls [342]. These calls have the same principal acoustic structure among the main rat strains (Wistar, Long–Evans, and Sprague Dawley) with only small differences [37].

Many experiments indicate that the positive emotional state expressed by the emission of 50 kHz calls contains an element of expectation and "wanting" [343]. It was, indeed, observed that the emission of 50 kHz vocalizations appeared prior to rewarding social interactions, such as in rough-and-tumble play, when seeking sexual contacts [344], or in anticipation of other incentive stimuli, such as rewarding physical activity in a running wheel [345]. Thus, hedonia should not be understood as a passive state of pleasant satisfaction (consummatory or post-consummatory state) but as an active state associated with the expectation of rewarding stimuli or the anticipation of additional rewarding stimuli. The state of hedonia is, therefore, a motorically active state, not only with the expectation of rewarding stimuli but also a state of actively looking for such stimuli, acquiring them and, at the same time, emitting honest signals to other conspecifics. This state is dopamine dependent and pharmacological activation of this system by psychostimulant agents, such as amphetamine or cocaine, always induced vigorous locomotor activity [291,346,347]. The magnitude of locomotor activity, however, is subject to individual differences, a basic level of spontaneous locomotor activity, or the intensity of the inborn response to novelty [348,349].

The direct physiological evidence for hedonia comes from self-stimulation behavior, during which rats volitionally deliver electrical stimulation to their own brains, or from place-preference behavior. Using electrical brain stimulation, all brain regions that induced emission of 50 kHz vocalizations by electrostimulation (e.g., nucleus accumbens, ventral pallidum, lateral preoptic area, lateral hypothalamus, ventral tegmental area) are also known from previous studies to support vigorous self-stimulation behavior [350]. Place-preference behavior was reported after amphetamine injections that induced emission of 50 kHz vocalizations [344], confirming its hedonic nature. Despite suggestions that 50 kHz calls might be an (anxious) indicator of negative reinforcement learning [351], a recent pharmacological study has confirmed that emission of amphetamine-induced 50 kHz vocalizations reflect a hedonic state that is resistant to anxiogenic agents and, therefore, does not reflect anxiety [352]. Moreover, rats can also learn self-injection of amphetamine directly into the shell of the nucleus accumbens, further indicating hedonic nature of this activation [353].

Emission of 50 kHz vocalizations that signal the hedonic state is perceived by receivers also as a positive and rewarding signal that can initiate a similar hedonic state in the recipients and prompt the rats to look for the cause of this behavior. This process or rapid generation of emotional arousal in the brains of receivers of vocalizations was termed ethotransmission, as a particularly fast and specific form of a broader category of behavioral transmission called emotional contagion [221]. It was even postulated that the emission of vocalizations directly targets the emotional systems of the listeners, impelling them to change their behavior [240]. Vocalization is an honest signal in rats as laughter is, in general, an honest signal in human spontaneous behavior [354]. Hence, rats showed an approach behavior to the source of the playback of 50 kHz calls as well as self-application behavior of 50 kHz calls [65,70,71,73] (for other details, see Section 2.2, above).

3.6.2. Interpretation of Pharmacological Studies Inducing 50 kHz Call Emission

Results of pharmacological studies provided further support for vocal expression of the hedonic state. Application of dopaminergic drugs (cocaine, heroin, amphetamine, methamphetamine, apomorphine, quinpirole, methylphenidate) into the terminal fields of the ascending mesolimbic dopaminergic system potentiated the physiological effects of this system and induced significant emission of 50 kHz calls over the control levels [289,290,355–358]. It may be speculated that with higher doses of the drugs, this potentiation resulted in stronger hedonia than that in physiological situations, and it created a state of euphoria. This pharmacologically induced euphoric state, which has some features of mania [359], is believed to be of the same nature as the hedonic state caused by rewarding self-stimulation because all euphorigenic drugs lowered the threshold for intracranial electrical self-stimulation [360,361].

Pharmacological studies of rat 50 kHz vocalizations appeared to be a useful approach to understand the rewarding and motivational properties of drugs of abuse and the development of drug addiction in humans [362,363]. The question arose as to what value the emission of 50 kHz vocalizations expresses. Is this pure hedonic value (pleasure and liking), motivational value (wanting and motivation of incentive salience), or prediction value (expecting by learning) [343]? All these values may have separate neural mechanisms [364]. It was initially postulated that the mesolimbic dopaminergic system is mostly responsible for "wanting", while the hedonic state ("liking") is associated with the opioid system [365,366]. However, the emission of frequency-modulated 50 kHz calls that was induced by rewarding cues was generally found to be signaling the "liking" state with intermixed "wanting", depending on the intensity of the motivational state of the animal [367]. Rats that attribute incentive salience to reward cues will have difficulty resisting them and were suggested to be prone to develop addiction [368].

The emission of 50 kHz vocalizations during dopamine-dependent emotional arousal (hedonia) is the activity with the signaling to conspecifics of all the aspects of positive expectation, wanting, and liking with an elevated level of locomotor activity at the same time [290,291]. Recent studies confirmed that brief optogenetic activation of the accumbens medium spiny neurons with D2 dopamine receptors increases the dopaminergic activity via effects on the ventral tegmental dopamine neurons and increases positive motivation [369] (although prolonged stimulation causes aversive effects [265]; see Section 3.2, above).

3.6.3. Morphine and Emission of 50 kHz Calls

Morphine has rewarding properties but also some other unique characteristics, so it warrants a separate subsection. Acute application of morphine, a mostly μ-opioid receptor agonist, did not elevate or induce 50 kHz ultrasonic vocalizations in rats, and even had decreasing effects on the emission of these calls after withdrawal [370–372]. However, morphine was reported as changing the acoustic features of some subtypes of 50 kHz calls and causing a strong place-preference response [370,371]. Significant place preference was observed after intracerebral injections of synthetic peptide, μ-opioid DAMGO, directly into the ventral tegmental area, the origin site of the ascending dopaminergic mesolimbic system [350]. It was also observed that some rats emitted significant numbers of 50 kHz calls after the DAMGO injection into the ventral tegmental area (DAMGO vocalizers) while other animals did not emit 50 kHz calls (DAMGO non-vocalizers). Rats that emitted significantly more 50 kHz calls after the drug than the control (DAMGO vocalizers) showed strong place preference while the animals that did not show any increase in calling failed to show place preference [350]. It seems that opioid system is involved in the hedonic state but in a different way than the dopamine system and this is particularly observed during the long-lasting effects of drugs on the emission of 50 kHz calls [371].

The intra-accumbens injection of morphine increased social play in rats and the response was antagonized by the antagonist, naloxone [373]. Moreover, application of morphine in a certain dose-range and time after application had a decreasing effect on locomotor activity as well as anxiolytic, analgesic, and pain-alleviating effects, including a decrease in the emission of 22 kHz calls that signaled anxiety after painful stimuli [374–376]. These observations may indicate two reward subsystems: one early and active, dopamine-dependent subsystem; and the other with less activity and limited or no calling—an opiate-dependent subsystem [377,378] that is active during the later phase of the rewarding process.

The process of seeking and obtaining positive stimuli (preparatory phase) and experiencing pleasurable stimuli (consummatory phase) is governed by a central process that was termed hedonesthesia more than 40 years ago. It was postulated that hedonesthesia is an active and critical process for positive motivated behavior [379,380]. In the light of current knowledge, hedonesthesia is the process of appetitive emotional arousal, driven by the ascending mesolimbic dopaminergic system as well as the animal's concomitant motor activity aiding in obtaining the positive stimuli. This process will involve many transmitters, for instance, norepinephrine at the cognitive arousal phase, dopamine at the positive emotional arousal phase, and possibly opioids at the later rewarding phase.

4. General Conclusions

The review summarized 22 functions of vocalizations, divided into eight groups, that play a role in rat behavior. Roughly, half of the functions are associated with negative emotional states and half with positive ones. The role of vocal communication is situation dependent and changes over a rat's life, from a basic, life-preservation role in infants and the development of social skills in play behavior, to the resolution of social conflicts and the organization of the social group in adults as well as defense against external threats and dangers. Different types of calls in different situations and at different stages of animal life may serve as a qualitative and quantitative measure of the functioning of the animal emotional system in physiological and pathological conditions. These basic animal emotional systems are homolog to basic human affective systems—both as to neurophysiological and neurochemical mechanisms—and rat expression of emotional arousal may be used in many preclinical models.

All the rat vocal expressions, regardless of their valence, are initiated by the mechanisms of emotional arousal and are emitted in biologically important situations. The term "arousal" is used here in the same sense as in the original discovery of the ascending reticular activating system [381–383], i.e., as a diffuse and extensive projection systems ascending from the brainstem and directly or indirectly changing ongoing activity in the entire brain.

Emotional arousal leads to the development of one of two opposite states differing in valence: the positive, hedonic, appetitive state or the negative, anhedonic, aversive state. These two arousal states are signaled by species-specific and valence-specific ultrasonic vocalizations that are emitted to influence the behavior of other conspecifics. Pharmacological studies have proven that these vocalizations reliably reflect emotional valence and point to sets of specific receptors responsible for the appetitive or aversive state, homolog to basic limbic human brain processes.

The appetitive state is initiated by ascending mesolimbic dopaminergic projections to some forebrain structure with a hot spot in the shell of the nucleus accumbens, and releases dopamine, while the aversive state is initiated by the ascending mesolimbic cholinergic system targeting many medial diencephalic and forebrain limbic structures with hot spots in the medial hypothalamic-preoptic area and lateral septum and the release of acetylcholine. Massive release of any of these two transmitters has the capacity to rapidly change the animal's state.

Large numbers of behavioral studies led to the conclusion that activity of the appetitive, dopaminergic system develops an active state of hedonia (pleasure in human terms) with the concurrent emission of 50 kHz vocalizations and an accompanying increase in motor activity to approach and acquire the appetitive stimuli, while the activity of the aversive, cholinergic system develops a defensive state of anxiety (displeasure) with the concurrent emission of 22 kHz calls, a decrease in motor activity and the avoidance of unpleasant stimuli.

Consistent congruence of many lines of investigation lead to the conclusion that the brain is equipped with two separate emotional arousal systems that prepare the animal for two opposite behavioral outcomes, and these systems work in parallel with the cognitive arousal. This review supports the hypothesis that all types of rat vocalizations, serving all biological functions, are driven by emotional arousal. Neural mechanisms initiating emotional arousal, positive or negative, are, therefore, common in fulfilling any of these functions.

The consistent association of 22 kHz and 50 kHz vocalizations with aversive or appetitive states, respectively, and the dual emotional arousal system makes these vocalizations particularly useful for numerous preclinical studies and models, particularly in physiological, psychological, neurological, psychiatric, and neurodevelopmental investigations. Therefore, rat ultrasonic vocalizations have been used in studies of human social psychopathologies [384], the screening of drugs for numerous conditions, particularly anxiolytic and antidepressant drugs [385–387], studies of schizophrenia [388], Parkinson's disease [389], bipolar disorder [390], post-traumatic stress disorder [391], alcohol use disorders [392,393], neurodevelopmental damages [394], immunity [395], affective component of pain [229], addiction [216,335], effects of malnutrition [396] and many other disorders and diseases. The present review should help in the interpretation of the results of these and future studies.

Funding: This article was not supported by external funding.

Institutional Review Board Statement: Not applicable for the review.

Acknowledgments: The author would like to express his thanks to Seal Li, Senior Managing Editor of *Brain Sciences* for the invitation and encouragement to write this review article. The author also directs his sincere thanks to his colleagues for the critical reading of this text, and particularly to Markus Wöhr for his constructive and useful comments.

Conflicts of Interest: The author declares no conflict of interest.

References

1. Bass, A.H.; Gilland, E.H.; Baker, R. Evolutionary Origins for Social Vocalization in a Vertebrate Hindbrain-spinal Compartment. *Science* **2008**, *321*, 417–421. [CrossRef]
2. Bass, A.H.; Chagnaud, B.P. Shared Developmental and Evolutionary Origins for Neural Basis of Vocal-acoustic and Pectoral-gestural Signaling. *Proc. Natl. Acad. Sci. USA* **2012**, *109* (Suppl. 1), 10677–10684. [CrossRef] [PubMed]
3. Bass, A.H. Central Pattern Generator for Vocalization: Is There a Vertebrate Morphotype? *Curr. Opin. Neurobiol.* **2014**, *28*, 94–100. [CrossRef]
4. Bass, A.H.; Baker, R. Phenotypic Specification of Hindbrain Rhombomeres and the Origins of Rhythmic Circuits in Vertebrates. *Brain Behav. Evol.* **1997**, *50* (Suppl. 1), 3–16. [CrossRef] [PubMed]
5. Chagnaud, B.P.; Baker, R.; Bass, A.H. Vocalization Frequency and Duration are Coded in Separate Hindbrain Nuclei. *Nat. Commun.* **2011**, *2*, 346. [CrossRef]
6. Yajima, Y.; Hayashi, Y. Ambiguous Motoneurons Discharging Synchronously with Ultrasonic Vocalization in Rats. *Exp. Brain Res.* **1983**, *50*, 359–366. [CrossRef]
7. Clack, J.A. Devonian Tetrapod Trackways and Trackmakers; A Review of the Fossils and Footprints. *Palaeogeogr. Palaeoclimatol. Palaeoecol.* **1997**, *130*, 227–250. [CrossRef]
8. Porges, S. The Polyvagal Theory: Phylogenetic Substrates of a Social Nervous System. *Int. J. Psychophysiol.* **2001**, *42*, 123–146. [CrossRef]
9. Simola, N.; Granon, S. Ultrasonic Vocalizations as a Tool in Studying Emotional States in Rodent Models of Social Behavior and Brain Disease. *Neuropharmacology* **2019**, *159*, 107420. [CrossRef]
10. Mainka, T.; Balint, B.; Gövert, F.; Kurvits, L.; van Riesen, C.; Kühn, A.A.; Tijssen, M.A.J.; Lees, A.J.; Müller-Vahl, K.; Bhatia, K.P.; et al. The Spectrum of Involuntary Vocalizations in Humans: A Video Atlas. *J. Mov. Disord.* **2019**, *34*, 1774–1791. [CrossRef] [PubMed]
11. Frederiksen, J.K.; Slobodchikoff, C.N. Referential Specificity in the Alarm Calls of the Black-tailed Prairie Dog. *Ethol. Ecol. Evol.* **2007**, *19*, 87–99. [CrossRef]
12. Slobodchikoff, C.N.; Paseka, A.; Verdolin, J.L. Prairie Dog Alarm Calls Encode Labels about Predator Colors. *Anim. Cogn.* **2009**, *12*, 435–439. [CrossRef] [PubMed]
13. Wilson-Henjum, G.E.; Job, J.R.; McKenna, M.F.; Shannon, G.; Wittemyer, G. Alarm Call Modification by Prairie Dogs in the Presence of Juveniles. *J. Ethol.* **2019**, *37*, 167–174. [CrossRef]
14. Rendall, D.; Owren, M.J.; Ryan, M.J. What do Animal Signals Mean? *Anim. Behav.* **2009**, *78*, 233–240. [CrossRef]
15. Brudzynski, S.M. Principles of Rat Communication: Quantitative Parameters of Ultrasonic Calls in Rats. *Behav. Genet.* **2005**, *35*, 85–92. [CrossRef]
16. Costantini, F.; D'Amato, F.R. Ultrasonic Vocalizations in Mice and Rats: Social Contexts and Functions. *Acta Zool. Sin.* **2005**, *52*, 619–633.
17. Portfors, C.V. Types and Functions of Ultrasonic Vocalizations in Laboratory Rats and Mice. *J. Am. Assoc. Lab. Anim. Sci.* **2007**, *46*, 28–34.
18. Brudzynski, S.M. Communication of Adult Rats by Ultrasonic Vocalization: Biological, Sociobiological, and Neuroscience Approaches. *ILAR J.* **2009**, *50*, 43–50. [CrossRef] [PubMed]
19. McLean, P.D. The Limbic System with Respect to Self-preservation and the Preservation of the Species. *J. Nerv. Ment. Dis.* **1958**, *127*, 1–11. [CrossRef] [PubMed]

20. Nauta, W.J. Hippocampal Projections and Related Neural Pathways to the Midbrain in the Cat. *Brain* **1958**, *81*, 319–340. [CrossRef] [PubMed]
21. Nauta, W.J. The Central Visceromotor System: A General Survey. In *Limbic System Mechanisms and Autonomic Function*; Hockman, C.H., Charles, C., Eds.; Thomas Publishers: Springfield, IL, USA, 1972; pp. 21–38.
22. Newman, J.D. Neural Circuits Underlying Crying and Cry Responding in Mammals. *Behav. Brain Res.* **2007**, *182*, 155–165. [CrossRef] [PubMed]
23. Hofer, M.A.; Masmela, J.R.; Brunelli, S.A.; Shair, H.N. Behavioral Mechanisms for Active Maternal Potentiation of Isolation Calling in Rat Pups. *Behav. Neurosci.* **1999**, *113*, 51–61. [CrossRef] [PubMed]
24. Hofer, M.A. Evolution of the Infant Separation Call: Rodent Ultrasonic Vocalization. In *Handbook of Mammalian Vocalization. An Integrative Neuroscience Approach*; Brudzynski, S.M., Ed.; Academic Press/Elsevier: Amsterdam, The Netherlands, 2010; pp. 29–35.
25. Scott, J.P. The Analysis of Social Organization in Animals. *Ecology* **1956**, *37*, 213–221. [CrossRef]
26. White, N.R.; Adox, R.; Reddy, A.; Barfield, R.J. Regulation of rat Maternal Behavior by Broadband Pup Vocalizations. *Behav. Neural Biol.* **1992**, *58*, 131–137. [CrossRef]
27. Gardner, C.R. Distress Vocalization in Rat Pups. A Simple Screening Method for Anxiolytic Drugs. *J. Pharmacol. Methods.* **1985**, *14*, 181–187. [CrossRef]
28. Iacobucci, P.; Colonnello, V.; Fuchs, T.; D'Antuono, L.; Panksepp, J. Differential Ultrasonic Indices of Separation Distress in the Presence and Absence of Maternal Cues in Infant Rats Bred for High and Low Positive Social Affect. *Acta Neuropsychiatr.* **2013**, *25*, 289–296. [CrossRef] [PubMed]
29. Allin, J.T.; Banks, E.M. Functional Aspects of Ultrasound Production by Infant Albino Rats (*Rattus norvegicus*). *Anim. Behav.* **1972**, *20*, 175–185. [CrossRef]
30. Ehret, G. Infant Rodent Ultrasounds–A Gate to the Understanding of Sound Communication. *Behav. Genet.* **2005**, *35*, 19–29. [CrossRef] [PubMed]
31. Farrell, W.J.; Alberts, J.R. Stimulus Control of Maternal Responsiveness to Norway Rat (*Rattus norvegicus*) Pup Ultrasonic Vocalizations. *J. Comp. Psychol.* **2002**, *116*, 297–307. [CrossRef]
32. Smotherman, W.P.; Bell, R.W.; Starzec, J.; Elias, J.; Zachman, T.A. Maternal Responses to Infant Vocalizations and Olfactory Cues in Rats and Mice 1. *Behav. Biol.* **1974**, *12*, 55–66. [CrossRef]
33. Smotherman, W.P.; Bell, R.W.; Hershberger, W.A.; Coover, G.D. Orientation to rat Pup Cues: Effects of Maternal Experiential History. *Anim. Behav.* **1978**, *26*, 265–273. [CrossRef]
34. Brudzynski, S.M.; Kehoe, P.; Callahan, M. Sonographic Structure of Isolation-induced Ultrasonic Calls of Rat Pups. *Dev. Psychobiol.* **1999**, *34*, 195–204. [CrossRef]
35. Spence, H.R.; Aslam, A.M.; Hofer, M.A.; Brunelli, S.A.; Shair, H.N. Vocal Coselection in Rat Pup Ultrasonic Vocalizations. *Ecol. Evol.* **2016**, *6*, 1922–1929. [CrossRef] [PubMed]
36. Schwarting, R.K.W.; Wöhr, M. Isolation-induced Ultrasonic Vocalizations in Pups: A Comparison between Long-Evans, Sprague-Dawley, and Wistar Rats. *Dev. Psychobiol.* **2018**, *60*, 534–543. [CrossRef]
37. Schwarting, R.K.W. Ultrasonic Vocalization In Juvenile and Adult Male Rats: A Comparison among Stocks. *Physiol. Behav.* **2018**, *191*, 1–11. [CrossRef] [PubMed]
38. Takahashi, L.K. Developmental Expression of Defensive Responses during Exposure to Conspecific Adults in Preweanling Rats (*Rattus norvegicus*). *J. Comp. Psychol.* **1992**, *106*, 69–77. [CrossRef]
39. Takahashi, L.K. Ontogeny of Behavioral Inhibition Induced by Unfamiliar Adult Male Conspecifics in Preweanling Rats. *Physiol. Behav.* **1992**, *52*, 493–498. [CrossRef]
40. Zimmerberg, B.; Rosenthal, A.J.; Stark, A.C. Neonatal Social Isolation Alters Both Maternal and Pup Behaviors in Rats. *Dev. Psychobiol.* **2003**, *42*, 52–63. [CrossRef]
41. Rilling, J.K.; Young, L.J. The Biology of Mammalian Parenting and Its Effect on Offspring Social Development. *Science* **2014**, *345*, 771–776. [CrossRef]
42. Klug, H.; Bonsall, M.B. What are the Benefits of Parental Care? The Importance of Parental Effects on Developmental Rate. *Ecol. Evol.* **2014**, *4*, 2330–2351. [CrossRef]
43. Wöhr, M.; Schwarting, R.K. Maternal Care, Isolation-induced Infant Ultrasonic Calling, and Their Relations to Adult Anxiety-related Behavior in the Rat. *Behav. Neurosci.* **2008**, *122*, 310–330. [CrossRef] [PubMed]
44. Blaze, J.; Scheuing, L.; Roth, T.L. Differential Methylation of Genes in the Medial Prefrontal Cortex of Developing and Adult Rats Following Exposure to Maltreatment or Nurturing Care during Infancy. *Dev. Neurosci.* **2013**, *35*, 306–316. [CrossRef]
45. Naito, H.; Inoue, M.; Makino, J. Ultrasonic Isolation Calls in Genetically High- and Low-emotional Rat Pups. *Exp. Anim.* **2000**, *49*, 289–294. [CrossRef]
46. Ise, S.; Ohta, H. Power Spectrum Analysis of Ultrasonic Vocalization Elicited by Maternal Separation in Rat Pups. *Brain Res.* **2009**, *1283*, 58–64. [CrossRef]
47. Kaidbey, J.H.; Ranger, M.; Myers, M.M.; Anwar, M.; Ludwig, R.J.; Schulz, A.M.; Barone, J.L.; Kolacz, J.; Welch, M.G. Early Life Maternal Separation and Maternal Behaviour Modulate Acoustic Characteristics of Rat Pup Ultrasonic Vocalizations. *Sci. Rep.* **2019**, *9*, 19012. [CrossRef]
48. Shair, H.N. Acquisition and Expression of a Socially Mediated Separation Response. *Behav. Brain Res.* **2007**, *182*, 180–192. [CrossRef] [PubMed]

49. Shair, H.N. Parental Potentiation of Vocalization as a Marker for Filial Bonds in Infant Animals. *Dev. Psychobiol.* **2014**, *56*, 1689–1697. [CrossRef] [PubMed]
50. Groenink, L.; Verdouw, P.M.; van Oorschot, R.; Olivier, B. Models of Anxiety: Ultrasonic Vocalizations of Isolated Rat Pups. *Curr. Protoc. Pharmacol.* **2008**. [CrossRef]
51. Brunelli, S.A.; Aviles, J.A.; Gannon, K.S.; Branscomb, A.; Shacham, S. PRX-00023, a Selective Serotonin 1A Receptor Agonist, Reduces Ultrasonic Vocalizations in Infant Rats Bred for High Infantile Anxiety. *Pharmacol. Biochem. Behav.* **2009**, *94*, 8–15. [CrossRef] [PubMed]
52. Mogi, K.; Takakuda, A.; Tsukamoto, C.; Ooyama, R.; Okabe, S.; Koshida, N.; Nagasawa, M.; Kikusui, T. Mutual Mother-infant Recognition in Mice: The Role of Pup Ultrasonic Vocalizations. *Behav. Brain Res.* **2017**, *325*, 138–146. [CrossRef]
53. Nagasawa, M.; Okabe, S.; Mogi, K.; Kikusui, T. Oxytocin and Mutual Communication in Mother-infant Bonding. *Front. Hum. Neurosci.* **2012**, *6*, 31. [CrossRef]
54. Schweinfurth, M.K.; Neuenschwander, J.; Engqvist, L.; Schneeberger, K.; Rentsch, A.K.; Gygax, M.; Taborsky, M. Do Female Norway Rats Form Social Bonds? *Behav. Ecol. Sociobiol.* **2017**, *71*, 98. [CrossRef]
55. Soltis, J. The Signal Functions of Early Infant Crying. *Behav. Brain Sci.* **2004**, *27*, 443–490. [CrossRef] [PubMed]
56. Panksepp, J. Emotional Causes and Consequences of Social-affective Vocalization. In *Handbook of Mammalian Vocalization. An Integrative Neuroscience Approach*; Brudzynski, S.M., Ed.; Academic Press/Elsevier: Amsterdam, The Netherlands, 2010; pp. 201–208.
57. Malinowski, B. The Problem of Meaning in Primitive Languages. In *The Meaning of Meaning. A Study of the Influence of Language upon Thought and of the Science of Symbolism*; Ogden, C.K., Richards, I.A., Eds.; Kegan Paul, Trench and Trubner: London, UK, 1923; pp. 296–336.
58. Eibl-Eibesfeldt, I. *Ethology. The Biology of Behavior*; Holt, Rinehart and Winston, Inc.: New York, NY, USA, 1970; p. 530.
59. Jordania, J. Times to Fight and Times to Relax: Singing and Humming at the Beginnings of Human Evolutionary History. *Kadmos* **2009**, *1*, 272–277. [CrossRef]
60. Schwarting, R.K.W.; Jegan, N.; Wöhr, M. Situational Factors, Conditions and Individual Variables Which Can Determine Ultrasonic Vocalizations in Male Adult Wistar Rats. *Behav. Brain Res.* **2007**, *182*, 208–222. [CrossRef] [PubMed]
61. Wöhr, M.; Houx, B.; Schwarting, R.K.W.; Spruijt, B. Effects of Experience and Context on 50-kHz Vocalizations in Rats. *Physiol. Behav.* **2008**, *93*, 766–776. [CrossRef]
62. Brudzynski, S.M. Pharmacology of Ultrasonic Vocalizations in Adult Rats: Significance, Call Classification and Neural Substrate. *Curr. Neuropharmacol.* **2015**, *13*, 180–192. [CrossRef] [PubMed]
63. Brudzynski, S.M.; Pniak, A. Social Contacts and Production of 50-kHz Short Ultrasonic Calls in Adult Rats. *J. Comp. Psychol.* **2002**, *116*, 73–82. [CrossRef]
64. Seffer, D.; Rippberger, H.; Schwarting, R.K.W.; Wöhr, M. Pro-social 50-kHz Ultrasonic Communication in Rats: Post-weaning but not Post-adolescent Social Isolation Leads to Social Impairments-phenotypic Rescue by Re-socialization. *Front. Behav. Neurosci.* **2015**, *9*, 102. [CrossRef] [PubMed]
65. Wöhr, M.; Seffer, D.; Schwarting, R.K.W. Studying Socio-affective Communication in Rats through Playback of Ultrasonic Vocalizations. *Curr. Protoc. Neurosci.* **2016**, *75*, 8.35.1–8.35.17. [CrossRef]
66. Rojas-Carvajal, M.; Brenes, J.C. Acute Stress Differentially Affects Grooming Subtypes and Ultrasonic Vocalisations in the Open-field and Home-cage Test in Rats. *Behav. Process.* **2020**, *176*, 104140. [CrossRef]
67. Wöhr, M. Ultrasonic Communication in Rats: Appetitive 50-kHz Ultrasonic Vocalizations as Social Contact Calls. *Behav. Ecol. Sociobiol.* **2018**, *72*, 14. [CrossRef]
68. Brudzynski, S.M.; Fletcher, N.H. Rat Ultrasonic Vocalization: Short-range Communication. In *Handbook of Mammalian Vocalization. An Integrative Neuroscience Approach*; Brudzynski, S.M., Ed.; Academic Press/Elsevier: Amsterdam, The Netherlands, 2010; pp. 69–76.
69. Credner, S.; Burda, H.; Ludescher, F. Acoustic Communication Underground: Vocalization Characteristics in Subterranean Social Mole-rats (*Cryptomys, sp.*, Bathyergidae). *J. Comp. Physiol.* **1997**, *180*, 245–255. [CrossRef]
70. Wöhr, M.; Schwarting, R.K.W. Ultrasonic Communication in Rats: Can Playback of 50-kHz Calls Induce Approach Behavior? *PLoS ONE* **2007**, *2*, e1365. [CrossRef] [PubMed]
71. Seffer, D.; Schwarting, R.K.W.; Wöhr, M. Pro-social Ultrasonic Communication in Rats: Insights from Playback Studies. *J. Neurosci. Methods.* **2014**, *234*, 73–81. [CrossRef]
72. Davidson, N.B.; Hurst, J.L. Testing the Potential of 50 kHz Rat Calls as a Species-specific Rat Attractant. *PLoS ONE* **2019**, *14*, e0211601. [CrossRef] [PubMed]
73. Burgdorf, J.; Kroes, R.A.; Moskal, J.R.; Pfaus, J.G.; Brudzynski, S.M.; Panksepp, J. Ultrasonic Vocalizations of Rats (*Rattus norvegicus*) during Mating, Play, and Aggression: Behavioral Concomitants, Relationship to Reward, and Self-administration of Playback. *J. Comp. Psychol.* **2008**, *122*, 357–367. [CrossRef] [PubMed]
74. Barnett, S.A. *The Rat. A Study in Behavior*, 2nd ed.; The University of Chicago Press: Chicago, IL, USA, 1967; p. 318.
75. Schweinfurth, M.K. The Social Life of Norway Rats (*Rattus norvegicus*). *eLife* **2020**, *9*, e54020. [CrossRef] [PubMed]
76. Glancy, J.; Groß, R.; Stone, J.V.; Wilson, S.P. A Self-organising Model of Thermoregulatory Huddling. *PLoS Comput. Biol.* **2015**, *11*, e1004283. [CrossRef]

77. Alberts, J.R. Huddling by Rat Pups: Multisensory Control of Contact Behavior. *J. Comp. Physiol. Psychol.* **1978**, *92*, 220–230. [CrossRef] [PubMed]
78. Schank, J.C.; Alberts, J.R. Self-organized Huddles of Rat Pups Modeled by Simple Rules of Individual Behavior. *J. Theor. Biol.* **1997**, *189*, 11–25. [CrossRef]
79. Bowen, M.T.; Keats, K.; Kendig, M.D.; Cakic, V.; Callaghan, P.D.; McGregor, I.S. Aggregation in Quads but not Pairs of Rats Exposed to Cat Odor or Bright Light. *Behav. Process.* **2012**, *90*, 331–336. [CrossRef]
80. Bowen, M.T.; Kevin, R.C.; May, M.; Staples, L.G.; Hunt, G.E.; McGregor, I.S. Defensive Aggregation (huddling) in *Rattus norvegicus* Toward Predator Odor: Individual Differences, Social Buffering Effects and Neural Correlates. *PLoS ONE* **2013**, *8*, e68483. [CrossRef]
81. Barnett, S.A. An Analysis of Social Behaviour in Wild Rats. *Proc. Zool. Soc.* **1958**, *130*, 107–152. [CrossRef]
82. Edmunds, M. *Defence in Animals. A Survey of Anti-predator Defences*; Longman Group, Ltd.: New York, NY, USA, 1974; p. 357.
83. Park, E.S.; Hinsz, V.B. "Strength and Safety in Numbers": A Theoretical Perspective on Group Influences on Approach and Avoidance Motivation. *Motiv. Emot.* **2006**, *30*, 135–142. [CrossRef]
84. Burke, C.J.; Kisko, T.M.; Swiftwolfe, H.; Pellis, S.M.; Euston, D.R. Specific 50-kHz Vocalizations are Tightly Linked to Particular Types of Behavior in Juvenile Rats Anticipating Play. *PLoS ONE* **2017**, *12*, e0175841. [CrossRef]
85. Kisko, T.M.; Braun, M.D.; Michels, S.; Witt, S.H.; Rietschel, M.; Culmsee, C.; Schwarting, R.K.W.; Wöhr, M. *Cacna1c* Haploinsufficiency Leads to Pro-social 50-kHz Ultrasonic Communication Deficits in Rats. *Dis. Model. Mech.* **2018**, *11*, dmm034116. [CrossRef]
86. Redecker, T.M.; Kisko, T.M.; Schwarting, R.K.W.; Wöhr, M. Effects of *Cacna1c* Haploinsufficiency on Social Interaction Behavior and 50-kHz Ultrasonic Vocalizations in Adult Female Rats. *Behav. Brain Res.* **2019**, *367*, 35–52. [CrossRef] [PubMed]
87. Boinski, S. Vocal Coordination of Troop Movement among White-faced Capuchin Monkeys, Cebus capucinus. *Am. J. Primatol.* **1993**, *30*, 85–100. [CrossRef] [PubMed]
88. Campbell, A.F.; Boinski, S. Use of Trill Vocalizations to Coordinate Troop Movement among White-faced Capuchins: A Second Field Test. *Behaviour* **1995**, *132*, 875–901. [CrossRef]
89. Cloutier, S.; Wahl, K.; Baker, C.; Newberry, R.C. The Social Buffering Effect of Playful Handling on Responses to Repeated Intraperitoneal Injections in Laboratory Rats. *J. Am. Assoc. Lab. Anim. Sci.* **2014**, *53*, 168–173.
90. Cloutier, S.; LaFollette, M.R.; Gaskill, B.N.; Panksepp, J.; Newberry, R.C. Tickling, a Technique for Inducing Positive Affect when Handling Rats. *J. Vis. Exp.* **2018**, *135*, e57190. [CrossRef] [PubMed]
91. Kikusui, T.; Winslow, J.T.; Mori, Y. Social Buffering: Relief from Stress and Anxiety. *Philos. Trans. R. Soc. Lond. B Biol. Sci.* **2006**, *361*, 2215–2228. [CrossRef] [PubMed]
92. Okabe, S.; Yoshida, M.; Takayanagi, Y.; Onaka, T. Activation of Hypothalamic Oxytocin Neurons Following Tactile Stimuli in Rats. *Neurosci. Lett.* **2015**, *600*, 22–27. [CrossRef] [PubMed]
93. Okabe, S.; Takayanagi, Y.; Yoshida, M.; Onaka, T. Gentle Stroking Stimuli Induce Affiliative Responsiveness to Humans in Male Rats. *Sci. Rep.* **2020**, *10*, 9135. [CrossRef]
94. Dellu, F.; Piazza, P.V.; Mayo, W.; Le Moal, M.; Simon, H. Novelty seeking in Rats–Biobehavioral Characteristics and Possible Relationship with the Sensation-seeking Trait in Man. *Neuropsychobiology* **1996**, *34*, 136–145. [CrossRef]
95. Blanchard, M.M.; Mendelsohn, D.; Stamp, J.A. The HR/LR Model: Further Evidence as an Animal Model of Sensation Seeking. *Neurosci. Biobehav. Rev.* **2009**, *33*, 1145–1154. [CrossRef]
96. Saito, Y.; Yuki, S.; Seki, Y.; Kagawa, H.; Okanoya, K. Cognitive Bias in Rats Evoked by Ultrasonic Vocalizations Suggests Emotional Contagion. *Behav. Process.* **2016**, *132*, 5–11. [CrossRef] [PubMed]
97. Robakiewicz, I.; Polak, M.; Rawska, M.; Alberski, D.; Polowy, R.; Wytrychiewicz, K.; Syperek, M.; Matysiak, J.; Filipkowski, R.K. Stimulus-seeking in Rats is Accompanied by Increased c-Fos Expression in Hippocampal CA1 as Well as Short 22 kHz and Flat 50 kHz Calls. *Acta Neurobiol. Exp.* **2019**, *79*, 309–317. [CrossRef]
98. Mun, H.S.; Lipina, T.V.; Order, J.C. Ultrasonic Vocalizations in Mice during Exploratory Behavior are Context-dependent. *Front. Behav. Neurosci.* **2015**, *9*, 316. [CrossRef]
99. Kisko, T.M.; Wöhr, M.; Pellis, V.C.; Pellis, S.M. From Play to Aggression: High-frequency 50-kHz Ultrasonic Vocalizations as Play and Appeasement Signals in Rats. *Curr. Top. Behav. Neurosci.* **2017**, *30*, 91–108. [CrossRef] [PubMed]
100. Müller-Schwarze, D. Ludic Behavior in Young Mammals. In *Brain Development and Behavior*; Sterman, M.B., McGinty, D.J., Adinolfi, A.M., Eds.; Academic Press: New York, NY, USA, 1971; pp. 229–249.
101. Knutson, B.; Burgdorf, J.; Panksepp, J. Anticipation of Play Elicits High-frequency Ultrasonic Vocalizations in Young Rats. *J. Comp. Psychol.* **1998**, *112*, 65–73. [CrossRef] [PubMed]
102. Lukas, M.; Wöhr, M. Endogenous Vasopressin, Innate Anxiety, and the Emission of Pro-social 50-kHz Ultrasonic Vocalizations during Social Play Behavior in Juvenile Rats. *Psychoneuroendocrinology* **2015**, *56*, 35–44. [CrossRef]
103. Himmler, B.T.; Kisko, T.M.; Euston, D.R.; Kolb, B.; Pellis, S.M. Are 50-kHz Calls Used as Play Signals in the Playful Interactions of Rats? I. Evidence from the Timing and Context of Their Use. *Behav. Process.* **2014**, *106*, 60–66. [CrossRef] [PubMed]
104. Kisko, T.M.; Himmler, B.T.; Himmler, S.M.; Euston, D.R.; Pellis, S.M. Are 50-kHz Calls Used as Play Signals in the Playful Interactions of Rats? II. Evidence from the Effects of Devocalization. *Behav. Process.* **2015**, *111*, 25–33. [CrossRef] [PubMed]
105. Thompson, B.; Leonard, K.C.; Brudzynski, S.M. Amphetamine-induced 50 kHz Calls from Rat Nucleus Accumbens: A Quantitative Mapping Study and Acoustic Analysis. *Behav. Brain Res.* **2006**, *168*, 64–73. [CrossRef]
106. Gzielo, K.; Potasiewicz, A.; Hołuj, M.; Litwa, E.; Popik, P.; Nikiforuk, A. Valproic Acid Exposure Impairs Ultrasonic Communication in Infant, Adolescent and Adult Rats. *Eur. Neuropsychopharmacol.* **2020**, *41*, 52–62. [CrossRef] [PubMed]

107. Burgdorf, J.; Panksepp, J. Tickling Induces Reward in Adolescent Rats. *Physiol. Behav.* **2001**, *72*, 167–173. [CrossRef]
108. Panksepp, J.; Burgdorf, J. "Laughing" Rats and the Evolutionary Antecedents of Human Joy? *Physiol. Behav.* **2003**, *79*, 533–547. [CrossRef]
109. Burgdorf, J.; Kroes, R.A.; Moskal, J.R. Rough-and-tumble Play Induces Resilience to Stress in Rats. *Neuroreport* **2017**, *28*, 1122–1126. [CrossRef] [PubMed]
110. Burgdorf, J.; Panksepp, J.; Brudzynski, S.M.; Kroes, R.; Moskal, J.R. Breeding for 50-kHz Positive Affective Vocalization in Rats. *Behav. Genet.* **2005**, *35*, 67–72. [CrossRef]
111. Popik, P.; Potasiewicz, A.; Pluta, H.; Zieniewicz, A. High-frequency Ultrasonic Vocalizations in Rats in Response to Tickling: The Effects of Restraint Stress. *Behav. Brain Res.* **2012**, *234*, 223–227. [CrossRef] [PubMed]
112. Shimoju, R.; Shibata, H.; Hori, M.; Kurosawa, M. Stroking Stimulation of the Skin Elicits 50-kHz Ultrasonic Vocalizations in Young Adult Rats. *J. Physiol. Sci.* **2020**, *70*, 41. [CrossRef] [PubMed]
113. Hori, M.; Shimoju, R.; Tokunaga, R.; Ohkubo, M.; Miyabe, S.; Ohnishi, J.; Murakami, K.; Kurosawa, M. Tickling Increases Dopamine Release in the Nucleus Accumbens and 50 kHz Ultrasonic Vocalizations in Adolescent Rats. *Neuroreport* **2013**, *24*, 241–245. [CrossRef]
114. Hinchcliffe, J.K.; Mendl, M.; Robinson, E.S.J. Rat 50 kHz Calls Reflect Graded Tickling-induced Positive Emotion. *Curr. Biol.* **2020**, *30*, R1034–R1035. [CrossRef]
115. Reinhold, A.S.; Sanguinetti-Scheck, J.I.; Konstantin Hartmann, K.; Brech, M. Behavioral and Neural Correlates of Hide-and-seek in Rats. *Science* **2019**, *365*, 1180–1183. [CrossRef]
116. Cooke, B.M.; Shukla, D. Double Helix: Reciprocity between Juvenile Play and Brain Development. *Dev. Cogn. Neurosci.* **2011**, *1*, 459–470. [CrossRef]
117. Panksepp, K. The Ontogeny of Play in Rats. *Dev. Psychobiol.* **1981**, *14*, 327–332. [CrossRef] [PubMed]
118. Berlyne, D.E. *Conflict, Arousal, and Curiosity*; McGraw-Hill Book Comp., Inc.: New York, NY, USA, 1960; p. 350.
119. Nakayama, H. Changes in the Affect of Infants before and after Episodes of Crying. *Infant. Behav. Dev.* **2013**, *36*, 507–512. [CrossRef]
120. Nakayama, H. Emergence of Amae Crying in Early Infancy as a Possible Social Communication Tool between Infants and Mothers. *Infant. Behav. Dev.* **2015**, *40*, 122–130. [CrossRef]
121. Łopuch, S.; Popik, P. Cooperative Behavior of Laboratory Rats (*Rattus norvegicus*) in an Instrumental Task. *J. Comp. Psychol.* **2011**, *125*, 250–253. [CrossRef]
122. Burke, C.J.; Kisko, T.M.; Euston, D.R.; Pellis, S.M. Do Juvenile Rats Use Specific Ultrasonic Calls to Coordinate Their Social Play? *Anim. Behav.* **2018**, *140*, 81–92. [CrossRef]
123. Barfield, R.J.; Auerbach, P.; Geyer, L.A.; McIntosh, T.K. Ultrasonic Vocalizations in Rat Sexual Behavior. *Am. Zool.* **1979**, *19*, 469–480. [CrossRef]
124. Pertsov, S.S.; Koplik, E.V.; Karkishchenko, N.N.; Sudakov, K.V. Ultrasonic Vocalization of Rats in Various Motivational and Emotional States. *Bull. Exp. Biol. Med.* **2012**, *153*, 804–807. [CrossRef] [PubMed]
125. Sales, G.D. Ultrasound and Aggressive Behaviour in Rats and Other Small Mammals. *Anim. Behav.* **1972**, *20*, 88–100. [CrossRef]
126. Lore, D.L.R. Aggression and Social Experience in Domesticated Rats. *J. Comp. Physiol. Psychol.* **1975**, *88*, 917–923. [CrossRef]
127. Sales, G.D. Strain Differences in the Ultrasonic Behavior of Rats (*Rattus norvegicus*). *Amer. Zool.* **1979**, *19*, 513–527. [CrossRef]
128. Takahashi, L.K.; Thomas, D.A.; Barfield, R.J. Analysis of Ultrasonic Vocalizations Emitted by Residents during Aggressive Encounters among Rats (*Rattus norvegicus*). *J. Comp. Psychol.* **1983**, *97*, 207–212. [CrossRef]
129. Thomas, D.A.; Takahashi, L.K.; Barfield, R.J. Analysis of Ultrasonic Vocalizations Emitted by Intruders during Aggressive Encounters among Rats (*Rattus norvegicus*). *J. Comp. Psychol.* **1983**, *97*, 201–206. [CrossRef] [PubMed]
130. Vivian, J.A.; Miczek, K.A. Morphine Attenuates Ultrasonic Vocalization during Agonistic Encounters in Adult Male Rats. *Psychopharmacology* **1993**, *111*, 367–375. [CrossRef]
131. Scott, J.P.; Fredericson, E. The Causes of Fighting in Mice and Rats. *Physiol. Zool.* **1951**, *24*, 273–309. [CrossRef]
132. Scott, J.P. Agonistic Behavior of Mice and Rats: A Review. *Am. Zool.* **1966**, *6*, 683–701. [CrossRef]
133. Marler, P. The Evolution of Communication. In *How Animals Communicate*; Sebeok, T.A., Ed.; Indiana University Press: Bloomington, IN, USA, 1977; pp. 45–69.
134. Rieger, N.S.; Marler, C.A. The Function of Ultrasonic Vocalizations during Territorial Defence by Pair-bonded Male and Female California Mice. *Anim. Behav.* **2018**, *135*, 97–108. [CrossRef]
135. McFarland, D. (Ed.) *The Oxford Companion to Animal Behaviour*, 2nd ed.; Oxford University Press: Oxford, UK, 1987; p. 685.
136. Takeuchi, H.; Kawashima, S. Ultrasonic Vocalizations and Aggressive Behavior in Male Rats. *Physiol. Behav.* **1986**, *38*, 545–550. [CrossRef]
137. Assini, R.; Sirotin, Y.B.; Laplagne, D.D. Rapid Triggering of Vocalizations Following Social Interactions. *Curr. Biol.* **2013**, *23*, R996–R997. [CrossRef]
138. Burke, C.J.; Kisko, T.M.; Pellis, S.M.; Euston, D.R. Avoiding Escalation from Play to Aggression in Adult Male Rats: The Role of Ultrasonic Calls. *Behav. Process.* **2017**, *144*, 72–81. [CrossRef]
139. White, N.R.; Barfield, R.J. Effects of Male Pre-ejaculatory Vocalizations on Female Receptive Behavior in the Rat (*Rattus norvegicus*). *J. Comp. Psychol.* **1990**, *104*, 140–146. [CrossRef] [PubMed]
140. White, N.R.; Barfield, R.J. Role of the Ultrasonic Vocalization of the Female Rat (*Rattus norvegicus*) in Sexual Behavior. *J. Comp. Psychol.* **1987**, *101*, 73–81. [CrossRef]

141. White, N.R.; Barfield, R.J. Playback of Female Rat Ultrasonic Vocalizations during Sexual Behavior. *Physiol. Behav.* **1989**, *45*, 229–233. [CrossRef]
142. Thomas, D.A.; Barfield, R.J. Ultrasonic Vocalization of the Female Rat (*Rattus norvegicus*) during Mating. *Anim. Behav.* **1985**, *33*, 720–725. [CrossRef]
143. Geyer, L.A.; Barfield, R.J. Influence of Gonadal Hormones and Sexual Behavior on Ultrasonic Vocalization in Rats: I. Treatment of Females. *J. Comp. Physiol. Psychol.* **1978**, *92*, 438–446. [CrossRef] [PubMed]
144. Ågmo, A.; Snoeren, E.M.S. Silent or Vocalizing Rats Copulate in a Similar Manner. *PLoS ONE* **2015**, *10*, e0144164. [CrossRef] [PubMed]
145. Chu, X.; Snoeren, E.; Ågmo, A. Functions of Vocalization in Sociosexual Behaviors in Rats (*Rattus norvegicus*) in a Seminatural Environment. *J. Comp. Psychol.* **2017**, *131*, 10–18. [CrossRef] [PubMed]
146. Börner, A.; Hjemdahl, R.; Götz, T.; Gillian, R.; Brown, G.R. Ultrasonic Vocalizations of Female Norway Rats (*Rattus norvegicus*) in Response to Social Partners. *J. Comp. Psychol.* **2016**, *130*, 76–80. [CrossRef]
147. Willadsen, M.; Seffer, D.; Schwarting, R.K.W.; Wöhr, M. Rodent Ultrasonic Communication: Male Prosocial 50-kHz Ultrasonic Vocalizations Elicit Social Approach Behavior in Female Rats (*Rattus norvegicus*). *J. Comp. Psychol.* **2014**, *128*, 56–64. [CrossRef]
148. McIntosh, T.K.; Barfield, R.J.; Thomas, D. Electrophysiological and Ultrasonic Correlates of Reproductive Behavior in the Male Rat. *Behav. Neurosci.* **1984**, *98*, 1100–1103. [CrossRef]
149. van der Poel, A.M.; Miczek, K.A. Long Ultrasonic Calls in Male Rats Following Mating, Defeat and Aversive Stimulation: Frequency Modulation and Bout Structure. *Behaviour* **1991**, *119*, 127–142. [CrossRef]
150. Barfield, R.J.; Geyer, L.A. Sexual Behavior: Ultrasonic Postejaculatory Song of the Male Rat. *Science* **1972**, *176*, 1349–1350. [CrossRef]
151. Anisko, J.J.; Suer, S.F.; McClintock, M.K.; Adler, N.T. Relation between 22-kHz Ultrasonic Signals and Sociosexual Behavior in Rats. *J. Comp. Physiol. Psychol.* **1978**, *92*, 821–829. [CrossRef]
152. Sachs, B.D.; Biały, M. Female Presence during Postejaculatory Interval Facilitates Penile Erection and 22-kHz Vocalization in Male Rats. *Behav. Neurosci.* **2000**, *114*, 1203–1208. [CrossRef] [PubMed]
153. Barfield, R.J.; Geyer, L.A. The Ultrasonic Postejaculatory Vocalization and the Postejaculatory Refractory Period of the Male Rat. *J. Comp. Physiol. Psychol.* **1975**, *88*, 723–734. [CrossRef]
154. Geyer, L.A.; Barfield, R.J. Regulation of Social Contact by the Female Rat during the Postejaculatory Interval. *Anim. Learn. Behav.* **1980**, *8*, 679–685. [CrossRef]
155. Biały, M.; Bogacki-Rychlik, W.; Kasarello, K.; Nikolaev, E.; Sajdel-Sułkowska, E.M. Modulation of 22-kHz Postejaculatory Vocalizations by Conditioning to New Place: Evidence for Expression of a Positive Emotional State. *Behav. Neurosci.* **2016**, *130*, 415–421. [CrossRef] [PubMed]
156. Thomas, D.A.; Howard, S.B.; Barfield, R.J. Male-produced Postejaculatory 22-kHz Vocalizations and the Mating Behavior of Estrous Female Rats. *Behav. Neural. Biol.* **1982**, *36*, 403–410. [CrossRef]
157. Biały, M.; Podobińska, M.; Barski, J.; Bogacki-Rychlik, W.; Sajdel-Sułkowska, E.M. Distinct Classes of Low Frequency Ultrasonic Vocalizations in Rats during Sexual Interactions Relate to Different Emotional States. *Acta Neurobiol. Exp.* **2019**, *79*, 1–12. [CrossRef]
158. Galef, B.G., Jr.; Wigmore, S.W. Transfer of Information Concerning Distant Foods: A Laboratory Investigation of the "Information-Centre" Hypothesis. *Anim. Behav.* **1983**, *31*, 748–758. [CrossRef]
159. Galef, B.G.; Kennett, D.J.; Wigmore, S.W. Transfer of Information Concerning Distant Foods in Rats: A Robust Phenomenon. *Anim. Learn. Behav.* **1984**, *12*, 292–296. [CrossRef]
160. Laland, K.N.; Plotkin, H.C. Social Transmission of Food Preferences among Norway Rats by Marking of Food Sites and by Gustatory Contact. *Anim. Learn. Behav.* **1993**, *21*, 35–41. [CrossRef]
161. Galef, B.G., Jr. Necessary and Sufficient Conditions for Communication of Diet Preferences by Norway Rats. *Anim. Learn. Behav.* **1990**, *18*, 347–351. [CrossRef]
162. Mayeaux, D.J.; Wallace, M.B.; Young, A.M. Effect of Nutrient Restriction on Social Transmission of Food Preferences Depends on Nutrient and Species. *J. Behav. Brain Sci.* **2014**, *4*, 590–601. [CrossRef]
163. Galef, B.G., Jr.; Jeimy, S. Ultrasonic Vocalizations and Social Learning of Food Preferences by Female Norway Rats. *Anim. Behav.* **2004**, *68*, 483–487. [CrossRef]
164. Monfils, M.H.; Agee, L.A. Insights from Social Transmission of Information in Rodents. *Genes Brain Behav.* **2019**, *18*, e12534. [CrossRef]
165. Moles, A.; D'amato, F.R. Ultrasonic Vocalization by Female Mice in the Presence of a Conspecific Carrying Food Cues. *Anim. Behav.* **2000**, *60*, 689–694. [CrossRef]
166. Takahashi, N.; Kashino, M.; Hironaka, N. Structure of Rat Ultrasonic Vocalizations and Its Relevance to Behavior. *PLoS ONE* **2010**, *5*, e14115. [CrossRef]
167. Schweinfurth, M.K.; Taborsky, M. Norway Rats (*Rattus norvegicus*) Communicate Need, Which Elicits Donation of Food. *J. Comp. Psychol.* **2018**, *132*, 119–129. [CrossRef] [PubMed]
168. Blanchard, R.J.; Blanchard, D.C. Antipredator Defensive Behaviors in a Visible Burrow System. *J. Comp. Psychol.* **1989**, *103*, 70–82. [CrossRef] [PubMed]
169. Blanchard, R.J.; Blanchard, D.C. Anti-predator Defense as a Model of Animal Fear and Anxiety. In *Fear and Defence*; Brain, P.F., Parmigiani, S., Blanchard, R.J., Mainardi, D., Eds.; Ettore Majorana International Life Sciences Series; Harwood Academic Publishers GmbH: Chur, Switzerland, 1990; Volume 8, pp. 89–108.
170. Blanchard, R.J.; Blanchard, D.C.; Rodgers, J.; Weiss, S.M. The Characterization and Modelling of Antipredator Defensive Behavior. *Neurosci. Biobehav. Rev.* **1990**, *14*, 463–472. [CrossRef]

171. Blanchard, R.J.; Blanchard, D.C.; Agullana, R.; Weiss, S.M. Twenty-two kHz Alarm Cries to Presentation of a Predator, by Laboratory Rats Living in Visible Burrow Systems. *Physiol. Behav.* **1991**, *50*, 967–972. [CrossRef]
172. Fendt, M.; Gonzalez-Guerrero, C.P.; Kahl, E. Observational Fear Learning in Rats: Role of Trait Anxiety and Ultrasonic Vocalization. *Brain Sci.* **2021**, *11*, 423. [CrossRef] [PubMed]
173. Litvin, Y.; Blanchard, C.D.; Blanchard, R.J. Rat 22 kHz Ultrasonic Vocalizations as Alarm Cries. *Behav. Brain Res.* **2007**, *182*, 166–172. [CrossRef]
174. Lee, J.-H.; Kimm, S.; Han, J.-S.; Choi, J.-S. Chasing as a Model of Psychogenic Stress: Characterization of Physiological and Behavioral Responses. *Stress* **2018**, *21*, 323–332. [CrossRef] [PubMed]
175. Dielenberg, R.A.; McGregor, I.S. Defensive Behavior in Rats towards Predatory Odors: A Review. *Neurosci. Biobehav. Rev.* **2001**, *25*, 597–608. [CrossRef]
176. Apfelbach, R.; Blanchard, C.D.; Blanchard, R.J.; Hayes, R.A.; McGregor, I.S. The Effects of Predator Odors in Mammalian Prey Species: A Review of Field and Laboratory Studies. *Neurosci. Biobehav. Rev.* **2005**, *29*, 1123–1144. [CrossRef] [PubMed]
177. Blanchard, D.C.; Blanchard, R.J.; Griebel, G. Defensive Responses to Predator Threat in the Rat and Mouse. *Curr. Protoc. Neurosci.* **2005**. [CrossRef]
178. Fendt, M.; Brosch, M.; Wernecke, K.E.A.; Willadsen, M.; Wöhr, M. Predator Odour but not TMT Induces 22-kHz Ultrasonic Vocalizations in Rats That Lead to Defensive Behaviours in Conspecifics upon Replay. *Sci. Rep.* **2018**, *8*, 11041. [CrossRef]
179. Groeber Travis, C.M.; Genovese, R.F. Cat-exposure Results in Significantly more Elicited Alarm Calls (22kHz Ultrasonic Vocalizations, USVs) Compared to Snake-, Ferret-, or Sham-exposure during a Rodent Model of Traumatic Stress. *FASEB J.* **2016**, *30*, 938–939. [CrossRef]
180. Thomson, A.P.D. A History of the Ferret. *J. Hist. Med.* **1951**, *6*, 471–480. [CrossRef]
181. Shelley, E.L.; Blumstein, D.T. The Evolution of Vocal Alarm Communication in Rodents. *Behav. Ecol.* **2005**, *16*, 69–177. [CrossRef]
182. Nyby, J.; Whitney, G. Ultrasonic Communication of Adult Myomorph Rodents. *Neurosci. Biobehav. Review.* **1978**, *2*, 1–14. [CrossRef]
183. Beason, R.C. What Can Birds Hear. In *21st Vertebrate Pest Conference*; Timm, R.M., Gorenzel, W.P., Eds.; University of California: Davis, CA, USA, 2004; pp. 92–96.
184. Parsons, M.H.; Banks, P.B.; Deutsch, M.A.; Munshi-South, J. Temporal and Space-use Changes by Rats in Response to Predation by Feral Cats in an Urban Ecosystem. *Front. Ecol. Evol.* **2018**, *6*, 146. [CrossRef]
185. Himsworth, C.G.; Zabek, E.; Tang, P.; Parsons, K.L.; Koehn, M.; Jardine, C.M.; Patrick, D.M. Bacteria Isolated from Conspecific Bite Wounds in Norway and Black Rats: Implications for Rat Bite-associated Infections in People. *Vector Borne Zoonotic Dis.* **2014**, *14*, 94–100. [CrossRef] [PubMed]
186. Tucker, V.A.; Cade, T.J.; Tucker, A.E. Diving Speeds and Angles of a Gyrafalcon (*Falco rusticolus*). *J. Exp. Biol.* **1998**, *201*, 2061–2070. [CrossRef] [PubMed]
187. Wagner, H.; Weger, M.; Klaas, M.; Schröder, W. Features of Owl Wings That Promote Silent Flight. *Interface Focus* **2017**, *7*, 20160078. [CrossRef]
188. Brudzynski, S.M.; Holland, G. Acoustic Characteristics of Air Puff-induced 22-kHz Alarm Calls in Direct Recordings. *Neurosci. Biobehav. Rev.* **2005**, *29*, 1169–1180. [CrossRef]
189. Brudzynski, S.M.; Chiu, E. Behavioural Responses of Laboratory Rats to Playback of 22 kHz Ultrasonic Calls. *Physiol. Behav.* **1995**, *57*, 1039–1044. [CrossRef]
190. Wöhr, M.; Willadsen, M.; Kisko, T.M.; Schwarting, R.K.W.; Fendt, M. Sex-dependent Effects of Cacna1c Haploinsufficiency on Behavioral Inhibition Evoked by Conspecific Alarm Signals in Rats. *Prog. Neuropsychopharmacol. Biol. Psychiat.* **2020**, *99*, 109849. [CrossRef]
191. Blanchard, R.J.; Flannelly, K.J.; Blanchard, D.C. Defensive Behavior of Laboratory and Wild *Rattus norvegicus*. *J. Comp. Psychol.* **1986**, *100*, 101–107. [CrossRef] [PubMed]
192. Brudzynski, S.M.; Ociepa, D. Ultrasonic Vocalization of Laboratory Rats in Response to Handling and Touch. *Physiol. Behav.* **1992**, *52*, 655–660. [CrossRef]
193. Kaltwasser, M.-T. Startle-inducing Acoustic Stimuli Evoke Ultrasonic Vocalization in the Rat. *Physiol. Behav.* **1990**, *48*, 13–17. [CrossRef]
194. Kaltwasser, M.-T. Acoustic Startle Induced Ultrasonic Vocalization in the Rat: A Novel Animal Model of Anxiety? *Behav. Brain Res.* **1991**, *43*, 133–137. [CrossRef]
195. Knapp, D.J.; Pohorecky, L.A. 1995. An Air-puff Stimulus Method for Elicitation of Ultrasonic Vocalizations in Rats. *J. Neurosci. Methods.* **1995**, *62*, 1–5. [CrossRef]
196. Tonoue, T.; Ashida, Y.; Makino, H.; Hata, H. Inhibition of Shock-elicited Ultrasonic Vocalization by Opioid Peptides in the Rat: A Psychotropic Effect. *Psychoneuroendocrinology* **1986**, *11*, 177–184. [CrossRef]
197. Brudzynski, S.M.; Ociepa, D.; Bihari, F. Comparison between Cholinergically and Naturally Induced Ultrasonic Vocalization in the Rat. *J. Psychiatry. Neurosci.* **1991**, *16*, 221–226. [PubMed]
198. De Vry, J.; Benz, U.; Schreiber, R.; Traber, J. Shock-induced Ultrasonic Vocalization in Young Adult Rats: A Model for Testing Putative Anti-anxiety Drugs. *Eur. J. Pharmacol.* **1993**, *249*, 331–339. [CrossRef]
199. Ardid, D.; Jourdan, D.; Eschalier, A.; Arabia, C.; Lé Bars, D. Vocalization Elicited by Activation of A Delta- and C-fibres in the Rat. *Neuroreport* **1993**, *5*, 105–108. [CrossRef] [PubMed]

200. Jourdan, D.; Ardid, D.; Chapuy, E.; Eschalier, A.; Lé Bars, D. Audible and Ultrasonic Vocalization Elicited by Single Electrical Nociceptive Stimuli to the Tail in the Rat. *Pain* **1995**, *63*, 237–249. [CrossRef]
201. Bali, A.; Jaggi, A.S. Electric Foot Shock Stress: A Useful Tool in Neuropsychiatric Studies. *Rev. Neurosci.* **2015**, *26*, 655–677. [CrossRef]
202. Szechtman, H.; Woody, E. Obsessive-compulsive Disorder as a Disturbance of Security Motivation. *Psychol. Rev.* **2004**, *111*, 111–127. [CrossRef]
203. Hinds, A.L.; Woody, E.Z.; Drandic, A.; Schmidt, L.A.; Van Ameringen, M.; Coroneos, M.; Szechtman, H. The Psychology of Potential Threat: Properties of the Security Motivation System. *Biol. Psychol.* **2010**, *85*, 331–337. [CrossRef] [PubMed]
204. Woody, E.Z.; Szechtman, H. Adaptation to Potential Threat: The Evolution, Neurobiology, and Psychopathology of the Security Motivation System. *Neurosci. Biobehav. Rev.* **2011**, *35*, 1019–1033. [CrossRef]
205. Brudzynski, S.M. Social Origin of Vocal Communication in Rodents. In *Biocommunication of Animals*, 1st ed.; Witzany, G., Ed.; Springer Science+Business Media: Dordrecht, The Netherlands, 2014.
206. Ma, S.T.; Maier, E.Y.; Ahrens, A.M.; Schallert, T.; Duvauchelle, C.L. Repeated Intravenous Cocaine Experience: Development and Escalation of Pre-drug Anticipatory 50-kHz Ultrasonic Vocalizations in Rats. *Behav. Brain. Res.* **2010**, *212*, 109–114. [CrossRef]
207. Tornatzky, W.; Miczek, K.A. Behavioral and Autonomic Responses to Intermittent Social Stress: Differential Protection by Clonidine and Metoprolol. *Psychopharmacology* **1994**, *116*, 346–356. [CrossRef]
208. Naito, H.; Inoue, M.; Suzuki, Y.; Tohei, A.; Watanabe, G.; Taya, K.; Makino, J. Ultrasonic Vocalization Responses in Genetically High- and Low-emotional Rats. *Exp. Anim.* **2001**, *50*, 285–291. [CrossRef] [PubMed]
209. Choi, E.A.; Leman, S.; Vianna, D.M.L.; Waite, P.M.E.; Carrive, P. Expression of Cardiovascular and Behavioural Components of Conditioned Fear to Context in T4 Spinally Transected Rats. *Auton. Neurosci.* **2005**, *120*, 26–34. [CrossRef]
210. Carrive, P. Dual Activation of Cardiac Sympathetic and Parasympathetic Components during Conditioned Fear to Context in the Rat. *Clin. Exp. Pharmacol. Physiol.* **2006**, *33*, 1251–1254. [CrossRef]
211. Demaestri, C.; Brenhouse, H.C.; Honeycutt, J.A. 22 kHz and 55 kHz Ultrasonic Vocalizations Differentially Influence Neural and Behavioral Outcomes: Implications for Modeling Anxiety via Auditory Stimuli in the Rat. *Behav. Brain Res.* **2019**, *360*, 134–145. [CrossRef]
212. Olszyński, K.H.; Polowy, R.; Małż, M.; Boguszewski, P.M.; Filipkowski, R.K. Playback of Alarm and Appetitive Calls Differentially Impacts Vocal, Heart-rate, and Motor Response in Rats. *iScience* **2020**, *23*, 101577. [CrossRef] [PubMed]
213. Barros, H.M.; Miczek, K.A. Withdrawal from Oral Cocaine in Rate: Ultrasonic Vocalizations and Tactile Startle. *Psychopharmacology* **1996**, *125*, 379–384. [CrossRef] [PubMed]
214. Covington, H.E., III, Miczek, K.A. Vocalizations during Withdrawal from Opiates and Cocaine: Possible Expressions of Affective Distress. *Eur. J. Pharmacol.* **2003**, *467*, 1–13. [CrossRef]
215. Miczek, K.A.; Vivian, J.A. Automatic Quantification of Withdrawal from 5-day Diazepam in Rats: Ultrasonic Distress Vocalizations and Hyperreflexia to Acoustic Startle Stimuli. *Psychopharmacology* **1993**, *110*, 379–382. [CrossRef]
216. Barker, D.J.; Simmons, S.J.; West, M.O. Ultrasonic Vocalizations as a Measure of Affect in Preclinical Models of Drug Abuse: A Review of Current Findings. *Curr. Neuropharmacol.* **2015**, *13*, 193–210. [CrossRef]
217. Simmons, S.J.; Barker, D.J.; West, M.O. Ultrasonic Vocalizations Capture Opposing Affective States during Drug Self-stimulation: Revisiting the Opponent-process Model of Addiction. In *Handbook of Ultrasonic Vocalization. A Window into the Emotional Brain*; Brudzynski, S.M., Ed.; Elsevier/Academic Press: Amsterdam, The Netherlands, 2018.
218. Klein, S.D.; Beacher, N.J.; Kulik, J.M.; Estrin, D.J.; Pawlak, A.P.; West, M.O. Emergence of Negative Affect as Motivation for Drug Taking in Rats Chronically Self-administering Cocaine. *Psychopharmacology* **2020**, *237*, 1407–1420. [CrossRef]
219. Frenois, F.; Le Moine, C.; Cador, M. The Motivational Component of Withdrawal in Opiate Addiction: Role of Associative Learning and Aversive Memory in Opiate Addiction from a Behavioral, Anatomical and Functional Perspective. *Rev. Neurosci.* **2005**, *16*, 255–276. [CrossRef]
220. Mogenson, G.J. *The Neurobiology of Behavior: An Introduction*, 1st ed.; Lawrence Erlbaum Associates, Inc. Publishers: Hillsdale, NJ, USA.
221. Brudzynski, S.M. Ethotransmission: Communication of Emotional States through Ultrasonic Vocalization in Rats. *Curr. Opin. Neurobiol.* **2013**, *23*, 310–317. [CrossRef] [PubMed]
222. Bell, R.W. Ultrasounds in Small Rodents: Arousal-produced and Arousal-producing. *Dev. Psychobiol.* **1974**, *7*, 39–42. [CrossRef] [PubMed]
223. Cuomo, V.; Cagiano, R. Ultrasonic Vocalization in Rodents: A New Potential Tool for Detecting Emotional and Motivational Changes Produced by Adverse Treatments. *Zentralbl. Bakteriol. Mikrobiol. Hyg. B* **1987**, *185*, 55–60.
224. Cuomo, V.; Cagiano, R.; De Salvia, M.A.; Mazzoccoli, M.; Persichella, M.; Renna, G. Ultrasonic Vocalization as an Indicator of Emotional State during Active Avoidance Learning in Rats. *Life Sci.* **1992**, *50*, 1049–1055. [CrossRef]
225. Knutson, B.; Burgdorf, J.; Panksepp, J. Ultrasonic Vocalizations as Indices of Affective States in Rats. *Psychol. Bull.* **2002**, *128*, 961–977. [CrossRef]
226. Wöhr, M.; Schwarting, R.K.W. Affective Communication in Rodents: Ultrasonic Vocalizations as a Tool for Research on Emotion and Motivation. *Cell Tissue Res.* **2013**, *354*, 81–97. [CrossRef] [PubMed]
227. Barker, D.J. Ultrasonic Vocalizations as an Index of Positive Emotional State. In *Handbook of Ultrasonic Vocalization: A Window into the Emotional Brain*; Brudzynski, S.M., Ed.; Elsevier/Academic Press: Amsterdam, The Netherlands, 2018; pp. 253–260.

228. Shair, N.H. Factors Affecting Emission of Infantile Vocalizations and Vocal Expression of Emotional States. In *Handbook of Ultrasonic Vocalization. A Window into the Emotional Brain*; Brudzynski, S.M., Ed.; Elsevier/Academic Press: Amsterdam, The Netherlands, 2018; pp. 157–168.
229. Burgdorf, J.S.; Ghoreishi-Haack, N.; Cearley, C.N.; Kroes, R.A.; Moskal, J.R. Rat Ultrasonic Vocalizations as a Measure of the Emotional Component of Chronic Pain. *Neuroreport* **2019**, *30*, 863–866. [CrossRef]
230. Burgdorf, J.S.; Brudzynski, S.M.; Moskal, J.R. Using Rat Ultrasonic Vocalization to Study the Neurobiology of Emotion: From Basic Science to the Development of Novel Therapeutics for Affective Disorders. *Curr. Opin. Neurobiol.* **2020**, *60*, 192–200. [CrossRef]
231. Roth, B.; Bruns, V. Postnatal Development of the Rat organ Of Corti. II. Hair Cell Receptors and Their Supporting Elements. *Anat. Embryol.* **1992**, *185*, 571–581. [CrossRef]
232. Berke, G.S.; Long, J.L. Functions of the Larynx and Production of Sound. In *Handbook of Mammalian Vocalization. An Integrative Neuroscience Approach*; Brudzynski, S.M., Ed.; Academic Press/Elsevier: Amsterdam, The Netherlands, 2010; pp. 419–426.
233. Shiba, K. Functions of Larynx in Breathing, Vocalization, and Airway Protective Reflexes. In *Handbook of Mammalian Vocalization. An Integrative Neuroscience Approach*; Brudzynski, S.M., Ed.; Academic Press/Elsevier: Amsterdam, The Netherlands, 2010; pp. 373–381.
234. Stewart, A.M.; Lewis, G.F.; Yee, J.R.; Kenkel, W.M.; Davila, M.I.; Carter, C.S.; Porges, S.W. Acoustic Features of Prairie Vole (*Microtus ochrogaster*) Ultrasonic Vocalizations Covary with Heart Rate. *Physiol. Behav.* **2015**, *138*, 94–100. [CrossRef] [PubMed]
235. Grafe, T.U.; Thein, J. Energetics of Calling and Metabolic Substrate Use during Prolonged Exercise in the European Treefrog *Hyla arborea*. *J. Comp. Physiol. B* **2001**, *171*, 69–76. [CrossRef]
236. Brudzynski, S.M. Ultrasonic Calls of Rats as Indicator Variables of Negative or Positive States: Acetylcholine-dopamine Interaction and Acoustic Coding. *Behav. Brain Res.* **2007**, *182*, 261–273. [CrossRef] [PubMed]
237. Young, B.J.; Leaton, R.N. Fear Potentiation of Acoustic Startle Stimulus-evoked Heart Rate Changes in Rats. *Behav. Neurosci.* **1994**, *108*, 1065–1079. [CrossRef] [PubMed]
238. Wang, Y.; Li, Z.; Tian, Z.; Wang, X.; Li, Y.; Qin, L. Emotional Arousal Modifies Auditory Steady State Response in the Auditory Cortex and Prefrontal Cortex of Rats. *Stress* **2019**, *22*, 492–500. [CrossRef] [PubMed]
239. Brudzynski, S.M. Communication of Emotions in Animals. In *Encyclopedia of Behavioral Neuroscience*; Koob, G.F., Le Moal, M., Thompson, R.F., Eds.; Elsevier/Academic Press: Amsterdam, The Netherlands, 2010; Volume 1, pp. 302–307.
240. Rendall, D.; Owren, M.J. Vocalizations as Tools for Influencing the Affect and Behavior of Others. In *Handbook of Mammalian Vocalization. An Integrative Neuroscience Approach*; Brudzynski, S.M., Ed.; Academic Press/Elsevier: Amsterdam, The Netherlands, 2010; pp. 177–185.
241. Newman, J.D.; Harris, J.C. The Scientific Contributions of Paul, D. MacLean (1913–2007). *J. Nerv. Ment. Dis.* **2009**, *197*, 3–5. [CrossRef] [PubMed]
242. Morgane, P.J.; Stern, W.C. Relationship of Sleep Neuroanatomical Circuits, Biochemistry, and Behavior. *Ann. N. Y. Acad. Sci.* **1972**, *193*, 95–111. [CrossRef]
243. Hammer, R.P., Jr.; Lindsay, R.D.; Scheibel, A.B. Development of the Brain Stem Reticular Core: An Assessment of Dendritic State and Configuration in the Perinatal Rat. *Brain Res.* **1981**, *227*, 179–190. [CrossRef]
244. Ramón-Moliner, E.; Nauta, W.J.H. The Isodendritic Core of the Brain Stem. *J. Comp. Neurol.* **1966**, *126*, 311–335. [CrossRef] [PubMed]
245. Hobson, J.A.; Scheibel, A.B. The Brainstem Core: Sensorimotor Integration and Behavioral State Control. *Neurosci. Res. Program. Bull.* **1980**, *18*, 1–173.
246. Venkatraman, A.; Edlow, B.L.; Immordino-Yang, M.H. The Brainstem in Emotion: A Review. *Front. Neuroanat.* **2017**, *11*, 15. [CrossRef] [PubMed]
247. Porges, S.W. Love: An Emergent Property of the Mammalian Autonomic Nervous System. *Psychoneuroendocrinology* **1998**, *23*, 837–861. [CrossRef]
248. Opeyemi, A.; Berzofsky, C.; Sharma, S.; Pitman, M.J. Development of the Rat Larynx: A Histological Study. *Laryngoscope* **2013**, *123*, 3093–3098. [CrossRef]
249. Pitman, M.J.; Berzofsky, C.E.; Opeyemi, A.; Sharma, S. Embryologic Innervation of the Rat Laryngeal Musculature–A Model for Investigation of Recurrent Laryngeal Nerve Reinnervation. *Laryngoscope* **2013**, *123*, 3117–3126. [CrossRef]
250. Hangai, K.; Kobayashi, Y.; Nonaka, S. Developmental Changes in Histochemical Properties of Intrinsic Laryngeal Muscles in Rats. *Auris Nasus Larynx* **1999**, *26*, 467–478. [CrossRef]
251. Nitschke, W. *Acoustic Behavior in the Rat. Research, Theory, and Applications*; Praeger Publishers: New York, NY, USA, 1982; p. 185.
252. Sales, G.; Pye, D. *Ultrasonic Communication by Animals*, 1st ed.; Chapman and Hall Ltd.: London, UK, 1974; p. 281.
253. Saito, Y.; Tachibana, R.O.; Okanoya, K. Acoustical Cues for Perception of Emotional Vocalizations in Rats. *Sci. Rep.* **2019**, *9*, 10539. [CrossRef]
254. Brudzynski, S.M. Vocalizations as Indicators of Emotional States in Rats and Cats. In *Evolution of Emotional Communication. From Sounds in Nonhuman Mammals to Speech and Music in Man*, 1st ed.; Altenmüller, E., Schmidt, S., Zimmermann, E., Eds.; Oxford University Press: Oxford, UK, 2013; pp. 75–91.
255. Klappenbach, M.; Nally, A.; Locatelli, F.F. Parallel Memory Traces are Built after an Experience Containing Aversive and Appetitive Components in the Crab *Neohelice*. *Proc. Natl. Acad. Sci. USA* **2017**, *114*, E4666–E4675. [CrossRef]
256. Silkstone, M.; Brudzynski, S.M. Intracerebral Injection of R-(-)-Apomorphine into the Nucleus Accumbens Decreased Carbachol-induced 22-kHz Ultrasonic Vocalizations in Rats. *Behav. Brain Res.* **2019**, *364*, 264–273. [CrossRef]

257. Silkstone, M.; Brudzynski, S.M. The Antagonistic Relationship between Aversive and Appetitive Emotional States in Rats as Studied by Pharmacologically-induced Ultrasonic Vocalization from the Nucleus Accumbens and Lateral Septum. *Pharmacol. Biochem. Behav.* **2019**, *181*, 77–85. [CrossRef] [PubMed]
258. Ungless, M.A.; Magill, P.J.J.; Bolam, J.P. Uniform Inhibition of Dopamine Neurons in the Ventral Tegmental Area by Aversive Stimuli. *Science* **2004**, *303*, 2040–2042. [CrossRef]
259. Brudzynski, S.M.; Mogenson, G.J. Inhibition of Amphetamine-induced Locomotor Activity by Injection of Carbachol into the Anterior Hypothalamic/Preoptic Area: Pharmacological and Electrophysiological Studies in the Rat. *Brain Res.* **1986**, *376*, 47–56. [CrossRef]
260. Silkstone, M.; Brudzynski, S.M. Dissimilar Interaction between Dopaminergic and Cholinergic Systems in the Initiation of Emission of 50-kHz and 22-kHz Vocalizations. *Pharmacol. Biochem. Behav.* **2020**, *188*, 172815. [CrossRef] [PubMed]
261. Omelchenko, N.; Sesack, S.R. Laterodorsal Tegmental Projections to Identified Cell Populations in the Rat Ventral Tegmental Area. *J. Comp. Neurol.* **2005**, *483*, 217–235. [CrossRef]
262. Omelchenko, N.; Sesack, S.R. Cholinergic Axons in the Rat Ventral Tegmental Area Synapse Preferentially onto Mesoaccumbens Dopamine Neurons. *J. Comp. Neurol.* **2006**, *494*, 863–875. [CrossRef]
263. Steidl, S.; Miller, A.D.; Blaha, C.D.; Yeomans, J.S. M_5 Muscarinic Receptors Mediate Striatal Dopamine Activation by Ventral Tegmental Morphine and Pedunculopontine Stimulation in Mice. *PLoS ONE* **2011**, *6*, e27538. [CrossRef] [PubMed]
264. Scardochio, T.; Clarke, P.B.S. Inhibition of 50-kHz Ultrasonic Vocalizations by Dopamine Receptor Subtype-selective Agonists and Antagonists in Adult Rats. *Psychopharmacology* **2013**, *226*, 589–600. [CrossRef]
265. Soares-Cunha, C.; de Vasconcelos, N.A.P.; Coimbra, B.; Domingues, A.V.; Silva, J.M.; Loureiro-Campos, E.; Gaspar, R.; Sotiropoulos, I.; Sousa, N.; Rodrigues, A.J. Nucleus Accumbens Medium Spiny Neurons Subtypes Signal Both Reward and Aversion. *Mol. Psychiatry* **2020**, *25*, 3241–3255. [CrossRef]
266. Schwarz, L.A.; Luo, L. Organization of the Locus Coeruleus-norepinephrine System. *Curr. Biol* **2015**, *25*, R1051–R1056. [CrossRef] [PubMed]
267. Mason, S.T.; Fibiger, H.C. Possible Behavioural Function for Noradrenaline–acetylcholine Interaction in Brain. *Nature* **1979**, *277*, 396–397. [CrossRef]
268. Aston-Jones, G. Brain Structures and Receptors Involved in Alertness. *Sleep Med.* **2005**, *6* (Suppl. 1), S3–S7. [CrossRef]
269. Wright, J.M.; Dobosiewicz, M.R.S.; Clarke, P.B.S. α- and β-Adrenergic Receptors Differentially Modulate the Emission of Spontaneous and Amphetamine-induced 50-kHz Ultrasonic Vocalizations in Adult Rats. *Neuropsychopharmacology* **2012**, *37*, 808–821. [CrossRef] [PubMed]
270. Grant, L.M.; Barth, K.J.; Muslu, C.; Kelm-Nelson, C.A.; Vaishali, P.; Bakshi, V.P.; Ciucci, M.R. Noradrenergic Receptor Modulation Influences the Acoustic Parameters of Pro-social Rat Ultrasonic Vocalizations. *Behav. Neurosci.* **2018**, *132*, 269–283. [CrossRef]
271. Lee, T.-H.; Sakaki, M.; Cheng, R.; Velasco, R.; Mather, M. Emotional Arousal Amplifies the Effects of Biased Competition in the Brain. *Soc. Cogn. Affect. Neurosci.* **2014**, *9*, 2067–2077. [CrossRef] [PubMed]
272. Kayama, Y.; Ito, S.; Koyama, Y.; Jodo, E. Tonic and Phasic Components of the Ascending Reticular Activating System. *Fukushima J. Med. Sci.* **1991**, *37*, 59–74. [PubMed]
273. Kayama, Y.; Koyama, Y. Control of Sleep and Wakefulness by Brainstem Monoaminergic and Cholinergic Neurons. *Acta Neurochir. Suppl.* **2003**, *87*, 3–6. [CrossRef] [PubMed]
274. Sarter, M.; Parikh, V.; Howe, W.M. Phasic Acetylcholine Release and the Volume Transmission Hypothesis: Time to Move on. *Nat. Rev. Neurosci.* **2009**, *10*, 383–390. [CrossRef] [PubMed]
275. Atzori, M.; Cuevas-Olguin, R.; Esquivel-Rendon, E.; Garcia-Oscos, F.; Salgado-Delgado, R.C.; Saderi, N.; Miranda-Morales, M.; Treviño, M.; Pineda, J.C.; Salgado, H. Locus Ceruleus Norepinephrine Release: A Central Regulator of CNS Spatio-Temporal Activation? *Front. Synaptic Neurosci.* **2016**, *8*, 25. [CrossRef]
276. Alcaro, A.; Huber, R.; Panksepp, J. Behavioral Functions of the Mesolimbic Dopaminergic System: An Affective Neuroethological Perspective. *Brain Res. Rev.* **2007**, *56*, 283–321. [CrossRef] [PubMed]
277. Yetnikoff, L.; Lavezzi, H.N.; Reichard, R.A.; Zahm, D.S. An Update on the Connections of the Ventral Mesencephalic Dopaminergic Complex. *Neuroscience* **2014**, *282*, 23–48. [CrossRef]
278. Mesulam, M.M.; Mufson, E.J.; Wainer, B.H.; Levey, A.I. Central Cholinergic Pathways in the Rat: An Overview Based on an Alternative Nomenclature (Ch1-Ch6). *Neuroscience* **1983**, *10*, 1185–1201. [CrossRef]
279. Charnay, Y. Brain Serotonergic Circuitries. *Dialogues Clin. Neurosci.* **2010**, *12*, 471–487. [CrossRef]
280. Blandina, P.; Munari, L.; Provensi, G.; Passani, M.B. Histamine Neurons in the Tuberomamillary Nucleus: A Whole Center or Distinct Subpopulations? *Front. Syst. Neurosci.* **2012**, *6*. [CrossRef]
281. Alexandre, C.; Andermann, M.L.; Scammell, T.E. Control of Arousal by the Orexin Neurons. *Curr. Opin. Neurobiol.* **2013**, *23*, 752–759. [CrossRef] [PubMed]
282. Inutsuka, A.; Yamanaka, A. The Physiological Role of Orexin/hypocretin Neurons in the Regulation of Sleep/wakefulness and Neuroendocrine Functions. *Front. Endocrinol.* **2013**, *4*, 18. [CrossRef] [PubMed]
283. Brudzynski, S.M. Ultrasonic Vocalization Induced by Intracerebral Carbachol in Rats: Localization and a Dose-response Study. *Behav. Brain Res.* **1994**, *63*, 133–143. [CrossRef]

284. Brudzynski, S.M. Medial Cholinoceptive Vocalization Strip in the Cat and Rat Brains: Initiation of Defensive Vocalizations. In *Handbook of Mammalian Vocalization: An Integrative Neuroscience Approach*; Brudzynski, S.M., Ed.; Elsevier/Academic Press: Amsterdam, The Netherlands, 2010; pp. 265–279.
285. Brudzynski, S.M.; Barnabi, F. Contribution of the Ascending Cholinergic Pathways in the Production of Ultrasonic Vocalization in the Rat. *Behav. Brain Res.* **1996**, *80*, 145–152. [CrossRef]
286. Brudzynski, S.M.; McLachlan, R.S.; Girvin, J.P. Cholinergically Mediated Reduction of Locomotor Activity from the Basal Forebrain of the Rat. *Exp. Neurol.* **1989**, *105*, 197–205. [CrossRef]
287. Bihari, A.; Hrycyshyn, A.W.; Brudzynski, S.M. Role of the Mesolimbic Cholinergic Projection to the Septum in the Production of 22 kHz Alarm Calls in Rats. *Brain Res. Bull.* **2003**, *60*, 263–274. [CrossRef]
288. Burgdorf, J.; Knutson, B.; Panksepp, J.; Ikemoto, S. Nucleus Accumbens Amphetamine Microinjections Unconditionally Elicit 50-kHz Ultrasonic Vocalizations in Rats. *Behav. Neurosci.* **2001**, *115*, 940–944. [CrossRef] [PubMed]
289. Wintink, A.J.; Brudzynski, S.M. The Related Roles of Dopamine and Glutamate in the Initiation of 50-kHz Ultrasonic Calls in Adult Rats. *Pharmacol. Biochem. Behav.* **2001**, *70*, 317–323. [CrossRef]
290. Brudzynski, S.M.; Silkstone, S.; Komadoski, M.; Scullion, K.; Duffus, S.; Burgdorf, J.; Kroes, R.A.; Moskal, J.R.; Panksepp, J. Effects of Intraaccumbens Amphetamine on Production of 50 kHz Vocalizations in Three Lines of Selectively Bred Long-Evans Rats. *Behav. Brain. Res.* **2011**, *217*, 32–40. [CrossRef] [PubMed]
291. Brudzynski, S.M.; Gibson, B.; Silkstone, M.; Burgdorf, J.; Kroes, R.A.; Moskal, J.R.; Panksepp, J. Motor and Locomotor Responses to Systemic Amphetamine in Three Lines of Selectively Bred Long-Evans Rats. *Pharmacol. Biochem. Behav.* **2011**, *100*, 119–124. [CrossRef]
292. Wang, Y.; Wan, B.; Huang, J.; Clarke, P.B.S. Effects of Nicotine, Nornicotine and Cotinine, Alone or in Combination, on Locomotor Activity and Ultrasonic Vocalization Emission in Adult Rats. *Psychopharmacology* **2020**, *237*, 2809–2822. [CrossRef]
293. Swalve, N.; Mulholland, M.M.; Li, M. Alterations of Acoustic Features of 50 kHz Vocalizations by Nicotine and Phencyclidine in Rats. *Behav. Pharmacol.* **2019**, *30*, 446–451. [CrossRef] [PubMed]
294. Remy, S.M.; Schreiber, R.; Dalmus, M.; De Vry, J. Somatodendritic 5-HT1A Receptors are Critically Involved in the Anxiolytic Effects of 8-OH-DPAT. *Psychopharmacology* **1996**, *125*, 89–91. [CrossRef] [PubMed]
295. Gołebiowska, J.; Hołuj, M.; Potasiewicz, A.; Piotrowska, D.; Kuziak, A.; Popik, P.; Homberg, J.R.; Nikiforuk, A. Serotonin Transporter Deficiency Alters Socioemotional Ultrasonic Communication in Rats. *Sci. Rep.* **2019**, *9*, 20283. [CrossRef]
296. Narita, M.; Nagumo, Y.; Hashimoto, S.; Narita, M.; Khotib, J.; Miyatake, M.; Sakurai, T.; Yanagisawa, M.; Nakamachi, T.; Shioda, S.; et al. Direct Involvement of Orexinergic Systems in the Activation of the Mesolimbic Dopamine Pathway and Related Behaviors Induced by Morphine. *J. Neurosci.* **2006**, *26*, 398–405. [CrossRef]
297. Morden, A. The Role of Orexin-A in Anxiety and the Emission of Ultrasonic Vocalizations in Rats. Master's Thesis, Psychology, Faculty of Graduate Studies, Brock University, St. Catharines, ON, Canada, 2013.
298. Mogi, K.; Nagasawa, M.; Kikusui, T. Developmental Consequences and Biological Significance of Mother-infant Bonding. *Prog Neuropsychopharmacol. Biol. Psychiatry* **2011**, *35*, 1232–1241. [CrossRef] [PubMed]
299. Walker, C.D.; Kudreikis, K.; Sherrard, A.; Johnston, C.C. Repeated Neonatal Pain Influences Maternal Behavior, but not Stress Responsiveness in Rat Offspring. *Brain Res. Dev. Brain Res.* **2003**, *140*, 253–261. [CrossRef]
300. Stark, R.A.; Harker, A.; Salamanca, S.; Pellis, S.M.; Li, F.; Gibb, R.L. Development of Ultrasonic Calls in Rat Pups Follows Similar Patterns Regardless of Isolation Distress. *Dev. Psychobiol.* **2020**, *62*, 617–630. [CrossRef]
301. Kim, H.; Bao, S. Experience-dependent Overrepresentation of Ultrasonic Vocalization Frequencies in the Rat Primary Auditory Cortex. *J. Neurophysiol.* **2013**, *110*, 1087–1096. [CrossRef]
302. Bao, S. Perceptual Learning in the Developing Auditory Cortex. *Eur. J. Neurosci.* **2015**, *41*, 718–724. [CrossRef] [PubMed]
303. Bao, S.; Chang, E.F.; Teng, C.-L.; Heiser, M.A.; Merzenich, M.M. Emergent Categorical Representation of Natural, Complex Sounds Resulting from the Early Post-natal Sound Environment. *Neuroscience* **2013**, *248*, 30–42. [CrossRef]
304. Craig, A.; Luo, N.L.; Beardsley, D.J.; Wingate-Pearse, N.; Walker, D.W.; Hohimer, A.R.; Back, S.A. Quantitative Analysis of Perinatal Rodent Oligodendrocyte Lineage Progression and Its Correlation with Human. *Exp. Neurol.* **2003**, *181*, 231–240. [CrossRef]
305. Brudzynski, S.M. Ontogenetic Development of Ultrasonic Alarm Calls in Long-Evans Rats. In Proceedings of the 5th Forum of Federation of European Neuroscience Societies (FENS), Vienna, Austria, 8–12 July 2006.
306. Ninomiya, Y.; Koyama, Y.; Kayama, Y. Postnatal Development of Choline Acetyltransferase Activity in the Rat Laterodorsal Tegmental Nucleus. *Neurosci. Lett.* **2001**, *308*, 138–140. [CrossRef]
307. Ninomiya, Y.; Kayama, Y.; Koyama, Y. Postnatal Development of Cholinergic Neurons in the Mesopontine Tegmentum Revealed by Histochemistry. *Int. J. Dev. Neurosci.* **2005**, *23*, 711–721. [CrossRef]
308. Kehoe, P.; Callahan, M.; Daigle, A.; Mallinson, K.; Brudzynski, S. The Effect of Cholinergic Stimulation on Rat Pup Ultrasonic Vocalizations. *Dev. Psychobiol.* **2001**, *38*, 92–100. [CrossRef]
309. Boulanger-Bertolus, J.; Rincón-Cortés, M.; Sullivan, R.M.; Mouly, A.-M. Understanding Pup Affective State through Ethologically Significant Ultrasonic Vocalization Frequency. *Sci. Rep.* **2017**, *7*, 13483. [CrossRef]
310. Fibiger, H.C.; Lytle, L.D.; Campbell, B.A. Cholinergic Modulation of Adrenergic Arousal in the Developing Rat. *J. Comp. Physiol. Psychol.* **1970**, *72*, 384–389. [CrossRef]
311. Jelen, P.; Soltysik, S.; Zagrodzka, J. 22-kHz Ultrasonic Vocalization in Rats as an Index of Anxiety but not Fear: Behavioral and Pharmacological Modulation of Affective State. *Behav. Brain Res.* **2003**, *141*, 63–72. [CrossRef]

312. Naito, H.; Nakamura, A.; Inoue, M.; Suzuki, Y. Effect of Anxiolytic Drugs on Air-puff-elicited Ultrasonic Vocalization in Adult Rats. *Exp. Anim.* **2003**, *52*, 409–414. [CrossRef] [PubMed]
313. Borta, A.; Wöhr, M.; Schwarting, R.K.W. Rat Ultrasonic Vocalization in Aversively Motivated Situations and the Role of Individual Differences in Anxiety-related Behavior. *Behav. Brain Res.* **2006**, *166*, 271–280. [CrossRef] [PubMed]
314. Schwarting, R.K.W.; Wöhr, M. On the Relationships between Ultrasonic Calling and Anxiety-related Behavior in Rats. *Braz. J. Med. Biol. Res.* **2012**, *45*, 337–348. [CrossRef] [PubMed]
315. Brudzynski, S.M. Emission of 22 kHz Vocalizations in Rats as an Evolutionary Equivalent of Human Crying: Relationship to Depression. *Behav. Brain Res.* **2019**, *363*, 1–12. [CrossRef]
316. Steimer, T. The Biology of Fear- and Anxiety-related Behaviors. *Dialogues Clin. Neurosci.* **2002**, *4*, 231–249. [CrossRef] [PubMed]
317. Perusini, J.N.; Fanselow, M.S. Neurobehavioral Perspectives on the Distinction between Fear and Anxiety. *Learn. Mem.* **2015**, *22*, 417–425. [CrossRef]
318. Schwarting, R.K.W. Ultrasonic Vocalization in Female Rats: A Comparison Among Three Outbred Stocks from Pups to Adults. *Physiol. Behav.* **2018**, *196*, 59–66. [CrossRef] [PubMed]
319. Kroes, R.A.; Burgdorf, J.; Otto, N.J.; Panksepp, J.; Moskal, J.R. Social Defeat, a Paradigm of Depression in Rats That Elicits 22 kHz Vocalizations, Preferentially Activates the Cholinergic Signaling Pathway in the Periaqueductal Gray. *Behav. Brain Res.* **2007**, *182*, 290–300. [CrossRef]
320. Jourdan, D.; Ardid, D.; Eschalier, A. Analysis of Ultrasonic Vocalisation Does Not Allow Chronic Pain to be Evaluated in Rats. *Pain* **2002**, *95*, 165–173. [CrossRef]
321. Oliveira, A.R.; Barros, H.M.T. Ultrasonic Rat Vocalizations during the Formalin Test: A Measure of the Affective Dimension of Pain? *Anesth. Analg.* **2006**, *102*, 832–839. [CrossRef]
322. Barroso, A.R.; Araya, E.I.; Pasquini de Souza, C.; Andreatini, R.; Chichorro, J.G. Characterization of Rat Ultrasonic Vocalization in the Orofacial Formalin Test: Influence of the Social Context. *Eur. Neuropsychopharmacol.* **2019**, *29*, 1213–1226. [CrossRef] [PubMed]
323. Jourdan, D.; Ardid, D.; Chapuy, E.; Le Bars, D.; Eschalier, A. Effect of Analgesics on Audible and Ultrasonic Pain-induced Vocalization in the Rat. *Life Sci.* **1998**, *63*, 1761–1768. [CrossRef]
324. Naito, H.; Okumura, T.; Inoue, M.; Suzuki, Y. Ultrasonic Vocalization Response Elicited in Adjuvant-induced Arthritic Rats as a Useful Method for Evaluating Analgesic Drugs. *Exp. Anim.* **2006**, *55*, 125–129. [CrossRef]
325. Calvino, B.; Besson, J.M.; Boehrer, A.; Depaulis, A. Ultrasonic Vocalization (22–28 kHz) in a Model of Chronic Pain, the Arthritic Rat: Effects of Analgesic Drugs. *Neuroreport* **1996**, *7*, 581–584. [CrossRef] [PubMed]
326. Borszcz, G. Contribution of the Ventromedial Hypothalamus to Generation of the Affective Dimension of Pain. *Pain* **2006**, *123*, 155–168. [CrossRef]
327. Naito, H.; Inoue, M. Quantitative Analysis of the Ultrasonic Vocalization Responses Elicited in Adjuvant-induced Arthritic Rats for Screening Analgesic Drugs. *Exp. Anim.* **2008**, *57*, 419–422. [CrossRef] [PubMed]
328. Portavella, M.; Depaulis, A.; Vergnes, M. 22–28 kHz Ultrasonic Vocalizations Associated with Defensive Reactions in Male Rats do not Result from Fear or Aversion. *Psychopharmacology* **1993**, *111*, 190–194. [CrossRef] [PubMed]
329. Endres, T.; Widmann, K.; Fendt, M. Are Rats Predisposed to Learn 22 kHz Calls as Danger-predicting Signals? *Behav. Brain Res.* **2007**, *185*, 69–75. [CrossRef]
330. Calub, C.A.; Furtak, S.C.; Brown, T.H. Revisiting the Autoconditioning Hypothesis for Acquired Reactivity to Ultrasonic Alarm Calls. *Physiol. Behav.* **2018**, *194*, 380–386. [CrossRef]
331. Kim, E.J.; Kim, E.S.; Covey, E.; Kim, J.J. Social Transmission of Fear in Rats: The Role of 22-kHz Ultrasonic Distress Vocalization. *PLoS ONE* **2010**, *5*, e15077. [CrossRef]
332. Inagaki, H.; Ushida, T. Changes in Acoustic Startle Reflex in Rats Induced by Playback of 22-kHz Calls. *Physiol. Behav.* **2017**, *169*, 189–194. [CrossRef]
333. Birchley, G. Opioid and Benzodiazepine Withdrawal Syndromes in the Paediatric Intensive Care Unit: A Review of Recent Literature. *Nurs. Crit. Care.* **2009**, *14*, 26–37. [CrossRef]
334. Ista, E.; van Dijk, M.; de Hoog, M.; Tibboel, D.; Duivenvoorden, H.J. Construction of the Sophia Observation Withdrawal Symptoms-scale (SOS) for Critically Ill Children. *Intensive Care Med.* **2009**, *35*, 1075–1081. [CrossRef] [PubMed]
335. Barker, D.J.; Root, D.H.; Ma, S.; Jha, S.; Megehee, L.; Pawlak, A.P.; West, M.O. Dose-dependent Differences in Short Ultrasonic Vocalizations Emitted by Rats during Cocaine Self-administration. *Psychopharmacology* **2010**, *211*, 435–442. [CrossRef] [PubMed]
336. Brudzynski, S.M.; Bihari, F.; Ociepa, D.; Fu, X.W. Analysis of 22 kHz Ultrasonic Vocalization in Laboratory Rats: Long and Short Calls. *Physiol. Behav.* **1993**, *54*, 215–221. [CrossRef]
337. Taylor, J.E.; Ficzere, B.; St. Louis, J.; Schoenfeld., T.J. Examining the Effects of Exercise on Frustration-Induced Anxiety-like Behavior in Rats. *PSI CHI J. Psychol. Res.* **2019**, *24*, 210–221. [CrossRef]
338. Bierzynska, M.; Bielecki, M.; Marchewka, A.; Debowska, W.; Duszyk, A.; Zajkowski, W.; Falkiewicz, M.; Nowicka, A.; Strelau, J.; Kossut, M. Effect of Frustration on Brain Activation Pattern in Subjects with Different Temperament. *Front. Psychol.* **2015**, *6*, 1989. [CrossRef] [PubMed]
339. Panksepp, J.; Burgdorf, J. 50-kHz Chirping (Laughter?) in Response to Conditioned and Unconditioned Tickle-induced Reward in Rats: Effects of Social Housing and Genetic Variables. *Behav. Brain Res.* **2000**, *115*, 25–38. [CrossRef]
340. Dorland, W.A.N. *The American Illustrated Medical Dictionary*, 19th ed.; W.B. Saunders Company: Philadelphia, PA, USA, 1942; p. 1647.

341. Rush, B. *Medical Inquiries and Observations upon the Disease of the Mind*, 5th ed.; Grigg and Elliot: Philadelphia, PA, USA, 1835; p. 365, Facsimile of the Original Book by Harvard College Library, 2003; Digitized by Google.
342. Burgdorf, J.; Panksepp, J.; Moskal, J.R. Frequency-modulated 50 kHz Ultrasonic Vocalizations: A Tool for Uncovering the Molecular Substrates of Positive Affect. *Neurosci. Biobehav. Rev.* **2011**, *35*, 1831–1836. [CrossRef] [PubMed]
343. Berridge, K.C.; Robinson, T.E. Liking, Wanting and The Iincentive-sensitization Theory of Addiction. *Am. Psychol.* **2016**, *71*, 670–679. [CrossRef] [PubMed]
344. Knutson, B.; Burgdorf, J.; Panksepp, J. High-frequency Ultrasonic Vocalizations Index Conditioned Pharmacological Reward in Rats. *Physiol. Behav.* **1999**, *66*, 639–643. [CrossRef]
345. Heyse, N.C.; Brenes, J.C.; Schwarting, R.K.W. Exercise Reward Induces Appetitive 50-kHz Calls in Rats. *Physiol. Behav.* **2015**, *147*, 131–140. [CrossRef]
346. Antoniou, K.; Kafetzopoulos, E.; Papadopoulou-Daifoti, Z.; Hyphantis, T.; Marselos, M. D-amphetamine, Cocaine and Caffeine: A Comparative Study of Acute Effects on Locomotor Activity and Behavioural Patterns in Rats. *Neurosci. Biobehav. Rev.* **1998**, *23*, 189–196. [CrossRef]
347. Mathews, I.Z.; Brudzynski, S.M.; McCormick, C.M. Heightened Locomotor-activating Effects of Amphetamine Administered into the Nucleus Accumbens in Adolescent Rats. *Int. J. Dev. Neurosci.* **2011**, *29*, 501–507. [CrossRef]
348. Alttoa, A.; Eller, M.; Herm, L.; Rinken, A.; Harro, J. Amphetamine-induced Locomotion, Behavioral Sensitization to Amphetamine, and Striatal D2 Receptor Function in Rats with High or Low Spontaneous Exploratory Activity: Differences in the Role of Locus Coeruleus. *Brain Res.* **2007**, *1131*, 138–148. [CrossRef]
349. Garcia, E.J.; Cain, M.E. Novelty Response and 50 kHz Ultrasonic Vocalizations: Differential Prediction of Locomotor and Affective Response to Amphetamine in Sprague-Dawley Rats. *Psychopharmacology* **2016**, *233*, 625–637. [CrossRef] [PubMed]
350. Burgdorf, J.; Wood, P.L.; Kroes, R.A.; Moskal, J.R.; Panksepp, J. Neurobiology of 50-kHz Ultrasonic Vocalizations in Rats: Electrode Mapping, Lesion, and Pharmacology Studies. *Behav. Brain Res.* **2007**, *182*, 274–283. [CrossRef]
351. Buck, C.L.; Malavar, J.C.; George, O.; Koob, G.F.; Vendruscolo, L.F. Anticipatory 50 kHz Ultrasonic Vocalizations are Associated with Escalated Alcohol Intake in Dependent Rats. *Behav. Brain Res.* **2014**, *271*, 171–176. [CrossRef]
352. Willadsen, M.; Best, L.M.; Wöhr, M.; Clarke, P.B.S. Effects of Anxiogenic Drugs on the Emission of 22- and 50-kHz Ultrasonic Vocalizations in Adult Rats. *Psychopharmacology* **2018**, *235*, 2435–2445. [CrossRef]
353. Hoebel, B.G.; Monaco, A.P.; Hernandez, L.; Aulisi, E.F.; Stanley, B.G.; Lenard, L. Self-injection of Amphetamine Directly into the Brain. *Psychopharmacology* **1983**, *81*, 158–163. [CrossRef] [PubMed]
354. Brosy, J.; Bangerter, A.; Sieber, J. Laughter in the Selection Interview: Impression Management or Honest Signal? *Eur. J. Work Organ Psychol.* **2020**. [CrossRef]
355. Brudzynski, S.M.; Silkstone, M.; Komadoski, M.; Scullion, K.; Duffus, S.; Burgdorf, J.; Kroes, R.A.; Moskal, J.R.; Panksepp, J. Quinpirole-induced 50 kHz Ultrasonic Vocalization in the Rat: Role of D2 and D3 Dopamine Receptors. *Behav. Brain Res.* **2012**, *226*, 511–518. [CrossRef]
356. Simola, N.; Fenu, S.; Costa, G.; Pinna, A.; Plumitallo, A.; Morelli, M. Pharmacological Characterization of 50-kHz Ultrasonic Vocalizations in Rats: Comparison of the Effects of Different Psychoactive Drugs and Relevance in Drug-induced Reward. *Neuropharmacology* **2012**, *63*, 224–234. [CrossRef] [PubMed]
357. Simola, N.; Costa, G. Emission of Categorized 50-kHz Ultrasonic Vocalizations in Rats Repeatedly Treated with Amphetamine or Apomorphine: Possible Relevance to Drug-induced Modifications in the Emotional State. *Behav. Brain Res.* **2018**, *347*, 88–98. [CrossRef] [PubMed]
358. Avvisati, R.; Contu, L.; Stendardo, E.; Michetti, C.; Montanari, C.; Scattoni, M.L.; Badiani, A. Ultrasonic Vocalization in Rats Self-administering Heroin and Cocaine in Different Settings: Evidence of Substance-specific Interactions between Drug and Setting. *Psychopharmacology* **2016**, *233*, 1501–1511. [CrossRef]
359. Wöhr, M. Measuring Mania-like Elevated Mood through Amphetamine-induced 50-kHz Ultrasonic Vocalizations in Rats. *Br. J. Pharmacol.* **2021**. [CrossRef]
360. Kornetsky, C.; Esposito, R.U. Euphorigenic Drugs: Effects on the Reward Pathways of the Brain. *Fed. Proc.* **1979**, *38*, 2473–2476. [PubMed]
361. Kornetsky, C.; Esposito, R.U.; McLean, S.; Jacobson, J.O. Intracranial Self-stimulation Thresholds: A Model for the Hedonic Effects of Drugs of Abuse. *Arch. Gen. Psychiatry* **1979**, *36*, 289–292. [CrossRef] [PubMed]
362. Taracha, E.; Kaniuga, E.; Chrapusta, S.J.; Maciejak, P.; Sliwa, L.; Hamed, A.; Krząścik, P. Diverging Frequency-modulated 50-kHz Vocalization, Locomotor Activity and Conditioned Place Preference Effects in Rats Given Repeated Amphetamine Treatment. *Neuropharmacology* **2014**, *83*, 128–136. [CrossRef] [PubMed]
363. Simola, N.; Morelli, M. Repeated Amphetamine Administration and Long-term Effects on 50-kHz Ultrasonic Vocalizations: Possible Relevance to the Motivational and Dopamine-Stimulating properties of the Drug. *Eur. Neuropsychopharmacol.* **2015**, *25*, 343–355. [CrossRef]
364. Smith, K.S.; Berridge, K.C.; Aldridge, J.W. Disentangling Pleasure from Incentive Salience and Learning Signals in Brain Reward Circuitry. *Proc. Natl. Acad. Sci. USA* **2011**, *108*, E255–E264. [CrossRef]
365. Berridge, K.C.; Robinson, T.E. What is the Role of Dopamine in Reward: Hedonic Impact, Reward Learning, or Incentive Salience? *Brain Res. Brain Res. Rev.* **1998**, *28*, 309–369. [CrossRef]

366. Olney, J.J.; Warlow, S.M.; Erin, E.; Naffziger, E.N.; Berridge, K.C. Current Perspectives on Incentive Salience and Applications to Clinical Disorders. *Curr. Opin. Behav. Sci.* **2018**, *22*, 59–69. [CrossRef] [PubMed]
367. Brenes, J.C.; Schwarting, R.K.W. Attribution and Expression of Incentive Salience are Differentially Signaled by Ultrasonic Vocalizations in Rats. *PLoS ONE* **2014**, *9*, e102414. [CrossRef] [PubMed]
368. Meyer, P.J.; Ma, S.T.; Robinson, T.E. A Cocaine Cue is More Preferred and Evokes More Frequency-modulated 50-kHz Ultrasonic Vocalizations in Rats Prone to Attribute Incentive Salience to a Food Cue. *Psychopharmacology* **2012**, *219*, 999–1009. [CrossRef] [PubMed]
369. Soares-Cunha, C.; Coimbra, B.; Domingues, A.V.; Vasconcelos, N.; Sousa, N.; Rodrigues, A.J. Nucleus Accumbens Microcircuit Underlying D2-MSN-driven Increase in Motivation. *eNeuro* **2018**, *5*. [CrossRef]
370. Wright, J.M.; Deng, L.; Clarke, P.B.S. Failure of Rewarding and Locomotor Stimulant Doses of Morphine to Promote Adult Rat 50-kHz Ultrasonic Vocalizations. *Psychopharmacology* **2012**, *224*, 477–487. [CrossRef]
371. Simola, N.; Frau, L.; Plumitallo, A.; Morelli, M. Direct and long-lasting effects elicited by repeated drug administration on 50-kHz ultrasonic vocalizations are regulated differently: Implications for the study of the affective properties of drugs of abuse. *Int. J. Neuropsychopharmacol.* **2014**, *17*, 429–441. [CrossRef]
372. Hamed, A.; Taracha, E.; Szyndler, J.; Krząścik, P.; Lehner, M.; Maciejak, P.; Skórzewska, A.; Płaźnik, A. The Effects of Morphine and Morphine Conditioned Context on 50 kHz Ultrasonic Vocalisation in Rats. *Behav. Brain Res.* **2012**, *229*, 447–450. [CrossRef]
373. Trezza, V.; Damsteegt, R.; Achterberg, E.J.M.; Vanderschuren, L.J.M.J. Nucleus Accumbens μ-opioid Receptors Mediate Social Reward. *J. Neurosci.* **2011**, *31*, 6362–6370. [CrossRef]
374. Van Ree, J.M.; Leys, A. Behavioral Effects of Morphine and Phencyclidine in Rats: The Influence of Repeated Testing before and after Single Treatment. *Eur. J. Pharmacol.* **1985**, *113*, 353–362. [CrossRef]
375. Babbini, M.; Davis, W.M. Time-dose Relationships for Locomotor Activity Effects of Morphine after Acute or Repeated Treatment. *Br. J. Pharmacol.* **1972**, *46*, 213–224. [CrossRef] [PubMed]
376. Hamed, A.; Boguszewski, P.M. Effects of Morphine and Other Opioid Ligands on Ultrasonic Vocalization in Rats. In *Handbook of Ultrasonic Vocalization. A Window into the Emotional Brain*; Brudzynski, S.M., Ed.; Elsevier/Academic Press: Amsterdam, The Netherlands, 2018.
377. Wiskerke, J.; Schetters, D.; van Es, I.E.; van Mourik, Y.; den Hollander, B.R.O.; Schoffelmeer, A.N.M.; Pattij, T. μ-Opioid Receptors in the Nucleus Accumbens Shell Region Mediate the Effects of Amphetamine on Inhibitory Control but not Impulsive Choice. *J. Neurosci.* **2011**, *31*, 262–272. [CrossRef] [PubMed]
378. De Jaeger, X.; Bishop, S.F.; Ahmad, T.; Lyons, D.; Ng, G.A.; Laviolette, S.R. The Effects of AMPA Receptor Blockade in the Prelimbic Cortex on Systemic and Ventral Tegmental Area Opiate Reward Sensitivity. *Psychopharmacology* **2013**, *225*, 687–695. [CrossRef] [PubMed]
379. Cytawa, J.; Trojniar, W. Hedonesthesia: The Nervous Process Determining Motivated Ingestive Behavior. *Acta Neurobiol. Exp.* **1978**, *38*, 139–151.
380. Cytawa, J.; Trojniar, W. The Pleasure System of the Brain and Its Neurotransmitters. *Pol. J. Pharmacol. Pharm.* **1979**, *31*, 283–292.
381. Moruzzi, G.; Magoun, H.W. Brain Stem Reticular Formation and Activation of the EEG. *Electroencephal. Clin. Neurophysiol.* **1949**, *1*, 455–473. [CrossRef]
382. Neylan, T.C. Physiology of Arousal: Moruzzi and Magoun's Ascending Reticular Activating System. *J. Neuropsychiatry Clin. Neurosci.* **1995**, *7*, 250. [CrossRef]
383. Wijdicks, E.F.M. The Ascending Reticular Activating System. *Neurocrit. Care* **2019**, *31*, 419–422. [CrossRef] [PubMed]
384. Peters, S.M.; Pothuizen, H.H.J.; Spruijt, B.M. Ethological Concepts Enhance the Translational Value of Animal Models. *Eur. J. Pharmacol.* **2015**, *759*, 42–50. [CrossRef]
385. Mällo, T.; Matrov, D.; Herm, L.; Kõiv, K.; Eller, M.; Rinken, A.; Harro, J. Tickling-induced 50-kHz Ultrasonic Vocalization is Individually Stable and Predicts Behaviour in Tests of Anxiety and Depression in Rats. *Behav. Brain Res.* **2007**, *184*, 57–71. [CrossRef] [PubMed]
386. Hodgson, R.A.; Guthrie, D.H.; Varty, G.B. Duration of Ultrasonic Vocalizations in the Isolated Rat Pup as a Behavioral Measure: Sensitivity to Anxiolytic and Antidepressant Drugs. *Pharmacol. Biochem. Behav.* **2008**, *88*, 341–348. [CrossRef] [PubMed]
387. Simola, N. Rat Ultrasonic Vocalizations and Behavioral Neuropharmacology: From the Screening of Drugs to the Study of Disease. *Curr. Neuropharmacol.* **2015**, *13*, 164–179. [CrossRef]
388. Potasiewicz, A.; Holuj, M.; Piotrowska, D.; Zajda, K.; Wojcik, M.; Popik, P.; Nikiforuk, A. Evaluation of Ultrasonic Vocalizations in a Neurodevelopmental Model of Schizophrenia during the Early Life Stages of Rats. *Neuropharmacology* **2019**, *146*, 28–38. [CrossRef] [PubMed]
389. Kelm-Nelson, C.A.; Brauer, A.F.L.; Ciucci, M.R. Vocal Training, Levodopa, and Environment Effects on Ultrasonic Vocalizations in a Rat Neurotoxin Model of Parkinson Disease. *Behav. Brain Res.* **2016**, *307*, 54–64. [CrossRef] [PubMed]
390. Wendler, E.; de Souza, C.P.; Dornellas, A.P.S.; Santos, L.E.; Ferreira, S.T.; Galduróz, J.C.F.; Wöhr, M.; Schwarting, R.K.W.; Andreatini, R. Mania-like Elevated Mood in Rats: Enhanced 50-kHz Ultrasonic Vocalizations after Sleep Deprivation. *Prog. Neuropsychopharmacol. Biol. Psychiatry* **2019**, *88*, 142–150. [CrossRef]
391. Chen, X.; Li, Y.; Li, S.; Kirouac, G.J. Early Fear as a Predictor of Avoidance in a Rat Model of Post-traumatic Stress Disorder. *Behav. Brain Res.* **2012**, *226*, 112–117. [CrossRef]

392. Mittal, N.; Maddox, W.T.; Schallert, T.; Duvauchelle, C.L. Rodent Ultrasonic Vocalizations as Biomarkers of Future Alcohol Use: A Predictive Analytic Approach. *Cogn. Affect. Behav. Neurosci.* **2018**, *18*, 88–98. [CrossRef]
393. Mittal, N.W.; Maddox, T.; Schallert, T.; Duvauchelle, C.L. Spontaneous Ultrasonic Vocalization Transmission in Adult, Male Long–Evans Rats is Age-dependent and Sensitive to EtOH Modulation. *Brain Sci.* **2020**, *10*, 89. [CrossRef] [PubMed]
394. Berg, E.L.; Ching, T.M.; Bruun, D.A.; Rivera, J.K.; Careaga, M.; Ellegood, J.; Lerch, J.P.; Wöhr, M.; Lein, P.J.; Silverman, J.L. Translational Outcomes Relevant to Neurodevelopmental Disorders Following Early Life Exposure of Rats to Chlorpyrifos. *J. Neurodev. Disord* **2020**, *12*, 40. [CrossRef] [PubMed]
395. Jouda, J.; Wöhr, M.; Del Rey, A. Immunity and Ultrasonic Vocalization in Rodents. *Ann. N. Y. Acad. Sci.* **2019**, *1437*, 68–82. [CrossRef] [PubMed]
396. Tonkiss, J.; Bonnie, K.E.; Hudson, J.L.; Shultz, P.L.; Duran, P.; Galler, J.R. Ultrasonic Call Characteristics of Rat Pups are Altered Following Prenatal Malnutrition. *Dev. Psychobiol.* **2003**, *43*, 90–101. [CrossRef] [PubMed]

brain sciences

MDPI

Article

The Effect of Vasopressin Antagonists on Maternal-Separation-Induced Ultrasonic Vocalization and Stress-Hormone Level Increase during the Early Postnatal Period

Bibiána Török [1,2], Anna Fodor [1,2], Sándor Zsebők [3], Eszter Sipos [1] and Dóra Zelena [1,4,*]

1 Institute of Experimental Medicine, 1085 Budapest, Hungary; torok.bibiana@koki.hu (B.T.); anna_@index.hu (A.F.); sipos.eszter@koki.hu (E.S.)
2 János Szentágothai School of Neurosciences, Semmelweis University, 1085 Budapest, Hungary
3 Centre for Ecological Research, Institute of Ecology and Botany, 2163 Vácrátót, Hungary; zsebok.s@gmail.com
4 Centre for Neuroscience, Szentágothai Research Centre, Institute of Physiology, Medical School, University of Pécs, 7622 Pécs, Hungary
* Correspondence: zelena@koki.hu

Abstract: In adults, vasopressin exerts an anxiogenic effect, but less is known about the perinatal period. As a sign of distress, rat pups emit ultrasonic vocalizations when they are separated from their mothers, known as maternal separation-induced ultrasonic vocalization (MS-USV). Previously, reduced MS-USV was reported in 7–8-day-old genetically vasopressin-deficient Brattleboro rats. Here, we aimed to examine the contributing vasopressin receptor (VR) subtypes using Wistar pups. MS-USV was recorded for 10 min, 30 min after vasopressin (V) 1aR, V1bR or V2R antagonist treatment (SR49059, SSR149415, SR121463B; 3, 10 and 30 mg/kg, intraperitoneal). Sedation was studied by the righting reflex and negative geotaxis, and finally, the stress hormone levels were measured by radioimmunoassay. The vasopressin-deficient pups showed decreased MS-USV and adrenocorticotropin levels even after a saline injection, with unchanged corticosterone levels. Thirty mg/kg of V1aR-antagonist increased the corticosterone levels. All V1bR antagonist doses decreased the MS-USV and adrenocorticotropin, while 10 + 10 mg/kg of V1aR and V1bR antagonists decreased MS-USV without influencing the stress hormones. Three mg/kg of V2R antagonist enhanced MS-USV, while 30 mg/kg increased the stress hormone levels. We confirmed that vasopressin deficiency already caused anxiolytic effects in pups. V1bRs are the most important player in connection with their adrenocorticotropin (ACTH)-regulatory role, but a combination of V1aR and V1bR antagonists might be also beneficial through other mechanisms, reducing the possibility of side effects. In contrast, antagonizing the V2Rs may be stressful due to an induction of imbalance in saltwater homeostasis.

Keywords: USV; maternal separation; pup; anxiety; vasopressin antagonists; righting reflex; negative geotaxis; ACTH; corticosterone; Brattleboro rat

Citation: Török, B.; Fodor, A.; Zsebők, S.; Sipos, E.; Zelena, D. The Effect of Vasopressin Antagonists on Maternal-Separation-Induced Ultrasonic Vocalization and Stress-Hormone Level Increase during the Early Postnatal Period. *Brain Sci.* **2021**, *11*, 444. https://doi.org/10.3390/brainsci11040444

Academic Editor: Stefan M. Brudzynski

Received: 2 March 2021
Accepted: 25 March 2021
Published: 30 March 2021

1. Introduction

Arginine vasopressin (AVP) plays an important role in saltwater homeostasis and the regulation of blood pressure as a peripheral hormone [1]. Besides these functions, it also works as a neuropeptide and contributes to the regulation of learning and memory, social behavior and emotionality. Most of the AVP is synthesized in the magnocellular neurons of the hypothalamic paraventricular (PVN) and supraoptic nuclei. It may have an effect on the stress axis by strengthening the function of corticotropin-releasing hormone (CRH) in the PVN [1]. Thus, AVP can increase the CRH-induced adrenocorticotropin (ACTH) levels, then finally lead to glucocorticoid (corticosterone in rodents) secretion from the adrenal gland [1,2]. This forms the hypothalamic–pituitary–adrenocortical (HPA) axis, the fundamental component of adaptation to stress. Disturbances of these mechanisms may contribute to the development of many diseases among them being anxiety. Many research studies supported the observation that the AVP level is positively correlated with the manifestation of anxiety symptoms in adulthood [3].

The perinatal period is less studied in relation to anxiety-like behavior, most probably because of the low number of available tests. One of the best tools seems to be maternal separation-induced ultrasonic vocalization (MS-USV) [4,5]. Our previous study indicated the contribution of AVP to anxiety and the emission of MS-USV using a genetic model: the natural AVP-deficient Brattleboro strain [6]. We found significantly less anxiety-like behavior and low ACTH levels compared with the wild type, both in adults [7] as well as in pups [6]. However, the contributing receptor subtype remained to be elucidated.

AVP has three vasopressin receptors (VR) in mammals: vasopressin (V) 1aR, V1bR and V2R, with different localizations in the body [8]. V1aRs can be found in the vessels' endothelium, and their most important function is to regulate vasoconstriction. We can also find this VR subtype in the limbic system (lateral septum, amygdala and hippocampal areas), the most important regions in the regulation of anxiety-related behaviors. V1bR signaling seems to also be important in anxiety [8]. Most of the V1bRs are present in the adenohypophysis, regulating the HPA axis. V1bR, as the main VR in the pituitary, was the main target of anxiety research in the 2000s [9]. However, among others, Bayerl et al. investigated the anxiogenic role of the VRs in the PVN of adult rats with the elevated plus maze test and found that V1aR, but not the V1bR antagonist, decreased anxiety-like behavior [10]. Moreover, in 2012, clinical studies by Griebel et al. showed that the nonpeptide V1bR antagonist SSR149415 may not be useful for the treatment of generalized anxiety disorder in adult patients [11]. V2Rs are localized in the kidney, regulating saltwater homeostasis. In the brain, its appearance is restricted to the cerebellum. Thus, its contribution to anxiety is questionable but cannot be excluded [12].

Here, we aimed to establish which VR subtype could be responsible for the observed behavioral (MS-USV) and HPA axis effects of the Brattleboro rat pups. As any stress, even a single saline injection, could influence the anxiety measured by MS-USV, we first confirmed that the effect of genetic AVP mutation was still detectable after the mild stress of a saline injection. Then, the effect of pharmacological antagonism by different VRs (V1aR-antagonist SR49059, V1bR-antagonist SSR149415 and V2R-antagonist SR121463B) was studied in increasing doses (3, 10 or 30 mg/kg) in comparison with vehicle injection in association with stress hormone changes (ACTH and corticosterone) [13]. Finally, a combination of V1aR and V1bR antagonists (10 + 10 mg/kg) was used.

2. Materials and Methods

2.1. Subjects

We used 7–8-day-old Brattleboro (n = 19) and Wistar (n = 149) male and female rat pups. There was no sex effect in either case, therefore we pulled results from two sexes together. The litter sizes of 7–10 pups, and 4–8 litters were included in each experiment, leading to 1–3 pups/treatment group/litter [14]. We chose this age group based on previous data [6]. Brattleboro rats were maintained at the Institute of Experimental Medicine in a colony started from breeder rats from Harlan, Indianapolis, IN, USA. Parental Wistar rats were purchased from Charles River (Budapest, Hungary) and were kept in the local animal facility. After birth, the litters were left undisturbed with their mothers until experimentation on postnatal day 7 or 8. The families were kept under standard laboratory conditions: a 12 h light–dark cycle with the lights on at 7:00 a.m., room temperature (20 ± 2 °C), 50–70% humidity and food and tap water ad libitum.The experiments were carried out between 10:00 a.m. and 2:00 p.m.

2.2. Measurements

2.2.1. Maternal Separation-Induced Ultrasonic Vocalization

We measured the weights of the pups, marked them with waterproof ink and gave them an intraperitoneal (ip) injection. Afterward, they were placed back with the litter and were left undisturbed for 30 min. Then, the rat pups were separated from their mother and littermates and placed in a 2 L empty glass beaker without bedding or heating. The experiment was carried out in an empty, closed, soundproof room. MS-USV was recorded for 10 min by an ultrasonic sensitive frequency division detector (CIEL Electronique, CDB205 R2) connected to a personal computer and used Audacity 2.0.5 free software. The detector was put on a platform 10 cm to the side from the top opening of a glass beaker. The recordings were later analyzed by a rat call counter developed by S. Zsebők [6]. The signals were filtered, and the power spectrum was analyzed, ranging from 30 kHz to 50 kHz. In previous studies, the large portions of MS-USV emitted by the 8-day-old rats were found from 30 kHz to 50 kHz [15,16]. The threshold value was set at a signal amplitude of 0.4 V to exclude background noise. The MS-USV duration and number of calls were studied in each group. The MS-USV duration was the sum of the emitted USV durations in seconds during the 10 min observation period, and the number of calls was the average number of emitted USVs per minute.

2.2.2. Testing Sedative Side Effects

At the end of 10 min of MS-USV detection, possible sedative side effects were evaluated.

For the righting reflex, the rat pups were placed on their back on a smooth, flat surface, and the latency necessary to reach the normal upright position with all four feet on the table was measured. The cut-off time was 15 s.

For the negative geotaxis, the offspring were placed on a 45° inclined foam rubber board with their nose pointing down. The animals had 30 s to rotate their body through 180°.

Both tests are widely used to assess neurobehavioral development and evaluated as being positive (the pup can conduct it within the given timeframe) or negative [17]. At this age, a control animal should be able to perform well, and a negative outcome could be judged as a sedative effect of the treatment.

2.2.3. Hormone Measurements

Right after the sedative tests, the rats were decapitated, and trunk blood was collected into ice-cold Eppendorf tubes and centrifuged at 3000 rpm for 30 min at −4 °C. The serum was stored at −20 °C until hormone measurements were conducted. From the serum samples, the ACTH and corticosterone concentrations were measured by a specific radioimmunoassay (RIA) without previous extraction. Both antibodies were developed in our institute as described elsewhere [18–20]. The detection limits were 4 fmol/mL for ACTH and 2.7 pmol/mL for corticosterone. The intraassay coefficients of variation were 4.7% for ACTH and 12.3% for corticosterone. From the Brattleboro rats (Experiment 1), we also collected the hypophysis of the pups to determine the AVP content. Pituitary samples were stored in 100 µL 0.1 N HCl at −20 °C and then homogenized by ultrasound and centrifuged. Then, the AVP content was measured from the hundredfold diluted supernatant using a specific RIA. The rabbit antibodies were donated by Dr. M. Vecsernyés (University of Debrecen, Debrecen, Hungary). The limit of detection was 1 pg AVP/assay tube. The intraassay coefficient of variation was 10.7%. All the samples from a particular experiment were assayed in the same RIA.

2.3. Experiments

2.3.1. Experiment 1: Genetic AVP Deficiency

Brattleboro rats born from heterozygous (di/+) mothers and homozygous diabetes insipidus (di/di) fathers were given ip injections with physiological saline 30 min before the test and placed back with their mothers. MS-USV was measured for 10 min, and at termination, trunk blood was collected for ACTH and corticosterone measurement, and hypophysis was collected for determination of the genotype (di/+ or di/di disposition based upon the AVP content) [17].

2.3.2. Experiment 2: Pharmacological AVP-Effect Deficiency

A (2A) V1aR antagonist (SR49059), (2B) V1bR antagonist (SSR149415) or (2C) V2R antagonist (SR121463B) was suspended in 0.4% Tween 80 (1 μL/g volume for every animal), then delivered intraperitoneally 30 min before MS-USV recording in three different doses: 3, 10 or 30 mg/kg (a generous gift from the Sanofi-Synthélabo company). In a further experimental series, a (2D) 10 mg/kg V1aR antagonist was mixed with a 10 mg/kg V1bR antagonist. The control treatment was the solvent in a 1 μL/g volume washed with 15 μL saline. MS-USV was measured for 10 min, and at the end, the righting reflex and negative geotaxis were evaluated within 2 min. At the termination of the experiments, trunk blood was collected for ACTH and corticosterone determination.

The specificity of these drugs was strongly confirmed by previous studies, and each of them effectively influenced their main target symptom. More precisely, SR49059 showed high affinity to V1aRs in the rat liver (Ki/inhibition constant = 1.6 ± 0.2 nmol/L) and human platelets, adrenals and myometrium (Ki ranging from 1.1 to 6.3 nmol/L). In vivo, SR49059 inhibited the pressor response to exogenous AVP in rats with a long duration of action [21].

SRR149415 had high affinity to V1bRs (hypophysis, human: Ki = 4.2 ± 1.1 nmol/L; rat: Ki = 3.7 ± 1.3 nmol/L). Its in vivo activity could be characterized as anxiolytic-like and stress relieving [13].

The human V2R bonds SR121463 with high affinity (Ki = 0.54 ± 0.09 nmol/L) [22]. SR121463 normalized serum Na^+ levels, abolished hyponatremia and restored normal urine excretion, urine osmolality and renal function in a rat model of cirrhosis [23].

2.4. Statistical Analysis

Data were expressed as mean ± SEM and analyzed using the STATISTICA 13.0 software package (StatSoft, Inc., Tulsa, OK, USA) by analysis of variance (ANOVA) using one-way ANOVA (factor = treatment). In case of the Levene's assumption being significant, the data were transformed logarithmically, and the analysis was conducted on the transformed data. Post hoc comparison was made by the Fisher's Least Significant Difference method, and the results were presented on the figures. Correlations were calculated by the Pearson method. The level of statistical significance was taken as $p < 0.05$.

3. Results

3.1. Genetic AVP Deficiency

The AVP-deficient Brattleboro rats (di/di) were less anxious, based upon their emitted MS-USV number of calls ($F_{(1,17)} = 7.600$; $p = 0.014$; Figure 1a) and duration ($F_{(1,17)} = 7.600$; mboxemphp = 0.014; Figure 1b) compared with their heterozygous littermates. In connection, their ACTH levels were significantly lower ($F_{(1,17)} = 15.380$; $p = 0.001$; Figure 1c) without significant alterations in corticosterone levels ($F_{(1,17)} = 1.915$; $p = 0.184$; Figure 1d).

The righting reflex ($F_{(1,17)} = 1.850$; $p = 0.193$) and negative geotaxis ($F_{(1,17)} = 0.042$; $p = 0.840$) values were comparable in the two genotypes (data not shown).

3.2. Pharmacological AVP Deficiency

3.2.1. V1aR Antagonist

The V1aR antagonist treatment decreased MS-USV only with the 30 mg/kg dose (number of calls: $F_{(3,42)} = 1.788$; $p = 0.164$; Fisher post hoc control vs. 30 mg/kg: $p = 0.026$; Figure 2a; duration: $F_{(3,42)} = 1.551$; $p = 0.215$; Fisher post hoc control vs. 30 mg/kg: $p = 0.040$; Figure 2b). The ACTH levels showed no alterations ($F_{(3,42)} = 1.219$; $p = 0.315$; Figure 2c), while the corticosterone levels were significantly higher in the group with the 30 mg/kg antagonist treatment ($F_{(3,42)} = 1.857$; $p = 0.152$; Fisher post hoc control vs. 30 mg/kg: $p = 0.030$; Figure 2d).

There was no difference between the groups in the latency of the righting reflex as well as the negative geotaxis (Table 1).

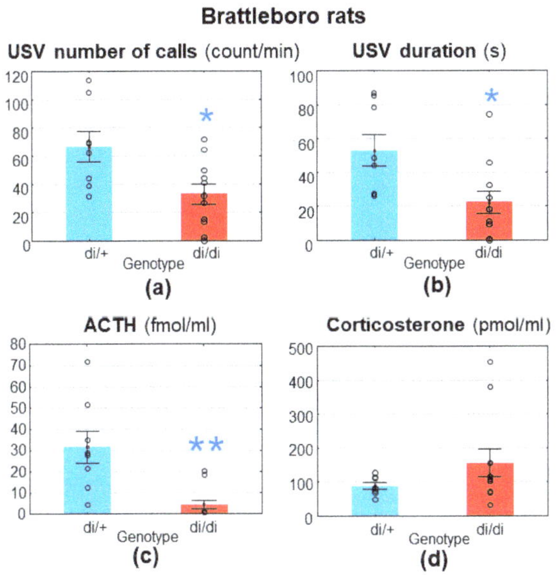

Figure 1. Brattleboro rats. The 7–8-day-old pups emitted reduced ultrasonic vocalization to maternal separation, measured for 10 min both in terms of (**a**) the number of calls (count/min) and (**b**) the duration (s) 30 min after a single intraperitoneal saline injection. Moreover, they showed reduced (**c**) adrenocorticotropin (ACTH, fmol/mL) elevation at the end of separation without significant changes in (**d**) corticosterone (pmol/mL) levels. di/+ = heterozygous; di/di = homozygous diabetes insipidus pups, without functional vasopressin; n = 8–11. * $p < 0.05$. ** $p < 0.01$ vs. di/+ control.

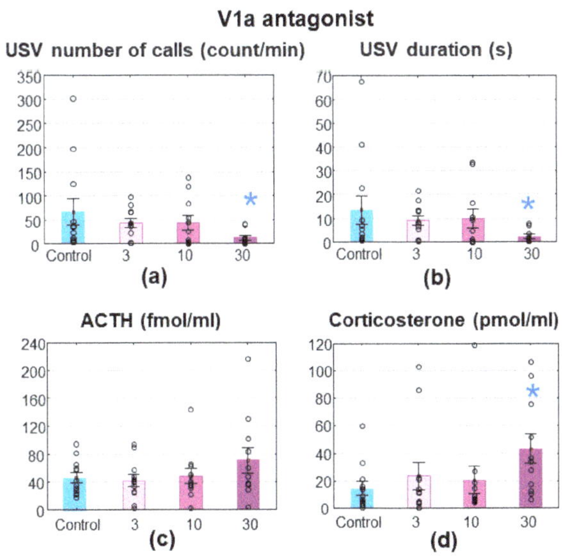

Figure 2. Wistar rats, treated with SR49059, a vasopressin (V) 1aR antagonist. The 7–8-day-old Wistar rat pups were treated intraperitoneally with a 3, 10 or 30 mg/kg V1aR antagonist 30 min before a 10 min maternal separation. The 30 mg/kg dose significantly reduced the emitted ultrasonic vocalization both in terms of (**a**) the number of calls and (**b**) the duration without changes in (**c**) adrenocorticotropin (ACTH, fmol/mL), but an elevation in (**d**) corticosterone (pmol/mL) levels. n = 12–14. * $p < 0.05$ vs. control.

Table 1. Righting reflex and negative geotaxis values.

Antagonist Treatment	Time (s)	Doses (mg/kg)			
		0	3	10	30
V1aR	Righting	2.333 ± 0.343	2.361 ± 0.230	2.758 ± 0.397	2.212 ± 0.187
	Neg.geo.	19.861 ± 2.687	20.750 ± 2.984	18.212 ± 2.735	16.485 ± 2.929
V1bR	Righting	1.303 ± 0.156	2.279 ± 0.669	1.947 ± 0.303	2.103 ± 0.266
	Neg.geo.	9.194 ± 0.963	8.487 ± 0.880	9.528 ± 1.215	10.100 ± 1.673
V2R	Righting	2.133 ± 0.218	2.267 ± 0.325	2.033 ± 0.195	2.250 ± 0.300
	Neg.geo.	11.967 ± 1.961	13.970 ± 2.707	15.267 ± 3.010	18.667 ± 2.567
V1aR + V1bR	Righting	1.741 ± 0.282		2.750 ± 0.552	
	Neg.geo.	8.296 ± 1.301		7.125 ± 0.900	

No significant alterations were discovered. Righting = righting reflex; Neg.geo. = negative geotaxis.

3.2.2. V1bR Antagonist

The V1bR antagonist treatment decreased the MS-USV number of calls ($F_{(3,43)} = 5.719$; $p = 0.002$; Figure 3a) and duration ($F_{(3,43)} = 4.470$; $p = 0.008$; Figure 3b), accompanied by reduced ACTH levels ($F_{(3,19)} = 3.008$; $p = 0.056$; Figure 3c) without changes in corticosterone levels ($F_{(3,19)} = 1.230$; $p = 0.326$; Figure 3d).

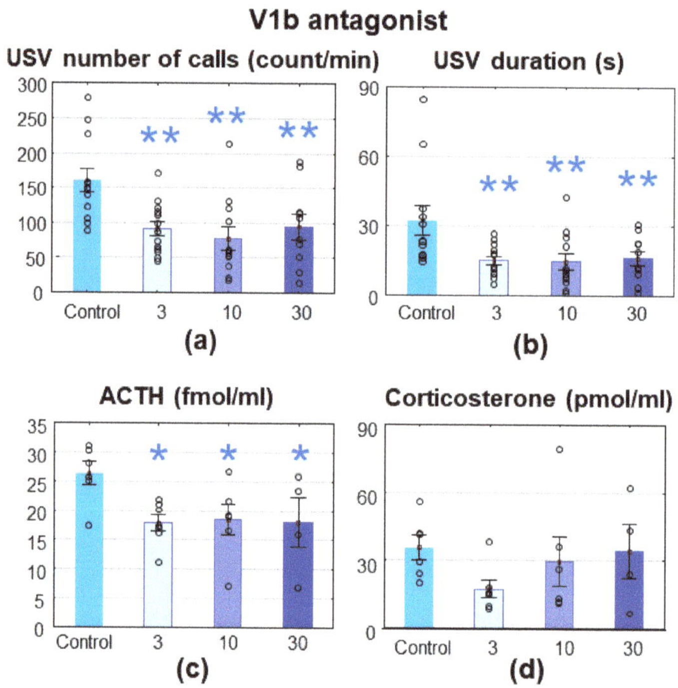

Figure 3. Wistar rats, treated with SSR149415, a a vasopressin (V) 1bR antagonist. The 7–8-day-old Wistar rat pups were treated intraperitoneally with 3, 10 or 30 mg/kg of the V1bR antagonist 30 min before a 10 min maternal separation. All doses significantly reduced the emitted ultrasonic vocalization both in terms of (**a**) the number of calls and (**b**) the duration and reduced (**c**) adrenocorticotropin (ACTH, fmol/mL) levels without affecting the (**d**) corticosterone (pmol/mL) values. n = 10–13. * $p < 0.05$. ** $p < 0.01$ vs. control.

There was no difference between the groups in the latency of the righting reflex and the negative geotaxis (Table 1) [24].

3.2.3. V2R Antagonist

The 3 mg/kg V2R antagonist enhanced MS-USV (number of calls: $F_{(3,35)} = 4.891$; $p = 0.006$; Fisher post hoc control vs. 3 mg/kg: $p = 0.041$; Figure 4a; duration: $F_{(3,35)} = 4.935$; $p = 0.006$; Fisher post hoc control vs. 3 mg/kg: $p = 0.057$; Figure 4b), while the higher doses had no effect on MS-USV. Both stress hormone levels were higher 45 min after a single 30 mg/kg V2R antagonist treatment compared with the control injection group (ACTH: $F_{(3,35)} = 13.321$; $p = 0.000$; Fisher post hoc control vs. 30 mg/kg: $p = 0.000$; Figure 4c; corticosterone: $F_{(3,35)} = 8.363$; $p = 0.000$; Fisher post hoc control vs. 30 mg/kg: $p = 0.000$; Figure 4d).

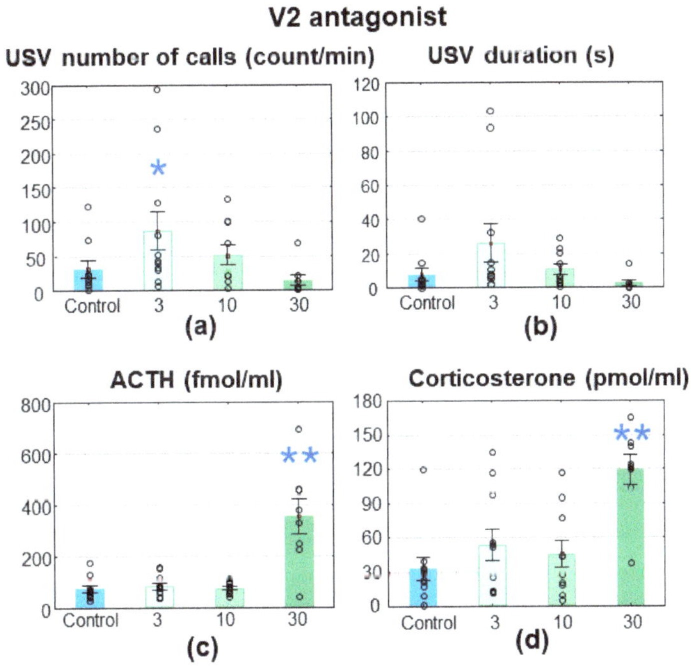

Figure 4. Wistar rats, treated with SR121463B, a a vasopressin (V) 2R antagonist. The 7–8-day-old Wistar rat pups were treated intraperitoneally with a 3, 10 or 30 mg/kg V2R antagonist 30 min before a 10 min maternal separation. The 3 mg/kg dose significantly enhanced the emitted ultrasonic vocalization in terms of (**a**) the number of calls, with a similar tendency in (**b**) the duration and 30 mg/kg elevated (**c**) adrenocorticotropin (ACTH, fmol/mL) and (**d**) corticosterone (pmol/mL) levels. $n = 9$–11; * $p < 0.05$, ** $p < 0.01$ vs. control.

There was no difference between the groups in the latency of the righting reflex or the negative geotaxis (Table 1).

3.2.4. V1aR + V1bR Antagonists

The combination of V1aR and V1bR antagonists effectively reduced MS-USV (number of calls: $F_{(1,15)} = 10.440$; $p = 0.006$; Figure 5a; duration: $F_{(1,15)} = 15.616$; $p = 0.001$; Figure 5b) without any effect on the stress hormones (ACTH: $F_{(1,15)} = 0.008$; $p = 0.931$; Figure 5c; corticosterone: $F_{(1,15)} = 0.001$; $p = 0.982$; Figure 5d). The same dose of the V1aR antagonist induced 34.3% and 26.8% nonsignificant reductions in the MS-USV number of calls and duration, respectively, while in the case of the V1bR-antagonist, 51.5% and 54.3% significant reductions were visible. The combination induced a 57.1% reduction in the MS-USV number of calls and a 68.5% reduction in duration.

There was no difference between the groups in the latency of the righting reflex or the negative geotaxis (Table 1).

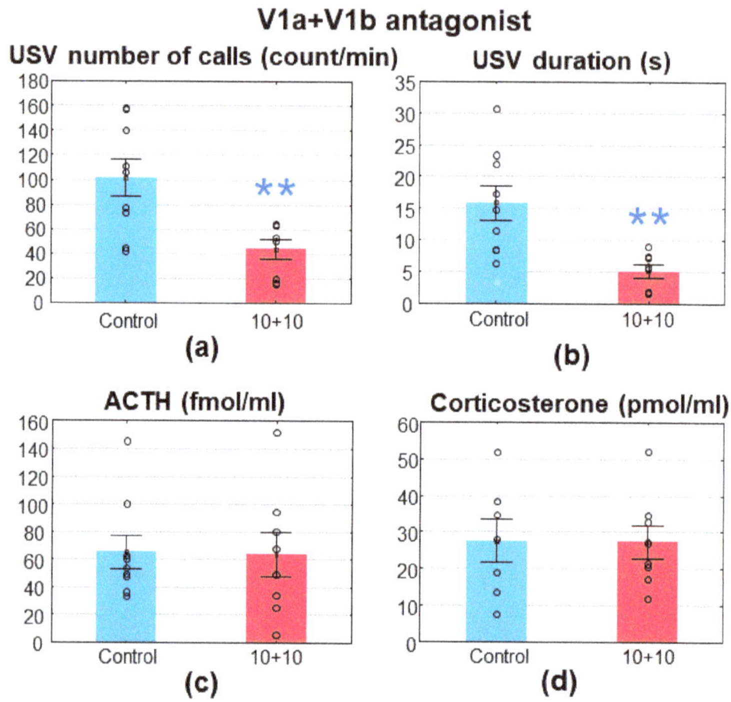

Figure 5. Wistar rats, treated with SR49059 (a vasopressin (V) 1aR) + SSR149415 (V1bR) antagonists. The 7–8-day-old Wistar rat pups were treated intraperitoneally with a mixture of 10 + 10 mg/kg V1aR + V1bR antagonists 30 min before a 10 min maternal separation. The combination significantly reduced the emitted ultrasonic vocalization both in terms of (**a**) the number of calls and (**b**) the duration without affecting the (**c**) adrenocorticotropin (ACTH, fmol/mL) or (**d**) corticosterone (pmol/mL) levels. n = 9–8. ** $p < 0.01$ vs. control.

3.3. Correlations

As could have been expected, the number of calls and duration of MS-USV positively correlated with each other in all experimental series. Interestingly, the same was true for ACTH and corticosterone correlation, except in the case of Brattleboro animals, where there was no correlation at all (data not shown). In the Brattleboro strain, the AVP content of the hypophysis showed a significant positive correlation with the MS-USV number of calls (r = 0.556; p = 0.017; Figure 6a) and duration (r = 0.541; p = 0.020). Moreover, in this case, the serum ACTH level also showed positive correlation with the number of emitted MS-USVs (r = 0.491; p = 0.038; Figure 6b). Interestingly, similar ACTH and MS-USV number of calls correlation was detected after V1bR antagonist treatment (r = 0.424; p = 0.044; Figure 6c).

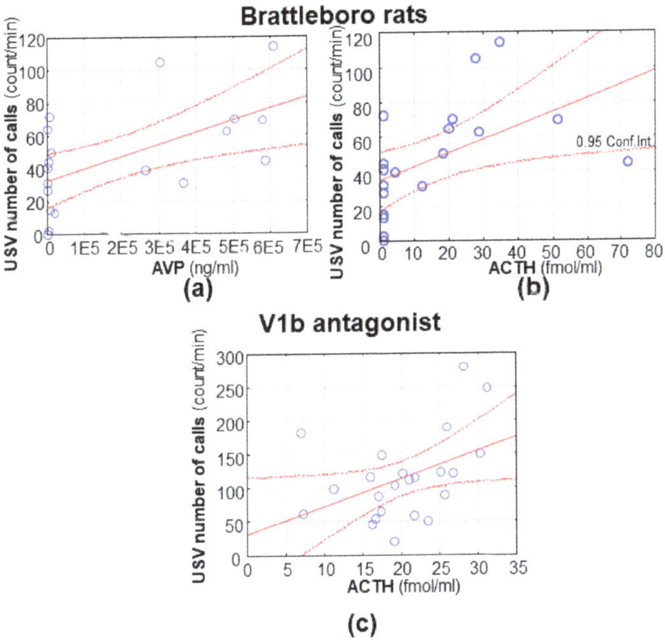

Figure 6. The most important correlations. In Brattleboro pups (both di/+ and di/di genotypes), (**a**) the hypophysis vasopressin (AVP) content positively correlated with the emitted number of ultrasonic vocalizations. A similar positive correlation was observable between the serum ACTH levels and the ultrasonic vocalization number both in (**b**) the Brattleboro pups and (**c**) after a vasopressin (V) 1bR antagonist treatment.

4. Discussion

Our results confirmed that genetic AVP deficiency already had an anxiolytic effect during the early postnatal age, which was not influenced by the mild stress of an ip saline injection. The positive correlation between the pituitary AVP content and MS-USV further confirmed the participation of this neuropeptide in the separation-induced vocalization. Pharmacological analysis showed that a high dose (30 mg/kg) of the V1aR antagonist and all studied doses of the V1bR antagonist reduced MS-USV, while the V2R antagonist elevated it in the smallest studied dose (3 mg/kg). The number of MS-USVs correlated positively with the ACTH levels in the case of the V1bR antagonist only, similar to the Brattleboro rats. None of the studied interventions influenced the latency of the righting reflex negative geotaxis, suggesting that they were without any sedative side effects.

As anxiety is a stress-related disorder, drugs influencing the HPA axis were the focus of interest for its treatment. Indeed, the first selective and orally active nonpeptide antagonist of V1bRs (SSR149415) in adult rodent models had anxiolytic and antidepressant-like effects [25]. We also found a strong anxiolytic effect of the V1bR antagonist in rat pups with all the studied doses (3, 10 or 30 mg/kg), which confirmed our previous results with 10 [6] and 30 mg/kg [5]. In a previous experiment, the same V1bR antagonist in the same doses showed only a tendency to reduce MS-USV [26]. However, the authors used 9–11-day-old animals (for age-dependent MS-USV, see Figure 1 in [6]) and 5 min of measurement, which might be responsible for the reduced sensitivity of their assay. Similarly, the tendency seen with another V1bR-antagonist, TASP0233278 [27], might be also attributed to their less sensitive assay, as they also used 5 min of measurement and older animals (21–30 g in contrast to our 16–20 g animals). In another study on mice, the shorter recording time (5 min) might also have been responsible for the possible small difference between wild type (WT) and V1bR knockout (KO) animals [28]. However, in these V1bR KO animals, the repeated MS was not able to induce any increase in the number of MS-USV, in contrast to their WT littermates, supporting some anxiolytic role of this VR subtype.

In our hands, the positive correlation between MS-USV and stress hormones, in the case of V1bR antagonist treatment, confirmed that this VR subtype is able to influence anxiety through the regulation of stress hormones. Indeed, although SSR149415 can penetrate into the brain, its half-life is relatively short (about 60 min) [24]. Thus, it is more likely that it acts on the pituitary V1b receptors influencing the HPA axis. Interestingly, it was not the end hormone of the axis (corticosterone in rodents), but the pituitary component where ACTH was implicated in this phenomenon. This was in line with the results found in the Brattleboro rats (see Figure 6b and [5,6]), where ACTH levels did not go parallel with the corticosterone levels. Although it is hard to separate the effect of ACTH from the effect of its downstream molecules (e.g., glucocorticoids, mineralocorticoids or adrenal androgens produced in the adrenal cortex), based upon ACTH administration, many extra-adrenal effects of ACTH have been suggested (e.g., cardiovascular, metabolic, motivational or memory influencing [29]). Moreover, chronic ACTH administration in rats induced depression-like behavioral changes [30]. ACTH-producing tumors were also associated with mood swings [31]; however, in this case, the role of other factors could not be entirely excluded. Additionally, other pathways may also contribute to the V1bR-induced anxiolysis, as previous studies showed altered V1bR protein levels in the rat hypothalamus in connection with anxiolytic treatment [32].

On the other hand, several data speak in favor of the role of V1aRs in anxiety [33]. In adult rodents, both the genetic (KO mice) and pharmacological (VR antagonism) blockade of the V1aRs showed anxiolytic effects [34]. Furthermore, in adult rats, overexpression of the V1aR gene in the lateral septum increased anxiety-related behavior [35]. Our data confirmed the anxiolytic effect of V1aR antagonism in pups, but only in the highest used dose (30 mg/kg). A 10 min observation period was sufficient to reveal the anxiolytic role of another V1aR antagonist (JNJ-17308616) in 11-day-old Sprague Dawley rat pups, too [36]. Interestingly, in this study, anxiolysis was detected only in the highest (100 mg/kg) dose, when JNJ-17308616 may influence the V2Rs as well. At the periphery, V1aRs might induce vasodilatation, confirmed also by lower blood pressure in the V1aR KO mice [37]. The drop in blood pressure might be stressful; thus, it was not surprising that the highest dose of the V1aR antagonist (30 mg/kg) stimulated the HPA axis. However, this stress could hardly explain the anxiolysis. Thus, we can conclude that, in our hands, V1aR antagonists should have a central effect. Although SR49059 cannot cross the blood–brain barrier in adults [38], the increased permeability of the blood–brain barrier [39] in pups can make its anxiolytic effect possible.

The decrease in the MS-USV number and duration after combined V1aR and V1bR antagonism was higher than after any antagonist treatment alone (for exact numbers, see Section 3.2.3), without any correlation with the stress hormones. Thus, it seems that during the early postnatal period, both V1bRs and V1aRs are involved in the development of anxiety, most probably through a central brain target other than the HPA axis.

Based upon a previous finding that a V1R antagonist was able antagonize AVP administration-induced MS-USV in 8–9-day-old Sprague Dawley rat pups [40], we did not truly expect an anxiolytic effect from the V2R antagonist. In our hands, even 3 mg/kg of the V2R antagonist was anxiogenic, while 30 mg/kg was highly stressful. As V2Rs are important in saltwater homeostasis, their antagonism may induce an imbalance, leading to the appearance of anxiety and high levels of stress hormones. Once again, the HPA axis parameters were clearly separated from the MS-USV behavioral measure.

The relevance of our study is supported by the presence of AVP already in rat embryos (first appearance of its binding at embryonic day 16) [40–42]. At birth, the hypothalamic level of AVP is comparable to adult levels, with the same ligand selectivity and affinity. In the brains of rat pups, VR1 [42] as well as V2R subtypes [43] were found, while V2Rs were expressed at the periphery [44].

We observed interesting strain differences. First, the length of an MS-USV bout (duration/number of calls) was substantially longer in the Brattleboro (87.335 ± 5.685 ms/bout) than in the Wistar rat strain (18.995 ± 2.142 ms/bout), despite similar MS-USV frequencies (no differences were found in this parameter between the treatment groups). The Brattleboro pups (17.768 ± 0.327 g) were smaller than the age-matched Wistar rats (19.718 ± 0.242 g), suggesting possible developmental differences. However, our previous study showed that the number of MS-USV calls was higher, not lower, in a heavier pup [5]. Further, in the controls, the corticosterone values were higher in the Brattleboro than in the Wistar pups. This is consistent with our previous results [6,45]. Moreover, an earlier study found that Long Evans animals (the origin of the Brattleboro strain) were more stress reactive than the Wistar strain [46]. This different stress sensitivity of the two strains can—at least partly—explain the observed strain differences.

5. Conclusions

All in all, we confirmed the involvement of both V1bRs and V1aRs in the anxiolytic effect of AVP without the contribution of V2Rs. HPA axis changes can only partly contribute to the observed anxiolysis. Taking into consideration the possible side effects, a mixed V1aR/V1bR antagonist might be more beneficial than either antagonist alone.

Author Contributions: Conceptualization, D.Z.; methodology, S.Z., who also developed and validated the software for USV detection and analysis; formal analysis, B.T., A.F. and D.Z.; investigation, B.T., A.F., E.S. and D.Z.; resources, D.Z.; writing—original draft preparation, B.T. and D.Z.; writing—review and editing, all co-authors; visualization, B.T. and D.Z. All authors have read and agreed to the published version of the manuscript.

Funding: This research was funded by the National Research, Development and Innovation Office (NKFIH) of Hungary, grant no. K120311 and K131406 to D.Z. and PD-115730 and K-129215 to S.Z.s.

Institutional Review Board Statement: The study was conducted according to the guidelines of the Declaration of Helsinki, in accordance with the European Communities Council Directive recommendations for the care and use of laboratory animals (2010/63/EU) and was approved by the Animal Welfare and Ethics Committee of Institute of Experimental Medicine (protocol code 22.1/3895/003/2009 and date of approval: 07/10/2009 and protocol code PEI/001/38-4/2013 and date of approval: 06/03/2013).

Informed Consent Statement: Not applicable.

Data Availability Statement: The data presented in this study are available on request from the corresponding author.

Acknowledgments: We are thankful to Éva Dobozi for her help in the radioimmunoassay.

Conflicts of Interest: The authors declare no conflict of interest. The funding agency had no role in the design of the study; in the collection, analyses, or interpretation of data; in the writing of the manuscript, or in the decision to publish the results.

References

1. Frank, E.; Landgraf, R. The vasopressin system–from antidiuresis to psychopathology. *Eur. J. Pharmacol.* **2008**, *583*, 226–542. [CrossRef] [PubMed]
2. Gibbs, D. Vasopressin and oxytocin: Hypothalamic modulators of the stress response: A review. *Psychoneuroendocrinology* **1986**, *11*, 131–139. [CrossRef]
3. Keck, M. Corticotropin-releasing factor, vasopressin and receptor systems in depression and anxiety. *Amino Acids* **2006**, *31*, 241–250. [CrossRef] [PubMed]
4. Hofer, M.A. Multiple regulators of ultrasonic vocalization in the infant rat. *Psychoneuroendocrinology* **1996**, *21*, 203–217. [CrossRef]
5. Zelena, D. The role of vasopressin in anxiety: What can we learn from studying ultrasonic vocalization? In *Handbook of Behavioral Neuroscience, A Window into the Emotional Brain*, 1st ed.; Brudzynski, S.M., Ed.; Elsevier/Academic Press: London, UK, 2018; Chapter 41; pp. 433–441.
6. Varga, J.; Fodor, A.; Klausz, B.; Zelena, D. Anxiogenic role of vasopressin during the early postnatal period: Maternal separation-induced ultrasound vocalization in vasopressin-deficient Brattleboro rats. *Amino Acids* **2015**, *47*, 2409–2418. [CrossRef] [PubMed]

7. Mlynarik, M.; Zelena, D.; Bagdy, G.; Makara, G.B.; Jezova, D. Signs of attenuated depression-like behavior in vasopressin deficient Brattleboro rats. *Horm. Behav.* **2007**, *51*, 395–405. [CrossRef] [PubMed]
8. Koshimizu, T.; Nakamura, K.; Egashira, N.M.H.; Nonoguchi, H.; Tanoue, A. Vasopressin V1a and V1b receptors: From molecules to physiological systems. *Physiol. Rev.* **2012**, *92*, 1813–1864. [CrossRef]
9. Shimazaki, T.; Iijima, M.; Chaki, S. The pituitary mediates the anxiolytic-like effects of the vasopressin V1B receptor antagonist, SSR149415, in a social interaction test in rats. *Eur. J. Pharmacol.* **2006**, *543*, 63–67. [CrossRef] [PubMed]
10. Bayerl, D.; Hönig, J.; Bosch, O. Vasopressin V1a, but not V1b, receptors within the PVN of lactating rats mediate maternal care and anxiety-related behaviour. *Behav. Brain Res.* **2016**, *305*, 318–322. [CrossRef]
11. Griebel, G.; Beeské, S.; Stahl, S. The vasopressin V(1b) receptor antagonist SSR149415 in the treatment of major depressive and generalized anxiety disorders: Results from 4 randomized, double-blind, placebo-controlled studies. *J. Clin. Psychiatry* **2012**, *73*, 1403–1411. [CrossRef]
12. Zelena, D. Vasopressin in health and disease with a focus on affective disorders. *Cent. Nerv. Syst. Agents Med. Chem.* **2012**, *12*, 286–303. [CrossRef] [PubMed]
13. Le Gal, C.S.; Wagnon, J.; Simiand, J.; Griebel, G.; Lacour, C.; Guillon, G.; Barberis, C.; Brossard, G.; Soubrie, P.; Nisato, D.; et al. Characterization of (2*S*,4*R*)-1-[5-chloro-1-[(2,4-dimethoxyphenyl)sulfonyl]-3-(2-methoxy-phenyl)-2-oxo- 2,3-dihydro-1*H*-indol-3-yl]-4-hydroxy-*N*,*N*-dimethyl-2-pyrrolidine carboxamide (SSR149415), a selective and orally active vasopressin V_{1b} receptor antagonist. *J. Pharmacol. Exp. Ther.* **2002**, *300*, 1122–1130. [CrossRef]
14. Williams, D.R.; Carlsson, R.; Burkner, P.C. Between-litter variation in developmental studies of hormones and behavior: Inflated false positives and diminished power. *Front. Neuroendocrinol.* **2017**, *47*, 154–166. [CrossRef]
15. Blumberg, M.S.; Sokoloff, G.; Kent, K.J. A developmental analysis of clonidine's effects on cardiac rate and ultrasound production in infant rats. *Dev. Psychobiol.* **2000**, *36*, 186–193. [CrossRef]
16. Allin, J.T.; Banks, E.M. Effects of temperature on ultrasound production by infant albino rats. *Dev. Psychobiol.* **1971**, *4*, 149–156. [CrossRef] [PubMed]
17. Zelena, D.; Mergl, Z.; Makara, G.B. Postnatal development in vasopressin deficient Brattleboro rats with special attention to the hypothalamo-pituitary-adrenal axis function: The role of maternal genotype. *Int. J. Dev. Neurosci.* **2009**, *27*, 175–183. [CrossRef] [PubMed]
18. Zelena, D.; Kiem, D.T.; Barna, I.; Makara, G.B. Alpha 2-adrenoreceptor subtypes regulate ACTH and beta-endorphin secretions during stress in the rat. *Psychoneuroendocrinology* **1999**, *24*, 333–343. [CrossRef]
19. Zelena, D.; Mergl, Z.; Foldes, A.; Kovacs, K.J.; Toth, Z.; Makara, G.B. Role of hypothalamic inputs in maintaining pituitary-adrenal responsiveness in repeated restraint. *Am. J. Physiol. Endocrinol. Metab.* **2003**, *285*, E1110–E1117. [CrossRef] [PubMed]
20. Zelena, D.; Mergl, Z.; Makara, G.B. Maternal genotype influences stress reactivity of vasopressin-deficient brattleboro rats. *J Neuroendocrinol.* **2003**, *15*, 1105–1110. [CrossRef] [PubMed]
21. Serradeil-Le Gal, C.; Wagnon, J.; Garcia, C.; Lacour, C.; Guiraudou, P.; Christophe, B.; Villanova, G.; Nisato, D.; Maffrand, J.P.; Le Fur, G.; et al. Biochemical and pharmacological properties of SR 49059, a new, potent, nonpeptide antagonist of rat and human vasopressin V1a receptors. *J. Clin. Investig.* **1993**, *92*, 224–231. [CrossRef] [PubMed]
22. Le Gal, C.S.; Raufaste, D.; Double-Cazanave, E.; Guillon, G.; Garcia, C.; Pascal, M.; Maffrand, J.P. Binding properties of a selective tritiated vasopressin V2 receptor antagonist, [H]-SR 121463. *Kidney Int.* **2000**, *58*, 1613–1622. [CrossRef]
23. Jimenez, W.; Gal, C.S.; Ros, J.; Cano, C.; Cejudo, P.; Morales-Ruiz, M.; Arroyo, V.; Pascal, M.; Rivera, F.; Maffrand, J.P.; et al. Long-term aquaretic efficacy of a selective nonpeptide V(2)-vasopressin receptor antagonist, SR121463, in cirrhotic rats. *J. Pharmacol. Exp. Ther.* **2000**, *295*, 83–90. [PubMed]
24. Hodgson, R.A.; Higgins, G.A.; Guthrie, D.H.; Lu, S.X.; Pond, A.J.; Mullins, D.E.; Guzzi, M.F.; Parker, E.M.; Varty, G.B. Comparison of the V1b antagonist, SSR149415, and the CRF1 antagonist, CP-154,526, in rodent models of anxiety and depression. *Pharmacol. Biochem. Behav.* **2007**, *86*, 431–440. [CrossRef] [PubMed]
25. Griebel, G.; Simiand, J.; Serradeil-Le Gal, C.; Wagnon, J.; Pascal, M.; Scatton, B.; Maffrand, J.P.; Soubrie, P. Anxiolytic- and antidepressant-like effects of the non-peptide vasopressin V1b receptor antagonist, SSR149415, suggest an innovative approach for the treatment of stress-related disorders. *Proc. Natl. Acad. Sci. USA* **2002**, *99*, 6370–6375. [CrossRef] [PubMed]
26. Iijima, M.; Chaki, S. Separation-induced ultrasonic vocalization in rat pups: Further pharmacological characterization. *Pharmacol. Biochem. Behav.* **2005**, *82*, 652–657. [CrossRef]
27. Iijima, M.; Yoshimizu, T.; Shimazaki, T.; Tokugawa, K.; Fukumoto, K.; Kurosu, S.; Kuwada, T.; Sekiguchi, Y.; Chaki, S. Antidepressant and anxiolytic profiles of newly synthesized arginine vasopressin V1B receptor antagonists: TASP0233278 and TASP0390325. *Br. J. Pharmacol.* **2014**, *171*, 3511–3525. [CrossRef] [PubMed]
28. Scattoni, M.L.; McFarlane, H.G.; Zhodzishsky, V.; Caldwell, H.K.; Young, W.S.; Ricceri, L.; Crawley, J.N. Reduced ultrasonic vocalizations in vasopressin 1b knockout mice. *Behav. Brain Res.* **2008**, *187*, 371–378. [CrossRef] [PubMed]
29. Zelena, D.; Makara, G.B. The role of adrenocorticotropin beyond the glucocorticoid horizon. In *Advances in Medicine and Biology*; Nova Publisher: Hauppauge, NY, USA, 2012; Volume 43, pp. 109–138.
30. Kitamura, Y.; Gomita, Y. Development of animal models of treatment-resistant depression in rats. *Nihon Shinkei Seishin Yakurigaku Zasshi* **2008**, *28*, 93–100.
31. McCaughey, E.S.; Walker, V.; Rolles, C.J.; Scheurmier, N.I.; Hale, A.C.; Rees, L.H. Ectopic ACTH production by a thymic carcinoid tumour. *Eur. J. Pediatr.* **1987**, *146*, 590–591. [CrossRef]

32. Kokras, N.; Sotiropoulos, I.; Pitychoutis, P.M.; Almeida, O.F.; Papadopoulou-Daifoti, Z. Citalopram-mediated anxiolysis and differing neurobiological responses in both sexes of a genetic model of depression. *Neuroscience* **2011**, *194*, 62–71. [CrossRef] [PubMed]
33. Ebner, K.; Wotjak, C.; Holsboer, F.; Landgraf, R.; Engelmann, M. Vasopressin released within the septal brain area during swim stress modulates the behavioural stress response in rats. *Eur. J. Neurosci.* **1999**, *11*, 997–1002. [CrossRef] [PubMed]
34. Egashira, N.; Tanoue, A.; Matsuda, T.; Koushi, E.; Harada, S.; Takano, Y.; Tsujimoto, G.; Mishima, K.; Iwasaki, K.; Fujiwara, M. Impaired social interaction and reduced anxiety-related behavior in vasopressin V1a receptor knockout mice. *Behav. Brain Res.* **2007**, *178*, 123–127. [CrossRef] [PubMed]
35. Bielsky, I.; Hu, S.; Ren, X.; Terwilliger, E.; Young, L. The V1a vasopressin receptor is necessary and sufficient for normal social recognition: A gene replacement study. *Neuron* **2005**, *47*, 503–513. [CrossRef] [PubMed]
36. Bleickardt, C.J.; Mullins, D.E.; Macsweeney, C.P.; Werner, B.J., Pond, A.J.; Guzzi, M.F.; Martin, F.D.; Varty, G.B.; Hodgson, R.A. Characterization of the V1a antagonist, JNJ-17308616, in rodent models of anxiety-like behavior. *Psychopharmacology* **2009**, *202*, 711–718. [CrossRef] [PubMed]
37. Koshimizu, T.A.; Nasa, Y.; Tanoue, A.; Oikawa, R.; Kawahara, Y.; Kiyono, Y.; Adachi, T.; Tanaka, T.; Kuwaki, T.; Mori, T.; et al. V1a vasopressin receptors maintain normal blood pressure by regulating circulating blood volume and baroreflex sensitivity. *Proc. Natl. Acad. Sci. USA* **2006**, *103*, 7807–7812. [CrossRef]
38. Tribollet, E.; Raufaste, D.; Maffrand, J.; Serradeil-Le Gal, C. Binding of the non-peptide vasopressin V1a receptor antagonist SR-49059 in the rat brain: An in vitro and in vivo autoradiographic study. *Neuroendocrinology* **1999**, *69*, 113–120. [CrossRef] [PubMed]
39. Schmitt, G.; Parrott, N.; Prinssen, E.; Barrow, P. The great barrier belief: The blood-brain barrier and considerations for juvenile toxicity studies. *Reprod. Toxicol.* **2017**, *72*, 129–135. [CrossRef] [PubMed]
40. Winslow, J.T.; Insel, T.R. Effects of central vasopressin administration to infant rats. *Eur. J. Pharmacol.* **1993**, *233*, 101–107. [CrossRef]
41. Altstein, M.; Gainer, H. Differential biosynthesis and posttranslational processing of vasopressin and oxytocin in rat brain during embryonic and postnatal development. *J. Neurosci.* **1988**, *8*, 3967–3977. [CrossRef]
42. Tribollet, E.; Goumaz, M.; Raggenbass, M.; Dubois-Dauphin, M.; Dreifuss, J.J. Early appearance and transient expression of vasopressin receptors in the brain of rat fetus and infant. An autoradiographical and electrophysiological study. *Brain Res. Dev. Brain Res.* **1991**, *58*, 13–24. [CrossRef]
43. Kato, Y.; Igarashi, N.; Hirasawa, A.; Tsujimoto, G.; Kobayashi, M. Distribution and developmental changes in vasopressin V2 receptor mRNA in rat brain. *Differentiation* **1995**, *59*, 163–169. [CrossRef] [PubMed]
44. Ammar, A.; Roseau, S.; Butlen, D. Postnatal ontogenesis of vasopressin receptors in the rat collecting duct. *Mol. Cell Endocrinol.* **1992**, *86*, 193–203. [CrossRef]
45. Varga, J.; Ferenczi, S.; Kovacs, K.J.; Garafova, A.; Jezova, D.; Zelena, D. Comparison of stress-induced changes in adults and pups: Is aldosterone the main adrenocortical stress hormone during the perinatal period in rats? *PLoS ONE* **2013**, *8*, e72313. [CrossRef]
46. Jurcovicova, J.; Vigas, M.; Klir, P.; Jezova, D. Response of prolactin, growth hormone and corticosterone secretion to morphine administration or stress exposure in Wistar-AVN and Long Evans rats. *Endocrinol. Exp.* **1984**, *18*, 209–214. [PubMed]

brain sciences

MDPI

Article

The Effect of Maternal Immune Activation on Social Play-Induced Ultrasonic Vocalization in Rats

Kinga Gzielo, Agnieszka Potasiewicz, Ewa Litwa, Diana Piotrowska, Piotr Popik and Agnieszka Nikiforuk *

Maj Institute of Pharmacology, Polish Academy of Sciences, Department of Behavioral Neuroscience and Drug Development, 12 Smętna Street, 31-343 Kraków, Poland; gzielo@if-pan.krakow.pl (K.G.); potasiew@if-pan.krakow.pl (A.P.); litwa@if-pan.krakow.pl (E.L.); dpiotrow@if-pan.krakow.pl (D.P.); nfpopik@cyf-kr.edu.pl (P.P.)

* Correspondence: nikifor@if-pan.krakow.pl; Tel.: +48-12-662-3374

Abstract: Prenatal maternal infection is associated with an increased risk of various neurodevelopmental disorders, including autism spectrum disorders (ASD). Maternal immune activation (MIA) can be experimentally induced by prenatal administration of polyinosinic:polycytidylic acid (poly I:C), a synthetic viral-like double-stranded RNA. Although this MIA model is adopted in many studies, social and communicative deficits, included in the first diagnostic criterion of ASD, are poorly described in the offspring of poly(I:C)-exposed dams. This study aimed to characterize the impact of prenatal poly(I:C) exposure on socio-communicative behaviors in adolescent rats. For this purpose, social play behavior was assessed in both males and females. We also analyzed quantitative and structural changes in ultrasonic vocalizations (USVs) emitted by rats during the play test. Deficits of social play behaviors were evident only in male rats. Males also emitted a significantly decreased number of USVs during social encounters. Prenatal poly(I:C) exposure also affected acoustic call parameters, as reflected by the increased peak frequencies. Additionally, repetitive behaviors were demonstrated in autistic-like animals regardless of sex. This study demonstrates that prenatal poly(I:C) exposure impairs socio-communicative functioning in adolescent rats. USVs may be a useful tool for identifying early autistic-like abnormalities.

Keywords: ultrasonic vocalization; maternal immune activation; autism; communication; rat; sex differences; social play; rats

Citation: Gzielo, K.; Potasiewicz, A.; Litwa, E.; Piotrowska, D.; Popik, P.; Nikiforuk, A. The Effect of Maternal Immune Activation on Social Play-Induced Ultrasonic Vocalization in Rats. *Brain Sci.* **2021**, *11*, 344. https://doi.org/10.3390/brainsci11030344

Academic Editors: Stefan M. Brudzynski and Jeffrey Burgdorf

Received: 26 January 2021
Accepted: 5 March 2021
Published: 9 March 2021

1. Introduction

Environmental conditions during prenatal life can influence fetal brain development. One of the most well-known factors linked to neurodevelopmental disorders is infection [1]. Plenty of studies and meta-analyses of human data have shown that maternal infection increases the risk of neuropsychiatric disorders in children [2]. The neurodevelopmental consequences of maternal immune activation (MIA) have also been confirmed in animal models [3]. One of the most frequently applied experimental MIA protocols is based on prenatal exposure of pregnant dams to polyinosinic:polycytidylic acid (poly(I:C)) [4]. Poly(I:C) is a synthetic analog of double-stranded RNA that mimics a viral infection. This MIA-based model has been widely used to explore the biochemical, neuroanatomical, and behavioral aspects of neuropsychiatric conditions such as autism spectrum disorders (ASD) [5].

ASD is a heterogeneous neurodevelopmental disorder that is considered one of the most severe public health problems because of its early onset, lifelong persistence, and significant limitations in everyday functioning [6]. According to DSM-5 (Diagnostic and Statistical Manual of Mental Disorders), the ASD symptoms are grouped into two main diagnostic criteria that are social/communicative deficits and restricted, repetitive patterns of behaviors, interests, or activities [7]. One of the most troublesome aspects of ASD is that the number of patients has strikingly increased in the last years [8]. Despite the severity of the problem, no specific and effective drug treatment is currently available for patients diagnosed with ASD, particularly in the social/communication domain [9]. Therefore, there is a need for preclinical research to better understand ASD etiopathology and to develop novel pharmacotherapeutic strategies.

Social and communicative deficits, included in the first diagnostic criterion of ASD, can also be modeled in preclinical paradigms [10]. Rodents, especially rats, are very social animals with a highly developed pattern of social behaviors, which are directly related to conspecifics [11]. Social play behavior is the first non-mother-directed social behavior in rats [12]. It emerges three weeks after birth and peaks between the 28th and 40th postnatal days. It is crucial to study social play behavior precisely during this period because some of the specific playful events decline when the animals become sexually mature. Playful interactions can be split into socio-positive behaviors such as grooming, pinning, pouncing, sniffing, and socio-negative behaviors, which are mostly associated with aggression, such as boxing, avoiding, and pushing away the conspecific [13,14].

During the rough-and-tumble play, rats also communicate using ultrasonic vocalizations (USVs) [15,16]. The USVs can be divided into two basic groups, such as 22-kHz "unhappy" calls and 50-kHz "happy" calls. The low frequency (22-kHz) calls, termed alarms, are emitted in response to unpleasant events, e.g., during losing aggressive encounters, facing predators, or experiencing anxiety. On the other hand, ultrasounds at about 50-kHz are associated with rewarding events and encourage and boost playing [15]. These "happy" calls are heterogeneous and may differ in pattern and duration [17]. They comprise flat (i.e., constant frequency) and frequency modulated ultrasounds. The frequency modulated calls are further divided into several categories, including "trills" that appear in spectrograms as rhythmic waves of ups and downs and are considered the most characteristic type of frequency-modulated calls. Thus, the rich repertoire of rats' juvenile play behaviors accompanied by a complex acoustic communication system may be employed to model early-onset socio-communicative deficits.

This study characterized poly(I:C)-induced socio-communicative deficits in early adolescence. To this aim, we analyzed social play behavior and recorded USVs emitted by juvenile rats during the social play test. We performed the detailed characteristics of the acoustic calls' features that provide a more comprehensive assessment of rats' USVs than by using purely quantitative measures. To the best of our knowledge, there is little data on ASD-related social behaviors during the adolescent period in a poly(I:C) model (detailed description in Section 4). None of the studies examined the potential outcome of poly(I:C) on rats' juvenile play. Even less is known about the impact of MIA on ultrasonic communication, and most existing studies are restricted to the assessment of neonatal USVs [18]. Based on previous studies using other rat ASD models [19,20], we hypothesized that juvenile play deficits and impaired vocalization should also occur due to poly(I:C) treatment.

To study another core symptom of ASD, i.e., repetitive behaviors, we employed a marble-burying test that is commonly used to score repetitive digging behavior [21]. Additionally, the number of repetitive/stereotypic-like movements was measured using activity meters.

Autism spectrum disorders affect males more frequently than females [22]. However, sex-specific differences in the manifestation of autistic features may lead to delayed or missed diagnoses in women [23]. It has also been recently shown that women are better at concealing their autism [24]. Nevertheless, the male prevalence in ASD is also reflected in animal studies that have been predominantly focused on males while often excluding females. Consequently, our understanding of sex-differential vulnerability to ASD may be incomplete. Therefore, we chose to compare the two sexes in this study.

2. Materials and Methods

2.1. Animals

Pregnant dams (Sprague-Dawley rats, $N = 18$) were obtained from Charles River (Sulzfeld, Germany) on gestation day (GD) 9–10. They were housed individually in polycarbonate cages: 26.5 (width) × 18 (height) × 42 (length) cm. On postnatal day (PND) 21, pups were weaned and separated by sex and litter into groups of 3–5 rats. Females and males were housed in different temperature-controlled (21 ± 1 °C) and humidity-controlled (40–50%) colony rooms under a 12/12 h light/dark cycle (lights on at 06:00 h). Food and water were available ad libitum. Behavioral testing was performed during the light phase of the light/dark cycle. The experiments were conducted in accordance with the European Guidelines for animal welfare (2010/63/EU) and were approved by the II Local Ethics Committee for Animal Experiments at the Maj Institute of Pharmacology, Polish Academy of Science, Krakow, Poland.

2.2. Poly(I:C) Administration

On GD 15, the dams were injected intraperitoneally (i.p.) with either physiological saline (vehicle) ($N = 9$) or poly(I:C) at a dose of 5 mg/kg ($N = 9$). Poly(I:C) (Sigma-Aldrich, Poznan, Poland) was dissolved in physiological saline. Both poly(I:C) and vehicle were administered at a volume of 2 mL/kg. The dose and time of administration of poly(I:C) were based on previous reports demonstrating autistic-like behaviors, including social and communicative abnormalities [25–27]. There were no effects of treatment on gestation length and litter size (average number ≈ 12), but there was a prevalence of females in poly(I:C) offspring (male/female ratio: ≈50:50 and ≈40:60 for control and poly(I:C) groups, respectively). The rats prenatally exposed to poly(I:C) appeared healthy and could not be physically or behaviorally distinguished from the controls. They also did not differ from control animals in body weight (Figure S1, Table S1).

In total, 64 males and 54 females were born from 9 vehicle-treated dams and 39 males and 60 females from 9 poly(I:C)-treated dams. In order to minimize the risk of a litter effect, offspring were randomly distributed across the experimental procedure, and none of the animals was tested more than twice. Detailed group characteristics are provided in the given method description.

2.3. Social Play Test (PND 32–35)

The test procedure was conducted in same-sex, same-treatment pairs, as previously described [20,28]. One day before the test, the rats were transported to the experimental room, weighed, and the backsides of one-half of the animals were marked with a Pentel permanent marker. Next, they were individually adapted to the test area for about 5 min. The dimly illuminated (15 Lux) test area consisted of a rectangular, polycarbonate cage (width × height × length: 38 × 20 × 59 cm) with approximately 2 cm of wood shavings covering the floor. On the testing day, each rat was isolated in a non-transparent plastic cage (width × height × length: 22 × 15 × 28 cm) for 2.5 h before the test. Then, two unfamiliar (various cages/litters) rats of matched body weights (±5 g) were placed in the test area, and their behaviors were recorded for 10 min using the Observer software (Noldus Information Technology, Wageningen, The Netherlands). After the test, the rats were returned to their home cages. The testing arena was subsequently emptied of wood shavings and thoroughly cleaned with water before being refilled with fresh bedding.

The social play behavior of each rat was separately analyzed by an experienced observer blind to the experimental conditions using the Observer software. We scored the number of episodes of pouncing (sniffing of the conspecific's neck, followed by rubbing movement) and pinning (upon contact with the nape, the recipient animal fully rotates to a supine position while the other subject stands over it) that were considered the main indices of social play behavior in rats [14]. Based on these scores, play responsiveness was calculated as the percentage of responses (being pinned) to play solicitation (pouncing).

Additionally, the total play duration represented the summed time of pouncing, pinning, and play-wrestling behaviors.

Since each animal in a pair yielded similar social behavior scores, the results are expressed as the summed score of each pair of animals. The numbers of pairs used in the analysis were: N = 17 (vehicle males), N = 16 (vehicle females), N = 11 (poly(I:C) males), N = 18 (poly(I:C) females). Two pairs (poly(I:C) males) were excluded from the analysis due to failure in the recording.

2.4. USV Recording

As previously described [20,28], the rats' vocalizations were recorded during the entire test session (i.e., 10 min) using a frequency response range of 2 kHz–200 kHz microphone (UltraSoundGate Condensor Microphone CM16/CMPA, Avisoft Bioacoustics, Berlin, Germany) suspended 25 cm above the floor of the test area. Microphone signals were fed into an UltraSoundGate 416H (Avisoft Bioacoustics, Berlin, Germany) before the analog signal was digitized with a sampling rate of 200 kHz and a 16-bit resolution. Acoustic data were recorded using Raven Pro Interactive Sound Analysis Software, version 1.5 (The Cornell Lab of Ornithology Bioacoustics Research Program, Ithaca, NY, USA). The calls were manually marked on the computer screen and counted by an experienced user, blind to the treatment, using the Raven Pro software (The Cornell Lab of Ornithology Bioacoustics Research Program, Ithaca, NY, USA). The spectrograms were generated with a fast Fourier transform (FFT)-length of 512 points and a time-window overlap of 75% (100% frame, Hamming window).

We analyzed: (a) the number of 50-kHz USVs (expressed as a total number of USVs emitted by a pair of rats) and the following 50-kHz USV features: (b) the peak frequency (the frequency in kHz at which maximal energy occurs within the selection, (c) the call duration (length of the call, measured in milliseconds), and (d) the bandwidth (the difference between the highest and lowest frequencies, a measure of frequency modulation, expressed in kHz). We also manually divided the calls (based on their acoustic call features) into the following general types: short calls, flat calls with a near-constant frequency, and frequency-modulated calls. The frequency-modulated calls were subsequently classified as trills, one-component calls (complex, ramp, and inverted-U calls), and multi-component calls (multi-step, step-up, step-down, and composite calls) [20,29]. Moreover, call subtypes were visualized using Kernel density plots depicting peak frequency versus call duration. The 22-kHz alarms were excluded from the analysis due to their negligible distribution (0–1.9%).

2.5. Locomotor and Repetitive/Stereotypic-Like Activity (PND 35–38)

Spontaneous locomotor activity was measured automatically in Opto-Varimex-4 Auto-Tracks (Columbus Instruments, Columbus, OH, USA) located in the sound attenuated and ventilated boxes. The Auto-Track System sensed the motion with a grid of infrared photocells (16 beams per x- and y-axis) surrounding the arena. The data collected every 5 min during a 30-min session are presented as the total distance traveled. Moreover, the number of repetitive/stereotypic-like movements was defined as the number of repeated breaks of the same beam. The number of animals in a given group was: N = 20 (vehicle males), N = 18 (vehicle females), N = 13 (poly(I:C) males), N = 22 (poly(I:C) females).

2.6. Marble-Burying Test (PND 35–38)

Clean cages (27 $\times$ 16.5 $\times$ 12.5 cm) were filled with a 4-cm layer of chipped wood bedding. Twenty-five green glass marbles (20 mm diameter) were gently laid on top of the bedding, equidistant from each other in a 5 $\times$ 5 arrangement. Animals were placed into the testing cage, and the number of marbles buried (>50% marble covered by bedding material) in 30 min was recorded. Additionally, the distance traveled was automatically measured using the Any-maze® tracking system (Stoelting Co., Wood Dale, IL, USA). The number of animals in a given group was: $N = 19$ (vehicle males), $N = 18$ (vehicle females), $N = 13$ (poly(I:C) males), $N = 22$ (poly(I:C) females).

2.7. Statistics

Data on social play behavior (i.e., pouncing episodes and total play duration), USV number, and acoustic call features (peak frequency, duration, and bandwidth) were analyzed by two-way ANOVAs with treatment (vehicle vs. poly(I:C)) and sex (male vs. female) as the between-subject factors. When there was a significant main effect of treatment or sex, we used the Tukey HSD post hoc test to assess overall differences between treatment or sex conditions. The Student's *t*-test was used for the planned comparisons between vehicle and poly(I:C) treatment conditions within a given sex.

In case data were not normally distributed (i.e., pinning episodes, latency to pinning, play responsiveness, and marble-burying test), differences between groups were analyzed using the Mann–Whitney U-test.

Data on the percentage distribution of call categories were arcsine-transformed and subjected to a repeated measure ANOVA with treatment and sex as between-subject factors and call type as a repeated measure. The Newman–Keuls multiple comparison test was used to analyze between-group differences in call type distribution. Locomotor activity data were analyzed by repeated measure ANOVAs with treatment and sex as between-subject factors and measurement time as a repeated measure. The sphericity was verified using Mauchly's test. If the assumption of sphericity was rejected, the corrected Greenhouse–Geisser value was utilized.

The effect size was estimated using partial eta squared (η_p^2). The normality of data distribution was evaluated by the Kolmogorov-Smirnov test. Statistical significance was set at $p < 0.05$. The statistical analyses were performed using Statistica 12.0 for Windows (StatSoft Inc, Tulsa, OK, USA).

3. Results

3.1. Social Play Behavior

While poly(I:C) animals did not differ from the vehicle-treated controls in the number of pouncing episodes (Figure 1a, Table S2), the treatment sex-dependently affected pinning behavior (Figure 1b,c). The number of pinning episodes ($U = 25$, $p = 0.0015$, Mann–Whitney U test, Figure 1b) and play responsiveness ($U = 28$, $p = 0.0022$, Mann–Whitney U test, Figure 1c) were reduced in poly(I:C) males compared to their controls. Moreover, the latency to pinning was increased in poly(I:C) males ($U = 43$, $p = 0.019$; Mann–Whitney U test, vehicle male: 172 ± 28 s, poly(I:C) male: 334 ± 526 s, vehicle female: 371 ± 46 s, poly(I:C) female: 295 ± 43 s). Poly(I:C) exposure also reduced total play durations in male rats ($t = 2.773$, $p = 0.012$, Student's *t*-test; ANOVA treatment × sex interaction: $F(1,58) = 6.95$, $p = 0.011$, Figure 1d, Table S2).

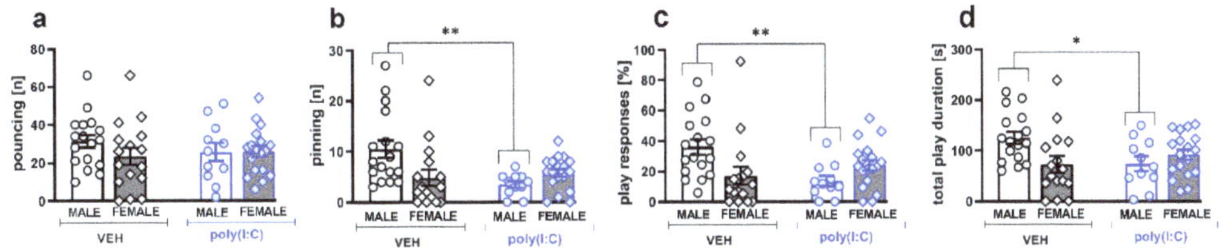

Figure 1. Poly (I:C) exposure reduced social play behavior in males. Data are presented as a mean ± SEM of the number of pouncing (**a**) and pinning episodes (**b**), the percentage of play responsiveness (**c**), and total play duration (**d**). Symbols: ** $p < 0.01$, * $p < 0.05$, a significant difference between vehicle- and poly(I:C)-exposed animals in a given sex group (Mann–Whitney U test (**b**,**c**) and Student's *t*-test (**d**)).

3.2. Ultrasonic Vocalizations

Poly(I:C) exposure also affected USV emission during social play. However, this effect was not sex-specific, as ANOVA analysis revealed significant effects of treatment on the number of USVs ($F(1,58) = 8.32$, $p = 0.005$, Figure 2a, Table S2) and peak frequency ($F(1,58) = 4.32$, $p = 0.042$, Figure 2b, Table S2), but insignificant treatment × sex interactions (Table S2). Poly(I:C)-exposed animals emitted a lower number of calls compared to the controls ($p = 0.002$, Tukey post hoc test following a significant treatment effect, Figure 2a, Table S2), and they tended to emit calls of a higher peak frequency ($p = 0.059$, Tukey post hoc test following a significant treatment effect, Figure 2b, Table S2). However, between-treatment comparisons within each sex group revealed a significant reduction only for the USV number in males ($t = 2.705$, $p = 0.012$, Student's t-test). There was no effect of poly(I:C) treatment on other acoustic parameters of calls, i.e., duration (Figure 2c, Table S2) and bandwidth (Figure 2d, Table S2).

Regardless of the treatment, females emitted less USVs ($p = 0.0003$, Tukey HSD post hoc test following a significant sex effect: $F(1,58) = 11.49$, $p = 0.0013$; Figure 2a, Table S2) and their calls were shorter ($p = 0.002$, Tukey HSD post hoc test following a significant sex effect: $F(1,58) = 10.12$, $p = 0.002$; Figure 2c, Table S2) and of a narrower frequency bandwidth ($p = 0.011$, Tukey HSD post hoc test following a significant sex effect: $F(1,58) = 6.28$, $p = 0.015$; Figure 2d, Table S2).

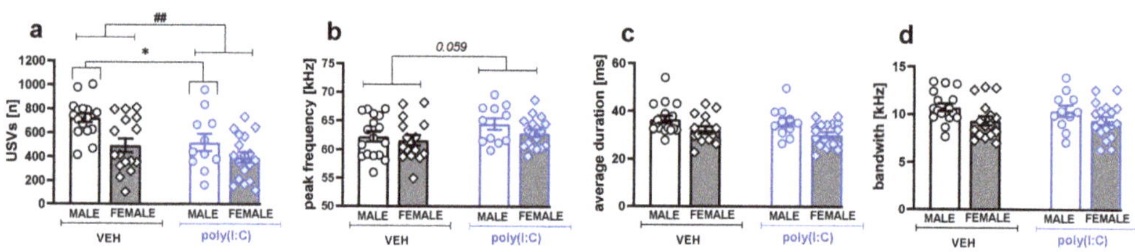

Figure 2. Poly (I:C) exposure decreased the number of USVs and changed their acoustic characteristics. Data are presented as a mean ± SEM of the total number (**a**), peak frequency (**b**), average duration (**c**), and bandwidth (**d**) of the calls. Symbols: $^{\#\#}$ $p < 0.01$, an overall difference between vehicle- and poly(I:C)-exposed animals (Tukey HSD post hoc test following a significant treatment effect), * $p < 0.05$, a significant difference between vehicle- and poly(I:C)-exposed animals in a given sex group (Student's t-test).

Call type characteristics. By analyzing the call type distribution, we observed that regardless of sex, poly(I:C) rats produced a higher proportion of trill calls ($p = 0.014$, Newman–Keuls post hoc test following a significant treatment × call type interaction: $F(4,232) = 2.963$, $p = 0.021$; Figure 3, Table S2). Analysis of individual call types revealed that the trills emitted by poly(I:C)-treated rats did not differ from those emitted by their controls in any of the acoustic parameters measured (Figure S2, Table S3). Between-treatment comparisons within each sex group did not reveal any significant differences in call distributions in either males or females.

Moreover, females emitted more of the short ($p = 0.013$) and one-component calls ($p = 0.023$), but fewer trills ($p = 0.006$) and multi-component calls ($p = 0.036$, Newman–Keuls post hoc test following a significant sex × call type interaction: $F(4,232) = 5.04$, $p = 0.0007$; Figure 3, Table S2).

Density plots. To identify clusters of USVs emitted by vehicle and poly(I:C) animals, we used density plots illustrating the distribution of individual calls depending on their peak frequencies versus call durations. Visual inspections showed that poly(I:C) did not affect the overall distribution of USVs (Figure 4). However, an enhanced representation of high-frequency calls in poly(I:C) animals corroborated the peak frequency increases as shown in Figure 2.

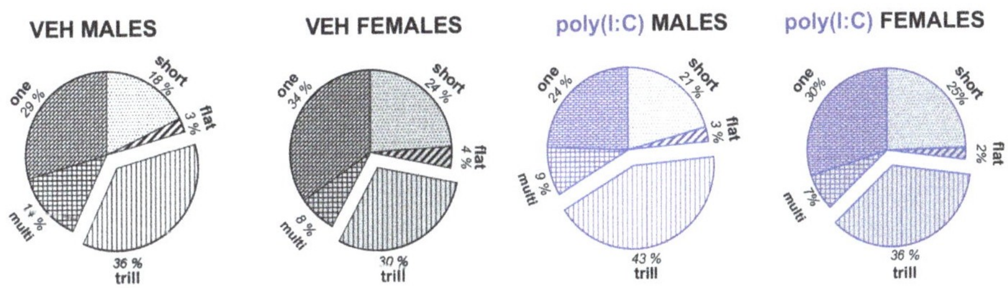

Figure 3. Poly(I:C) changed the percentage distribution of call categories within the tested groups. There was a significant overall difference between vehicle- and poly(I:C)-exposed animals in the distribution of trill calls ($p < 0.05$, Newman–Keuls post hoc test following a significant treatment × call category interaction).

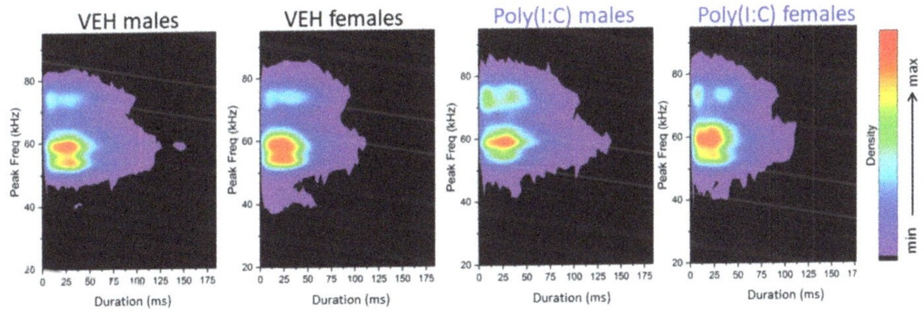

Figure 4. Density plots illustrating the distribution of individual calls depending on their peak frequencies versus durations in vehicle- and poly(I:C) exposed rats. The densities were visualized in a color-coded way. Poly(I:C) did not change the overall USV cluster distributions, but an enhanced representation of high-frequency calls was observed for poly(I:C) animals.

3.3. Locomotor Activity and Stereotypic-Like Movements

Irrespective of sex, poly(I:C) exposure enhanced rats' locomotor activity ($p = 0.0012$, Tukey HSD post hoc test following a significant treatment effect: $F(1,69) = 5.91$, $p = 0.018$; Figure 5a, Table S2) and increased the number of stereotypic movements ($p = 0.0006$, Tukey HSD post hoc test following a significant treatment effect: $F(1,69) = 12.47$, $p = 0.0007$; Figure 5b, Table S2). Moreover, there was an overall effect of sex on locomotor activity, as the distance traveled was increased in females ($p = 0.0001$, Tukey HSD post hoc test following a significant sex effect: $F(1,69) = 5.91$, $p = 0.012$; Figure 5a, Table S2).

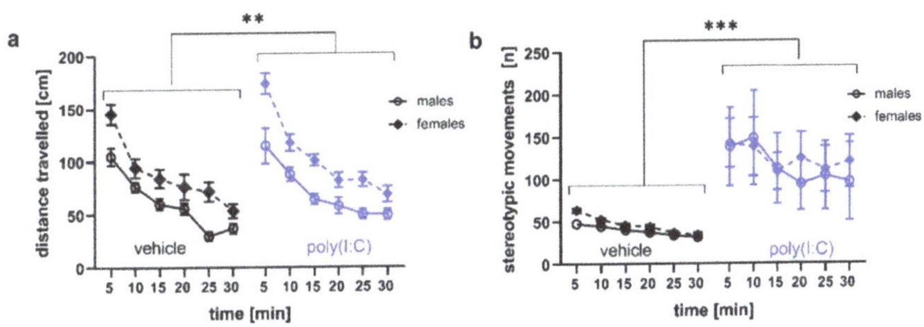

Figure 5. Poly(I:C) exposure elevated locomotor activity and stereotypic-like movements. Data are presented as a mean ± SEM of the distance traveled (**a**) and stereotypic-like movements (**b**) assessed during a 30-min test session. Symbols: ** $p < 0.01$, *** $p < 0.001$; a significant overall difference between vehicle- and poly(I:C)-exposed animals (Tukey HSD post hoc test following a significant treatment effect).

3.4. Marble-Burying

Poly(I:C) exposure significantly increased the number of buried marbles in females (U = 110, $p = 0.0054$, Mann–Whitney U test, Figure 6a). There was no effect of treatment on the distance traveled during the test session (Figure 6b, Table S2), suggesting that changes in locomotor activity were not a confounding factor when scoring buried marbles.

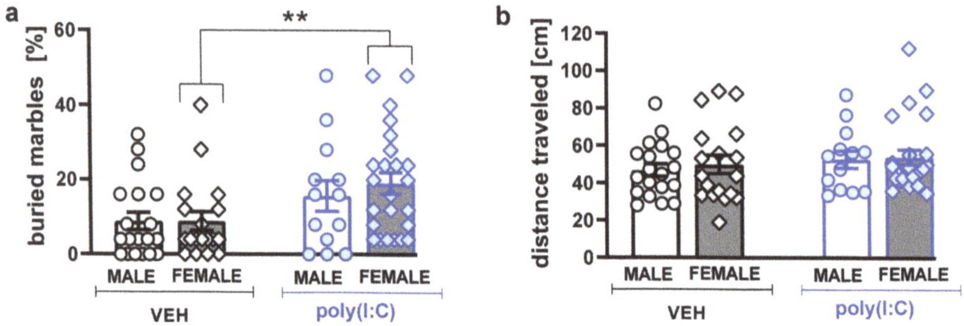

Figure 6. Poly(I:C) exposure increased repetitive-like marble-burying behavior. Data are presented as the percentage of buried marbles (**a**) and distance traveled (**b**). Symbol: ** $p < 0.01$, a significant difference between vehicle- and poly(I:C)-exposed animals in a given sex group (Mann–Whitney U test).

4. Discussion

This study demonstrates that prenatal poly(I:C) exposure impairs socio-communicative functioning in adolescent rats. Deficits of social play behaviors were evident only in male rats. While poly(I:C) treatment reduced USV emission during the social encounter irrespectively of sex, significant reductions were demonstrated only in males. Acoustic call parameters were also slightly affected, as reflected by the increased peak frequencies of the emitted calls. Additionally, repetitive behaviors were demonstrated in autistic-like animals, regardless of sex.

Prenatal exposition to poly(I:C) altered the pattern of social play behavior in males. While poly(I:C) males did not differ from controls in play-soliciting behaviors (i.e., pouncing), their responsiveness to play solicitation was decreased, as reflected by the reduced frequency of pinning and increased latency to pinning. To the best of our knowledge, no studies have examined the impact of prenatal poly(I:C) administration on juvenile rats' social play. Most of the published reports in this area have utilized the three-chamber test, which provides an easily quantifiable sociability measure reflected as a tendency to spend more time in the compartment with an unfamiliar conspecific than in the empty compartment. In line with the social deficits observed in the current experiments, poly(I:C) exposure affected sociability in adolescent [30,31] or adult [32,33] mice and in rats during early adulthood [34] (but see also [35,36] for the opposite results). The above-cited research reports were mainly carried out on males. Females were included in only a few studies demonstrating reductions of females' sociability assessed in a three-chamber test [25,31] or no effect [36,37]. However, deficits in juvenile rats' play behavior were noted in another MIA model based on administering the bacterial endotoxin lipopolysaccharide (LPS) [38]. Interestingly, prenatal LPS exposure reduced the frequency of juvenile play behavior exclusively in males [39]. Similar sex-specific impairments were observed in the current poly(I:C) studies that may corroborate the higher prevalence of ASD in males. It has been recently suggested that these protective effects may be attributed to sex differences in immune responses [40]. Accordingly, pro-inflammatory factors were increased in both sexes, but anti-inflammatory factors were decreased in males and increased in females.

However, considering the well-described sexual dimorphism in rough-and-tumble play, it is not surprising that females engage in fewer play behaviors than males. Consequently, a floor effect due to a relatively low level of playful interactions in females may not allow for the demonstration of further decrements. In line with this assumption, social deficits in other behavioral measures were noted in poly(I:C) females [25,31]. Interestingly, our previous study, demonstrating vocalization deficits in valproic acid (VPA)-exposed females [20], suggested that USV assessment in females may be a more sensitive measure of autistic-like disturbances than other indices of juvenile social behavior. Likewise, poly(I:C) treatment reduced USV emission during the social encounter irrespectively of sex (a significant treatment effect in the absence of sex × treatment interaction). However, significant reductions were demonstrated only in males (between-group comparisons), suggesting that communicative deficits in poly(I:C) females may be less pronounced than those previously demonstrated in VPA females.

The demonstrated vocalization deficit in poly(I:C)-exposed rats is per se interesting as there is a limited number of data on USVs in MIA models, and most of them focus on isolation-induced USVs in pups. Several studies have reported decreased USV emission in rat and mouse pups from poly(I:C)-exposed mothers, but others also noted enhanced isolation-induced vocalization (reviewed in [18]). The observed vocalization deficits did not result from altered maternal care [41]. While these discrepancies may arise from diverse poly(I:C) administration schedules, one may conclude that early communication abnormalities occur in this model. Less is known about MIA effects on vocalization during the latter stages of development. For example, adult male rats prenatally exposed to poly(I:C) emitted an increased number of aversive 22-kHz USVs during fear conditioning [26] that further supports the link between immune activation and socio-emotional communication. Moreover, adult poly(I:C) male mice emitted fewer calls during social interactions with either males or females [32,33]. However, other reports did not observe poly(I:C) effects on male mice's USV emission during direct interactions with females [35] or when exposed to female urine [40]. While we are unaware of any studies concerned with juvenile USVs in the poly(I:C) model, social play-induced USVs were decreased in LPS exposed male offspring [38]. The effects of LPS on juvenile social play behavior were also assessed by the recording of USVs emitted in response to manual tickling by an experimenter [42]. However, the latter study demonstrated only greater USV variability in LPS-exposed adolescent male and female rats [42]. The literature mentioned above, together with the present findings, indicate that vocalization abnormalities occur in MIA models. Nevertheless, the paucity of female data hinders a definite conclusion on sex differences in this measure.

There are limited data available on the qualitative analysis of USVs in MIA models. For example, calls emitted by poly(I:C) exposed mice in response to a social encounter were characterized by a shortened average duration [32]. However, this parameter was not affected when analyzing social play-evoked USVs in LPS-treated rats [38]. While we also did not observe significant differences in call durations between poly(I:C)- and vehicle-exposed animals, poly(I:C) exposure increased USV peak frequencies. Interestingly, poly(I:C) exposed pups also emitted calls of an elevated frequency when separated from their mothers [27]. Similarly, our previous study employing a valproic acid model of ASD demonstrated higher frequency calls in neonatal and adolescent rats [20]. Hence, it is tempting to interpret these atypical call features as a signature of autistic-like ultrasonic communication.

Moreover, call type categorizations revealed that poly(I:C) mice produced significantly fewer two-step frequency and chevron calls than controls [33]. The changes in call distribution were also observed in our study, as poly(I:C) rats emitted an increased proportion of trills. Interestingly, we found a similar tendency in VPA-exposed rats [20]. Considering that trills prevail before playful actions [43,44], their increased percentage may reflect the enhanced anticipatory aspects of social play. Indeed, poly(I:C) rats seem to be equally motivated as controls to initiate play, as revealed by unchanged play soliciting behaviors, but their reduced play responsiveness may suggest that their trills do not efficiently stimulate play behavior. In line with this assumption, we analyzed the acoustic feature of trills; however, in contrast to previous findings in a VPA model, no differences were seen between poly(I:C) and control rats. Alternatively, altered distribution of USV subtypes may result from an aberrant organization of social behaviors. Indeed, the temporal analysis of USVs emitted during social play revealed a close connection between specific call categories and particular play events [45]. Unfortunately, the lack of the temporal correlation analysis between the USV call sequence and specific behavior limits our ability to draw definitive conclusions regarding the current study. Finally, the trill emission may also be related to general arousal [46], which would agree with the increased locomotor activity in poly(I:C) rats.

The neurobehavioral origin of the USV deficit may be related to the altered processing of sensory information. Impaired processing and integration of auditory information is common in autistic children and may lead to delayed language development [47]. Interestingly, deficient sensory processing has also been reported in MIA models [25]. Besides, a growing body of evidence has shown that alterations in the periaqueductal gray, a brainstem nucleus engaged in the production of vocalizations and indirect modulation (via motor neurons in the brainstem) of vocal fold vibration, may be the leading cause of impaired communication in mammals [48,49]. Further studies are necessary to answer whether sensory or motor dysfunctions lead to USV alterations and consequently which brain regions (ascending auditory pathway or the periaqueductal gray) are involved in the MIA-induced vocal communication impairment.

Besides social and communication deficits, ASD symptoms include restricted, repetitive, and stereotyped patterns of behavior. Common motor stereotypies displayed by rodents are repetitive digging behaviors that are assessed in the marble-burying test. In line with previously published data [32,33], poly(I:C) exposure increased the number of buried marbles. However, it should be noticed that the higher number of marbles buried by the poly(I:C) offspring may not necessarily result from goal-directed digging behavior but, for example, from enhanced circling behavior (Figure S3). Possible alterations in anxiety levels in poly(I:C)-exposed animals (e.g., [32]) may also affect the outcome of this test. Nevertheless, the locomotor pattern of poly(I:C) rats was also characterized by stereotypic-like movements.

In contrast to social deficits, repetitive behaviors were exhibited to a greater extent in females. Similarly, VPA prenatal exposure induced autistic-like behaviors in a sex-specific manner [50]. While females were less susceptible to VPA-evoked socio-emotional deficits than males, they also demonstrated stereotypic behavior [50]. These results further support the necessity of the inclusion of females in ASD studies.

5. Conclusions

The current study demonstrated that the poly(I:C)-based model recapitulates ASD-related behavioral abnormalities. Similar to human ASD patients, affected offspring exhibit social/communicative impairments and repetitive behaviors. The finding that vocalization deficits occur in the poly(I:C)-based MIA model supports the utility of USVs as a measure of autistic-like socio-communicative abnormalities. The question of whether these changes persist beyond the adolescent period requires further study.

Supplementary Materials: The following are available online at https://www.mdpi.com/2076-3425/11/3/344/s1, Figure S1. Body weight, Figure S2. The acoustic characteristics of trill calls, Figure S3. Poly (I:C) treatment increased circling behaviors in females, Table S1. Body weight—ANOVA results, Table S2. The results of the ANOVA analysis, Table S3. The acoustic characteristics of trill calls—ANOVA results.

Author Contributions: Conceptualization, A.N., K.G.; methodology, A.N., K.G., A.P.; formal analysis, A.N., K.G.; investigation, K.G., E.L., D.P.; writing—original draft preparation, A.N., K.G.; writing—review and editing, A.N., K.G., P.P.; visualization data presentation, A.N., K.G.; supervision, A.N.; funding acquisition, A.N. All authors have read and agreed to the published version of the manuscript.

Funding: This research was funded by the National Science Centre (Kraków, Poland); grant no. 2016/23/B/NZ7/01131 (to A.N.) and supported by statutory funds of Maj Institute of Pharmacology, Polish Academy of Sciences (Kraków, Poland).

Institutional Review Board Statement: Animal experiments and were approved by the II Local Ethics Commission at Maj Institute of Pharmacology, Polish Academy of Sciences (permission number: 203/2017, approval date: 06.07.2017).

Data Availability Statement: Data are contained within the article and Supplementary Materials.

Conflicts of Interest: The authors declare no conflict of interest.

References

1. Scola, G.; Duong, A. Prenatal maternal immune activation and brain development with relevance to psychiatric disorders. *Neuroscience* **2017**, *346*, 403–408. [CrossRef] [PubMed]
2. Estes, M.L.; McAllister, A.K. Immune mediators in the brain and peripheral tissues in autism spectrum disorder. *Nat. Rev. Neurosci.* **2015**, *16*, 469–486. [CrossRef]
3. Ji-Xu, A.; Vincent, A. Maternal Immunity in Autism Spectrum Disorders: Questions of Causality, Validity, and Specificity. *J. Clin. Med.* **2020**, *9*, 2590. [CrossRef] [PubMed]
4. Meyer, U. Prenatal Poly(I:C) Exposure and Other Developmental Immune Activation Models in Rodent Systems. *Biol. Psychiatry* **2014**, *75*, 307–315. [CrossRef]
5. Haddad, F.L.; Patel, S.V.; Schmid, S. Maternal Immune Activation by Poly I:C as a preclinical Model for Neurodevelopmental Disorders: A focus on Autism and Schizophrenia. *Neurosci. Biobehav. Rev.* **2020**, *113*, 546–567. [CrossRef] [PubMed]
6. Lord, C.; Brugha, T.S.; Charman, T.; Cusack, J.; Dumas, G.; Frazier, T.; Jones, E.J.H.; Jones, R.M.; Pickles, A.; State, M.W.; et al. Autism spectrum disorder. *Nat. Rev. Dis. Primers* **2020**, *6*, 5. [CrossRef]
7. Association, A.P. *Diagnostic and Statistical Manual of Mental Disorders (DSM-5®)*; American Psychiatric Publishing: Arlington, VA, USA, 2013.
8. Christensen, D.L.; Maenner, M.J.; Bilder, D.; Constantino, J.N.; Daniels, J.; Durkin, M.S.; Fitzgerald, R.T.; Kurzius-Spencer, M.; Pettygrove, S.D.; Robinson, C.; et al. Prevalence and Characteristics of Autism Spectrum Disorder Among Children Aged 4 Years—Early Autism and Developmental Disabilities Monitoring Network, Seven Sites, United States, 2010, 2012, and 2014. *MMWR Surveill. Summ.* **2019**, *68*, 1–19. [CrossRef]
9. LeClerc, S.; Easley, D. Pharmacological therapies for autism spectrum disorder: A review. *Pharm. Ther.* **2015**, *40*, 389–397. [PubMed]
10. Wohr, M.; Scattoni, M.L. Behavioural methods used in rodent models of autism spectrum disorders: Current standards and new developments. *Behav. Brain Res.* **2013**, *251*, 5–17. [CrossRef]
11. Schweinfurth, M.K. The social life of Norway rats (*Rattus norvegicus*). *eLife* **2020**, *9*, e54020. [CrossRef]
12. Panksepp, J. The ontogeny of play in rats. *Dev. Psychobiol.* **1981**, *14*, 327–332. [CrossRef]
13. Vanderschuren, L.J.M.J.; Trezza, V. What the Laboratory Rat has Taught us about Social Play Behavior: Role in Behavioral Development and Neural Mechanisms. In *The Neurobiology of Childhood*; Andersen, S.L., Pine, D.S., Eds.; Springer: Berlin/Heidelberg, Germany, 2014; pp. 189–212. [CrossRef]
14. Vanderschuren, L.J.M.J.; Achterberg, E.J.M.; Trezza, V. The neurobiology of social play and its rewarding value in rats. *Neurosci. Biobehav. Rev.* **2016**, *70*, 86–105. [CrossRef]
15. Pellis, S.M.; Burke, C.J.; Kisko, T.M.; Euston, D.R. Chapter 11—50-kHz Vocalizations, Play and the Development of Social Competence. In *Handbook of Behavioral Neuroscience*; Brudzynski, S.M., Ed.; Elsevier: Amsterdam, The Netherlands, 2018; Volume 25, pp. 117–126.
16. Simola, N.; Brudzynski, S.M. Chapter 17—Repertoire and Biological Function of Ultrasonic Vocalizations in Adolescent and Adult Rats. In *Handbook of Behavioral Neuroscience*; Brudzynski, S.M., Ed.; Elsevier: Amsterdam, The Netherlands, 2018; Volume 25, pp. 177–186.
17. Wright, J.M.; Gourdon, J.C.; Clarke, P.B.S. Identification of multiple call categories within the rich repertoire of adult rat 50-kHz ultrasonic vocalizations: Effects of amphetamine and social context. *Psychopharmacology* **2010**, *211*, 1–13. [CrossRef]
18. Jouda, J.; Wohr, M.; Del Rey, A. Immunity and ultrasonic vocalization in rodents. *Ann. N. Y. Acad. Sci.* **2019**, *1437*, 68–82. [CrossRef]
19. Raza, S.; Himmler, B.T.; Himmler, S.M.; Harker, A.; Kolb, B.; Pellis, S.M.; Gibb, R. Effects of prenatal exposure to valproic acid on the development of juvenile-typical social play in rats. *Behav. Pharmacol.* **2015**, *26*, 707–719. [CrossRef]
20. Gzielo, K.; Potasiewicz, A.; Hołuj, M.; Litwa, E.; Popik, P.; Nikiforuk, A. Valproic acid exposure impairs ultrasonic communication in infant, adolescent and adult rats. *Eur. Neuropsychopharmacol.* **2020**, *41*, 52–62. [CrossRef]
21. Thomas, A.; Burant, A.; Bui, N.; Graham, D.; Yuva-Paylor, L.A.; Paylor, R. Marble burying reflects a repetitive and perseverative behavior more than novelty-induced anxiety. *Psychopharmacology* **2009**, *204*, 361–373. [CrossRef]
22. Loomes, R.; Hull, L.; Mandy, W.P.L. What Is the Male-to-Female Ratio in Autism Spectrum Disorder? A Systematic Review and Meta-Analysis. *J. Am. Acad Child. Adolesc. Psychiatry* **2017**, *56*, 466–474. [CrossRef]

23. Lai, M.-C.; Baron-Cohen, S. Identifying the lost generation of adults with autism spectrum conditions. *Lancet Psychiatry* **2015**, *2*, 1013–1027. [CrossRef]
24. Hull, L.; Lai, M.-C.; Baron-Cohen, S.; Allison, C.; Smith, P.; Petrides, K.V.; Mandy, W. Gender differences in self-reported camouflaging in autistic and non-autistic adults. *Autism* **2019**, *24*, 352–363. [CrossRef]
25. Lins, B.R.; Marks, W.N.; Zabder, N.K.; Greba, Q.; Howland, J.G. Maternal Immune Activation during Pregnancy Alters the Behavior Profile of Female Offspring of Sprague Dawley Rats. *eNeuro* **2019**, *6*. [CrossRef] [PubMed]
26. Yee, N.; Schwarting, R.K.; Fuchs, E.; Wohr, M. Increased affective ultrasonic communication during fear learning in adult male rats exposed to maternal immune activation. *J. Psychiatr. Res.* **2012**, *46*, 1199–1205. [CrossRef]
27. Potasiewicz, A.; Gzielo, K.; Popik, P.; Nikiforuk, A. Effects of prenatal exposure to valproic acid or poly(I:C) on ultrasonic vocalizations in rat pups: The role of social cues. *Physiol. Behav.* **2020**, *225*, 113113. [CrossRef]
28. Potasiewicz, A.; Holuj, M.; Piotrowska, D.; Zajda, K.; Wojcik, M.; Popik, P.; Nikiforuk, A. Evaluation of ultrasonic vocalizations in a neurodevelopmental model of schizophrenia during the early life stages of rats. *Neuropharmacology* **2019**, *146*, 28–38. [CrossRef]
29. Golebiowska, J.; Hołuj, M.; Potasiewicz, A.; Piotrowska, D.; Kuziak, A.; Popik, P.; Homberg, J.R.; Nikiforuk, A. Serotonin transporter deficiency alters socioemotional ultrasonic communication in rats. *Sci. Rep.* **2019**, *9*, 20283. [CrossRef]
30. Vuillermot, S.; Luan, W.; Meyer, U.; Eyles, D. Vitamin D treatment during pregnancy prevents autism-related phenotypes in a mouse model of maternal immune activation. *Mol. Autism* **2017**, *8*, 9. [CrossRef]
31. Aavani, T.; Rana, S.A.; Hawkes, R.; Pittman, Q.J. Maternal immune activation produces cerebellar hyperplasia and alterations in motor and social behaviors in male and female mice. *Cerebellum* **2015**, *14*, 491–505. [CrossRef]
32. Hsiao, E.Y.; McBride, S.W.; Hsien, S.; Sharon, G.; Hyde, E.R.; McCue, T.; Codelli, J.A.; Chow, J.; Reisman, S.E.; Petrosino, J.F.; et al. Microbiota Modulate Behavioral and Physiological Abnormalities Associated with Neurodevelopmental Disorders. *Cell* **2013**, *155*, 1451–1463. [CrossRef]
33. Malkova, N.V.; Yu, C.Z.; Hsiao, E.Y.; Moore, M.J.; Patterson, P.H. Maternal immune activation yields offspring displaying mouse versions of the three core symptoms of autism. *Brain Behav. Immun.* **2012**, *26*, 607–616. [CrossRef]
34. Lins, B.R.; Hurtubise, J.L.; Roebuck, A.J.; Marks, W.N.; Zabder, N.K.; Scott, G.A.; Greba, Q.; Dawicki, W.; Zhang, X.; Rudulier, C.D.; et al. Prospective Analysis of the Effects of Maternal Immune Activation on Rat Cytokines during Pregnancy and Behavior of the Male Offspring Relevant to Schizophrenia. *eNeuro* **2018**. [CrossRef]
35. Vigli, D.; Palombelli, G.; Fanelli, S.; Calamandrei, G.; Canese, R.; Mosca, L.; Scattoni, M.L.; Ricceri, L. Maternal Immune Activation in Mice Only Partially Recapitulates the Autism Spectrum Disorders Symptomatology. *Neuroscience* **2020**. [CrossRef]
36. Gray, A.; Tattoli, R.; Dunn, A.; Hodgson, D.M.; Michie, P.T.; Harms, L. Maternal immune activation in mid-late gestation alters amphetamine sensitivity and object recognition, but not other schizophrenia-related behaviours in adult rats. *Behav. Brain Res.* **2019**, *356*, 358–364. [CrossRef]
37. Ruskin, D.N.; Murphy, M.I.; Slade, S.L.; Masino, S.A. Ketogenic diet improves behaviors in a maternal immune activation model of autism spectrum disorder. *PLoS ONE* **2017**, *12*, e0171643. [CrossRef]
38. Kirsten, T.B.; Taricano, M.; Maiorka, P.C.; Palermo-Neto, J.; Bernardi, M.M. Prenatal lipopolysaccharide reduces social behavior in male offspring. *Neuroimmunomodulation* **2010**, *17*, 240–251. [CrossRef] [PubMed]
39. Taylor, P.V.; Veenema, A.H.; Paul, M.J.; Bredewold, R.; Isaacs, S.; de Vries, G.J. Sexually dimorphic effects of a prenatal immune challenge on social play and vasopressin expression in juvenile rats. *Biol. Sex Differ.* **2012**, *3*, 15. [CrossRef]
40. Carlezon, W.A.; Kim, W.; Missig, G.; Finger, B.C.; Landino, S.M.; Alexander, A.J.; Mokler, E.L.; Robbins, J.O.; Li, Y.; Bolshakov, V.Y.; et al. Maternal and early postnatal immune activation produce sex-specific effects on autism-like behaviors and neuroimmune function in mice. *Sci. Rep.* **2019**, *9*, 16928. [CrossRef] [PubMed]
41. Morais, L.H.; Felice, D.; Golubeva, A.V.; Moloney, G.; Dinan, T.G.; Cryan, J.F. Strain differences in the susceptibility to the gut-brain axis and neurobehavioural alterations induced by maternal immune activation in mice. *Behav. Pharmacol.* **2018**, *29*, 181–198. [CrossRef]
42. Cieslik, M.; Gassowska-Dobrowolska, M.; Jesko, H.; Czapski, G.A.; Wilkaniec, A.; Zawadzka, A.; Dominiak, A.; Polowy, R.; Filipkowski, R.K.; Boguszewski, P.M.; et al. Maternal Immune Activation Induces Neuroinflammation and Cortical Synaptic Deficits in the Adolescent Rat Offspring. *Int. J. Mol. Sci.* **2020**, *21*, 4097. [CrossRef]
43. Knutson, B.; Burgdorf, J.; Panksepp, J. Anticipation of play elicits high-frequency ultrasonic vocalizations in young rats. *J. Comp. Psychol.* **1998**, *112*, 65–73. [CrossRef]
44. Himmler, B.T.; Kisko, T.M.; Euston, D.R.; Kolb, B.; Pellis, S.M. Are 50-kHz calls used as play signals in the playful interactions of rats? I. Evidence from the timing and context of their use. *Behav. Process.* **2014**, *106*, 60–66. [CrossRef]
45. Burke, C.J.; Kisko, T.M.; Swiftwolfe, H.; Pellis, S.M.; Euston, D.R. Specific 50-kHz vocalizations are tightly linked to particular types of behavior in juvenile rats anticipating play. *PLoS ONE* **2017**, *12*, e0175841. [CrossRef]
46. Mulvihill, K.G.; Brudzynski, S.M. Non-pharmacological induction of rat 50 kHz ultrasonic vocalization: Social and non-social contexts differentially induce 50 kHz call subtypes. *Physiol. Behav.* **2018**, *196*, 200–207. [CrossRef] [PubMed]
47. Baum, S.H.; Stevenson, R.A.; Wallace, M.T. Behavioral, perceptual, and neural alterations in sensory and multisensory function in autism spectrum disorder. *Prog. Neurobiol.* **2015**, *134*, 140–160. [CrossRef] [PubMed]
48. Tschida, K.; Michael, V.; Takatoh, J.; Han, B.-X.; Zhao, S.; Sakurai, K.; Mooney, R.; Wang, F. A Specialized Neural Circuit Gates Social Vocalizations in the Mouse. *Neuron* **2019**, *103*, 459–472.e4. [CrossRef] [PubMed]

49. Kelm-Nelson, C.A.; Gammie, S. Gene expression within the periaqueductal gray is linked to vocal behavior and early-onset parkinsonism in Pink1 knockout rats. *BMC Genom.* **2020**, *21*, 625. [CrossRef] [PubMed]
50. Melancia, F.; Schiavi, S.; Servadio, M.; Cartocci, V.; Campolongo, P.; Palmery, M.; Pallottini, V.; Trezza, V. Sex-specific autistic endophenotypes induced by prenatal exposure to valproic acid involve anandamide signalling. *Br. J. Pharmacol.* **2018**, *175*, 3699–3712. [CrossRef] [PubMed]

brain sciences

MDPI

Article

Effects of Ethanol Exposure during Lactation on Ultrasonic Vocalizations of Rat Pups upon Their Isolation: Increase in Pup Distress Calls †

Mohd. Ashik Shahrier [1,*] and Hiromi Wada [2]

1 Department of Psychology, Faculty of Biological Sciences, Rajshahi University, Rajshahi 6205, Bangladesh
2 Department of Psychology, Faculty of Humanities and Human Sciences, Hokkaido University, Kita 10, Nishi 7, Kita-Ku, Sapporo 060-0810, Japan; wada@let.hokudai.ac.jp
* Correspondence: ashikpsy@ru.ac.bd

† This report is based on the dissertation submitted to Hokkaido University in partial fulfillment of the requirements for the Ph.D. degree of Mohd. Ashik Shahrier.

Abstract: Recording ultrasonic vocalizations (USVs) is a highly sensitive tool to study the dam–pup social relationships, and USV recordings have been used to study the effects of ethanol on pups. Gestational effects of ethanol on the emission of USVs in rat pups have been studied in our previous research. In the present study, the effects of ethanol given to dams during lactation on the acoustic parameters of USVs emitted by isolated pups were examined. Ethanol was administered to dams from postnatal days (PNDs) 5–21. From PNDs 11–21, the high- and low-ethanol-treated dams were exposed to ethanol-containing water (v/v) at concentrations of 30% and 15%, respectively. Tap water without ethanol (0%) was provided to the control dams. The pups in all three ethanol-treated groups were separated from the dam and littermates on PNDs 4, 8, 12, and 16, and USVs produced by the pups were recorded for 5 min. It was found that elevated distress USVs with longer duration and higher percentage of frequency modulations were displayed by the pups from the high-ethanol dams. Alterations in USVs were particularly evident in the pups with a reduced body weight at PND 12. This effect might be because high-ethanol dams showed significantly lower intake of higher ethanol-containing water, and consequently, produced lower amount of milk, as well as exhibited poor maternal care. Insufficient maternal care and malnutrition resulted in pup growth retardation and increased mortality rate in the high-ethanol group, which were not observed in the low-ethanol or control pups. Accordingly, the pups in the high-ethanol group experienced elevated negative emotionality during isolation from their dam and increased emission of USVs. Longer duration and increased frequency modulation of pup USVs are expected to be noticed by the dam and to initiate/increase proper maternal care. It is concluded that ethanol given to lactating mothers has more serious consequences on pup development than the gestational ethanol exposure, and has more harmful effects on pups.

Keywords: dam–pup interaction; ethanol; pup isolation; rat pup; ultrasonic vocalization

Citation: Shahrier, M.A.; Wada, H. Effects of Ethanol Exposure during Lactation on Ultrasonic Vocalizations of Rat Pups upon Their Isolation: Increase in Pup Distress Calls. *Brain Sci.* **2021**, *11*, 1249. https://doi.org/10.3390/brainsci11091249

Academic Editors: Stefan M. Brudzynski and Jeffrey Burgdorf

Received: 3 August 2021
Accepted: 19 September 2021
Published: 21 September 2021

1. Introduction

A popular earlier notion suggested that due to the lactogenic properties alcoholic beverages facilitate milk let-down and rectify milk insufficiency in lactating women [1,2]. Recently, this notion has changed [3,4], though confusion remains as to its positive or negative effects [5,6]. Alcohol consumed by lactating mothers diffuses into the breast milk, and the sensory qualities of milk in intoxicated mothers are changed during breastfeeding [2]. Altered sensory properties of breast milk result in altered sucking patterns reflected through longer nipple attachment in infants. Consequently, 20% less consumption of breast milk in infants during the 3 to 4 h following their mothers' consumption of alcoholic beverages was observed [7,8]. Moreover, maternal ethanol exposure in the nursing context was found to disrupt the sleep–wake patterns [9,10] and psychomotor development of infants [11]. Alcohol exposure through breast milk also caused dose-dependent reductions in cognitive abilities of children at ages 6 to 7 years [12].

However, the consequences of lactational alcohol exposure in human infants cannot be tested experimentally due to obvious ethical reasons. In human studies, alcohol consumption during lactation, its availability in breast milk, and infant growth are highly comorbid with the ingestion of substances such as tobacco, marijuana, or other psychoactive drugs and with factors such as nutritional status, drinking during pregnancy, prenatal care, socio-economic status of the family, and genetic variations, which cannot be controlled outside of clinical studies. Animal model studies, therefore, play an important role in examining the effects of lactational alcohol exposure on mother–infant interactions, allowing control of confounding factors such as genetic variations, nutritional status, and multidrug use, as well as timing, dose, duration, and pattern of exposure.

In animal model studies, the early postnatal period (PNDs 4–12) in rats was characterized by a brain growth spurt [13], and high doses of ethanol exposure during this period caused the cell death of specific neuronal populations and resulted in neurobehavioral disorders [14,15]. During this critical period, rat pups whose dams were exposed to ethanol for at least 4 days required more time to attach to the nipple [16] and consumed decreased amount of milk [17]. Notably, ethanol administration from lactation days 5–12 caused growth deficits [18] and a high dose exposure (3 g/kg/day) during the early postnatal period resulted in altered suckling behavior, motor impairments and decreased perioral responsiveness in rat pups [19]. Pups can encode the ethanol-related memories of negative emotional valence while interacting with ethanol intoxicated dam. This was evidenced by experiments showing that pups of ethanol-intoxicated dam associated ethanol odor with novel textures and were found to avoid these textures consistently in locational preference tactile tests [20]. In addition, ethanol exposure during lactation promoted hyperactivity and impaired the gait patterns in pups [21,22].

In dam–pup social interactions, ultrasonic vocalizations (USVs) produced by rat pups play an adaptive role in activating maternal responses towards the endangered pups. In rats, maternal attachment is critical in regulating pups' physiological homeostasis. Disruptions of the dam–pup relationships resulted in alterations of the pup's heart rate, circadian rhythm, temperature, and so on [23,24]. Consequently, neonatal rat pups emit USVs in the range of 30 to 65 kHz with an approximate average frequency of 40 kHz [25–27] that are directed to their dam and were termed distress calls. Distress USVs produced by pups promote search and retrieval behaviors in the dam and elicit pup-directed behaviors such as pup grooming, suppressing, or decreasing maternal biting and cannibalism. Thus, distress USVs have biological significance in triggering appropriate maternal responses, and thus, increase the chances of pup survival [28,29].

Exposure to early life stressors has been found to change the isolation-induced USVs in pups. Environmental temperature during the first neonatal week [27], and a variety of teratogens [30,31] and other social cues [32] during the second neonatal week play a key role in the emission of isolation-induced USVs in pups. Familiar odors, tactile stimulation, the presence of the dam, and littermates decreased the emission of USVs indicating reduced anxiety [33,34]. Conversely, stressful experimental situations increased the emission of USVs reflecting elevations of anxiety [35,36]. Exposure to prenatal insults such as prenatal stress [37], nitrogen [38], electromagnetic fields [39], and methylmercury [40] caused alterations in the acoustic parameters of isolation-induced USVs in pups. Ethanol as an acute stressor has been found to modulate the distress USVs in pups during development. Prenatal ethanol exposure has been reported to decrease [41] or increase [42] the isolation-induced USVs in pups with no alterations reported in one earlier research [43]. In our recent study [44], ethanol exposure (30%, v/v) from gestational days 8–20 elevated the number, frequencies, and amplitudes of USVs reflecting more negative emotionality in high-ethanol pups compared to their low-ethanol (15%, v/v) and control (0%) counterparts. Lactational ethanol exposure, on the other hand, disrupted the typical responses to isolation that were reflected through increased latency to vocalize and decreased number of USVs in pups [30,45]. Contrary to this, after interaction with an ethanol-intoxicated dam during lactation, rat pups displayed high levels of isolation-induced USVs indicating heightened behavioral distress [20]. Thus, it is evident that the isolation-induced USVs in pups were altered following exposure to ethanol during the critical developmental period.

Maternal care such as licking and grooming [46] are induced by USV-emitting pups and lactational ethanol exposure has been found to cause serious modifications in these behaviors as well. Administration of ethanol into the drinking water during lactation impaired the mammary gland development in dams [18,47] and resulted in significant changes in maternal nutritional status and milk production, initiating impaired litter growth [48]. Similar to these results, a study [49] with maternal ethanol consumption (5 g/kg/day) during lactation (PNDs 4–10) resulted in impaired maternal behavior that was reflected in longer latency to retrieve the pups. Furthermore, exposure to ethanol reduced the rectal temperature in dams and caused malnourishment of pups as reflected through the incapability to facilitate maternal attachment [50]. The presence of ethanol as a novel chemosensory cue in milk [51] and/or the excretion of ethanol through respiration, urine, salivation, and perspiration [52] influenced the infantile perception of ethanol-induced physiological and behavioral changes in the dam and altered the social interactions between the dam and the pup [20,53].

The acoustic parameters of isolation-induced USVs in pups differed significantly in number, duration, frequency, amplitude, frequency modulations, and bandwidth [54]. The USV repetition rate ranges between 80 and 90 USVs/min, and the call duration varies between 80 and 150 ms [40,55,56]. Therefore, all acoustic parameters facilitating sound localization by the dam are present in the pups' isolation calls. The expression of distressful states in rat pups through the emission of USV has been more frequently studied by recording and measuring the duration, frequency, and amplitude of calls along with the number of USVs. Duration, frequency, and amplitude of USVs are crucial acoustic parameters for dam–pup social interactions and these parameters are considered as signals for the dams to identify each pup [57]. However, very few studies have considered acoustic parameters in research to clarify the effects of lactational ethanol exposure on dam–pup social interactions. In this study, which is a continuation of our previous studies [44,58], we examine the number, duration, frequency, amplitude, and frequency modulated USVs in pups exposed to ethanol via breast milk, and to ethanol-altered dam behavior, and predict that ethanol exposure during lactation will change the acoustic parameters of distress USVs of pups upon their isolation.

2. Materials and methods

2.1. Animals

Wistar-derived pregnant rats were purchased from Japan SLC Inc. (Hamamatsu, Japan) at gestational day (GD) 13. They were housed in standard maternity cages lined with wood shavings. On GD 13, rats were assigned to the high-ethanol, low-ethanol, and control groups based on the doses of ethanol administered to the six dams in each group. They had ad libitum access to rat chow MF (Oriental Yeast Ltd., Sapporo, Japan) and tap water before and after the ethanol-exposure period.

The day of parturition was considered as postnatal day (PND) 0. The total number of pups born from the high-ethanol, low-ethanol and control dams were 59, 65, and 57, respectively. On PND 4, litters were culled to eight pups (four males and four females) per dam to ensure that the experimental conditions in all three groups were standardized. Accordingly, the number of pups was 48 in each of the three groups.

Ethanol exposure period in dams ranged from PNDs 5 to 21 and the experimental timeline for the recording of USVs in pups covered PNDs between 4 and 16. The high- and low-ethanol dams were treated with ethanol (purity = 99.5%; Kanto Chemical Co., Inc., Tokyo, Japan) in tap water along with free access to standard diet (rat chow) throughout the exposure period. On PNDs 5–7, 10% and 5%, on PNDs 8–10, 20% and 10% and on

PNDs 11–21, 30% and 15% ethanol (*v*/*v*) in tap water were administered to the high- and low-ethanol dams, respectively. Thus, dams from both groups had ad libitum access to ethanol mixture as the only source of drinking between PNDs 5 and 21 and their pups were exposed to ethanol via the breast milk of their dam. The control dams were exposed to pure tap water throughout the exposure period. Ethanol mixture was prepared in a 200 mL drinking bottle and each bottle was provided to each of the high- and low-ethanol dams. The same content of pure tap water without ethanol was provided to the control dams. Ethanol mixture and the tap water for dams in all three groups were refreshed every day from PNDs 5 to 21. The pattern, doses and days of ethanol administration were adopted from relevant previous studies [18,44,47,58–63].

The holding room for the dams and pups was maintained under controlled temperature (22 ± 2 °C), humidity (50% ± 10%), and artificial 12 h light/dark (light: 20:00–08:00 and dark: 08:00–20:00) conditions. The research design was approved by the animal ethics committee of Hokkaido University, and rats used in this study were maintained and treated in accordance with the guidelines for Care and Use of Laboratory Animals, Hokkaido University.

2.2. Apparatus

The Sonotrack system (version 2.4.0; Metris, Hoofddorp, The Netherlands) was employed to digitally record and analyze the USVs of rat pups. The Sonotrack system was installed on a personal computer and linked with an ultrasonic microphone affixed in a sound-proof chamber.

2.3. USV Recording

A total of 30 pups was used for the USV recording in the present study. USVs were recorded from two pups (one male and one female) randomly selected from the litters of each dam. Thus, 10 pups from five dams in each of the three ethanol groups were used for the USV recording procedure. The pups were repeatedly tested for USV recording on PNDs 4, 8, 12, and 16. Each pup was individually isolated from the dam and littermates in the holding room and placed in a translucent cup (13 cm bottom diameter, 15 cm top diameter, and 15 cm height) for transfer to the experiment room for USV recording. The pup was left alone in the sound-proof dark box. The first 5 min was the period of habituation followed by another 5 min of USV recording. The ultrasonic microphone was positioned at a height of 20 cm from the bottom of the translucent cup. After recording, the body weight of the pup was measured and then it was returned to the dam and littermates. Thus, the pup was isolated from the dam and littermates for 10 min on each day of the recording.

USVs were derived during the dark period of the light–dark cycle with maintained temperature (17–21 °C) and humidity (45–67%) conditions. The other pups were used for play fighting studies in juvenile period in another project.

2.4. Statistical Analyses

The automatic mode setting of the Sonotrack system selected and analyzed the recorded USVs produced by the pups. The resolution time was 1 ms, and to reduce background noise, low and high cut-off frequencies for analyses were set to 30 and 90 kHz, respectively. The Sonotrack system determined the lowest frequency in a periodic waveform of USV at every 1 ms interval and derived fundamental frequencies. Fundamental frequencies of USVs between <30 and ≥70 kHz at the start and end points, respectively, initiated the re-analyses of USV data with the manual selection mode due to the possibility of artifacts. USVs that satisfied all of the following criteria were selected for statistical analyses [44].

i. The fundamental frequencies at both start and end points were ≥30 and <70 kHz.
ii. The duration was ≥20 ms.
iii. The mean fundamental frequency was <90 kHz.

iv. The bandwidth between the maximum and minimum fundamental frequencies was <60 kHz.

Because vocalization-related organs of the pups are immature, they cannot produce the USVs with wider bandwidth and longer durations as juvenile and adult rats can. Acoustic parameters of selected USVs considered for statistical analyses included the number, duration, fundamental frequency, bandwidth, and amplitude. Duration, amplitude, and fundamental frequencies were determined through the average USVs of PNDs 4, 8, and 12 for 10 pups in each of the three groups. Bandwidth was the range between the maximum and minimum fundamental frequencies of USVs. Moreover, the percentage of USVs with frequency modulations was calculated as (total number of USVs with frequency modulations/total number of USVs) $\times$ 100. Frequency modulation was defined as the USV with a bandwidth $\geq$5 kHz [64]. All USVs, including the frequency modulated ones, were the 40 kHz distress-type calls emitted by pups ranging from 40 to 55 kHz.

The acoustic parameters of the USVs were analyzed using a three-way analysis of variance (ANOVA) defined by two between-group factors, concentration (high/low/control) and sex (male/female), and one within-subject factor, age (PNDs 4/8/12/16). USV data reported in the figures were the average USVs produced by 10 pups (5 males and 5 females) in each ethanol group. On PND 16, no other acoustic parameters except the number of USVs were possible to be analyzed with ANOVA, because no USVs were produced by many pups regardless of ethanol exposure. The body weights of pups were analyzed via three-way ANOVA comprising two between-group factors of concentration and sex and one within-group factor of age. Ethanol consumption in dams was subjected to a one-way ANOVA comprising one between-group factor of concentration. When a factor was found to be significant, multiple comparisons were performed using Ryan's method. Statistical analyses were conducted through ANOVA 4 on the website (http://www.hju.ac.jp/_kiriki/anova4/about.html, accessed on 17 June 2021).

3. Results

3.1. Daily Consumption of Ethanol-Containing Water

The daily intake of ethanol-containing water significantly varied across the groups (F (2, 15) = 177.508, $p < 0.001$). As shown in Figure 1, the daily intake of ethanol-containing water for the high-ethanol dams on PNDs 5–7 (10% ethanol), PNDs 8–10 (20% ethanol), and PNDs 11–21 (30% ethanol) was 29.33, 13.56, and 16.33 mL/day, respectively. For the low-ethanol dams, the daily consumption on PNDs 5–7 (5% ethanol), 8–10 (10% ethanol), and 11–21(15% ethanol) was 37.56, 27.67, and 30.36 mL/day, respectively. The daily consumption of pure tap water for the control dams on PNDs 5–7, 8–10, and 11–21 was 42.34, 38.89, and 47.62 mL/day, respectively. For high-ethanol dams, the daily consumption of ethanol-containing water was consistent between PNDs 8–10 and 11–21 but was significantly lower than the daily consumption of PNDs 5–7 ($p < 0.001$). Thus, it is evident that high-ethanol dams drank significantly less water with higher content of ethanol.

3.2. Landmarks of Physical Growth

Lactational ethanol exposure adversely affected the physical growth of pups from the group with the higher content of ethanol. Almost all landmarks of physical growth were delayed in the high-ethanol pups. The time of incisor eruption, onset of body hair, and eye openings were evidently delayed compared to those of the control pups (Table 1). More importantly, 16 of the high-ethanol pups died before weaning, whereas none of the low-ethanol and control pups were found dead. The low-ethanol and the control pups were timely weaned on PND 22, but the high-ethanol pups were weaned on PND 30 because growth retardation threatened their survival.

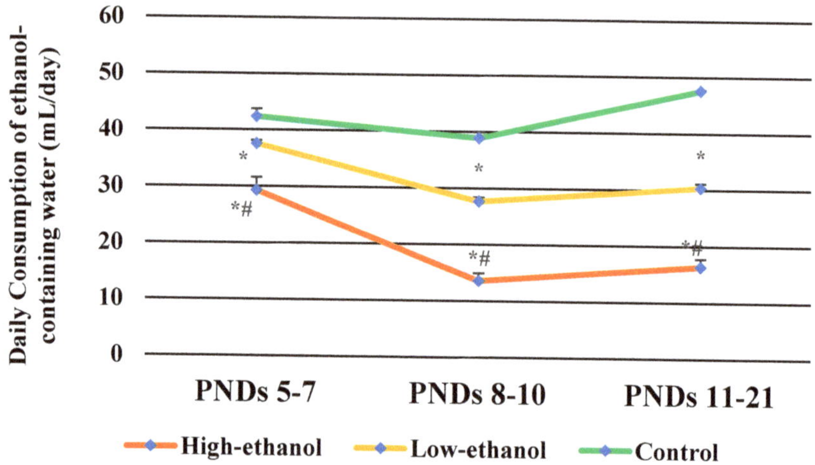

Figure 1. Daily consumption of ethanol-containing water in rat dams on PNDs 5–7, 8–10 and 11–21; Data are expressed as the mean ± SEM; * $p < 0.01$ compared with that in the control dams; # $p < 0.001$ compared with that in the low-ethanol dams; the number of dams was 6 in each group.

Table 1. Effects of lactational ethanol exposure on landmarks of physical growth in rat pups.

Pups	Incisor Eruption (PNDs)	Body Hair (PNDs)	Eye Opening (PNDs)	Timing of Death (PND)	Weaning
High-ethanol	15.094 ± 0.222	11.844 ± 0.191	20 ± 0.174	18 ± 0.224 (n = 16)	PND 30
Low-ethanol	11 ± 0.179	8.75 ± 0.113	16.708 ± 0.168	No death	PND 22
Control	9.271 ± 0.142	8.167 ± 0.124	15 ± 0.139	No death	PND 22

Note: The sample size for the high-ethanol, low-ethanol and control pups is 32, 48, and 48, respectively. Physical landmarks of sixteen dead pups in the high-ethanol group have not been incorporated in Table 1. Data are expressed as the mean ± SEM.

3.3. Body Weights of Dams

The body weights of dams significantly varied across the groups (F (2, 15) = 66.912, $p < 0.001$). Dams significantly lost body weights as the concentrations of ethanol increased, exhibiting lowest body weights in the high-ethanol group followed by the low-ethanol and the control groups ($p < 0.05$). A two-way interaction between ethanol concentration and the timing of exposure (PNDs) was significant (F (4, 30) = 85.453, $p < 0.001$, Figure 2A). The average body weights of high-ethanol dams on PNDs 8–10 and 11–21 were significantly reduced compared to the low-ethanol and control dams ($p < 0.05$).

3.4. Body Weights of Pups

The body weights of individual pup were measured just after the USV recording on PNDs 4, 8, 12, and 16. The body weights significantly varied across the groups (F (2, 24) = 86.371, $p < 0.001$). Reductions in body weights were evident as the concentration of ethanol increased, showing lowest body weights in the high-ethanol pups followed by the low-ethanol and the control pups ($p < 0.05$). A two-way interaction effect between ethanol and age was significant (see Figure 2B) (F (6, 72) = 131.024, $p < 0.001$). On PNDs 12 and 16, the high-ethanol pups significantly lost their body weights compared to the low-ethanol and control pups ($p < 0.05$). An age-dependent elevation in body weights was significant in all three ethanol groups (F (3, 72) = 1670.920, $p < 0.001$). Sex differences in body weights of the pups were not significant (data not shown).

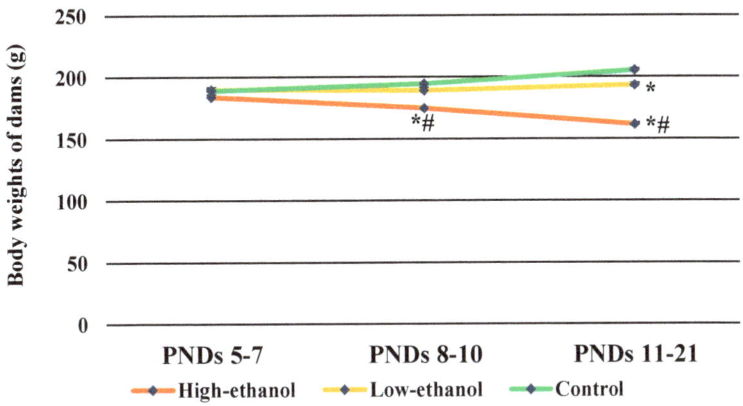

(A)

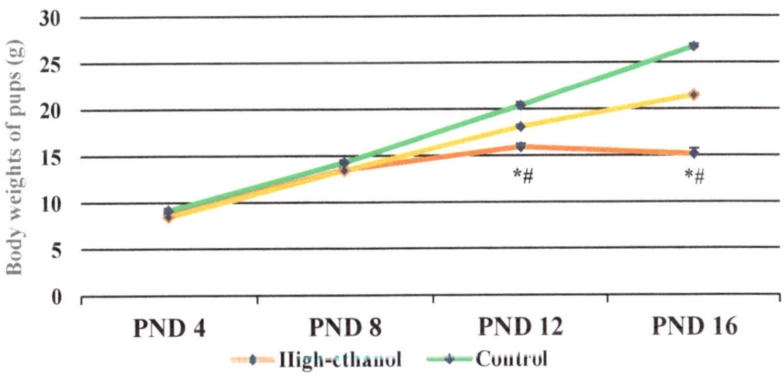

(B)

Figure 2. (**A**) Effects of ethanol exposure during lactation on body weights of dams; (**B**) Effects of ethanol exposure during lactation on body weights of rat pups; Data are expressed as the mean $\pm$ SEM; * $p < 0.05$ compared with that in the control dams and control pups, # $p < 0.05$ compared with that in the low-ethanol dams and low-ethanol pups; the number of dams was 6 in each of the three groups; the number of pups was 10 (5 males and 5 females) in each of the three groups.

3.5. Number of USVs

An elevated number of USVs was evident by the main effects of ethanol (F (2, 24) = 4.789, $p < 0.05$). Production of the greater number of USVs was prominent in the high-ethanol pups than the control pups ($p < 0.05$). The significant interaction between ethanol and age (F (6, 72) = 2.318, $p < 0.05$, Figure 3A) reflected the elevations of USVs on PND 12 in both the high- and low-ethanol pups compared to the control pups ($p < 0.05$). The main effect of age on the number of calls (F (3, 72) = 42.041, $p < 0.001$) exhibited that the highest number of USVs was produced on PND 8 followed by those on PNDs 4, 12, and 16 ($p < 0.05$). Sex effects on the number of emitted USVs were not significant (data not shown).

3.6. Mean Duration of USVs

The duration of USVs significantly varied across the groups (F (2, 24) = 3.494, $p < 0.05$, Figure 3B). Longer calls were produced by pups from the high-ethanol group than in the control group ($p < 0.05$). Sex of the pups had no significant effects on the mean duration of USVs.

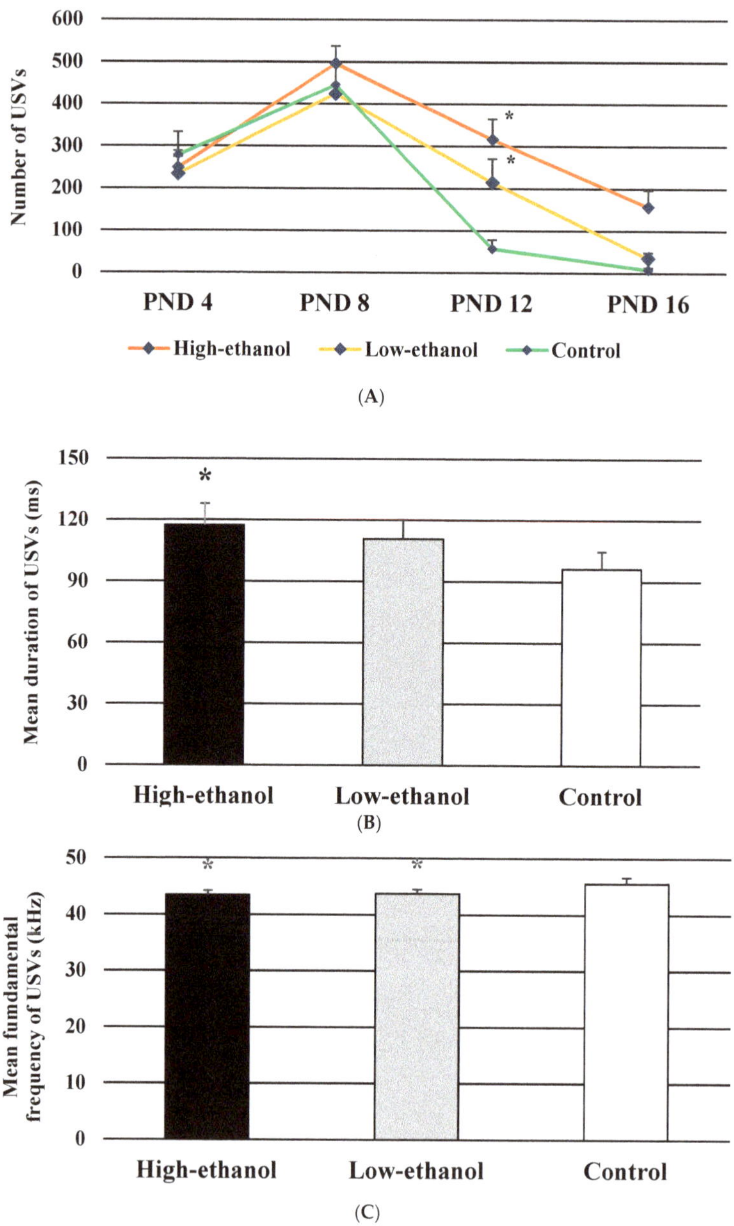

Figure 3. *Cont.*

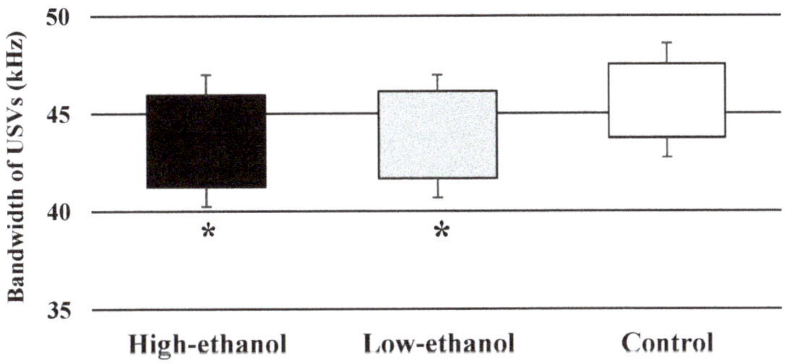

(D)

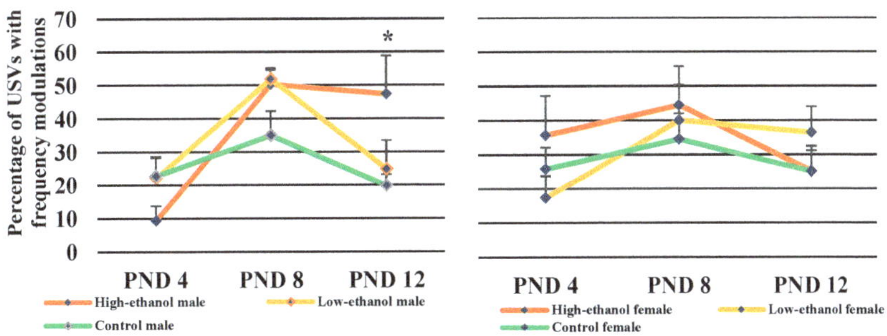

(E)

Figure 3. (**A**) Effects of ethanol exposure during lactation on the number of USVs in rat pups; data are expressed as the mean ± SEM; * $p < 0.05$ compared with that in the control pups; the number of pups was 10 (5 males and 5 females) in each of the three groups. (**B**) Effects of ethanol-exposure during lactation on the mean duration of USVs in rat pups; data are expressed as the mean ± SEM; * $p < 0.05$ compared with that in the control pups; the number of pups was 10 (5 males and 5 females) in each of the three groups. (**C**) Effects of ethanol-exposure during lactation on the mean fundamental frequency of USVs in rat pups; data are expressed as the mean ± SEM; * $p < 0.05$ compared with that in the control pups; the number of pups was 10 (5 males and 5 females) in each of the three groups. (**D**) Effects of ethanol-exposure during lactation on the bandwidth of USVs in rat pups; the lower and upper segments of bandwidth indicate the minimum and maximum fundamental frequencies, respectively; data are expressed as the mean ± SEM; * $p < 0.05$ compared with that in the control pups; the number of pups was 10 (5 males and 5 females) in each of the three groups. (**E**) Effects of ethanol-exposure during lactation on the percentage of USVs with frequency modulations in rat pups; data are the mean ± SEM; * $p < 0.05$ compared with that in the control male pups; the number of pups was 5 in each of the six groups.

3.7. Mean Fundamental Frequency of USVs

Ethanol had significant group effects on the fundamental frequency of USVs (F (2, 24) = 4.226, $p < 0.05$, Figure 3C). Lower fundamental frequencies were produced by the high- and low-ethanol pups than the control pups ($p < 0.05$). Sex differences in the mean fundamental frequency of pup USVs were not significant.

3.8. Bandwidth of USVs

Ethanol exposure had no significant effects on the bandwidth of USVs. However, significant effects were evident in minimum fundamental frequencies of USVs (F (2, 24) = 7.625, $p < 0.005$, Figure 3D). Decreased minimum fundamental frequencies were noticed in the high- and low-ethanol pups compared to the control pups ($p < 0.05$). Minimum fundamental frequencies of USVs significantly varied across the age of the pups (F (2, 48) = 6.203, $p < 0.005$). With the advancement of age, the minimum fundamental frequencies were decreased on PNDs 8 and 12 compared with those on PND 4 ($p < 0.05$). Maximum fundamental frequencies of USVs were not affected by ethanol exposure. Sex had no significant effects on USV bandwidth as well as on the minimum and maximum fundamental frequencies of USVs.

3.9. Percentage of USVs with Frequency Modulations

A three-way interaction of ethanol, sex, and age on the percentage of USVs with frequency modulations was significant (F (4, 48) = 3.624, $p < 0.05$, Figure 3E), exhibiting that the high-ethanol male pups produced greater percentage of USVs with frequency modulations than the control male pups on PND 12 ($p < 0.05$). A significant age effect was observed (F (2, 48) = 15.453, $p < 0.001$) as reflected through elevations in the percentage of frequency modulated USVs on PND 8 compared with those on PNDs 4 and 12 ($p < 0.05$).

3.10. Amplitude of USVs

The amplitude of USVs had no significant main effect or interaction effect of ethanol and sex. However, the age effects were significant (F (2, 48) = 7.079, $p < 0.005$) indicating greater amplitudes on PND 8 than those on PNDs 4 and 12 ($p < 0.05$) (data not shown).

In summary, these findings suggest that the gradually elevated exposure of ethanol concentrations in the high-ethanol dams on PNDs 5–7, 8–10 and 11–21 resulted in lower daily consumption of ethanol-containing water with higher content of ethanol. Accordingly, a significant loss in body weight of both the high-ethanol dams and their pups was noticed and USV acoustic parameters upon pups' isolation from the dams were altered, showing elevations in pups' distress calls with longer duration and higher percentage of frequency modulations.

4. Discussion

Ethanol consumed by rat dams is secreted into breast milk and readily transmitted into pups, exerting direct impacts on them. For dams, ethanol exposure during lactation reduced the breast milk production [17,48] and induced poor maternal care in building nest, retrieving pups, and crouching for breast feeding [65,66]. The pups from those ethanol intoxicated dams exhibited suckling deficits [19] and, in turn, suffered from malnutrition contributing to retarded growth [67,68]. Growth retardation due to ethanol exposure (2.5 g/kg/day on PNDs 3–13) affected the motor coordination in pups necessary for nipple attachment, which was reflected through decreased perioral responsiveness and longer latencies to attach to the nipples [69,70]. Growth retardation might also affect the vocalization related organs of ethanol-exposed pups during lactation and alter their communication with dams.

4.1. Physical Growth

The pups of the high-ethanol group in the present study carried the same body weights on PNDs 12–16 indicating growth retardation, whereas the body weights of control pups were steadily increasing on each PND. Other physical landmarks like body hair growth, incisor eruptions, and eye-openings were also delayed in high-ethanol pups compared with the low-ethanol and control pups. Due to growth retardation, weaning of the high-ethanol pups was delayed up to PND 30. In the present study, consumption of 30% ethanol-containing water resulted in lower intake (16.33 mL/day) by the high-ethanol dams, whereas consumption of 15% ethanol-containing water resulted in higher intake (30.36 mL/day) by the low-ethanol dams from PNDs 11 to 21. Growth retardation was highly evident in high-ethanol pups, whereas low-ethanol pups had no such evidence. Therefore, higher ethanol-containing water (30%) with lower intake (16.33 mL/day) was more detrimental than lower ethanol-containing water (15%) with greater intake (30.36 mL/day).

The parameters of physical growth of high-ethanol pups in the present study were found to be highly consistent with the results of earlier studies [19], where lactational ethanol exposure at a dose of 6 g/kg/day on PNDs 4–12 resulted in retarded body weights in pups compared to those of other treatment groups. Moreover, delayed eye openings and incisor eruptions altered the sensory modalities related to nipple attachment and contributed to the suckling deficits in ethanol-exposed (6 g/kg/day) pups. Combined gestational and lactational exposure to 20% ethanol concentrations resulted in lower intake of ethanol in dams and reductions in the body weights of pups [47]. Lower intake of higher ethanol-containing water (20%, v/v) made the dams dehydrated, reflected through deficiencies in maternal nutritional status and milk production. Consequently, sucking pups of high-ethanol exposed dams were malnourished and exhibited decreased body weights [18] similar to those of other studies with higher ethanol concentrations (25% ethanol) [48]. Repeated states of ethanol intoxication (1 and 2 g/kg/day) either in the early (PNDs 5–8) or mid-stage (PNDs 9–12) of lactation resulted in decreased breast milk production in rat dams and growth retardation in the pups [71,72]. In addition, ethanol (2.5 g/kg/day) administered to dams on PNDs 3–13 impaired the capability of the dams to retrieve the pups and disrupted the fixed action patterns relevant to maternal care [50]. Insufficient maternal care, in turn, led to lower body weights, increased mortality rates, and other growth retardation in pups observed in the present study. All these results suggest that the effects of ethanol on pups could be caused by its effects on the dams. Exposure to higher ethanol-containing water caused lower water consumption, resulted in decreased body weights, and presumably dehydration and less milk produced in the dams. Resulting inappropriate maternal care to pups was evident, so they were unsuccessful in sucking behaviors and displayed growth retardation and malnutrition up to the point of death.

4.2. USV Acoustic Parameters

The main purpose of the study was to determine the acoustic parameters of USVs of rat pups upon their isolation from the dam, because few studies have been carried out on the dam–pup interactions in the USV domain and comparing them with results of behavioral studies. In our study, increased numbers of USVs with longer duration and frequency modulations were prominent in high-ethanol pups compared with the control pups. These USV alterations were particularly evident on PND 12 along with the inhibition of body weight gains in the high-ethanol pups. Consistent with this is the study in [20], where intragastric ethanol administration on PNDs 3, 5, 7, 9, 11, and 13 at doses of 2.5 g/kg/day to rat dams resulted in increased USVs in pups on PNDs 3 and 9, as indexed by behavioral distress. Conversely, in [45,73] researchers administered ethanol (6 g/kg/day) intragastrically to rat pups on PNDs 4–10 and PNDs 1–7. The rat pups in both studies had longer latencies to produce the first vocalization, and produced decreased number of USVs. Similarly, intragastric ethanol administration (3 g/kg/day) to rat pups on PNDs 4–10 resulted in lower number of USVs on PND 14 [30]. However, intragastric administration of ethanol could have direct adverse effects on the organs and tissues of the vocal tract and larynx, and potentially affect production of USV. From the findings of the present study, it is reasonable to argue that the decreased body weights of the high-ethanol dams were caused by dehydration, and consequently, lowered amount of milk production and altered maternal care of their stressed pups.

Regardless of the effects of pup undernutrition following disrupted maternal care, significant elevations in the number of distress calls in both the high- and low-ethanol pups compared to the control pups could be because of ethanol itself. Importantly, the body weights of low-ethanol dams and their pups did not differ with those of the controls in the present study. Other acoustic parameters of USVs except for the number of distress calls created no differences between the low-ethanol and control pups. Thus, lower ethanol-containing water with greater intake in low-ethanol dams did not cause dramatic undernutrition, and promoted sufficient maternal care to their pups.

4.3. Comparisons between Gestational and Lactational Exposure

The effects of gestational ethanol exposure on USVs in rat pups have been demonstrated in earlier relevant studies [41,42,63]. These studies reported that prenatal exposure to ethanol reduces the number of USVs [41,63], except in one study [42] where perinatal exposure increased the number of USVs in pups. However, the pattern, dose, and duration of ethanol exposure as well as the acoustic parameters of USVs investigated in these studies varied grossly or seemed insufficient to draw a valid conclusion. Towards this end, in our recent research [44], ethanol was administered to pregnant rats on gestational days 8–20 via drinking water. On GDs 8–10, 10% and 5%, on GDs 11–13, 20% and 10%, and on GDs 14–20, 30% and 15% ethanol-containing water (v/v) was administered to the rats of the high- and low-ethanol groups, respectively. A higher number of distress USVs, elevated fundamental frequencies, and higher amplitudes were displayed by the high-ethanol pups (30%) compared to both the low-ethanol (15%) and control (0%) pups. Distress USVs emitted by high-ethanol pups indicated their negative emotionality upon maternal isolation [44]. However, pups in all three groups underwent similar body weight gains from infancy to adulthood, and other growth indices such as the time of ear and eye openings, body hair growth, and incisor eruptions, were not delayed in gestational exposure. They were weaned on PND 21 regardless of ethanol administration and the survival rate of pups in all three groups was 100% [44]. In gestational exposure, ethanol intake by pregnant dams is readily transmitted into the blood circulation and crosses the placental barrier to reach into the fetuses. However, gestational period showed limited effects on maternal behavior. After parturition, the dams were not exposed to ethanol anymore and they were able to ensure proper maternal behavior. Consequently, the pups from all ethanol groups grew up normally and weaned on PND 21 [44].

Contrary to the effects of gestational exposure, lactational high-ethanol exposure in the present study resulted in growth retardation, as reflected through delayed eye openings and incisor eruptions in pups. In addition to that, pups also showed signs of malnutrition. Therefore, it is likely that the high-ethanol pups were not able to receive sufficient breast milk from the dams and suffered from malnutrition leading to growth retardation. Consistent with this is the observation that the high-ethanol dams were unable to ensure proper maternal behavior such as retrieving the pups and adopting a crouching posture due to ethanol intoxication. Accordingly, pups of the high-ethanol group sought more maternal care and produced greater number of USVs demonstrating distressful emotional states. Disruptions in maternal care after ethanol consumption during lactation included the incapability of ethanol-consumed dams to retrieve the pups, to facilitate licking or grooming, and to adopt a crouching posture as revealed in previous studies [49,51,66]. It was also shown that time spent with the pups was reduced and time required for building nest was increased in ethanol-exposed dams and litter fragmentation was elevated [74]. Previous research [75] revealed that pups produce more distress USVs when they are isolated from the nest and number of USVs reached its peak when the pups were alone. Being isolated from the dam in the present study, pups of the high-ethanol group produced greater number of USVs but with longer duration and frequency modulations, indicating elevated negative emotionality. Adverse effects of lactational ethanol exposure on pups has been revealed in past studies as well. Researchers [19,69] found longer latencies for ethanol-exposed pups to attach to nipples, less time spent in suckling, and retarded body weight gains.

We randomly selected 10 pups from the 48 high-ethanol pups and used them for USV recording. All these 10 pups survived and weaned. However, 16 out of the remaining 38 high-ethanol pups died before weaning. Previous studies [54,74] reported that distress calls produced by pups serve as a biological signal leading to arousal and caregiving behavior of the dams and importantly, that rat dams gave more maternal care to isolated pups compared to pups staying in the nest all the time, and spent more time with the isolated pups and frequently showed licking and grooming behaviors. All acoustic parameters contribute to localize the pups by the dam [76] and facilitate appropriate maternal care, hence in the present study, all crucial acoustic parameters of USVs were examined. The high-ethanol pups in the present study produced distress USVs with frequency modulations upon maternal isolation, and continuation of USV emission after reunion with ethanol-intoxicated dam provoked arousal in the dam to facilitate maternal care, although it appeared insufficient for many endangered pups. Altogether, it is well-evident that lactational high-ethanol exposure has more serious consequences than the gestational one and induces more harmful effects in pups.

4.4. Central Mechanisms of Ethanol Exposure and Negative Emotionality

The tegmental structures in the midbrain are considered as the direct initiators of USVs in rats [77–79]. Negative emotionality is closely related to the enhanced neuronal activity in laterodorsal tegmental nucleus (LDT) [80] but not in pups [81]. LDT neurons play a functional role in negative emotion-induced USVs upon maternal isolation [82]. LDT is considered as one of the sources of GABAergic input to the ventral tegmental area (VTA) [83]. GABAergic innervation to the VTA originates from the rostromedial tegmental nucleus (RMTg) along with other brain regions. As an essential inhibitory afferent to midbrain dopamine neurons, the RMTg is activated during ethanol exposure [84–86] and the inhibitory RMTg GABA input to VTA dopaminergic (DA) neurons plays a role in regulating negative emotionality [87]. In the present study, the high-ethanol pups exhibited USV acoustic alterations reflected through a greater number of distress USVs with longer duration and a higher percentage of frequency modulations. Acoustic alterations in USVs indicate prominent elevations of negative emotionality and this might be due to the ethanol-induced excitation of RMTg neurons exhibited through the elevated distress calls in pups. However, the expression of the RMTg GABA input to the VTA after ethanol exposure and the resulting distressful states are different in adult rats compared to rat pups [88,89] and thus, further relevant studies on rat pups will be required to settle this issue.

In conclusion, all these results suggest that ethanol-induced distressful emotional states in pups might be because of ethanol effects on dams and/or due to the central nervous system effects of ethanol on pups. In the present study, elevations in the number of distress USVs in the high- and low-ethanol pups could be because of the ethanol-induced activation of RMTg that transmitted inhibitory GABA input to VTA DA neurons to exhibit negative emotionality in pups. Another explanation is that lower water consumption with higher ethanol content resulted in undernutrition and less milk produced in the high-ethanol dams. Accordingly, their pups were malnourished and exhibited unsuccessful suckling behaviors. Resulting malnutrition and growth retardation in the high-ethanol pups forced them to produce more distress calls with longer durations and frequency modulations upon isolation from the dams, as indicative of elevated negative emotionality.

Author Contributions: Conceptualization, M.A.S. and H.W.; methodology, M.A.S. and H.W.; software, H.W.; validation, M.A.S.; formal analysis, M.A.S.; investigation, M.A.S.; resources, H.W.; data curation, M.A.S.; writing—original draft preparation, M.A.S.; writing—review and editing, H.W.; visualization, H.W.; supervision, H.W.; project administration, H.W.; funding acquisition, H.W. All authors have read and agreed to the published version of the manuscript.

Funding: This research received no external funding.

Institutional Review Board Statement: The study was conducted according to the guidelines of the Care and Use of Laboratory Animals, Hokkaido University, and approved by The Institutional Animal Care and Use Committee of HOKKAIDO UNIVERSITY (protocol code: 17-0090 and date of approval: 18 August 2017).

Acknowledgments: This work was supported by the Graduate Grant Program of Graduate School of Humanities and Human Sciences, Hokkaido University, Japan.

Conflicts of Interest: The authors declare no conflict of interest.

References

1. Grossman, E.R. Beer, breast-feeding, and the wisdom of old wives. *JAMA* **1988**, *259*, 1016. [CrossRef]
2. Mennella, J.A. The transfer of alcohol to human milk Sensory implications and effects on mother-infant interaction. In *Alcohol and Alcoholism: Effects on Brain and Development*; Hanningan, J.H., Spear, L.P., Spear, N.E., Goodlett, C.R., Eds.; Lawrence Erlbaum: Mahwah, NJ, USA, 1999; pp. 177–198.
3. Mennella, J.A. Regulation of milk intake after exposure to alcohol in mothers' milk. *Alcohol. Clin. Exp. Res.* **2001**, *25*, 590–593. [CrossRef]
4. Mennella, J.A. Short-term effects of maternal alcohol consumption on lactational performance. *Alcohol. Clin. Exp. Res.* **1998**, *22*, 1389–1392. [CrossRef] [PubMed]
5. Giglia, R.C.; Binns, C.W. Patterns of alcohol intake of pregnant and lactating women in Perth, Australia. *Drug Alcohol Rev.* **2007**, *26*, 493–500. [CrossRef] [PubMed]
6. Mennella, J.A. Alcohol use during lactation: The folklore versus the science. In *Current Issues in Clinical Lactation*; Auerbach, K.G., Ed.; Jones and Burtlett Publishers: Sudbury, MA, USA, 2002.
7. Mennella, J.A.; Beauchamp, G.K. The transfer of alcohol to human milk: Effects on flavor and the infant's behavior. *N. Engl. J. Med.* **1991**, *325*, 981–985. [CrossRef] [PubMed]
8. Mennella, J.A.; Beauchamp, G.K. Beer, breast feeding and folklore. *Dev. Pychobiol.* **1993**, *26*, 459–466. [CrossRef]
9. Mennella, J.A.; Garcia-Gomez, P.L. Sleep disturbances after acute exposure to alcohol in mothers' milk. *Alcohol* **2001**, *25*, 153–158. [CrossRef]
10. Mennella, J.A.; Gerrish, C.J. Effects of exposure to alcohol in mother's milk on infant sleep. *Pediatrics* **1998**, *101*, E21–E25. [CrossRef]
11. Little, R.E.; Anderson, K.W.; Ervin, C.H.; Worthington-Roberts, B.; Clarren, S.K. Maternal alcohol use during breast-feeding and infant mental and motor development at one year. *N. Engl. J. Med.* **1989**, *321*, 425–430. [CrossRef]
12. Gibson, L.; Porter, M. Drinking or smoking while breastfeeding and later cognition in children. *Pediatrics* **2018**, *142*, e20174266. [CrossRef]
13. Maier, S.E.; West, J.R. Alcohol and nutritional control treatments during neurogenesis in rat brain reduce total neuron number in locus coeruleus, but not in cerebellum or inferior olive. *Alcohol* **2003**, *30*, 67–74. [CrossRef]
14. Chen, W.J.A.; West, J.R. Alcohol-induced brain damage during development: Potential risk factors. In *Alcohol and Alcoholism: Effects on Brain and Development*; Hannigan, J.H., Spear, L.P., Spear, N.E., Goodlett, C.R., Eds.; Lawrence Erlbaum: Mahwah, NJ, USA, 1999; pp. 17–38.
15. Goodlett, C.R.; Eilers, A.T. Alcohol-induced Purkinje cell loss with a single binge exposure in neonatal rats: A stereological study of temporal windows of vulnerability. *Alcohol. Clin. Exp. Res.* **1997**, *21*, 738–744. [CrossRef] [PubMed]
16. Subramanian, M.G.; Chen, X.G.; Bergeski, B.A. Pattern and duration of the inhibitory effect of alcohol administered acutely on suckling-induced prolactin in lactating rats. *Alcohol. Clin. Exp. Res.* **1990**, *14*, 771–775. [CrossRef] [PubMed]
17. Subramanian, M.G. Evaluation of lactational parameters after alcohol administration for four days during early or midlactation in the rat. *Alcohol. Clin. Exp. Res.* **1997**, *21*, 799–803. [CrossRef]
18. Tavares do Carmo, M.G.; Oller do Nascimento, C.M.; Martin, A.; Herrera, E. Ethanol intake during lactation impairs milk production in rats and affects growth and metabolism of suckling pups. *Alcohol* **1999**, *18*, 71–76. [CrossRef]
19. Barron, S.; Kelly, S.J.; Riley, E.P. Neonatal alcohol exposure alters suckling behavior in neonatal rat pups. *Pharmacol. Biochem. Behav.* **1991**, *39*, 423–428. [CrossRef]
20. Molina, J.C.; Pepino, M.Y.; Johnson, J.; Spear, N.E. The infant rat learns about alcohol through interaction with an intoxicated mother. *Alcohol. Clin. Exp. Res.* **2000**, *24*, 428–437. [CrossRef] [PubMed]
21. Melcer, T.; Gonzalez, S.; Barron, S.; Riley, E. Hyperactivity in preweanling rats following postnatal alcohol exposure. *Alcohol* **1994**, *11*, 41–45. [CrossRef]
22. Meyer, L.; Kotch, L.; Riley, E. Alterations in gait following ethanol exposure during the brain growth spurt in rats. *Alcohol. Clin. Exp. Res.* **1990**, *14*, 23–27. [CrossRef]
23. Kuhn, C.M.; Pauk, J.; Schanberg, S.M. Endocrine responses to mother–infant separation in developing rats. *Dev. Psychobiol.* **1990**, *23*, 395–410. [CrossRef]
24. Zimmerberg, B.; Shartrand, A.M. Temperature- dependent effects of maternal separation on growth, activity, and amphetamine sensitivity in the rat. *Dev. Psychobiol.* **1992**, *25*, 213–226. [CrossRef]
25. Ise, S.; Ohta, H. Power spectrum analysis of ultrasonic vocalization elicited by maternal separation in rat pups. *Brain Res.* **2009**, *1283*, 58–64. [CrossRef] [PubMed]
26. Portfors, C.V. Types and functions of ultrasonic vocalizations in laboratory rats and mice. *J. Am. Assoc. Lab. Anim. Sci.* **2007**, *46*, 28–34.
27. Schwarting, R.K.W.; Wohr, M. On the relationships between ultrasonic calling and anxiety-related behavior in rats. *Braz. J. Med. Biol. Res.* **2012**, *45*, 337–348. [CrossRef] [PubMed]
28. Brunelli, S.A.; Shair, H.N.; Hofer, M.A. Hypothermic vocalizations of rat pups (Rattus norvegicus) elicit and direct maternal search behavior. *J. Comp. Psychol.* **1994**, *108*, 298–303. [CrossRef] [PubMed]

29. Hofer, M.A. Hidden regulators in attachment, separation, and loss. *Monogr. Soc. Res. Child Dev.* **1994**, *59*, 192–207. [CrossRef] [PubMed]
30. Barron, S.; Gilbertson, R. Neonatal ethanol exposure but not neonatal cocaine selectively reduces specific isolation-induced vocalization waveforms in rats. *Behav. Genet.* **2005**, *35*, 93–102. [CrossRef]
31. Winslow, J.T.; Insel, T.R.; Trullas, R.; Skolnick, P. Rat pup isolation calls are reduced by functional antagonists of the NMDA receptor. *Eur. J. Pharmacol.* **1990**, *190*, 11–21. [CrossRef]
32. Hofer, M.A.; Shair, H. Sensory processes in the control of isolation-induced ultrasonic vocalization by 2-week-old rats. *J. Comp. Physiol. Psychol.* **1980**, *94*, 271. [CrossRef]
33. Carden, S.E.; Hernandez, N.; Hofer, M.A. The isolation and companion comfort responses of 7- and 3-day-old rat pups are modulated by drugs active at the opioid receptor. *Behav. Neurosci.* **1996**, *110*, 324–330. [CrossRef]
34. Hofer, M.A.; Masmela, J.R.; Brunelli, S.A.; Shair, H.N. Behavioral mechanism for active maternal potentiation of isolation calling in rat pups. *Behav. Neurosci.* **1999**, *113*, 51–61. [CrossRef]
35. Hennessy, M.B. Hypothalamic–pituitary–adrenal responses to brief social separation. *Neurosci. Biobehav. Rev.* **1997**, *21*, 11–29. [CrossRef]
36. Hofer, M.A.; Brunelli, S.A.; Shair, H.N. Potentiation of isolation-induced vocalization by brief exposure of rat pups to maternal cues. *Dev. Psychobiol.* **1994**, *27*, 503–517. [CrossRef]
37. Morgan, K.N.; Thayer, J.E.; Frye, C.A. Prenatal stress suppresses rat pup ultrasonic vocalization and myoclonic twitching in response to separation. *Dev. Psychobiol.* **1999**, *34*, 205–215. [CrossRef]
38. Di Giovanni, V.; Cagiano, R.; Carratu, M.R.; De Salvia, M.A.; Giustino, A.; Cuomo, V. Alterations in the ontogeny of rat pup ultrasonic vocalization produced by pre-natal exposure to nitrogen dioxide. *Psychopharmacology* **1994**, *116*, 423–427. [CrossRef] [PubMed]
39. Cobb, B.L.; Jauchem, J.R.; Mason, P.A.; Dooley, M.P.; Miller, S.A.; Ziriax, J.M.; Murphy, M.R. Neural and behavioral teratological evaluation of rats exposed to ultrawideband electromagnetic fields. *Bioelectromagnetics* **2000**, *21*, 524–537. [CrossRef]
40. Elsner, J.; Suter, D.; Alder, S. Microanalysis of ultrasound vocalizations of young rats: Assessment of the behavioral teratogenicity of methylmercury. *Neurotoxicol. Teratol.* **1990**, *12*, 7–14. [CrossRef]
41. Kehoe, P.; Shoemaker, W. Opioid-dependent behaviors in infant rats: Effects of prenatal exposure to ethanol. *Pharmacol. Biochem. Behav.* **1991**, *39*, 389–394. [CrossRef]
42. Marino, M.D.; Cronise, K.; Lugo, J.N., Jr.; Kelly, S.J. Ultrasonic vocalizations and maternal-infant interactions in a rat model of fetal alcohol syndrome. *Dev. Psychobiol.* **2002**, *41*, 341–351. [CrossRef] [PubMed]
43. Zimmerberg, B.; McDonald, B.C. Prenatal alcohol exposure influences the effects of neuroactive steroids on separation-induced ultrasonic vocalizations in rat pups. *Pharmacol. Biochem. Behav.* **1996**, *55*, 541–547. [CrossRef]
44. Shahrier, M.A.; Wada, H. Effects of prenatal ethanol exposure on acoustic characteristics of ultrasonic vocalizations in rat pups. *Neurotoxicology* **2018**, *69*, 29–36. [CrossRef]
45. Barron, S.; Segar, T.M.; Yahr, J.S.; Baseheart, B.J.; Willford, J.A. The effects of neonatal ethanol and/or cocaine exposure on isolation-induced ultrasonic vocalizations. *Pharmacol. Biochem. Behav.* **2000**, *67*, 1–9. [CrossRef]
46. Brouette-Lahlou, I.; Vernet-Maury, E.; Vigouroux, M. Role of pups' ultrasonic calls in a particular maternal behavior in Wistar rat: Pups' anogenital licking. *Behav. Brain Res.* **1992**, *50*, 147–154. [CrossRef]
47. Murillo-Fuentes, L.; Artillo, R.; Carreras, O.; Murillo, L. Effects of maternal chronic alcohol administration in the rat: Lactation performance and pup's growth. *Eur. J. Nutr.* **2001**, *40*, 147–154. [CrossRef] [PubMed]
48. Vilaro, S.; Viñas, O.; Remesar, X.; Herrera, E. Effects of chronic ethanol consumption on lactational performance in rat: Mammary gland and milk composition and pups' growth and metabolism. *Pharmacol. Biochem. Behav.* **1987**, *27*, 333–339. [CrossRef]
49. Wilson, J.H.; Kelly, S.J.; Wilson, M.A. Early postnatal alcohol exposure in rats: Maternal behavior and estradiol levels. *Physiol. Behav.* **1996**, *59*, 287–293. [CrossRef]
50. Pepino, M.Y.; Land, C.I.; Campbell, J.; Spear, N.E.; Molina, J. Infantile experience with an alcohol-intoxicated mother: Responsiveness to isolation and to specific social interactions (abstract). *Alcohol. Clin. Exp. Res.* **2000**, *24*, 87A.
51. Pepino, M.Y.; Abate, P.; Spear, N.E.; Molina, J.C. Disruption of maternal behavior by alcohol intoxication in the lactating rat: A behavioral and metabolic analysis. *Alcohol. Clin. Exp. Res.* **2002**, *26*, 1205–1214. [CrossRef]
52. Kalant, A. Pharmacokinetics of ethanol: Absorption, distribution and elimination. In *The Pharmacology of Alcohol and Alcohol Dependence*; Begleiter, H., Kissin, B., Eds.; Oxford University Press: New York, NY, USA, 1996; pp. 15–58.
53. Pepino, M.Y.; Spear, N.E.; Molina, J.C. Nursing experiences with an alcohol-intoxicated dam counteract appetitive conditioned responses toward alcohol. *Alcohol. Clin. Exp. Res.* **2001**, *25*, 118–124. [CrossRef]
54. Brudzynski, S.M.; Kehoe, P.; Callahan, M. Sonographic structure of isolation-induced ultrasonic calls of rat pups. *Dev. Psychobiol.* **1999**, *34*, 195–204. [CrossRef]
55. Brunelli, S.A.; Keating, C.C.; Hamilton, N.A.; Hofer, M.A. Development of ultrasonic vocalization responses in genetically heterogenous National Institute of Health (N:NIH) rats: I. Influence of age, testing experience, and associated factors. *Dev. Psychobiol.* **1996**, *29*, 507–516. [CrossRef]
56. Goodwin, G.A.; Molina, V.A.; Spear, L.P. Repeated exposure of rat pups to isolation attenuates isolation-induced ultrasonic vocalization rates: Reversal with naltrexone. *Dev. Psychobiol.* **1994**, *27*, 53–64. [CrossRef]

57. Hahn, M.; Benno, R.; Schanz, N.; Phadia, E. The effects of prenatal cocaine exposure and genotype on the ultrasonic calls of infant mice. *Pharmacol. Biochem. Behav.* **2000**, *67*, 729–738. [CrossRef]
58. Shahrier, M.A.; Wada, H. Effects of prenatal ethanol exposure on acoustic characteristics of play fighting-induced ultrasonic vocalizations in juvenile rats. *Neurotoxicology* **2020**, *79*, 25–39. [CrossRef]
59. Garcia-Moreno, L.M.; Cimadevilla, J.M. Acute and chronic ethanol intake: Effects on spatial and non-spatial memory in rats. *Alcohol* **2012**, *46*, 757–762. [CrossRef]
60. Macieira, M.S.; Almeida, W.G.; Silva, E.A.; Schenberg, L.C.; Nakamura-Palacios, E.M. Alcohol dependence induced in rats by semivoluntary intermittent intake. *Braz. J. Med. Biol. Res.* **1997**, *30*, 1107–1111. [CrossRef] [PubMed]
61. Nio, E.; Kogure, K.; Yae, T.; Onodera, H. The effects of maternal ethanol exposure on neurotransmission and second messenger systems: A quantitative auto radiographic study in the rat brain. *Dev. Brain Res.* **1991**, *62*, 51–60. [CrossRef]
62. Salami, M.; Aghanouri, Z.; Rashidi, A.A.; Keshavarz, M. Prenatal Alcohol Exposure and Dysfunction of Hippocampal Formation in Cognition. *Int. J. Reprod. Med.* **2004**, *2*, 43–50.
63. Wellmann, K.A.; George, F.; Brnouti, F.; Mooney, S.M. Docosahexaenoic acid partially ameliorates deficits in social behavior and ultrasonic vocalizations caused by prenatal ethanol exposure. *Behav. Brain Res.* **2015**, *286*, 201–211. [CrossRef]
64. Reno, J.M.; Marker, B.; Cormack, L.K.; Schallert, T.; Duvauchelle, C.L. Automating ultrasonic vocalization analyses: The WAAVES program. *J. Neurosci. Methods* **2013**, *219*, 155–161. [CrossRef]
65. Bond, N.W. Postnatal alcohol exposure in the rat: Its effects on avoidance conditioning, Hebb-Williams maze performance, maternal behavior, and pup development. *Physiol. Psychol.* **1980**, *8*, 437–443. [CrossRef]
66. Pueta, M.; Abate, P.; Haymal, O.B.; Spear, N.E.; Molina, J.C. Ethanol exposure during late gestation and nursing in the rat: Effects upon maternal care, ethanol metabolism and infantile milk intake. *Pharmacol. Biochem. Behav.* **2008**, *91*, 21–31. [CrossRef]
67. Martin, J.C.; Martin, D.C.; Sigman, G.; Radow, B. Offspring survival, development, and operant performance following maternal ethanol consumption. *Dev. Psychobiol.* **1977**, *10*, 435–446. [CrossRef]
68. Subramanian, M.G. Effects of chronic alcohol administration on lactational performance in the rat. *Alcohol* **1994**, *12*, 137–143. [CrossRef]
69. Cheslock, S.J.; Varlinskaya, E.I.; Silveri, M.M.; Petrov, E.S.; Spear, L.P.; Spear, N.E. Acute effects of ethanol and the first suckling episode in the newborn rat. *Alcohol. Clin. Exp. Res.* **2000**, *24*, 996–1002. [CrossRef] [PubMed]
70. Ponce, L.F.; Pautassi, R.M.; Spear, N.E.; Molina, J.C. Nursing from an ethanol-intoxicated dam induces short-and long-term disruptions in motor performance and enhances later self-administration of the drug. *Alcohol. Clin. Exp. Res.* **2004**, *28*, 1039–1050. [CrossRef]
71. Heil, S.H.; Subramanian, M.G. Chronic alcohol exposure and lactation: Extended observations. *Alcohol* **2000**, *21*, 127–132. [CrossRef]
72. Subramanian, M.G. Alcohol inhibits suckling-induced oxytocin release in the lactating rat. *Alcohol* **1999**, *19*, 51–55. [CrossRef]
73. Wellmann, K.; Lewis, B.; Barron, S. Agmatine reduces ultrasonic vocalization deficits in female rat pups exposed neonatally to ethanol. *Neurotoxicol. Teratol.* **2010**, *32*, 158–163. [CrossRef]
74. Bond, N.W. Effects of postnatal alcohol exposure on maternal nesting behavior in the rat. *Physiol. Psychol.* **1979**, *7*, 396–398. [CrossRef]
75. Zimmerberg, B.; Rosenthal, A.J.; Stark, A.C. Neonatal social isolation alters both maternal and pup behaviors in rats. *Dev. Psychobiol.* **2003**, *42*, 52–63. [CrossRef]
76. Brudzynski, S.M. Principles of Rat Communication: Quantitative Parameters of Ultrasonic Calls in Rats. *Behav. Genet.* **2005**, *35*, 85–92. [CrossRef]
77. Brudzynski, S.M.; Kadishevitz, L.; Fu, X.W. Mesolimbic component of the ascending cholinergic pathways: Electrophysiological pharmacological study. *J. Neurophysiol.* **1998**, *79*, 1675–1686. [CrossRef]
78. Brudzynski, S.M. (Ed.) . *Handbook of Mammalian Vocalization: An Integrative Neuroscience Approach*; Academic Press: London, UK, 2010; pp. 69–76. [CrossRef]
79. Brudzynski, S.M. Ethotransmission: Communication of emotional states through ultrasonic vocalization in rats. *Curr. Opin. Neurobiol.* **2013**, *23*, 310–317. [CrossRef]
80. Brudzynski, S.M.; Iku, A.; Harness, A. Activity of cholinergic neurons in the laterodorsal tegmental nucleus during emission of 22kHz vocalization in rats. *Behav. Brain Res.* **2011**, *225*, 276–283. [CrossRef]
81. Kehoe, P.; Callahan, M.; Daigle, A.; Mallinson, K. The Effect of Cholinergic Stimulation on Rat Pup Ultrasonic Vocalizations. *Dev. Psychobiol.* **2001**, *38*, 92–100. [CrossRef]
82. Middlemis-Brown, J.E.; Johnson, E.D.; Blumberg, M.S. Separable brainstem and forebrain contributions to ultrasonic vocalizations in infant rats. *Behav. Neurosci.* **2005**, *119*, 1111–1117. [CrossRef]
83. Wang, H.L.; Morales, M. Pedunculopontine and laterodorsal tegmental nuclei contain distinct populations of cholinergic, glutamatergic and GABAergic neurons in the rat. *Eur. J. Neurosci.* **2009**, *29*, 340–358. [CrossRef] [PubMed]
84. Glover, E.J.; McDougle, M.J.; Siegel, G.S.; Jhou, T.C.; Chandler, L.J. Role for the rostromedial tegmental nucleus in signaling the aversive properties of alcohol. *Alcohol. Clin. Exp. Res.* **2016**, *40*, 1651–1661. [CrossRef]
85. Fu, R.; Chen, X.; Zuo, W.; Li, J.; Kang, S.; Zhou, L.H.; Siegel, A.; Bekker, A.; Ye, J.-H. Ablation of mu opioid receptor- expressing GABA neurons in rostromedial tegmental nucleus increases ethanol consumption and regulates ethanol-related behaviors. *Neuropharmacology* **2016**, *107*, 58–67. [CrossRef] [PubMed]

86. Sheth, C.; Furlong, T.M.; Keefe, K.A.; Taha, S.A. Lesion of the rostromedial tegmental nucleus increases voluntary ethanol consumption and accelerates extinction of ethanol-induced conditioned taste aversion. *Psychopharmacology* **2016**, *233*, 3737–3749. [CrossRef] [PubMed]
87. Jhou, T.C.; Fields, H.L.; Baxter, M.G.; Saper, C.B.; Holland, P.C. The rostromedial tegmental nucleus (RMTg), a GABAergic afferent to midbrain dopamine neurons, encodes aversive stimuli and inhibits motor responses. *Neuron* **2009**, *61*, 786–800. [CrossRef] [PubMed]
88. Davis, A.M.; Penschuck, S.; Fritschy, J.M.; McCarthy, M.M. Developmental switch in the expression of GABAA receptor subunits $\alpha1$ and $\alpha2$ in the hypothalamus and limbic system of the rat. *Dev. Brain Res.* **2000**, *119*, 127–138. [CrossRef]
89. Fritschy, J.M.; Paysan, J.; Enna, A.; Mohler, H. Switch in the expression of rat GABAA-receptor subtypes during postnatal development: An immunohistochemical study. *J. Neurosci* **1994**, *14*, 5302–5324. [CrossRef] [PubMed]

brain sciences

MDPI

Review

Ultrasonic Vocalizations Emission across Development in Rats: Coordination with Respiration and Impact on Brain Neural Dynamics

Julie Boulanger-Bertolus [1,*] and Anne-Marie Mouly [2,*]

1 Department of Anesthesiology, Center for Consciousness Science, University of Michigan, Ann Arbor, MI 48109-5048, USA
2 Lyon Neuroscience Research Center, INSERM U1028, CNRS UMR 5292, University Lyon 1, 69366 Lyon, France
* Correspondence: jboulangerbertolus@gmail.com (J.B.-B.); annemarie.mouly@cnrs.fr (A.-M.M.)

Abstract: Rats communicate using ultrasonic vocalizations (USV) throughout their life when confronted with emotionally stimulating situations, either negative or positive. The context of USV emission and the psychoacoustic characteristics of the vocalizations change greatly between infancy and adulthood. Importantly, the production of USV is tightly coordinated with respiration, and respiratory rhythm is known to influence brain activity and cognitive functions. This review goes through the acoustic characteristics and mechanisms of production of USV both in infant and adult rats and emphasizes the tight relationships that exist between USV emission and respiration throughout the rat's development. It further describes how USV emission and respiration collectively affect brain oscillatory activities. We discuss the possible association of USV emission with emotional memory processes and point out several avenues of research on USV that are currently overlooked and could fill gaps in our knowledge.

Keywords: ultrasonic vocalizations; cognitive development; respiration; brain oscillations

Citation: Boulanger-Bertolus, J.; Mouly, A.-M. Ultrasonic Vocalizations Emission across Development in Rats: Coordination with Respiration and Impact on Brain Neural Dynamics. *Brain Sci.* **2021**, *11*, 616. https://doi.org/10.3390/brainsci11050616

Academic Editor: Stefan M. Brudzynski

Received: 21 April 2021
Accepted: 8 May 2021
Published: 11 May 2021

1. Introduction

Rats emit both audible and ultrasonic vocalizations (USV) throughout their life. USV are, by definition, emitted at a frequency higher than 19–20 kHz and are thought to be at the core of rat communication. The directionality of high frequencies facilitates the localization of the sender animal through interaural difference in the receiver while increasing their attenuation by distance, humidity, or obstacles, making them ideal for interindividual communication of small prey animals living in burrows [1]. USV are emitted when the animal is confronted with emotionally stimulating situations, but these actual situations vary between infancy and adulthood. Their psychoacoustic characteristics also change as the rat's physiology and cognition matures. This review describes the link between USV emission, respiration, and brain dynamics throughout the rat's development. Of note, other rodents emit USV, but their psychoacoustics, context of emission, or even link with physiology varies greatly between species [2,3], so this review is limited to rat USV.

2. Ethological and Anatomical Considerations for The Emission of USV

2.1. Vocalizing at the Different Ages of Life

Ecologically, both adult and infant rats emit USV, albeit in different contexts. However, even when emitted in response to the same artificial adverse stimulus, infant and adult USV present different characteristics [4].

2.1.1. Adults

Adult rats emit two types of USV in emotionally distinct contexts. The first type of USV has a frequency of 18–32 kHz, a duration of 300–3400 ms, and shows little frequency modulation. They are referred to as 22-kHz USV and are emitted in response to antagonistic interactions with conspecifics [5,6], social isolation [7], defense against a predator [8,9], aversive stimuli [6,10–12], withdrawal from drugs, or after intracerebral stimulation [13–16] or mating [6,17]. As such, they are interpreted as signaling a negative emotional state or social withdrawal from an attacker or conspecific. This is further supported by the correlation between their emission rate and the intensity of the aversive stimuli [11,12], the reduction of their emission rate by systemic injection of anxiolytics [18,19], and their increase by anxiogenic drugs [19]. Within this type of USV, two subtypes can be identified: short (<300 ms) and long USV [20]. It has been suggested they could reflect distinct negative emotional states, with the short USV signaling distress without an identified source of danger and the long USV signaling distress with an identified cause [21].

A second type of adult USV has a frequency of 35–80 kHz, with or without frequency modulation, and is commonly referred to as 50-kHz USV. These USV are observed in positive social interactions, such as play [22–25] or mating [26–30], in anticipation of food intake or intracerebral stimulation of the reward system, or injection of addictive drugs [13,31–33], and are reduced by some aversive situations [13,24,34]. Therefore, they have been suggested to reflect a positive emotional state akin to human joy and laughter [35]. However, some studies also report them when rats are interacting with a complex environment [36], during short social isolation [37–39], during morphine withdrawal [15], or during aggressive social encounters [16,40–42]. In these contexts, the 50-kHz USV emission is more difficult to reconcile with a positive emotional state. Different subtypes of 50-kHz USV have been identified that can be differentially modulated by experimental interventions and could reflect different emotional states [31,43–46]. It has been suggested that non-frequency-modulated 50-kHz USV serve a social coordinating role both for social contact and food intake. On the other end, frequency-modulated USV would reflect a highly motivated state, with the inclusion of trills measuring the intensity of the positive effect [21].

2.1.2. Infants

USV are also emitted by infant rats from the day after birth [47,48]. Although often called 40-kHz USV, mainly because they have often been recorded using filtering in a narrow frequency band around 40 kHz [47,49–59], these vocalizations span frequencies from 30 to over 100 kHz [4,60]. Ethologically, they are mostly emitted when the pups are isolated from the dam and nest [47,54] or when they are cold [49,51,56,61,62]. Additionally, they have been shown to be emitted during the extinction of operant appetitive conditioning [50], during exposure to an odor associated with gastric malaise [59], and in response to mild foot shocks [4]. The exact underlying physiological or emotional states responsible for the emission of infantile USV are still unclear. There seems to be a link with the animal's anxiety levels, as injection of anxiolytic or anxiogenic drugs modulates USV emission [43,52,63], and some of the situations eliciting USV, such as maternal isolation, are thought to be anxiogenic. However, USV are emitted in very small amounts by pups younger than 1 week if they are isolated at a temperature equivalent to that of the nest [49,61,64], suggesting USV observed at that age in the absence of careful control of the temperature are a reaction to the lowering of the pup's inner body temperature. Moreover, other factors affect the number of USV emitted. For example, a satiated state, the presence of the mother or littermates on the floor of the recording cage, or the presence of a threat all decrease USV production, despite having diverse consequences on the pup's anxiety level [54,56,65–67]. By contrast, reisolating a pup after a short reunion with its mother potentiates the number of USV emitted, even though the mother inhibits the HPA axis of the pup (i.e., its stress response) [67–70]. Overall, while these infantile USV are usually referred to as "distress calls", such interpretation seems to be an oversimplification and deserves further investigation.

In addition to this great variability in USV emission and our uncertainty regarding their physiological meaning and ethological purpose, the method of analysis of infantile

USV varies a lot, thus increasing the challenge of comparing the results across studies. Indeed, initial studies of infantile USV have simply investigated the amount of emitted calls and reported presence of clicks—described as very short sounds heard through a bat detector—in addition to calls [47,54,56]. However, a seminal article by Brudzynski et al. in 1999 reported a great variety in the duration, frequency, bandwidth, and sonographic structure of infantile USV in rats, classifying them in 10 categories according to their shape and duration [60]. This classification has been further expanded, and subsequent articles added new categories, eventually attempting to observe the effect of various environmental factors on each call category [71–75]. Other studies measured the average duration, frequency, or bandwidth of the emitted USV to investigate the effect of an intervention [53,55,75–77].

More recently, we suggested that infantile USV might be categorized depending on their frequency rather than their shape [4]. Using mild foot shock aversive stimulation, we showed that USV could be split in two classes. The first type presents a frequency centered around 40 kHz and a duration around 200 ms, while the second type, akin to the clicks described in the earlier literature [47,56], is much shorter, being 21 ms on average and presenting a frequency of 66 kHz on average (Figure 1A). These two kinds of infant USV can either co-occur or be emitted separately. Next, we questioned if these infant USV were also emitted in a more naturalistic context when pups received rough treatment from the mother. To do so, rat pups were isolated one at a time in an unfamiliar bedding-free plexiglass cage, after which the mother was introduced. In this new environmental context, the mother spent most of her time exploring the cage, occasionally stepping on the pup, rarely nursing it, and frequently roughly transporting it. This treatment enhanced 40-kHz USV while leaving 66-kHz USV unchanged. Preliminary observations further suggested that rough handling in transport, which is a source of painful stimulation for the pup, was the most efficient stimulus for enhancing 40-kHz USV (Figure 1B).

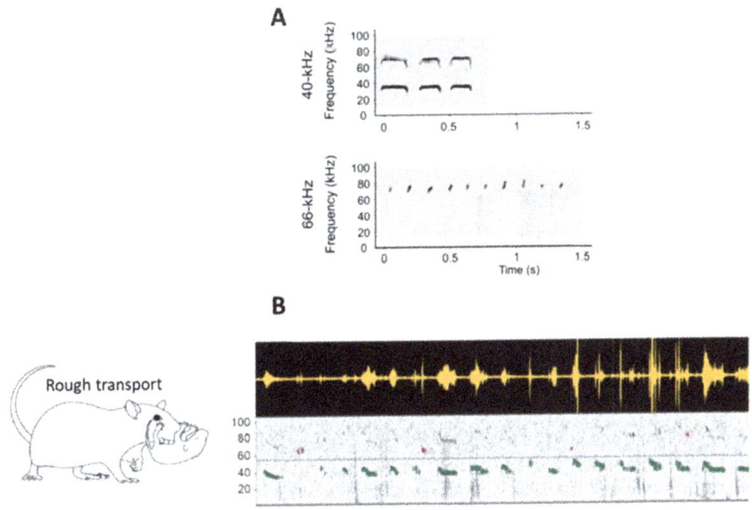

Figure 1. (**A**) Examples of individual sonograms of the two types of infant USV at 40 kHz and 66 kHz. (**B**) Examples of infant USV emitted in a naturalistic situation, here during rough transport by the mother. The top panel represents the raw USV signal, and the bottom panel represents the associated spectrogram (frequency in kHz as a function of time). The light horizontal gray line on the spectrogram represents the separation between 40-kHz (green) and 66-kHz USV (purple). Rough transport of the pup by the dam specifically enhanced the emission of 40-kHz USV (adapted from Boulanger-Bertolus et al. [4]).

This bimodal distribution of infant USV has been also described by others, and evidence suggests that they are differentially impacted by various interventions, such as breeding for high or low emission of 50-kHz USV in adulthood, chronic modulation of monoamines levels, or maternal presence or withdrawal [4,78,79]. Furthermore, while not explicitly described as the evidence of two categories of USV, the bimodal distribution is also visible in the data of other studies. Finally, several studies reported modulations of the average frequency and duration of USV following various interventions without reporting the distribution of the USV frequencies and durations. Therefore, in these studies, the intervention-induced modulations of the average USV frequency and duration could be the consequence of altered ratios of 40-kHz or 66-kHz USV [77,80].

Overall, the ecological relevance of different infantile USV is still unclear, regardless of the parameters analyzed. While a unification of the standards to analyze infantile USV would seem appealing, every categorization presented here has provided interesting results. However, requesting every study to analyze their USV using all these different categorizations might increase the risk of observing differences by chance (false positive increased by multiple comparisons) and likely reduce the reproducibility of the results, especially considering that USV emission is extremely variable between individuals and litters [81]. An alternative option would be for researchers to share their raw USV recordings along with their conditions of collection in a database akin to *mouseTube* to facilitate later meta-analyses [82].

2.2. Organs and Mechanisms Controlling USV Emission

Audible vocalizations in rats, as with human vocalizations, are the result of vibration of the vocal folds. Such audible vocalizations are usually constituted of a fundamental frequency and multiple harmonics. On the contrary, USV consist of a sound of a single frequency at a time with very little to no harmonics. The first investigation of the mechanisms of production of USV in rats dates back to the work of Laurence H. Robert in the early 1970s, showing that the frequency of USV, but not of audible vocalizations, was dramatically altered when rats were breathing light gases, suggesting a mode of production different from vibrations of the vocal folds [83]. Looking for the anatomical structure responsible for USV production, studies showed that sectioning the nerves innervating the larynx impairs the emission of USV [84–86], and endoscopic observations of the vocal folds during USV emission showed that they do not vibrate but are tightly opposed, leaving only a 1–2 mm opening in the back [87]. Furthermore, measuring laryngeal muscle activity suggested that they control the sound features of the emitted USV by controlling the glottal shape [88,89]. These observations led researchers to suggest that USV are produced through a hole tone, whistle-like mechanism in the larynx akin to a teakettle whistle. A hole tone whistle consists of two holes. The diameter and length of the first hole contribute to controlling the fundamental frequency of the whistle. The air flow is disturbed by passing through the first hole, and as this disturbed airflow passes through the second hole, these instabilities create vortices responsible for the sound wave [90]. For USV emission, researchers suggested that the first hole was formed at the back of the tightly opposed vocal folds, and the second hole constituted of the epiglottis and the basis of the tongue, approximately 1–1.5 mm higher in the vocal track of an adult rat [89]. This hypothesis was largely favored until recently; by studying the excised larynges of mice, Mahrt et al. showed that ultrasonic sounds comparable to USV could still be generated when the epiglottis was removed, therefore occurring without the second constriction, which is inconsistent with this hole tone whistle hypothesis [91]. The authors suggested that USV could instead be produced by an air jet from the glottis impinging on the planar wall, formed by the planar inner laryngeal wall as shown in Figure 2A (right side).

Riede et al. [92] proposed an alternative mechanism in which the exiting jet from the glottis would instead be projected onto the alar cartilage (Figure 2A, left side). In that model, sounds are generated through an edge tone whistle mechanism, with the ventral pouch formed between the vocal folds and the alar edge working as a resonator. Riede et al. [92]

suggested that the impinging jet hypothesis is unlikely, as they suggest, using laryngeal airway reconstruction, that the planar surface necessary for its functioning does not exist in the rat larynx. Interestingly, a very recent study by Håkansson et al. [93] (same laboratory as [91]) confirmed the impinging jet hypothesis in the rat using in vitro larynx physiology and individual-based 3D airway reconstructions with fluid dynamics simulations. They further showed that filling the ventral pouch using aluminum spheres did not prevent the production of USV.

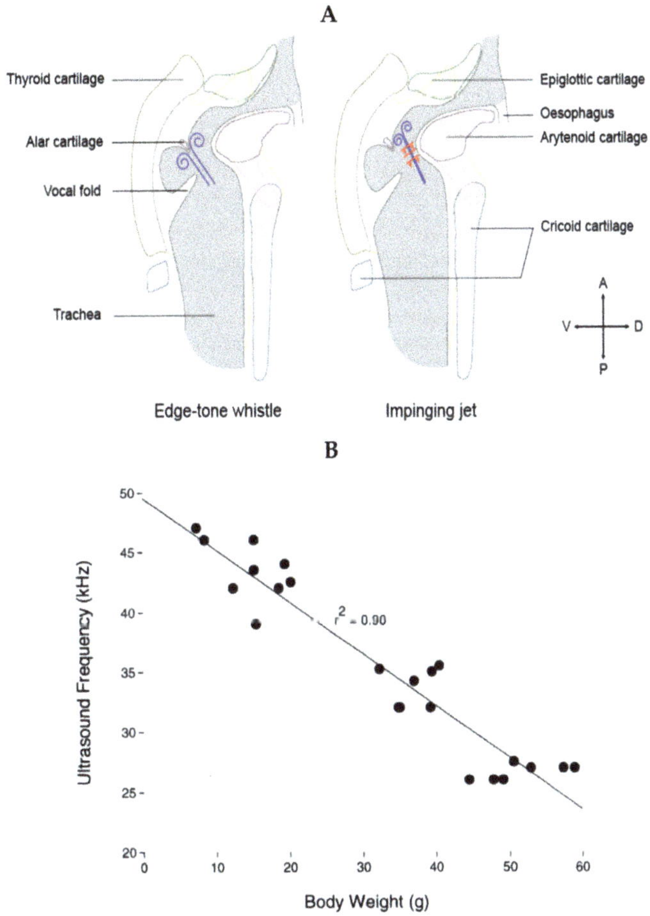

Figure 2. (**A**) Schematic of the proposed mechanisms for USV generation, with the edge tone whistle on the left and the impinging jet on the right. Constriction of the vocal folds leaves a small hole dorsally, through which the expiratory flow is constricted. The glottal exit jet is then either projected onto the alar edge formed by the alar cartilage (left) [92] or impinged against the thyroid wall (right) [91]. In the edge tone whistle mechanism, the ventral pouch would function as a resonator, while in the impinged jet mechanism, the upstream feedback travels back toward the glottis, generating the acoustic wave. Drawing of the larynx created from [3,94,95]. (**B**) In both models, the frequency and modulation of the USV depends on the laryngeal shape, itself determined both by the anatomy of the animal and the activity of the laryngeal muscles controlling the shape of the glottis. Accordingly, as the animal grows, the fundamental frequency of its longest USV lowers. Figure reproduced with permission from [96].

Importantly, in both models, the frequency of the emitted USV is then dependent on the distance between the glottis and the thyroid cartilage; the greater the distance, the lower the frequency of the USV. This is consistent with the observed decrease in the fundamental frequency of the rat's longest USV as it grows (Figure 2B). Furthermore, the USV frequency would also depend on the exit speed of the jet and the diameter of the jet, which the animal can modulate using its respiration and laryngeal muscles to produce frequency-modulated USV, such as trills. Finally, it is important to note that the impinging jet mechanism produces jumps between modal frequencies (usually described as steps in USV research) that are artifactual in USV production. Therefore, the rats might not control the magnitude of the jumps, but could still control their timing by subtle changes in the expiratory flow rate or glottal configuration (Dr. Elemans, personal communication, and [93]). This is particularly important when trying to categorize 50-kHz adult USV or infant USV and highlights the need to better understand the mechanism of USV production.

2.3. *Change in Larynx Size Changes the Sound*

As the rat grows in the first weeks of its life, going from about 6 g at birth to about 150 g at 60 days old, its laryngeal shape changes, and the frequency of the USV it emits changes accordingly (Figure 2B) [96]. Furthermore, these first weeks of life witness dramatic changes in the morphology and physiology of the rat that could all alter their USV production. Rats are born deaf, with their ear canals physically closed and opening at around 14 days of life. This change affects acoustic representation in their cortex, including the representation of ultrasonic frequencies [80,97,98]. The rat also transitions from a milk-based diet to a solid diet, thus undergoing a modification of the activity of the mandibular muscles [99]. The respiration also changes as the rat grows. Its central control switches from being controlled by feedback from the pulmonary branch of the vagal nerve to a central control by the Kölliker-Fuse nuclei in the pons [100]. The respiratory motor control by the phrenic nerve also changes from a short, rapid onset burst to a long-duration discharge with a ramp [101]. This modification in the control of the respiration is associated with modifications to how the respiratory pattern is affected by the rat's emotional state. For example, the respiratory response to an odor that has been associated with a foot shock (i.e., that became aversive) changes from a simple increase of the respiratory frequency to more modulated increases and decreases of the respiratory frequency that reflect the anticipation of a foot shock [102]. All these anatomical and physiological modifications are likely to affect the animal's USV emission. How similar bodily changes affect the production of vocalizations has been investigated in other species, such as the marmoset [103], but a detailed analysis of changes in the mechanism of USV production is still lacking for the rat.

2.4. *USV Emission Is Tighly Associated with the Respiratory Cycle*

As mentioned earlier, USV are emitted during the expiratory phase of the respiratory cycle [12,64,104–106]. In the 22-kHz USV of adult rats, the length of the expiratory phase is correlated with the length of the USV and is significantly longer than the silent expiratory phases [4,12], suggesting that emitting USV lengthens the expiratory phase, and it is possible that the USV duration is limited by the physiological need to breathe [107]. Similarly, longer lower-frequency infantile USV are also correlated in length with the expiratory phase of the respiration [4]. Higher-frequency infantile USV and 50-kHz adult USV are usually much shorter than the expiratory phase in which they are emitted. However, evidence suggests that emission of 50-kHz USV still prolongs the respiratory cycle [2,4]. Importantly, the emission of USV affects the expiratory air flow. During 22-kHz USV, for example, the expiration is characterized by a drastically reduced flow rate [4,12,106]. In addition, 50-kHz USV are also emitted during a low-pressure phase following exhalation onset [2]. In the rest of this article, we gather evidence suggesting that these modifications of the respiratory output associated with USV emission are likely to influence brain function and cognitive processes [108].

3. Respiration Influences Brain Activity and Cognitive Functions

During nasal respiration, odorant molecules enter the nasal cavity and rhythmically stimulate olfactory receptor neurons during inhalation. This rhythmical stimulation drives oscillations time-locked to breathing cycles in the olfactory pathways, as first reported by Lord Adrian [109] and confirmed by many other studies since then [110–115]. Interestingly, olfactory receptor neurons have mechanosensitive properties and respond to changes in pressure caused by the nasal airflow [116], thus allowing entrainment of neural activity in the olfactory pathways in the absence of odor stimulation [117]. In line with these results, Fontanini and Bower suggested that "slow-wave oscillations in the cerebral cortex as a whole, including the neocortex, might be entrained and coordinated by entry of air into the nostrils" [118]. Interestingly, a number of recent papers on rodents have highlighted that, in addition to its impact on olfactory regions, nasal breathing entrains respiration-locked oscillations in several non-olfactory brain areas, such as the whisker barrel cortex [119,120], the hippocampus [121–123], or the prefrontal cortex [124–126]. In a recent study, Tort et al. [127] investigated respiration-coupled oscillations throughout the brains of freely moving mice exhibiting a broad range of respiratory frequencies and found that they could be detected in several neocortical regions, from prefrontal to visual areas and also in subcortical structures such as the thalamus, amygdala, and ventral hippocampus.

Importantly, beside entraining brain oscillatory activity at the respiratory rhythm, nasal respiration also modulates the amplitude of higher frequency oscillations. This was first demonstrated in the olfactory pathways and more specifically in the olfactory bulb, where odorant stimulation was shown to induce prominent oscillatory activity in the beta (10-35 Hz) and gamma (40-80 Hz) ranges, which alternate during a respiratory cycle [128–131]. Recent studies showed that the ability of respiratory rhythm to modulate the amplitude of fast oscillations was also observed in non-olfactory structures. Ito et al. [119] were the first to show in awake, head-fixed mice that the power of the gamma oscillations in the whisker barrel cortex was modulated in phase with breathing, a phenomenon pertaining to phase–amplitude coupling. Biskamp et al. [124] extended this finding by showing that gamma activity in the prefrontal cortex, a key associational brain region, was paced by the respiration cycle. Zhong et al. [125] further documented that respiratory rhythm modulates gamma activity in a region- and state-specific manner.

Respiration-locked oscillations and respiration-locked modulations of gamma power also occur in humans. Indeed, Herrero et al. [132], using direct intracranial recordings in humans, correlated neuronal activity with the breathing cycle and showed that the recorded signal tracked the breathing cycle across a widespread network of cortical and limbic structures. More recently, Kluger and Gross [133] used magnetoencephalography (MEG) to assess the potential influence of the respiration depth and respiration phase on the human motor system. They found coherence within the beta band to be cyclically modulated by the respiration phase. Zelano et al. [115], performing intracranial recordings from the piriform cortex, amygdala, and hippocampus in chronically implanted epileptic patients, showed that the power of oscillatory activity in different frequency bands was modulated by respiration. Importantly, this modulation was dependent on the nasal airflow, because it disappeared when patients were breathing through the mouth and no longer through the nose.

These findings suggest that the breathing rhythm, such as slow oscillatory rhythms (e.g., the theta rhythm), could help coordinate neural activity across distant brain regions by supporting the formation and synchronization of co-active cell assemblies [134,135]. In both animals and humans, oscillatory activity in the gamma frequency range has been involved in several cognitive functions, among which are attention [136–138], sensory perception [139–141], and short-term and long-term memory [138,142–146]. Therefore, the respiratory rhythm, through its modulation of gamma oscillatory activity, is in a good position to modulate cognitive processes.

A few recent studies have addressed this question and highlighted that some cognitive functions are modulated by the phase of the respiratory cycle. For example, Perl et al. [147] reported that participants spontaneously inhaled at non-olfactory cognitive task onset, and this resulted in improved performance accuracy in a visuospatial task. Nakamura et al. [148] used a delayed matching-to-sample visual recognition task where a test cue phase-locked to the respiratory cycle was given and showed that subjects exhibited increased reaction time and reduced accuracy when their retrieval processes encompassed the expiration to inspiration transition. Zelano et al. [115] showed that the respiratory phase has a significant influence on emotion discrimination and recognition memory. Indeed, subjects exhibited higher performances in recognizing fearful expressions and retrieving visual object memories, when target stimuli were presented during nasal inspiration rather than during expiration. In addition, the authors reported that cognitive performance significantly declined when the subjects were breathing through the mouth instead of the nose, suggesting that the route of breathing played a critical role. Similar observations were made by Arshamian et al. [149], who examined the effect of respiration on the consolidation of episodic odor memory and showed that breathing through the nose, compared with the mouth, during consolidation enhanced recognition memory, suggesting that nasal respiration is important during the critical period where memories are reactivated and strengthened.

Overall, the respiratory rhythm is transmitted from the nose to the brain through the mechanical stimulation of the olfactory neuroreceptors by the nasal airflow. In the brain, this rhythm both entrains respiration-locked oscillations and modulates higher-frequency oscillations, such as gamma oscillations, in a wide network of structures. A growing body of data suggest that through its action on brain oscillatory activity, respiration might be able to modulate some cognitive functions. Since, as described above, USV emission alters the respiratory rhythm, one can wonder if it can also modulate brain activity.

4. USV Emission Impacts Brain Activity

As described above, in aversive situations, such as exposure to a predator or foot shock, rats emit 22-kHz USV [69]. In the fear-conditioning paradigm, the most studied index of fear response in rats is freezing. Recent studies have shown that the expression of freezing temporally coincides with the development of sustained 4-Hz oscillations in the prefrontal–amygdala circuits, which organize the spiking activity of local neuronal populations [150,151]. Importantly, this slow oscillation is distinct from the theta rhythm and predicts the onset and offset of freezing. Interestingly, recent work has shown that freezing-related 4-Hz oscillation in the median prefrontal cortex (mPFC) is correlated with the animal's respiratory rate, and disruption of the olfactory inputs to the mPFC significantly reduces the 4-Hz oscillation in this structure but leads to prolonged freezing periods [126]. These results indicate that the olfactory inputs can modulate rhythmic activity in the PFC and freezing behavior.

While the neural circuit involved in USV production [69,152–154] and the correlates of USV perception in the brain of conspecific receivers [155–157] are well-documented, the effect of USV production on the sender animal's brain activity has been largely overlooked. Yet, such information is needed to better understand how the different components of fear response collectively modulate a rat's brain neural dynamics. Importantly, as mentioned above, 22-kHz USV emission drastically slows down the animal's respiratory rate [4,12,106], potentially disrupting the respiration-related brain rhythm described above.

In a recent study [158], we investigated the consequences of USV emission on brain oscillatory activities in the fear neural network of the vocalizing animal and assessed to what extent these changes were related to changes in the breathing rhythm (Figure 3).

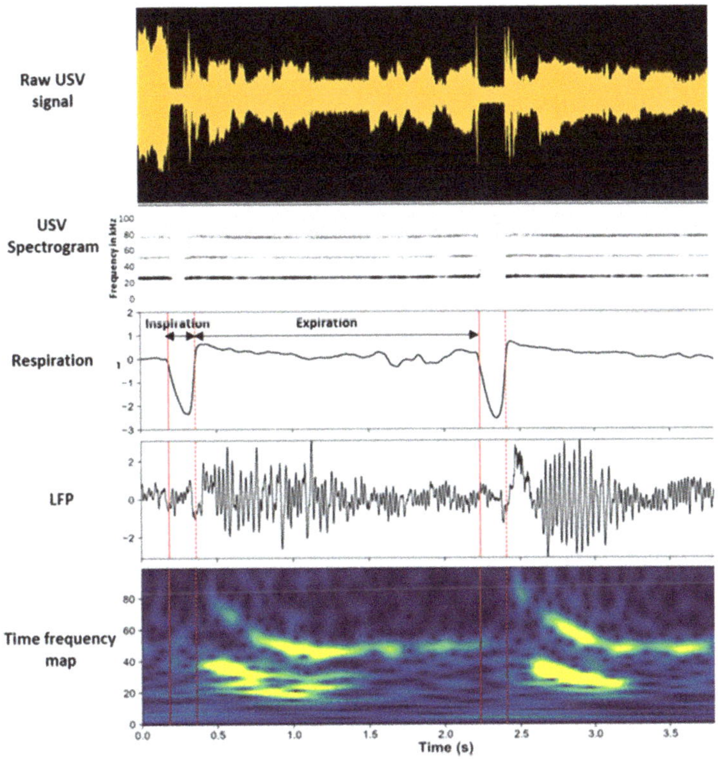

Figure 3. Modulation of beta and gamma oscillation power by the phase of the respiratory cycle. Individual traces represent, from the top, the raw USV calls, USV spectrogram, respiratory signal, raw local field potential (LFP) signal recorded in the piriform cortex (PIR), and its time frequency map (y-axis: LFP signal frequency in Hz; x-axis: time in milliseconds). Adapted from [158], visual abstract.

To address these questions, we trained rats in an odor fear conditioning paradigm, in which an odor signaled the arrival of a foot shock a few seconds later. Fear conditioning is a widely used task in the literature of fear memories in animals, and this paradigm readily induces USV in rats. Training was conducted in an experimental cage allowing the monitoring of ultrasonic vocalizations, overt behavior, and respiration. During training, we collected local field potentials reflecting the activity of populations of neurons in the basolateral amygdala (BLA), the mPFC, and the olfactory piriform cortex (PIR). We compared the brain oscillatory activity power in different frequency bands during sequences of USV calls versus sequences without USV (silent sequences). We showed that during USV emission, the activity power in the delta (0–5 Hz), beta (15–40 Hz), and gamma (40–80 Hz) bands increased in the recorded network, while the theta (5–15 Hz) activity power decreased (Figure 4A).

We then assessed the relationship between the frequency of oscillatory activity in the delta and theta ranges and breathing frequency (which varied between 0.5 and 10 Hz), as well as the impact of USV emission on this coordination. We showed that during silent respiratory sequences, delta oscillatory activity was coupled with the breathing rhythm in the recorded network. This coupling faded away during USV (Figure 4B).

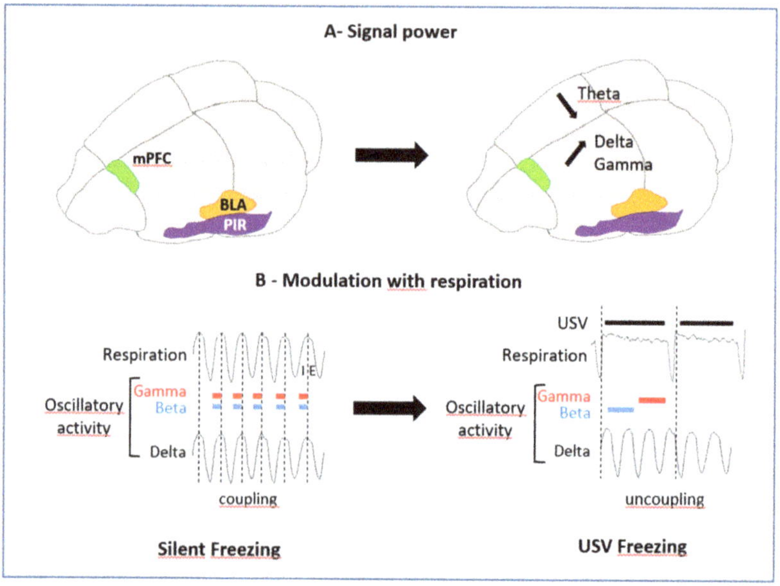

Figure 4. Effect of USV emission on oscillatory activity power and respiration. (**A**) Signal power. USV emission coincides with a decrease in theta power and an increase in delta and gamma power. (**B**) Modulation with respiration. During silent freezing, the delta activity frequency covaries with the nasal respiratory frequency. In addition, power in the beta and gamma bands is modulated in phase with respiration, with a higher beta and gamma activity power during inspiration (I) than expiration (E). During USV emission, a deep slow-down of respiratory frequency is observed, with the uncoupling between the delta frequency and respiratory frequency. Furthermore, a reorganization of beta and gamma activity power during the respiratory cycle occurs, with increased power during the first half of the expiration phase and increased gamma power during the second half of expiration (adapted from [108], Figure 10). mPFC: medial prefrontal cortex; BLA: basolateral amygdala; and PIR: piriform cortex.

Finally, we assessed whether the amplitudes of the beta and gamma oscillations were modulated by the phase of the respiratory cycle (inspiration versus expiration). We showed that during silent sequences, the beta and gamma activity power was strongly modulated by the phase of the respiratory cycle. The emission of USV was associated with drastic changes in the course of time of this modulation (Figure 4B).

We proposed that the deep slow-down of the respiratory rate added to the reduction of airflow through the nose during USV calls [2,88] might be responsible for the loss of coupling between the nasal rhythm and delta oscillation. This would result in an increase in the brain's delta oscillations power and an enhancement of the beta and gamma activity power during exhalation. We suggested that the window of a USV call, with its associated changes in nasal airflow, triggers a specific combination of brain oscillatory activities that might enhance plasticity at critical nodes of the network and ultimately strengthen long-term fear memories. Interestingly, we showed that the amount of ultrasonic vocalization during training was a good predictor of the animals' learned fear response, measured 24 h later. The higher the number of ultrasonic vocalizations in training, the stronger the learned fear response. Hence, USV calls might result in a differential gating of information within the fear neural network, thus potentially modulating later fear memory and expression.

It is important to point out that while we described important differences in the emission of USV between infancy and adulthood, the consequences of USV emission on the infant brain remains entirely unknown at this point. Aversive events that induce USV have drastically different consequences at these two stages of development, and their memories also differentially affected the animal's subsequent behavior [159,160]. The 22-kHz rat vocalizations present an evolutionary counterpart to human crying [161], and human infant crying and rat pup USV have been suggested to share some similarities [72]. Understanding the influence of USV, and therefore possibly cries, on the memorization of aversive events and brain processes in general could have important consequences for clinical care management of neonates.

5. Conclusions

Vocalization requires precise coordination of phonation, articulation, and respiration and involves a wide neural network spanning from the forebrain to the brainstem [69,162]. The present review goes through the acoustic characteristics and mechanisms of production of USV both in infant and adult rats and emphasizes the tight relationships existing between USV emission and respiration. It also provides new insights on how USV and respiratory rhythm collectively influence coordinated brain activity within the neural networks underlying defensive and emotional states. Better knowledge of the impact of ultrasonic vocalizations on brain neural dynamics is particularly relevant for rodent models of human neuropsychiatric disorders, for which socio-affective communication is severely impaired [163,164].

Author Contributions: J.B.-B. and A.-M.M. contributed equally to the writing of the manuscript and agreed to its published version. All authors have read and agreed to the published version of the manuscript.

Funding: J.B.-B. was funded in majority by the Department of Anesthesiology at the University of Michigan during the writing of this article. Additional funds were provided by Pôle Emploi. A.-M.M. was funded by the Centre National de la Recherche Scientifique.

Institutional Review Board Statement: Not applicable.

Informed Consent Statement: Not applicable.

Data Availability Statement: Not applicable.

Conflicts of Interest: The authors declare no conflict of interest.

References

1. Branchi, I.; Santucci, D.; Alleva, E. Ultrasonic vocalisation emitted by infant rodents: A tool for assessment of neurobehavioural development. *Behav. Brain Res.* **2001**, *125*, 49–56. [CrossRef]
2. Sirotin, Y.B.; Costa, M.E.; Laplagne, D.A. Rodent ultrasonic vocalizations are bound to active sniffing behavior. *Front. Behav. Neurosci.* **2014**, *8*, 399. [CrossRef] [PubMed]
3. Borgard, H.L.; Baab, K.; Pasch, B.; Riede, T. The Shape of Sound: A Geometric Morphometrics Approach to Laryngeal Functional Morphology. *J. Mamm. Evol.* **2019**, *27*, 577–590. [CrossRef]
4. Boulanger-Bertolus, J.; Rincón-Cortés, M.; Sullivan, R.M.; Mouly, A.-M. Understanding pup affective state through ethologically significant ultrasonic vocalization frequency. *Sci. Rep.* **2017**, *7*, 1–12. [CrossRef]
5. Lore, R.; Flannelly, K.; Farina, P. Ultrasounds produced by rats accompany decreases in intraspecific fighting. *Aggress. Behav.* **1976**, *2*, 175–181. [CrossRef]
6. Miczek, K.A.; Van Der Poel, A.M. Long Ultrasonic Calls in Male Rats Following Mating, Defeat and Aversive Stimulation: Frequency Modulation and Bout Structure. *Behaviour* **1991**, *119*, 127–142. [CrossRef]
7. Francis, R.L. 22-kHz calls by isolated rats. *Nature* **1977**, *265*, 236–238. [CrossRef]
8. Blanchard, R.; Blanchard, D.; Agullana, R.; Weiss, S.M. Twenty-two kHz alarm cries to presentation of a predator, by laboratory rats living in visible burrow systems. *Physiol. Behav.* **1991**, *50*, 967–972. [CrossRef]
9. Brudzynski, S.M.; Ociepa, D. Ultrasonic vocalization of laboratory rats in response to handling and touch. *Physiol. Behav.* **1992**, *52*, 655–660. [CrossRef]
10. Kaltwasser, M.T. Startle-inducing acoustic stimuli evoke ultrasonic vocalization in the rat. *Physiol. Behav.* **1990**, *48*, 13–17. [CrossRef]
11. Wöhr, M.; Borta, A.; Schwarting, R.K. Overt behavior and ultrasonic vocalization in a fear conditioning paradigm: A dose–response study in the rat. *Neurobiol. Learn. Mem.* **2005**, *84*, 228–240. [CrossRef]
12. Hegoburu, C.; Shionoya, K.; Garcia, S.; Messaoudi, B.; Thevenet, M.; Mouly, A.-M. The RUB Cage: Respiration–Ultrasonic Vocalizations–Behavior Acquisition Setup for Assessing Emotional Memory in Rats. *Front. Behav. Neurosci.* **2011**, *5*, 25. [CrossRef]
13. Burgdorf, J.; Knutson, B.; Panksepp, J. Anticipation of rewarding electrical brain stimulation evokes ultrasonic vocalization in rats. *Behav. Neurosci.* **2000**, *114*, 320–327. [CrossRef]

14. Covington, H.E.; Miczek, K.A. Vocalizations during withdrawal from opiates and cocaine: Possible expressions of affective distress. *Eur. J. Pharmacol.* **2003**, *467*, 1–13. [CrossRef]
15. Vivian, J.A.; Miczek, K.A. Ultrasounds during morphine withdrawal in rats. *Psychopharmacology* **1991**, *104*, 187–193. [CrossRef]
16. Vivian, J.A.; Miczek, K.A. Diazepam and gepirone selectively attenuate either 20–32 or 32–64 kHz ultrasonic vocalizations during aggressive encounters. *Psychopharmacology* **1993**, *112*, 66–73. [CrossRef]
17. Barfield, R.J.; Auerbach, P.; Geyer, L.A.; McIntosh, T.K. Ultrasonic Vocalizations in Rat Sexual Behavior. *Am. Zoöl.* **1979**, *19*, 469–480. [CrossRef]
18. Cuomo, V.; Cagiano, R.; De Salvia, M.A.; Maselli, M.; Renna, G.; Racagni, G. Ultrasonic vocalization in response to unavoidable aversive stimuli in rats: Effects of benzodiazepines. *Life Sci.* **1988**, *43*, 485–491. [CrossRef]
19. Jelen, P.; Soltysik, S.; Zagrodzka, J. 22-kHz ultrasonic vocalization in rats as an index of anxiety but not fear: Behavioral and pharmacological modulation of affective state. *Behav. Brain Res.* **2003**, *141*, 63–72. [CrossRef]
20. Brudzynski, S.M.; Bihari, F.; Ociepa, D.; Fu, X.-W. Analysis of 22 kHz ultrasonic vocalization in laboratory rats: Long and short calls. *Physiol. Behav.* **1993**, *54*, 215–221. [CrossRef]
21. Brudzynski, S.M. Pharmacology of Ultrasonic Vocalizations in adult Rats: Significance, Call Classification and Neural Substrate. *Curr. Neuropharmacol.* **2015**, *13*, 180–192. [CrossRef] [PubMed]
22. Burgdorf, J.; Panksepp, J. Tickling induces reward in adolescent rats. *Physiol. Behav.* **2001**, *72*, 167–173. [CrossRef]
23. Burgdorf, J.; Kroes, R.A.; Moskal, J.R.; Pfaus, J.G.; Brudzynski, S.M.; Panksepp, J. Ultrasonic vocalizations of rats (Rattus norvegicus) during mating, play, and aggression: Behavioral concomitants, relationship to reward, and self-administration of playback. *J. Comp. Psychol.* **2008**, *122*, 357–367. [CrossRef] [PubMed]
24. Knutson, B.; Burgdorf, J.; Panksepp, J. Anticipation of play elicits high-frequency ultrasonic vocalizations in young rats. *J. Comp. Psychol.* **1998**, *112*, 65–73. [CrossRef] [PubMed]
25. Panksepp, J.; Burgdorf, J. 50-kHz chirping (laughter?) in response to conditioned and unconditioned tickle-induced reward in rats: Effects of social housing and genetic variables. *Behav. Brain Res.* **2000**, *115*, 25–38. [CrossRef]
26. Sewell, G.D. Ultrasound in Adult Rodents. *Nature* **1967**, *215*, 512. [CrossRef]
27. Mcintosh, T.K.; Barfield, R.J.; Geyer, L.A. Ultrasonic vocalisations facilitate sexual behaviour of female rats. *Nature* **1978**, *272*, 163–164. [CrossRef]
28. McIntosh, T.K.; Barfield, R.J. The temporal patterning of 40–60 kHz ultrasonic vocalizations and copulation in the rat (Rattus norvegicus). *Behav. Neural Biol.* **1980**, *29*, 349–358. [CrossRef]
29. White, N.R.; Barfield, R.J. Effects of male pre-ejaculatory vocalizations on female receptive behavior in the rat (Rattus norvegicus). *J. Comp. Psychol.* **1990**, *104*, 140–146. [CrossRef]
30. White, N.R.; Cagiano, R.; Moises, A.U.; Barfield, R.J. Changes in mating vocalizations over the ejaculatory series in rats (Rattus norvegicus). *J. Comp. Psychol.* **1990**, *104*, 255–262. [CrossRef]
31. Buck, C.L.; Vendruscolo, L.F.; Koob, G.F.; George, O. Dopamine D1 and μ-opioid receptor antagonism blocks anticipatory 50 kHz ultrasonic vocalizations induced by palatable food cues in Wistar rats. *Psychopharmacology* **2013**, *231*, 929–937. [CrossRef]
32. Takahashi, N.; Kashino, M.; Hironaka, N. Structure of Rat Ultrasonic Vocalizations and Its Relevance to Behavior. *PLoS ONE* **2010**, *5*, e14115. [CrossRef]
33. Knutson, B.; Burgdorf, J.; Panksepp, J. High-Frequency Ultrasonic Vocalizations Index Conditioned Pharmacological Reward in Rats. *Physiol. Behav.* **1999**, *66*, 639–643. [CrossRef]
34. Knutson, B.; Burgdorf, J.; Panksepp, J. Ultrasonic vocalizations as indices of affective states in rats. *Psychol. Bull.* **2002**, *128*, 961–977. [CrossRef]
35. Panksepp, J.; Burgdorf, J. Laughing Rats? Playful Tickling Arouses High Frequency Ultrasonic Chirping in Young Rodents. In *Toward a Science of Consciousness III*; Hameroff, S.R., Kaszniak, A.W., Chalmers, D.J., Eds.; MIT Press: Cambridge, MA, USA, 1999; pp. 231–244.
36. Perez-Sepulveda, J.; Flagel, S.; Garcia-Fuster, M.; Slusky, R.; Aldridge, J.; Watson, S.; Akil, H. Differential impact of a complex environment on positive affect in an animal model of individual differences in emotionality. *Neuroscience* **2013**, *248*, 436–447. [CrossRef]
37. Brudzynski, S.M.; Pniak, A. Social contacts and production of 50-kHz short ultrasonic calls in adult rats. *J. Comp. Psychol.* **2002**, *116*, 73–82. [CrossRef]
38. Wintink, A.J.; Brudzynski, S.M. The related roles of dopamine and glutamate in the initiation of 50-kHz ultrasonic calls in adult rats. *Pharmacol. Biochem. Behav.* **2001**, *70*, 317–323. [CrossRef]
39. Wöhr, M.; Houx, B.; Schwarting, R.K.; Spruijt, B. Effects of experience and context on 50-kHz vocalizations in rats. *Physiol. Behav.* **2008**, *93*, 766–776. [CrossRef]
40. Sales, G.D. Ultrasound and aggressive behaviour in rats and other small mammals. *Anim. Behav.* **1972**, *20*, 88–100. [CrossRef]
41. Haney, M.; Miczek, K.A. Ultrasounds emitted by female rats during agonistic interactions: Effects of morphine and naltrexone. *Psychopharmacology* **1994**, *114*, 441–448. [CrossRef]
42. Thomas, D.A.; Takahashi, L.K.; Barfield, R.J. Analysis of ultrasonic vocalizations emitted by intruders during aggressive encounters among rats (Rattus norvegicus). *J. Comp. Psychol.* **1983**, *97*, 201–206. [CrossRef]
43. Simola, N. Rat Ultrasonic Vocalizations and Behavioral Neuropharmacology: From the Screening of Drugs to the Study of Disease. *Curr. Neuropharmacol.* **2015**, *13*, 164–179. [CrossRef]

44. Wright, J.M.; Gourdon, J.C.; Clarke, P.B.S. Identification of multiple call categories within the rich repertoire of adult rat 50-kHz ultrasonic vocalizations: Effects of amphetamine and social context. *Psychopharmacology* **2010**, *211*, 1–13. [CrossRef]
45. Wright, J.M.; Dobosiewicz, M.R.S.; Clarke, P.B.S. α- and β-Adrenergic Receptors Differentially Modulate the Emission of Spontaneous and Amphetamine-Induced 50-kHz Ultrasonic Vocalizations in Adult Rats. *Neuropsychopharmacology* **2011**, *37*, 808–821. [CrossRef]
46. Garcia, E.J.; McCowan, T.J.; Cain, M.E. Harmonic and frequency modulated ultrasonic vocalizations reveal differences in conditioned and unconditioned reward processing. *Behav. Brain Res.* **2015**, *287*, 207–214. [CrossRef]
47. Noirot, E. Ultrasounds in young rodents. II. Changes with age in albino rats. *Anim. Behav.* **1968**, *16*, 129–134. [CrossRef]
48. Hofer, M.A. Multiple regulators of ultrasonic vocalization in the infant rat. *Psychoneuroendocrinology* **1996**, *21*, 203–217. [CrossRef]
49. Allin, J.T.; Banks, E.M. Effects of temperature on ultrasound production by infant albino rats. *Dev. Psychobiol.* **1971**, *4*, 149–156. [CrossRef] [PubMed]
50. Amsel, A.; Radek, C.C.; Graham, M.; Letz, R. Ultrasound emission in infant rats as an indicant of arousal during appetitive learning and extinction. *Science* **1977**, *197*, 786–788. [CrossRef]
51. Blumberg, M.S.; Efimova, I.V.; Alberts, J.R. Ultrasonic vocalizations by rat pups: The primary importance of ambient temperature and the thermal significance of contact comfort. *Dev. Psychobiol.* **1992**, *25*, 229–250. [CrossRef]
52. Gardner, C. Distress vocalization in rat pups a simple screening method for anxiolytic drugs. *J. Pharmacol. Methods* **1985**, *14*, 181–187. [CrossRef]
53. Hodgson, R.; Guthrie, D.; Varty, G. Duration of ultrasonic vocalizations in the isolated rat pup as a behavioral measure: Sensitivity to anxiolytic and antidepressant drugs. *Pharmacol. Biochem. Behav.* **2008**, *88*, 341–348. [CrossRef]
54. Hofer, M.A.; Shair, H. Ultrasonic vocalization during social interaction and isolation in 2-week-old rats. *Dev. Psychobiol.* **1978**, *11*, 495–504. [CrossRef]
55. Ise, S.; Ohta, H. Power spectrum analysis of ultrasonic vocalization elicited by maternal separation in rat pups. *Brain Res.* **2009**, *1283*, 58–64. [CrossRef]
56. Oswalt, G.L.; Meier, G.W. Olfactory, thermal, and tactual influences on infantile ultrasonic vocalization in rats. *Dev. Psychobiol.* **1975**, *8*, 129–135. [CrossRef]
57. Shair, H.N.; Smith, J.A.; Welch, M.G. Cutting the vagus nerve below the diaphragm prevents maternal potentiation of infant rat vocalization. *Dev. Psychobiol.* **2011**, *54*, 70–76. [CrossRef]
58. Takahashi, L.K.; Turner, J.G.; Kalin, N.H. Development of stress-induced responses in preweanling rats. *Dev. Psychobiol.* **1991**, *24*, 341–360. [CrossRef]
59. Weber, M.; Paxinos, G.; Richardson, R. Conditioned changes in ultrasonic vocalizations to an aversive olfactory stimulus are lateralized in 6-day-old rats. *Dev. Psychobiol.* **2000**, *37*, 121–128. [CrossRef]
60. Brudzynski, S.M.; Kehoe, P.; Callahan, M. Sonographic structure of isolation-induced ultrasonic calls of rat pups. *Dev. Psychobiol.* **1999**, *34*, 195–204. [CrossRef]
61. Okon, E.E. The effect of environmental temperature on the production of ultrasounds by isolated non handled albino mouse pups. *J. Zool.* **1970**, *162*, 71–83. [CrossRef]
62. Okon, E.E. The temperature relations of vocalization in infant Golden hamsters and Wistar rats. *J. Zool.* **1971**, *164*, 227–237. [CrossRef]
63. Insel, T.R.; Hill, J.L.; Mayor, R.B. Rat pup ultrasonic isolation calls: Possible mediation by the benzodiazepine receptor complex. *Pharmacol. Biochem. Behav.* **1986**, *24*, 1263–1267. [CrossRef]
64. Hofer, M.A.; Shair, H.N. Ultrasonic vocalization by rat pups during recovery from deep hypothermia. *Dev. Psychobiol.* **1992**, *25*, 511–528. [CrossRef] [PubMed]
65. Hofer, M.A.; Brunelli, S.A.; Shair, H.N. Ultrasonic vocalization responses of rat pups to acute separation and contact comfort do not depend on maternal thermal cues. *Dev. Psychobiol.* **1993**, *26*, 81–95. [CrossRef]
66. Nelson, E.E.; Alberts, J.R. Gastric saline infusion reduces ultrasonic vocalizations and brown fat activity in suckling rat pups. *Dev. Psychobiol.* **2002**, *40*, 160–167. [CrossRef]
67. Hofer, M.A.; Shair, H.N.; Brunelli, S.A. Ultrasonic Vocalizations in Rat and Mouse Pups. In *Current Protocols in Neuroscience*; John Wiley & Sons, Inc.: Hoboken, NJ, USA, 2001; ISBN 978-0-471-14230-0.
68. Hofer, M.A.; Masmela, J.R.; Brunelli, S.A.; Shair, H.N. The ontogeny of maternal potentiation of the infant rats' isolation call. *Dev. Psychobiol.* **1998**, *33*, 189–201. [CrossRef]
69. Schwarting, R.; Wöhr, M. On the relationships between ultrasonic calling and anxiety-related behavior in rats. *Braz. J. Med Biol. Res.* **2012**, *45*, 337–348. [CrossRef]
70. Kraebel, K.S.; Brasser, S.M.; Campbell, J.O.; Spear, L.P.; Spear, N.E. Developmental differences in temporal patterns and potentiation of isolation-induced ultrasonic vocalizations: Influence of temperature variables. *Dev. Psychobiol.* **2002**, *40*, 147–159. [CrossRef]
71. Tonkiss, J.; Bonnie, K.E.; Hudson, J.L.; Shultz, P.L.; Durán, P.; Galler, J.R. Ultrasonic call characteristics of rat pups are altered following prenatal malnutrition. *Dev. Psychobiol.* **2003**, *43*, 90–101. [CrossRef]
72. Zeskind, P.S.; McMurray, M.S.; Garber, K.A.; Neuspiel, J.M.; Cox, E.T.; Grewen, K.M.; Mayes, L.C.; Johns, J.M. Development of Translational Methods in Spectral Analysis of Human Infant Crying and Rat Pup Ultrasonic Vocalizations for Early Neurobehavioral Assessment. *Front. Psychiatry* **2011**, *2*, 56. [CrossRef]

73. Gulia, K.K.; Patel, N.; Radhakrishnan, A.; Kumar, V.M. Reduction in Ultrasonic Vocalizations in Pups Born to Rapid Eye Movement Sleep Restricted Mothers in Rat Model. *PLoS ONE* **2014**, *9*, e84948. [CrossRef]
74. Hashimoto, H.; Moritani, N.; Katou, M.; Nishiya, T.; Kromkhun, P.; Yokosuka, M.; Tanaka, M.; Saito, T.R. Ontogenetic Changes of Ultrasonic Vocalizations Emitted from Infant Rats. *Exp. Anim.* **2007**, *56*, 315–318. [CrossRef]
75. Saucier, D.M.; Ehresman, C.A.; Keller, A.J.; Armstrong, E.; Elderkin, A.; Yager, J.Y. Hypoxia ischemia affects ultrasonic vocalization in the neonatal rat. *Behav. Brain Res.* **2008**, *190*, 243–247. [CrossRef]
76. Kehoe, P.; Callahan, M.; Daigle, A.; Mallinson, K.; Brudzynski, S. The Effect of Cholinergic Stimulation on Rat Pup Ultra-sonic Vocalizations. *Dev. Psychobiol.* **2001**, *38*, 92–100. [CrossRef]
77. Potasiewicz, A.; Gzielo, K.; Popik, P.; Nikiforuk, A. Effects of prenatal exposure to valproic acid or poly(I:C) on ultrasonic vocalizations in rat pups: The role of social cues. *Physiol. Behav.* **2020**, *225*, 113113. [CrossRef]
78. Harmon, K.; Cromwell, H.; Burgdorf, J.; Moskal, J.; Brudzynski, S.; Kroes, R.; Panksepp, J. Rats selectively bred for low levels of 50 kHz ultrasonic vocalizations exhibit alterations in early social motivation. *Dev. Psychobiol.* **2008**, *50*, 322–331. [CrossRef]
79. Blazevic, S.; Merkler, M.; Persic, D.; Hranilovic, D. Chronic postnatal monoamine oxidase inhibition affects affiliative behavior in rat pups. *Pharmacol. Biochem. Behav.* **2017**, *153*, 60–68. [CrossRef]
80. Kim, H.; Bao, S. Experience-dependent overrepresentation of ultrasonic vocalization frequencies in the rat primary auditory cortex. *J. Neurophysiol.* **2013**, *110*, 1087–1096. [CrossRef]
81. Rieger, M.A.; Dougherty, J.D. Analysis of within Subjects Variability in Mouse Ultrasonic Vocalization: Pups Exhibit Inconsistent, State-Like Patterns of Call Production. *Front. Behav. Neurosci.* **2016**, *10*, 182. [CrossRef]
82. Torquet, N.; De Chaumont, F.; Faure, P.; Bourgeron, T.; Ey, E. mouseTube—A database to collaboratively unravel mouse ultrasonic communication. *F1000Research* **2016**, *5*, 2332. [CrossRef]
83. Roberts, L. The rodent ultrasound production mechanism. *Ultrasonics* **1975**, *13*, 83–88. [CrossRef]
84. Roberts, L.H. Evidence for the laryngeal source of ultrasonic and audible cries of rodents. *J. Zool.* **1975**, *175*, 243–257. [CrossRef]
85. Nunez, A.A.; Pomerantz, S.M.; Bean, N.; Youngstrom, T.G. Effects of laryngeal denervation on ultrasound production and male sexual behavior in rodents. *Physiol. Behav.* **1985**, *34*, 901–905. [CrossRef]
86. Wetzel, D.M.; Kelley, D.B.; Campbell, B.A. Central control of ultrasonic vocalizations in neonatal rats: I. Brain stem motor nuclei. *J. Comp. Physiol. Psychol.* **1980**, *94*, 596–605. [CrossRef]
87. Sanders, I.; Weisz, D.J.; Yang, B.Y.; Fung, K.; Amirali, A. The Mechanism of Ultrasonic Vocalisation in the Rat. *Soc. Neurosci. Abstr.* **2001**, *27*, 88.19.
88. Riede, T. Subglottal pressure, tracheal airflow, and intrinsic laryngeal muscle activity during rat ultrasound vocalization. *J. Neurophysiol.* **2011**, *106*, 2580–2592. [CrossRef]
89. Riede, T. Stereotypic Laryngeal and Respiratory Motor Patterns Generate Different Call Types in Rat Ultrasound Vocalization. *J. Exp. Zool. Part A Ecol. Genet. Physiol.* **2013**, *319*, 213–224. [CrossRef]
90. Henrywood, R.H.; Agarwal, A. The aeroacoustics of a steam kettle. *Phys. Fluids* **2013**, *25*, 107101. [CrossRef]
91. Mahrt, E.; Agarwal, A.; Perkel, D.; Portfors, C.; Elemans, C.P. Mice produce ultrasonic vocalizations by intra-laryngeal planar impinging jets. *Curr. Biol.* **2016**, *26*, R880–R881. [CrossRef]
92. Riede, T.; Borgard, H.L.; Pasch, B. Laryngeal airway reconstruction indicates that rodent ultrasonic vocalizations are produced by an edge-tone mechanism. *R. Soc. Open Sci.* **2017**, *4*, 170976. [CrossRef]
93. Håkansson, J.; Jiang, W.; Xue, Q.; Zheng, X.; Ding, M.; Agarwal, A.A.; Elemans, C.P.H. Aerodynamics and Motor Control of Ultrasonic Vocalizations for Social Communication in Mice and Rats. *bioRxiv* **2021**. [CrossRef]
94. Roberts, L.H. The functional anatomy of the rodent larynx in relation to audible and ultrasonic cry production. *Zool. J. Linn. Soc.* **1975**, *56*, 255–264. [CrossRef]
95. Alli, O.; Berzofsky, C.; Sharma, S.; Pitman, M.J. Development of the rat larynx: A histological study. *Laryngoscope* **2013**, *123*, 3093–3098. [CrossRef] [PubMed]
96. Blumberg, M.S.; Sokoloff, G.; Kent, K.J. A Developmental Analysis of Clonidine's Effects on Cardiac Rate and Ultrasound Production in Infant Rats. *Dev. Psychobiol.* **2000**, *36*, 186–193. [CrossRef]
97. Mowery, T.M.; Kotak, V.C.; Sanes, D.H. Transient Hearing Loss Within a Critical Period Causes Persistent Changes to Cellular Properties in Adult Auditory Cortex. *Cereb. Cortex* **2014**, *25*, 2083–2094. [CrossRef]
98. De Villers-Sidani, E.; Chang, E.F.; Bao, S.; Merzenich, M.M. Critical Period Window for Spectral Tuning Defined in the Primary Auditory Cortex (A1) in the Rat. *J. Neurosci.* **2007**, *27*, 180–189. [CrossRef]
99. Westneat, M.W.; Hal, W.G. Ontogeny of feeding motor patterns in infant rats: An electromyographic analysis of suckling and chewing. *Behav. Neurosci.* **1992**, *106*, 539–554. [CrossRef]
100. Dutschmann, M.; Mörschel, M.; Rybak, I.A.; Dick, T.E. Learning to breathe: Control of the inspiratory-expiratory phase transition shifts from sensory- to central-dominated during postnatal development in rats. *J. Physiol.* **2009**, *587*, 4931–4948. [CrossRef]
101. Paton, J.F.R.; Richter, D.W. Maturational changes in the respiratory rhythm generator of the mouse. *Pflügers Arch. Eur. J. Physiol.* **1995**, *430*, 115–124. [CrossRef]
102. Boulanger Bertolus, J.; Hegoburu, C.; Ahers, J.L.; Londen, E.; Rousselot, J.; Szyba, K.; Thévenet, M.; Sullivan-Wilson, T.A.; Doyère, V.; Sullivan, R.M.; et al. Infant rats can learn time intervals before the maturation of the striatum: Evidence from odor fear conditioning. *Front. Behav. Neurosci.* **2014**, *8*, 176. [CrossRef]

103. Zhang, Y.S.; Takahashi, D.Y.; Liao, D.A.; Ghazanfar, A.A.; Elemans, C.P.H. Vocal state change through laryngeal development. *Nat. Commun.* **2019**, *10*, 1–12. [CrossRef]
104. Anderson, J.W. The Production of Ultrasonic Sounds by Laboratory Rats and Other Mammals. *Science* **1954**, *119*, 808–809. [CrossRef]
105. Roberts, L.H. Correlation of respiration and ultrasound production in rodents and bats. *J. Zool.* **1972**, *168*, 439–449. [CrossRef]
106. Frysztak, R.J.; Neafsey, E.J. The Effect of Medial Frontal Cortex Lesions on Respiration, "Freezing," and Ultrasonic Vocalizations during Conditioned Emotional Responses in Rats. *Cereb. Cortex* **1991**, *1*, 418–425. [CrossRef]
107. Riede, T.; Schaefer, C.P.; Stein, A. Role of deep breaths in ultrasonic vocal production of Sprague-Dawley rats. *J. Neurophysiol.* **2020**, *123*, 966–979. [CrossRef]
108. Dupin, M.; Garcia, S.; Messaoudi, B.; Doyère, V.; Mouly, A.-M. Respiration and brain neural dynamics associated with interval timing during odor fear learning in rats. *Sci. Rep.* **2020**, *10*, 1–18. [CrossRef]
109. Adrian, E.D. Olfactory reactions in the brain of the hedgehog. *J. Physiol.* **1942**, *100*, 459–473. [CrossRef]
110. Fontanini, A.; Bower, J.M. Variable Coupling Between Olfactory System Activity and Respiration in Ketamine/Xylazine Anesthetized Rats. *J. Neurophysiol.* **2005**, *93*, 3573–3581. [CrossRef]
111. Buonviso, N.; Amat, C.; Litaudon, P. Respiratory Modulation of Olfactory Neurons in the Rodent Brain. *Chem. Senses* **2005**, *31*, 145–154. [CrossRef]
112. Courtiol, E.; Hegoburu, C.; Litaudon, P.; Garcia, S.; Fourcaud-Trocmé, N.; Buonviso, N. Individual and synergistic effects of sniffing frequency and flow rate on olfactory bulb activity. *J. Neurophysiol.* **2011**, *106*, 2813–2824. [CrossRef]
113. Esclassan, F.; Courtiol, E.; Thévenet, M.; Garcia, S.; Buonviso, N.; Litaudon, P. Faster, Deeper, Better: The Impact of Sniffing Modulation on Bulbar Olfactory Processing. *PLoS ONE* **2012**, *7*, e40927. [CrossRef]
114. Rojas-Líbano, D.; Frederick, D.E.; Egaña, J.I.; Kay, L.M. The olfactory bulb theta rhythm follows all frequencies of diaphragmatic respiration in the freely behaving rat. *Front. Behav. Neurosci.* **2014**, *8*, 214. [CrossRef]
115. Zelano, C.; Jiang, H.; Zhou, G.; Arora, N.; Schuele, S.; Rosenow, J.; Gottfried, J.A. Nasal Respiration Entrains Human Limbic Oscillations and Modulates Cognitive Function. *J. Neurosci.* **2016**, *36*, 12448–12467. [CrossRef]
116. Grosmaitre, X.; Santarelli, L.C.; Tan, J.; Luo, M.; Ma, M. Dual functions of mammalian olfactory sensory neurons as odor detectors and mechanical sensors. *Nat. Neurosci.* **2007**, *10*, 348–354. [CrossRef]
117. Adrian, E.D. The role of air movement in olfactory stimulation. *J. Physiol.* **1951**, *114*, 4–5.
118. Fontanini, A.; Bower, J.M. Slow-waves in the olfactory system: An olfactory perspective on cortical rhythms. *Trends Neurosci.* **2006**, *29*, 429–437. [CrossRef]
119. Ito, J.; Roy, S.K.; Liu, Y.; Cao, Y.; Fletcher, M.L.; Lu, L.; Boughter, J.D.; Grun, S.; Heck, D.H. Whisker barrel cortex delta oscillations and gamma power in the awake mouse are linked to respiration. *Nat. Commun.* **2014**, *5*, 3572. [CrossRef]
120. Moore, J.D.; Deschênes, M.; Furuta, T.; Huber, D.; Smear, M.C.; Demers, M.; Kleinfeld, D. Hierarchy of orofacial rhythms revealed through whisking and breathing. *Nat. Cell Biol.* **2013**, *497*, 205–210. [CrossRef]
121. Yanovsky, Y.; Ciatipis, M.; Draguhn, A.; Tort, A.B.L.; Brankačk, J. Slow Oscillations in the Mouse Hippocampus Entrained by Nasal Respiration. *J. Neurosci.* **2014**, *34*, 5949–5964. [CrossRef]
122. Lockmann, A.L.V.; Laplagne, D.A.; Leão, R.N.; Tort, A.B.L. A Respiration-Coupled Rhythm in the Rat Hippocampus Independent of Theta and Slow Oscillations. *J. Neurosci.* **2016**, *36*, 5338–5352. [CrossRef]
123. Chi, V.N.; Müller, C.; Wolfenstetter, T.; Yanovsky, Y.; Draguhn, A.; Tort, A.B.L.; Brankačk, J. Hippocampal Respiration-Driven Rhythm Distinct from Theta Oscillations in Awake Mice. *J. Neurosci.* **2016**, *36*, 162–177. [CrossRef]
124. Biskamp, J.; Bartos, M.; Sauer, J.-F. Organization of prefrontal network activity by respiration-related oscillations. *Sci. Rep.* **2017**, *7*, srep45508. [CrossRef] [PubMed]
125. Zhong, W.; Ciatipis, M.; Wolfenstetter, T.; Jessberger, J.; Müller, C.; Ponsel, S.; Yanovsky, Y.; Brankačk, J.; Tort, A.B.L.; Draguhn, A. Selective entrainment of gamma subbands by different slow network oscillations. *Proc. Natl. Acad. Sci. USA* **2017**, *114*, 4519–4524. [CrossRef] [PubMed]
126. Moberly, A.H.; Schreck, M.; Bhattarai, J.P.; Zweifel, L.S.; Luo, W.; Ma, M. Olfactory inputs modulate respiration-related rhythmic activity in the prefrontal cortex and freezing behavior. *Nat. Commun.* **2018**, *9*, 1–10. [CrossRef]
127. Tort, A.B.L.; Ponsel, S.; Jessberger, J.; Yanovsky, Y.; Brankačk, J.; Draguhn, A. Parallel detection of theta and respiration-coupled oscillations throughout the mouse brain. *Sci. Rep.* **2018**, *8*, 1–14. [CrossRef]
128. Buonviso, N.; Amat, C.; Litaudon, P.; Roux, S.; Royet, J.-P.; Farget, V.; Sicard, G. Rhythm sequence through the olfactory bulb layers during the time window of a respiratory cycle. *Eur. J. Neurosci.* **2003**, *17*, 1811–1819. [CrossRef]
129. Cenier, T.; Amat, C.; Litaudon, P.; Garcia, S.; De Micheaux, P.L.; Liquet, B.; Roux, S.; Buonviso, N. Odor vapor pressure and quality modulate local field potential oscillatory patterns in the olfactory bulb of the anesthetized rat. *Eur. J. Neurosci.* **2008**, *27*, 1432–1440. [CrossRef]
130. Fourcaud-Trocmé, N.; Briffaud, V.; Thévenet, M.; Buonviso, N.; Amat, C. In vivo beta and gamma subthreshold oscillations in rat mitral cells: Origin and gating by respiratory dynamics. *J. Neurophysiol.* **2018**, *119*, 274–289. [CrossRef]
131. Manabe, H.; Mori, K. Sniff rhythm-paced fast and slow gamma-oscillations in the olfactory bulb: Relation to tufted and mitral cells and behavioral states. *J. Neurophysiol.* **2013**, *110*, 1593–1599. [CrossRef]
132. Herrero, J.L.; Khuvis, S.; Yeagle, E.; Cerf, M.; Mehta, A.D. Breathing above the brain stem: Volitional control and attentional modulation in humans. *J. Neurophysiol.* **2018**, *119*, 145–159. [CrossRef]

133. Kluger, D.S.; Gross, J. Depth and phase of respiration modulate cortico-muscular communication. *NeuroImage* **2020**, *222*, 117272. [CrossRef]
134. Heck, D.H.; McAfee, S.S.; Liu, Y.; Babajani-Feremi, A.; Rezaie, R.; Freeman, W.J.; Wheless, J.W.; Papanicolaou, A.C.; Ruszinkó, M.; Sokolov, Y.; et al. Breathing as a Fundamental Rhythm of Brain Function. *Front. Neural Circuits* **2017**, *10*, 115. [CrossRef]
135. Lisman, J.E.; Jensen, O. The Theta-Gamma Neural Code. *Neuron* **2013**, *77*, 1002–1016. [CrossRef]
136. Fries, P.; Reynolds, J.H.; Rorie, A.E.; DeSimone, R. Modulation of Oscillatory Neuronal Synchronization by Selective Visual Attention. *Science* **2001**, *291*, 1560–1563. [CrossRef]
137. Tallon-Baudry, C. The roles of gamma-band oscillatory synchrony in human visual cognition. *Front. Biosci.* **2009**, *14*, 321–332. [CrossRef]
138. Jensen, O.; Colgin, L.L. Cross-frequency coupling between neuronal oscillations. *Trends Cogn. Sci.* **2007**, *11*, 267–269. [CrossRef]
139. Cardin, J.A.; Carlén, M.; Meletis, K.; Knoblich, U.; Zhang, F.; Deisseroth, K.; Tsai, L.-H.; Moore, C.I. Driving Fast-Spiking Cells Induces Gamma Rhythm and Controls Sensory Responses. *Nature* **2009**, *459*, 663–667. [CrossRef]
140. Engel, A.K.; Fries, P.; Singer, W. Dynamic predictions: Oscillations and synchrony in top–down processing. *Nat. Rev. Neurosci.* **2001**, *2*, 704–716. [CrossRef]
141. Gross, J.; Schnitzler, A.; Timmermann, L.; Ploner, M. Gamma Oscillations in Human Primary Somatosensory Cortex Reflect Pain Perception. *PLoS Biol.* **2007**, *5*, e133. [CrossRef]
142. Osipova, D.; Takashima, A.; Oostenveld, R.; Fernández, G.; Maris, E.; Jensen, O. Theta and Gamma Oscillations Predict Encoding and Retrieval of Declarative Memory. *J. Neurosci.* **2006**, *26*, 7523–7531. [CrossRef]
143. Sederberg, P.B.; Schulze-Bonhage, A.; Madsen, J.R.; Bromfield, E.B.; Litt, B.; Brandt, A.; Kahana, M.J. Gamma Oscillations Distinguish True From False Memories. *Psychol. Sci.* **2007**, *18*, 927–932. [CrossRef]
144. Van Vugt, M.K.; Schulze-Bonhage, A.; Litt, B.; Brandt, A.; Kahana, M.J. Hippocampal Gamma Oscillations Increase with Memory Load. *J. Neurosci.* **2010**, *30*, 2694–2699. [CrossRef]
145. Howard, M.W.; Rizzuto, D.S.; Caplan, J.B.; Madsen, J.R.; Lisman, J.; Aschenbrenner-Scheibe, R.; Schulze-Bonhage, A.; Kahana, M.J. Gamma Oscillations Correlate with Working Memory Load in Humans. *Cereb. Cortex* **2003**, *13*, 1369–1374. [CrossRef]
146. Tallon-Baudry, C.; Bertrand, O.; Peronnet, F.; Pernier, J. Induced γ-Band Activity during the Delay of a Visual Short-Term Memory Task in Humans. *J. Neurosci.* **1998**, *18*, 4244–4254. [CrossRef]
147. Perl, O.; Ravia, A.; Rubinson, M.; Eisen, A.; Soroka, T.; Mor, N.; Secundo, L.; Sobel, N. Human non-olfactory cognition phase-locked with inhalation. *Nat. Hum. Behav.* **2019**, *3*, 501–512. [CrossRef]
148. Nakamura, N.H.; Fukunaga, M.; Oku, Y. Respiratory modulation of cognitive performance during the retrieval process. *PLoS ONE* **2018**, *13*, e0204021. [CrossRef]
149. Arshamian, A.; Iravani, B.; Majid, A.; Lundström, J.N. Respiration Modulates Olfactory Memory Consolidation in Humans. *J. Neurosci.* **2018**, *38*, 10286–10294. [CrossRef]
150. Dejean, C.; Courtin, J.; Karalis, N.; Chaudun, F.; Wurtz, H.; Bienvenu, T.C.M.; Herry, C. Prefrontal neuronal assemblies temporally control fear behaviour. *Nature* **2016**, *535*, 420–424. [CrossRef]
151. Karalis, N.; Dejean, C.; Chaudun, F.; Khoder, S.; Rozeske, R.R.; Wurtz, H.; Bagur, S.; Benchenane, S.B.K.; Sirota, N.K.A.; Courtin, J.; et al. 4-Hz oscillations synchronize prefrontal–amygdala circuits during fear behavior. *Nat. Neurosci.* **2016**, *19*, 605–612. [CrossRef]
152. Brudzynski, S.M. Ethotransmission: Communication of emotional states through ultrasonic vocalization in rats. *Curr. Opin. Neurobiol.* **2013**, *23*, 310–317. [CrossRef] [PubMed]
153. Tschida, K.; Michael, V.; Takatoh, J.; Han, B.-X.; Zhao, S.; Sakurai, K.; Mooney, R.; Wang, F. A Specialized Neural Circuit Gates Social Vocalizations in the Mouse. *Neuron* **2019**, *103*, 459.e4–472.e4. [CrossRef] [PubMed]
154. Michael, V.; Goffinet, J.; Pearson, J.; Wang, F.; Tschida, K.; Mooney, R. Circuit and synaptic organization of forebrain-to-midbrain pathways that promote and suppress vocalization. *eLife* **2020**, *9*. [CrossRef] [PubMed]
155. Sadananda, M.; Wöhr, M.; Schwarting, R.K. Playback of 22-kHz and 50-kHz ultrasonic vocalizations induces differential c-fos expression in rat brain. *Neurosci. Lett.* **2008**, *435*, 17–23. [CrossRef]
156. Roberts, P.D.; Portfors, C.V. Responses to Social Vocalizations in the Dorsal Cochlear Nucleus of Mice. *Front. Syst. Neurosci.* **2015**, *9*, 172. [CrossRef]
157. Parsana, A.J.; Moran, E.E.; Brown, T.H. Rats learn to freeze to 22-kHz ultrasonic vocalizations through autoconditioning. *Behav. Brain Res.* **2012**, *232*, 395–399. [CrossRef]
158. Dupin, M.; Garcia, S.; Boulanger-Bertolus, J.; Buonviso, N.; Mouly, A.-M. New Insights from 22-kHz Ultrasonic Vocalizations to Characterize Fear Responses: Relationship with Respiration and Brain Oscillatory Dynamics. *eNeuro* **2019**, *6*, 65. [CrossRef]
159. Bertolus, J.B.; Mouly, A.-M.; Sullivan, R.M. Ecologically relevant neurobehavioral assessment of the development of threat learning. *Learn. Mem.* **2016**, *23*, 556–566. [CrossRef]
160. Boulanger-Bertolus, J.; White, A.M.; Debiec, J. Enduring Neural and Behavioral Effects of Early Life Adversity in Infancy: Consequences of Maternal Abuse and Neglect, Trauma and Fear. *Curr. Behav. Neurosci. Rep.* **2017**, *4*, 107–116. [CrossRef]
161. Brudzynski, S.M. Emission of 22 kHz vocalizations in rats as an evolutionary equivalent of human crying: Relationship to depression. *Behav. Brain Res.* **2019**, *363*, 1–12. [CrossRef]
162. Jürgens, U. The Neural Control of Vocalization in Mammals: A Review. *J. Voice* **2009**, *23*, 1–10. [CrossRef]

163. Scattoni, M.L.; Crawley, J.; Ricceri, L. Ultrasonic vocalizations: A tool for behavioural phenotyping of mouse models of neurodevelopmental disorders. *Neurosci. Biobehav. Rev.* **2009**, *33*, 508–515. [CrossRef]
164. Simola, N.; Granon, S. Ultrasonic vocalizations as a tool in studying emotional states in rodent models of social behavior and brain disease. *Neuropharmacology* **2019**, *159*, 107420. [CrossRef]

brain sciences

MDPI

Review

Biological and Acoustic Sex Differences in Rat Ultrasonic Vocalization

Charles Lenell [1,2], Courtney K. Broadfoot [1,3], Nicole E. Schaen-Heacock [1,3] and Michelle R. Ciucci [1,3,*]

1 Department of Surgery, University of Wisconsin Madison, Madison, WI 53792, USA; Lenell@wisc.edu (C.L.); ckuehn2@wisc.edu (C.K.B.); nschaen@wisc.edu (N.E.S.-H.)
2 Communicative Sciences and Disorders, New York University, New York, NY 10001, USA
3 Department of Communication Sciences and Disorders, University of Wisconsin Madison, Madison, WI 53706, USA
* Correspondence: ciucci@surgery.wisc.edu; Tel.: +1-608-263-0192

Abstract: The rat model is a useful tool for understanding peripheral and central mechanisms of laryngeal biology. Rats produce ultrasonic vocalizations (USVs) that have communicative intent and are altered by experimental conditions such as social environment, stress, diet, drugs, age, and neurological diseases, validating the rat model's utility for studying communication and related deficits. Sex differences are apparent in both the rat larynx and USV acoustics and are differentially affected by experimental conditions. Therefore, the purpose of this review paper is to highlight the known sex differences in rat USV production, acoustics, and laryngeal biology detailed in the literature across the lifespan.

Keywords: ultrasonic vocalization; larynx; female; male; sex differences

check for updates

Citation: Lenell, C.; Broadfoot, C.K.; Schaen-Heacock, N.E.; Ciucci, M.R. Biological and Acoustic Sex Differences in Rat Ultrasonic Vocalization. *Brain Sci.* **2021**, *11*, 459. https://doi.org/10.3390/brainsci11040459

Academic Editors: Stefan M. Brudzynski and Jeffrey Burgdorf

Received: 11 February 2021
Accepted: 1 April 2021
Published: 4 April 2021

1. Introduction

Both male and female rats produce ultrasonic vocalizations (USVs) in a variety of contexts that hold communicative intent [1–7]. Adult rat USVs can be categorized into two primary types based on affective state and mean frequencies: 1) alarm USVs which are produced during negative affective states with a mean frequency near 22 kHz, and 2) 50 kHz USVs which are produced during positive affective states [1,6,8–10]. Pups produce USVs with an average frequency of 40 kHz to receive care from their dam (female parent) [8,9,11]. Sexual dimorphism is apparent in all three major categories of USVs. Therefore, the purpose of this review paper is to highlight the known sex differences in rat USV production and acoustics as well as laryngeal biological differences between sexes. All ages were included in this review of the literature. This review is focused on sexual dimorphism of the rat larynx and USVs; however, sex differences exist in other rodent species' USVs (e.g., hamsters [12] and mice [13–16]) and sexual dimorphism of USVs is also mediated by sex differences within the central nervous system, not just the larynx [17–19]. Nevertheless, the rat model is widely used to study vocal communication in a variety of contexts such as social environment [1,5,7,9,20–23], neurogenic disorders [24–29], aging [30–35], and pharmacology [36–44], justifying the need for a comprehensive review of the literature attuned to sex differences.

2. Review of Sex Differences

2.1. Sexual Dimorphism of the Vocal Fold

Rat USVs are produced using a complex orchestration of the respiratory, laryngeal, and resonatory systems [45]. The whistle-like vocalization is produced by airflow passing through glottal and supraglottal spaces, and the configuration of these spaces can be altered by subglottic pressure and intrinsic laryngeal muscle activity [45–48]. Laryngeal motor innervation is primarily by the nucleus ambiguus through two divisions of the vagus nerve: superior and recurrent laryngeal nerves [49–53]. Intrinsic laryngeal muscles such as the cricothyroid and thyroarytenoid elongate the glottis and shorten/close the vocal folds to regulate the glottal geometry for specific USV types [46]. Several studies have demonstrated that vocal fold approximation/configuration is critical for production and modulation of rat USVs, making rat vocal folds a targeted investigation in voice-related research [46–48,54–56]. Additionally, rat vocal folds, like human vocal folds, are composed of a body (thyroarytenoid muscles) and cover (lamina propria, macula flavae, and epithelium) [57,58] and can produce audible vocalizations with vocal fold vibration in the frequency range 1–6 kHz [45,59,60]. However, because conspecific communication occurs by USV rather than audible vocalizations, this paper discusses USV only.

Understanding how hormones affect vocal structure and function (USV) is crucial for advancing science and clinical practice. As such, this paper reviews sex differences in laryngeal biology, USV production rates, and USV acoustics (Figure 1). More specifically, the summary of the known sex differences in the rat larynx was organized into intrinsic laryngeal muscles and vocal fold mucosa, and sexual dimorphism of the USV production and acoustics sections were summarized by primary USV category with rat strain and age reported as needed (Figure 1). To ensure that this review encompassed pertinent articles reporting sex differences in rat USVs, we created a PubMed search using the following search terms: ((USV) OR (ultrasonic vocalization) AND (female)) NOT (mouse). Authors then read the methods of the 540 article results and included all articles that compared rat USVs between sexes with significant findings within the results. The majority of articles were excluded for the following reasons: did not evaluate rat USVs, did not compare sexes, and did not include both sexes (Figure 2).

2.1.1. Sex Differences in Intrinsic Laryngeal Muscles

Intrinsic laryngeal muscles are necessary to produce USVs, but few studies have examined sexual dimorphism within rat intrinsic laryngeal muscles [61]. Existing studies have primarily focused on evaluating the thyroarytenoid (TA) muscles, the primary muscles of the vocal folds [61,62]. The myofiber types of the lateral thyroarytenoid (LTA) and medial thyroarytenoid (MTA) muscles are similar between sexes, but the overall muscle areas of the LTA and MTA are larger in male rats [61]. Additionally, the individual minimum feret diameter of the myofibers of the LTA muscle are also larger in male rats [61]. Studies have not investigated sex differences in muscle fiber types and sizes in the other intrinsic laryngeal muscles such as the superior cricoarytenoid, lateral cricoarytenoid, posterior cricoarytenoid, cricothyroid, and alar muscles. Therefore, a lack of information exists regarding the potential sexual dimorphism of intrinsic laryngeal muscles.

Neuromuscular junctions (NMJs) of the TA muscles are also uniquely sexually dimorphic [62]. Female rats have more acetylcholine receptor fragments in the NMJs of the TA muscles but not the other intrinsic laryngeal muscles [62]. The study's authors hypothesized that this sexually dimorphic NMJ feature would result in higher synaptic strength and was likely mediated by higher estrogen levels of female rats [62]. However, a recent study did not find NMJ morphological differences between ovariectomized (elimination of ovarian hormones) and control female rats, suggesting that the NMJ of the TA muscle may not be ovarian hormone dependent, and that sex differences in the TA muscles may be more likely influenced by male hormones (androgens) rather than female hormones such as estrogens [63].

Because few investigations have evaluated sexually dimorphic neuromuscular parameters of the laryngeal mechanism and even fewer studies have evaluated the effects of sex hormones on these parameters, how sex differences in the underlying neuromuscular laryngeal mechanisms influence acoustic differences in USV is unknown. Therefore, future studies characterizing the extent of sex differences and influence of sex hormones on the neuromuscular proprieties of the laryngeal mechanism are warranted.

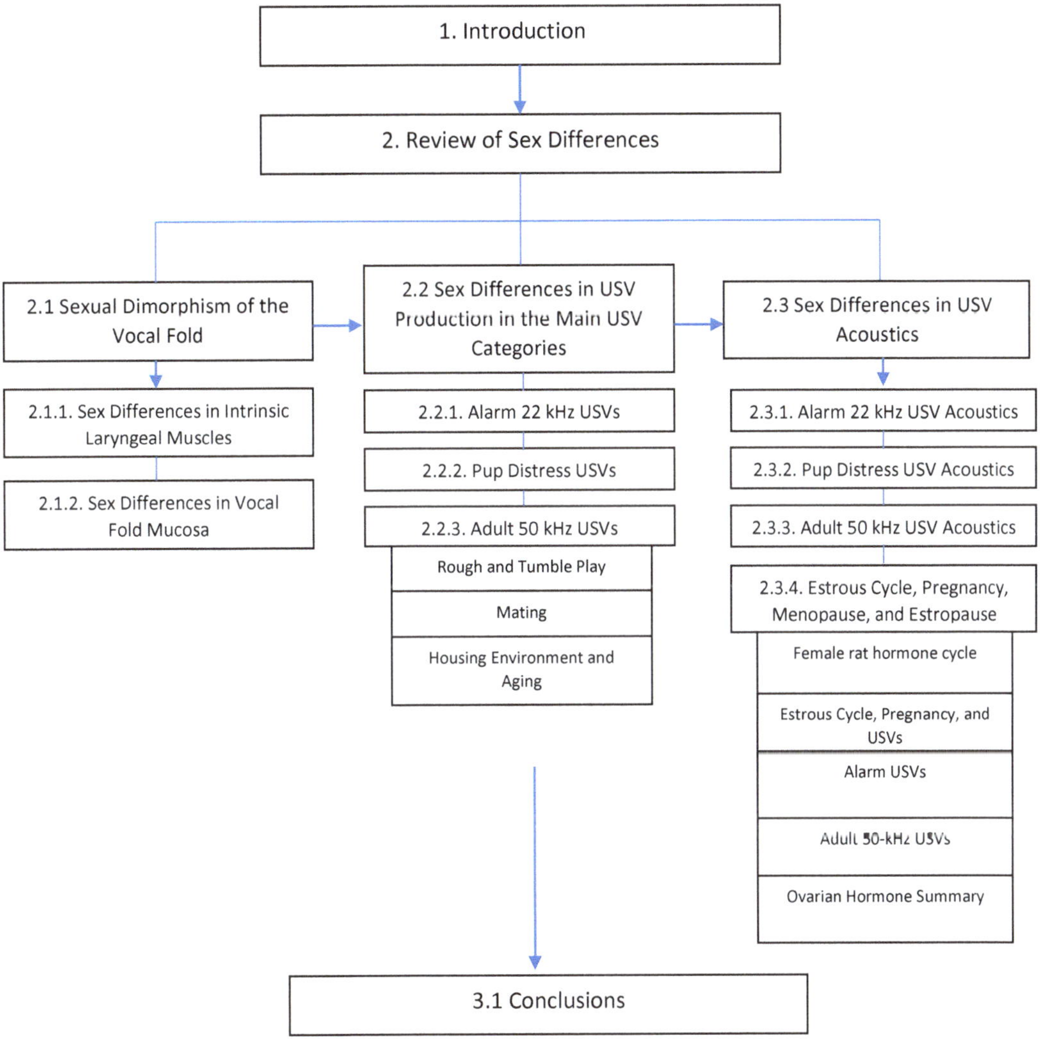

Figure 1. Organization of the manuscript.

2.1.2. Sex Differences in Vocal Fold Mucosa

Both male and female rats have sex hormone receptors within the vocal folds, indicating that sex hormones bind within the vocal fold and potentially modulate physiological effects [64]. In general, rat studies have demonstrated that ovarian hormones more drastically affect properties of the vocal fold mucosa than androgens.

In female rats, sex hormones are critical to homeostasis of the pre-menopausal vocal fold mucosal tissues [64–66]. Several studies have demonstrated that removal of the ovaries (elimination of ovarian hormones) results in the remodeling of the vocal fold mucosa including the following: decreased cellular layers of the epithelium, increased edema of the lamina propria, and decreased collagen I, hyaluronic acid, and elastin of the lamina propria [64–66]. Although ovariectomy procedures have demonstrated vocal fold mucosal remodeling, orchiectomy procedures in male rats have not resulted in significant remodeling [64]. Therefore, the vocal fold mucosa appears to be differentially regulated between sexes with female rats having hormone-dependent mucosa, whereas the vocal fold mucosa of male rats does not seem to be affected by hormone status.

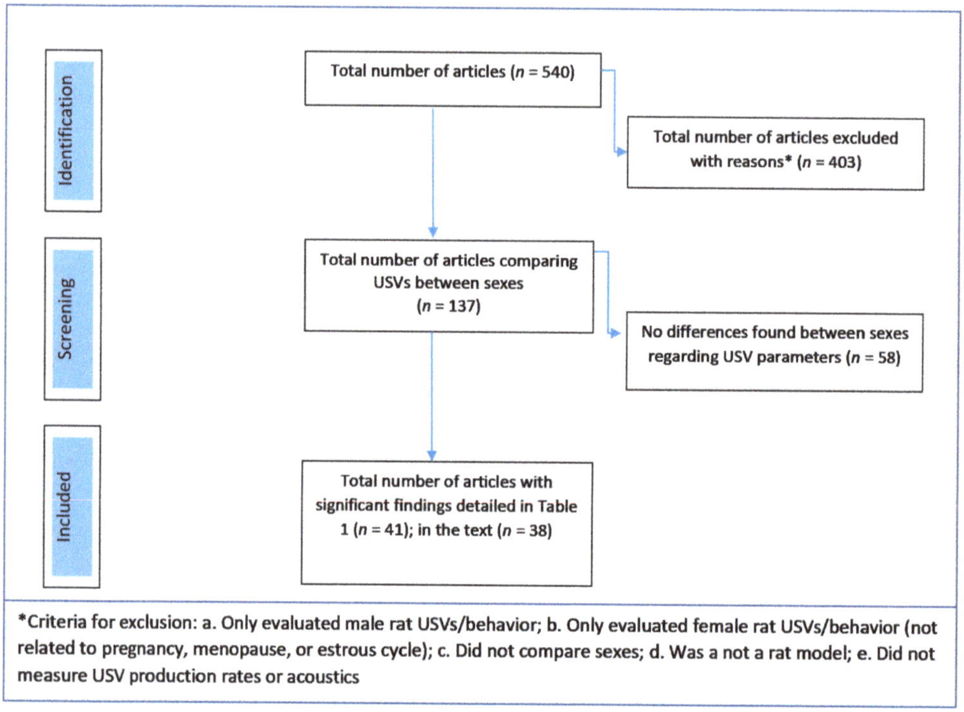

Figure 2. Number of articles reviewed and included in this article regarding sex differences in the rat USV with a PubMed search.

Pregnancy has also been shown to affect female rat vocal fold mucosa [67,68]. Pregnancy has been associated with the following histological changes in the vocal fold mucosa: increased edema, increased glycosaminoglycans, mast cell emergence, and increased cellularity of the lamina propria [67,68]. In addition, pregnancy has been found to change biomolecules within the vocal fold mucosa [67]. Specifically, pregnant rats had lower expression of nuclear factor-kappa B (a protein transcription factor related to immune response) and higher expression of mucin 5 subtype AC (the major mucin of the upper airway) [67]. Furthermore, progesterone levels were negatively correlated to the nuclear factor-kappa B, but estradiol levels were not correlated to either biomolecule [67]. Therefore, progesterone may activate transcriptional signaling responsible for mucosal changes during pregnancy and may consequently affect USV acoustics.

2.2. Sex Differences in USV Production in the Main USV Categories

Rat USVs can be broadly categorized into three main categories: adult alarm 22 kHz, adult 50 kHz, and pup distress USVs (Figure 3). Within these broad categorizes, USV production rates are different between sexes. The following section will describe the sex differences in production of these USV types and subsequent sections will focus more specifically on acoustic differences.

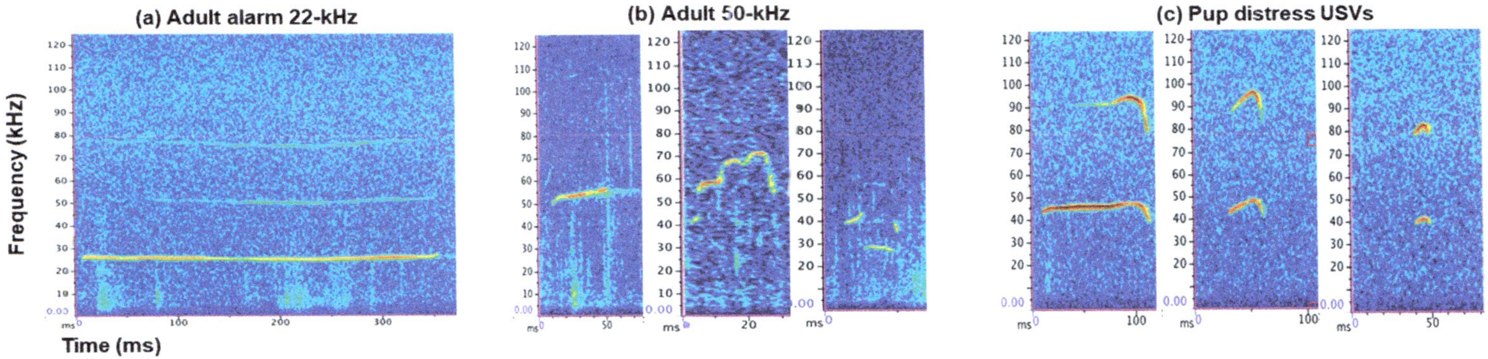

Figure 3. Spectrograms of the three main categories of rat USVs: adult alarm 22 kHz (**a**), adult 50 kHz (**b**), and pup distress USVs (**c**). We have included three subtypes of 50 kHz USVs (**b**) and three subtypes of pup distress USVs (**c**).

2.2.1. Alarm 22 kHz USVs

Alarm 22 kHz USVs are vocalizations produced during negative affective states with long durations (0.5–3.0 s), low frequencies (~22 kHz), and narrow bandwidths (1–4 kHz) [6,69]. Further, Blanchard identified six subtypes of alarm USVs during the presence of a predator: horizontal, linear ascending, linear descending, U-shaped, negatively accelerated ascending, and negatively accelerated descending vocalizations [70]. Sex differences in alarm 22 kHz USVs have been evaluated in response to both predators (a live cat) and laboratory experimental stressors [69]. In 2018, Inagaki reviewed sex differences in rat alarm 22 kHz USVs [69]. To complement this review, we will summarize the thematic sex differences in 22 kHz USVs incorporating more recent literature.

In general, female rats produce longer overall duration of alarm USVs in response to predators but shorter overall duration of alarm USVs in response to experimental stressors [69]; however, production of alarm USVs are influenced by both strain and sex [71].

For example, in response to fear conditioning training, Long–Evans female rats produced greater overall duration of alarm 22 kHz USVs than male Long–Evans rats; however, the opposite was true for Sprague–Dawley rats [71]. Additionally, following fear conditioning training, Sprague–Dawley male rats produced more alarm USVs than female Sprague–Dawley rats during contextual and auditory conditioned stimuli, whereas male and female Long–Evans rats had similar alarm USV productions in response to conditioned stimuli [71]. Likewise, another study that evaluated the effects of serotonin transporter deficiency in a fear condition found that female rats produced fewer alarm USVs than male rats [72].

Within strain, rats can be categorized as high vocalizing or low vocalizing, [73] and stressors have been found to differentially affect sexes of high- and low-vocalizing rats. Wistar rats exposed to chronic experimental stressors (variable lights, small cages, tail pinch, etc.) produced significantly increased levels of 22 kHz USVs for low-vocalizing male rats and high-vocalizing female rats [74]. Therefore, strain, sex, and vocalizing category all contribute to differences in alarm USV rates for rats.

Early life stress also has been shown to differentially impact alarm USV productions between sexes. A study that evaluated the effects of brief and prolonged maternal separation demonstrated that brief maternal separation attenuated fear conditioning (reduced alarm USV production and freezing behavior) in both male and female Sprague–Dawley rats; however, in general, male rats produced greater overall duration of alarm USVs than female rats during fear conditioning [75]. Neonatal maternal separation resulted in changes to 22 kHz USV production in adulthood, with fewer 22kHz USVs in response to a stressor for female Sprague–Dawley rats but more 22 kHz USVs in Sprague–Dawley males, demonstrating an opposite-sex effect [76,77].

Playback of alarm USVs to adult rats also affects behavior differently between sexes. A playback of 22 kHz alarm USVs resulted in more long-lasting behavioral inhibition in female rats than male rats [78]. This finding highlights that although rats may have hormone-mediated, sexually dimorphic USV rates and different acoustic characteristics, these differences may not be directly related to laryngeal differences but rather differences in behavior mediated by the central nervous system.

Hormones likely contribute to the sexual dimorphism observed in alarm USV production and differentially affect alarm USV productions. For example, in response to an air puff, female Wistar rats produce shorter overall duration of alarm USVs than males on both proestrus and diestrus phases of the estrous cycle [79]. This difference was hypothesized to be due to testosterone level differences between sexes and tested in subsequent experiments. In response to an air puff, castrated male Wistar rats produced shorter overall duration of alarm USVs than sham-operated or castrated male rats with a testosterone implant [80]. Additionally, because other anxiety responses (freezing and defecation) were not reduced in the castrated male rats, the lower alarm USV emissions did not indicate a reduced startle response [80]. The reduced emission rate is more likely indicative of a reduction in dominant behaviors [80]. Nevertheless, in response to an air puff, alarm USV production of ovariectomized female Wistar rats did not differ between female rats with testosterone implants and female rats with cholesterol implants [79]. Furthermore, alarm USV productions were similar between female rats in diestrus 1 (low hormones) and proestrus (high hormones), indicating no role of ovarian hormones in alarm USV productions. Therefore, while male alarm USVs may be influenced by sex hormones, female alarm USVs may not be. Nevertheless, these results were found in one anxiety context with one strain of rat and should not be assumed to be identical in all anxiety contexts or all rat strains.

Taken together, these results demonstrate that fearful conditions differentially affect male and female rat behavior including their production of alarm USVs. Nevertheless, more research is warranted to evaluate the patterns of hormonal influence across strain and sexes before definitive conclusions can be made regarding sexual dimorphism of alarm USV productions.

One uniquely male 22 kHz USV subtype is the post-ejaculation vocalization [81,82]. This extended vocalization is produced by the male approximately 30 seconds post-ejaculation and continues, repeatedly, for approximately 75% of the entire post-ejaculatory refractory period [81]. Such USVs are characterized as being highly frequency modulated, specifically within the medial terminal segments of the USV [83]. All spontaneous copulatory behavior ceases during this refractory period [84]. While such copulatory behaviors are associated with 22 kHz vocalizations, alarm USVs have also been observed in other aspects surrounding mating. For example, prior to mounting, male rats produce 22 kHz USVs as the rat approaches ejaculation, particularly if the female is non-receptive to male mounting and/or if the male was unsuccessful [81]. It is hypothesized, in the copulatory context, that this USV subtype represents a "de-arousal" mechanism or a type of "motivational cut-off" [85]. This may enforce mating separation between the male and female rats while still maintaining social contact [81]. Given the few studies available and the several-decade gap in published studies, more research into this unique vocalization is certainly warranted.

2.2.2. Pup Distress USVs

Rat pups produce USVs ranging in mean frequency between 30 and 65 kHz when separated from their mother and associated litter. These USVs are generally referred to as distress USVs [36]. The duration of distress USVs is variable (ranging between average durations of 80 and 150 ms) and starts with very short USVs at younger ages with increasing duration with maturation [86]. The distress USVs are unique in that the timeframe in which pups produce them is relatively short (~18 days of age). While the increase in call rate may be correlated with a heightened state of anxiety upon separation, both duration and frequency may reflect developmental changes pre- and post-weaning [87]. Pups are completely reliant on the mother for survival prior to weaning, supporting the hypothesis that these distress USVs are produced in the context of separation/isolation, and are consequently important for pup survival [36,88].

The pup USVs have also been described as occurring in the frequency range 40–70 kHz, further being classified into 2 groups: 40 kHz/300 ms and 66 kHz/21 ms [88]. These classes have specific relationships with both respiration and behavior and are produced during pup movement [88]. While 40 kHz distress USVs have been observed in the context of isolation, they can also be elicited in a more naturalistic setting when mothers engage in rough handling with the pups [88]. In contrast, 66 kHz are not related to the behavioral conditions [88]. In terms of respiration, distress USVs alter the length of expiration, lasting for the entirety of the expiration cycle, whereas 66 kHz USVs do not alter the respiratory signal in pups when mild foot shocks were administered [88].

Pup distress USVs can also occur following a drop in ambient temperature [9]. This behavior is thought to relate to a pup's dependence on the dam for survival, as pups cannot regulate their own body temperature. Additional research investigating the role of anxiogenic drugs on pup USVs supports the hypothesis that these USVs are correlated with presumptive distressed/anxious states, as administration of said drugs (e.g., selective serotonin reuptake inhibitors) lead to a reduction or complete blocking of the USVs [9,89–91]. Studies assessing selective breeding as well as breeding over several generations further support this hypothesis [92,93]. Pups bred to produce high rates of isolation-induced USVs for multiple generations demonstrated increased anxiety-related behaviors in adulthood. Additionally, rats who were selectively bred to demonstrate an anxious phenotype in adulthood produced more distress USVs as pups compared to a less anxious phenotype [92]. In summary, pups produce unique distress vocalizations prior to weaning that seem important to their survival and are influenced by strain and genetic lines.

In the context of sex differences, current research suggests preferential retrieval of male pups by the dam which may be attributed to sex differences in pup distress USVs [17]. Male pups produce significantly more distress USVs with a lower mean frequency and lower amplitude that results in preferential retrieval of the dam [17]. Research has shown from postnatal day (P)-2 to 3 through P-12 to 13, males pups tend to produce distress USVs more frequently than female pups, which results in the dam retrieving and returning male pups preferentially to female pups [94]. Therefore, sexual dimorphism of the pup distress USVs may result in the female rat prioritizing male pup survival.

Although few studies have investigated sex differences in distress USVs in typical/normal rat pups, many studies have investigated how drugs, neurological disorders, endocrine disruptors, diet, and environmental condition differentially affect male and female pup distress USVs. Table 1 summarizes major sex differences found in USV rate and acoustics for experimental models organized by model, age, and strain. While not all ages are prior to weaning (~P21), most summarized studies in this table measure USV rates and/or acoustics within this timeframe (Table 1). Because the sex differences are not uniform across studies or models, the articles are summarized individually. While this table highlights major USV sex differences present in experimental rat models (particularly rat pups), not all experiments find sex differences in USV production or acoustics.

Table 1. Summary of sex differences found in USV acoustics for experimental models.

Model	Sub Model	Age	Strain	Recording Duration	Major Sex Difference(s) in USV Acoustics
Drug exposure	Prenatal cannabis	P10	Wistar	15 s	Male pups produced fewer distress USVs during isolation, whereas females did not [95].
	Prenatal alcohol	P40–42	Wistar	10 min	For males, high ethanol exposure resulted in more 22 kHz and fewer 50 kHz USVs during play, whereas ethanol exposure did not affect female USV production during play [96].
		~P38–P48	LE	12 min	Prenatal exposure to alcohol decreased the mean frequency and total duration of 50 kHz USVs during same-sex social interaction for male rats, but not female rats [97].
		P28 P42	LE	10 min	At P28, during play female control whisker clipped rats produced more 22 kHz USVs than other female groups. At P42, during play male rats overall had more 50 kHz USVs than female rats [98].

Table 1. *Cont.*

Model	Sub Model	Age	Strain	Recording Duration	Major Sex Difference(s) in USV Acoustics
	Postnatal alcohol	P15	SD	6 min	Neonatal alcohol exposure significantly reduced distress USV rate for both sexes and significantly increased USV latency in female pups. Agmatine reduced these deficits, in female but not male pups [99].
		P25, P35, P110–P120	SD	45 min	Adult alcohol-treated males produced more 22 kHz USVs following initial handling which was suppressed with the startle stimulus than male rats receiving water. Alcohol did not affect female 22 kHz USV rate. Overall, male rats had a greater number of 22 kHz USVs in response to startle stimulus [100].
		~2–5 mo.	NS	4 h	Female rats produced more 50 kHz USVs than male rats during experimental conditions. EtOH males produce 50 kHz USVs with a higher mean frequency and greater power than EtOH females. EtOH males produced 22 kHz USVs with a lower mean frequency, reduced bandwidth, and longer duration than EtOH females [101].
	Cocaine	P90	SD	15 min	During foot shock procedure males had a dramatic increase in 22 kHz USVs and decrease in 50 kHz USVs. Male rats also had longer duration of 22 kHz USVs [102].
		P1, P14, P21	SD	5 min	At P1, both male and female pups with prenatal cocaine exposure (PCE) produced fewer distress USVs than saline or untreated pups and male pups with PCE produced fewer USVs with at least one observable harmonic than male saline or untreated pups. At P21, male PCE rats produced more USVs with longer overall total duration of USVs than female PCE rats [103].
		P10, P11	SD	5 min	Male pups produced more distress USVs than female pups during the final 2 min of a 5 min isolation test [104].
	Morphine	P130–P288	LE	45 min	In the presence of a cat, both male and female rats produced fewer 22 kHz USVs when exposed to morphine. Additionally, both control and morphine females produced significantly more 22 kHz USVs with longer total duration than male counterparts [105].
	Oxycodone	P3 P6 P9 P12	SD	3 min	Isolation distress USVs peaked in production rate at P9 for males and P6 and P9 for females [106].

Table 1. *Cont.*

Model	Sub Model	Age	Strain	Recording Duration	Major Sex Difference(s) in USV Acoustics
	Fluoxetine	P6	SERT	3 min	Fluoxetine reduced the total duration of distress USVs for male pups but did not affect female USV total duration [107].
	Diazepam	P3–P18	Wistar	3 min	Overall, male pups in all experimental conditions produced more distress USVs than females [108].
	Trimethylolpropane phosphate (TMPP)	P8, P14	LE	1 min	Males with prenatal TMPP treatment produced more distress USVs than control males, control females, and TMPP females [109].
Neurological disorder models	Shank3 deficiency	P7	Shank3	3 min	Fewer distress USVs were observed in male Shank3 −/− pups but not females [110].
	Pax6	P7	rSey2/+	5 min	Female rSey2/+ rat pups produce fewer distress USV from wild-type female pups, which was not observed in male rat pups [111].
	Valproic acid	P9 P31–P32 P65–P70	SD	5 min 10 min	In general, female rats had shorter duration of 50 kHz USVs during isolation, same-sex play, and same-sex social interaction than male rats. Female rats also had fewer 50 kHz USVs in same-sex social interaction [112].
	Valproic acid, chlorpyrifos	P7	Wistar	3 min	In isolation, male pups produced more distress USVs [113].
	Valproic acid, poly(I:C)	P6	SD	3 min	In the poly(I:C) condition, male pups produced more distress USVs than females [114].
	Cacna1c	P32–P34	Cacna1c	5 min	For control animals, female rats produced fewer overall 50 kHz USVs during same-sex play, with fewer step USVs and more trill USVs, than males. Female rat USVs also had a higher peak frequency. For experimental animals, female rats produced a similar rate of 50 kHz USVs during play as male control animals, whereas experimental male animals had reduced 50 kHz USV production during play [115,116].
	MAM	P60	SD	10 min	During same-sex social interaction, MAM exposure decreased the total number of 50 kHz USVs and increased the percentage of short USVs and decreased the percentage of frequency-modulated USVs for both sexes. However, control females had fewer frequency modulated USVs than control males, whereas it was opposite for MAM groups [117].

Table 1. *Cont.*

Model	Sub Model	Age	Strain	Recording Duration	Major Sex Difference(s) in USV Acoustics
		P8 P30 P31–P32	SD	3 min 3 min 10 min	At P8, males pups produced distress USVs with a lower frequency and reduced bandwidth than females. At P30, males produced tickle-induced 50 kHz USVs with a higher center frequency than females. At P31-P32, during same-sex play, males produced more USVs with greater bandwidth than females [118].
	AVP	P34 P44	Brattleboro	10 min	Males produced more trill 50 kHz USVs during same-sex play than females [119].
		P34–37	Brattleboro	10 min	Males produced more 50 kHz USVs than females during same-sex play [120].
	PD	2–8 mo	Pink1-/-	90 s	Pink-/- female rats did not have as many 50 kHz USV deficits as Pink1-/- male rats in a mating context [121].
	SE	P15 P16 P21	Wistar	5 min	SE male pups had a decrease in USV latency than control pups, which was not observed in female pups [122].
	Liposaccharide (LPS)	P11	Wistar	5 min	Prenatal LPS exposure caused male pups to produce fewer distress USVs, but this was not observed with female pups [123].
	Ischemic brain injury	P12	Wistar	3 min	Overall, ischemic pups produced fewer distress USVs than control pups with male ischemic pups experiencing greater reductions in USV call subcategories than female ischemic pups [25].
Endocrine disruption	A1221 VIN	P80–P100 d	SD	5 min	In a mating paradigm, VIN males produced fewer 50 kHz USVs than control males. A1221 produced 50 kHz USVs with reduced power, bandwidth, and lower frequency. Experimental female USVs were unaffected [124].
	A1221 estradiol	P60	SD	10 min	For female rats, estradiol treatment decreased the number of step 50 kHz USVs following opposite-sex exposure. For male rats, A1221 treatment increased the number of rise and step 50 kHz USVs following opposite-sex exposure [125].
	A1221	P30–39	SD	5 min–4 h	PBCs affected the number of 50 kHz USVs for female rats but not male rats during same-sex play [126].
Diet and environmental stressors	High-fat diet	P7 P13	LE	10 min	Female pups on the high-fat diet produced more 1-sweep distress USVs, whereas male pups on the high-fat diet produced more 2-sweep distress USVs when compared to sex-matched controls [127].

Table 1. *Cont.*

Model	Sub Model	Age	Strain	Recording Duration	Major Sex Difference(s) in USV Acoustics
	Maternal separation	P60	SD	15 min, 3 h	Brief maternal separation in pups resulted in changes in 22 kHz USVs in adulthood with fewer 22 kHz USVs in females but not in males, when compared to controls [75].
		P70–P90	SD	NS	After prenatal isolation, adult male rats produced 22 kHz USVs with greater duration compared to female rats [128].
		P120	LE	17 min	Maternal separation resulted in males producing more 22 kHz USVs in response to startle stimulus but did not affect female startle-induced 22 kHz USVs [129].
	Heat-induced convulsions	P10, P12	LE	2 min	Males produced more distress USVs (more category 5 and 6 USVs) compared to females [130].
	Moderate and extreme cold	P7–8	SD	70 min	Male pups produced more distress USVs than female pups; both sexes increased distress USVs in the presence of extreme cold temperature [131].
	Heat, light, and restraint stressors	P1 P6 P10 P14	SD	6 min	At P6, males produced more distress USVs than females during the first 2 min following maternal separation [132].
	corticotropin-releasing factor (CRF)	P6 P10 P14	SD	6 min	Overall, male pups produced more distress USVs than females [133].

A1221 = Aroclor 1221, AVP = arginine vasopressin, CRF = corticotropin-releasing factor, etOH = ethanol alcohol, LE = Long–Evans, MAM = methylazoxymethanol acetate, LPS = Liposaccharide, NS = not specified, P = postnatal day, PBCs = polychlorinated biphenyls, PCE = prenatal cocaine exposure, PD = Parkinson's Disease, poly(I:C) = polyinosinic-polycytidylic acid, SD= Sprague–Dawley, SE = status epilepticus, SERT = serotonin transporter deficient, TMPP = trimethylolpropane phosphate, USV = ultrasonic vocalization, and VIN = vinclozolin.

2.2.3. Adult 50 kHz USVs

Both male and female rats produce 50 kHz USVs during various social contexts such as rough-and-tumble play, mating, and in isolation. These USVs are critical to the communicative intent of the rats and often impact the behavior of conspecifics [134–136]. Because 50 kHz USVs are often observed in appetitive situations and during physical interactions, features of these USVs have been investigated in different social contexts. These 50 kHz USVs can be subcategorized based on spectral features such as duration and frequency modulation. Wright et al. introduced 14 categories of 50 kHz USVs [37]; however, many studies use simpler categorization such as flat vs frequency modulated [137]. Although there is no current consensus on rat 50 kHz USV subcategories, rats are known to produce a variety of these vocalizations, and recent advancements in the efficiency of USV data analysis will lend to greater cross-institutional collaboration to better elucidate USV subcategories and their communicative relevance [138]. Furthermore, these social USVs such as alarm and pup USVs are also influenced by rat strain [139].

Rough and Tumble Play

During social interactions, such as social play, both young and aged rats will frequently produce 50 kHz vocalizations [7,140,141]. These vocalizations are thought to be produced to promote playful contact with peer rats and function as play signals [22]. During rough-

and-tumble play, 50 kHz USVs often co-occur with attack-like, play behaviors [23]. These interactions, however, can transition beyond play fighting and into more serious fighting behavior [142]. Specifically, when pairs of unfamiliar adult males were exposed to each other, there was an increased risk to escalate into aggressive behavior if one partner is devocalized, or unable to communicate with USVs [143]. Thus, the importance of communication during rough-and-tumble play is critical to prevent this escalation [22]. Therefore, vocal communication during rough-and-tumble play has been investigated to explore the social ecological value it provides. To further explore their utility, sex differences in USV production have been identified. In rough-and-tumble play, male rats produce a greater amount of 50 kHz USVs when compared to female rats [143]. This sex difference has been attributed to males desire to engage in more rough play [142], but further exploration is warranted.

Mating

In addition to play-based social interactions, 50 kHz USVs are produced in mating contexts to initiate approach behaviors of mating partners [4]. Research findings are mixed regarding the role of USVs in mating, but one theory suggests that male 50 kHz USVs are prosocial in nature and elicit female copulation behaviors [144–147]. Other results suggest that female USVs do not provide mating incentive for male counterparts and instead support that male rats will show sexual interest in the female rat regardless of the presence of USVs [148–150], and male USVs did not influence female USV production [146]. However, the presence of an estrus (sexually receptive) female rat significantly increases male vocalizations [151]. Additionally, female vocalization rates were significantly increased during peak periods of sexual receptivity (during estrus), in contrast with male-only vocalizations, which were not linked to sexual receptivity [152,153]. In regards to female USVs, it has been found that female rats produced higher proportion of frequency modulated USVs to male peers compared to female peers, and females produced a higher total number of USVs when in the presence of male rats that have not been castrated [154]. These findings suggest that female vocalizations are also influential in motivating sexual interactions and may indicate communicative intent during opposite-sex encounters.

Housing Environment and Aging

Both housing environment and age influence USV production rates and acoustics in male rats. Social isolation influences USV production rates and acoustics depending on the length of time of isolation. Wöhr et al. found that male rats exposed to a brief period of social isolation produced more than twice the amount of USVs when compared to other rats, likely due to increased social motivation after isolating [5]. However, after longer periods of isolation (2–6 months), studies have found that socially isolated male rats produced fewer 50 kHz USVs with lower amplitude in response to a female rat than socially-housed rats [155,156]. Thus, it appears that while short-term social isolation may increase USV production rates in social situations, long-term social isolation decreases the number and amplitude of social USVs. The inclusion of female rats in future studies is warranted to understand how sexes may respond differently to social isolation in adulthood.

In male rats, aging has been shown to change USV acoustics with older rats producing fewer 50 kHz USVs with reduced peak frequency, frequency bandwidth, and amplitude in a mating context [31,34,35]. These changes to USV acoustics co-occur with non-muscular and neuromuscular changes in the larynx. Some of these changes include the following: reduced hyaluronic acid, reduced elastin densities, and increased collagen densities of the vocal fold [35]; motorneuron loss of the nucleus ambiguus [34]; deinnervation-like changes to the neuromuscular junction of the thyroarytenoid muscles [31,157–160]; reductions in muscle-twitch functions of the thyroarytenoid muscles [161]; and alterations to intrinsic laryngeal myofiber structures [33,158,161,162]. Although many of these changes may contribute to functional age-related deficits observed in rat larynx, as previously mentioned, USV production is a complex orchestration of many muscular subsystems that which simultaneously under age-related changes and cannot be explained by a single muscular system such as the larynx. Importantly, most of the studies investigating age-related changes to the larynx have been completed solely with male rats prohibiting any insight to sex differences in the aging rat larynx.

2.3. Sex Differences in USV Acoustics

2.3.1. Alarm 22 kHz USV Acoustics

Although studies have confirmed differences between male and female alarm USV productions, most studies have focused on solely the number or overall duration of alarm USVs produced in fear contexts and have overlooked acoustic features or alarm subtypes. In Blanchard's anti-predator USV study, female rats produce more frequent alarm USVs with a higher mean frequency and shorter duration [70]. In the response to a predator, male rats primarily produced negatively accelerated descending frequency alarm USVs, whereas female rats primarily produced linear descending alarm USVs. Another study that evaluated the effects of serotonin transporter deficiency in a fear condition found that female rats produced fewer overall alarm USVs than male rats with a higher frequency modulation and longer USV duration than males [72]. This finding complemented other studies that found higher frequencies of 22 kHz alarm USVs of female rats [70,105,163,164].

In summary, although alarm emission rates have been revealed to be different between sexes, sex differences in the subtypes and acoustic parameters of alarm USVs are relatively unexplored. The sex differences may be differentially regulated by the endocrine system with male rat alarm USVs being affected by sex hormones and female rat alarm USVs having less hormone dependence.

2.3.2. Pup Distress USV Acoustics

As previously described, pups produce distress USVs during approximately the first 3 weeks of life in response to separation/isolation from the dam. This distress signal functions as both a social and survival act and signals the dam to retrieve and care for the pup. Because male pups produce more distress USVs, and these USVs tend to be lower in both mean frequency and amplitude compared to female pups, dams tend to respond to the male pups producing these USVs more so than female pups [17,36]. This sexual dimorphism may be mediated by the *FOXP2* gene, with a general reduction in *FOXP2* protein observed in females compared to males [17]. Although few studies have investigated the sexual dimorphism of distress USVs of typical rat pups, several studies in Table 1 highlight sex differences in distress USVs in various experimental models.

2.3.3. Adult 50 kHz USV Acoustics

While USV acoustic properties have not been extensively studied between sexes, there have been a small number of studies exploring the difference between male and female rat vocalizations in terms of specific acoustic parameters. One study explored the impact of social situations on vocalizations between sexes, which revealed that female rats produced a higher proportion of frequency modulated 50 kHz USVs when exposed to a male partner compared to a female partner [154]. This suggested that female USVs may be indicative of sexual motivation. Other studies found that during rough-and-tumble play, levels of 50 kHz vocalizations was decreased as a result of Cacna1c haploinsufficiency, a gene implicated in social signal processing, which was more robustly noted in males than females [78]. Additionally, studies have explored acoustical parameter differences in male and female rats in isolation. Specifically, during isolation, the mean frequency of 50 kHz USVs was significantly lower in males than in females [61]. While there is work to be carried out in acoustic analysis exploring sex differences, these studies highlight the need for the inclusion of male and female rats into experimental studies as we continue to learn how social settings impact vocalizations produced by both sexes.

2.3.4. Estrous Cycle, Pregnancy, Menopause, and Estropause Effects on USV Acoustics

Female Rat Hormone Cycle

The female rat, like other mammals, has an ovarian hormone cycle that begins following sexual maturation (puberty); undergoes cyclical regulation by the hypothalamus, ovaries, and pituitary gland; is ceased during pregnancy; and finally undergoes age-related dysregulation and subsequently infertility [165]. Nevertheless, the estrous cycle and age-related changes in the rat are uniquely different from other mammals.

In brief, female rats reach sexual maturity at ~3 months of age [166]. The typical estrous cycle of a female rat lasts between 4 and 5 days beginning with proestrus, which is ~14 h and has both high estradiol and progesterone levels (female rat might be receptive during proestrus), estrus (the sexually receptive state) which is ~24–48 h and has low estradiol and progesterone levels, and finally diestrus states which hormone levels begin to rise and the female rat refuses copulation [166,167]. At ~9 months of age female rats will begin to experience irregular cycling for ~1 month and enter estropause that has three stages: constant estrus, persistent diestrus with irregular cycling, and persistent diestrus [165,168]. An important distinction of rat estropause is that rats continue to secrete low-moderate ovarian hormones during constant estrus and elevated ovarian hormone levels during persistent diestrus, which contrasts to humans who experience dramatic loss of ovarian hormones at menopause [165,168]. Because of this difference in ovarian hormone status, the effects of menopause are studied using ovariectomy procedures in the rat model, which more closely mimics menopause of humans by eliminating the primary production of ovarian hormones via removal of the ovaries.

The subsequent sections will summarize the known effects of the ovarian cycle of the female rat on USV production and acoustics. The scant knowledge will be evident in the review.

Estrous Cycle, Pregnancy, and USVs

Few studies have investigated the effects of the hormone cycle or pregnancy on USV acoustics. Furthermore, to our knowledge no studies to date have investigated the effects of estropause stages on USV production or acoustics. In the following sections, both the USV production rates and acoustics will be described according to USV type: alarm 22 kHz USVs and 50 kHz USVs.

Alarm USVs

Few studies have evaluated the effects of the estrous cycle on alarm 22 kHz USVs. One study evaluated the total duration of alarm USVs produced after a puff of air and found no differences between proestrus and diestrus stages [79]. Nevertheless, although the overall duration of alarm USVs did not differ between the two evaluated estrous states, the entire cycle was not evaluated, and this analysis did not include acoustics. Therefore, the effects of the estrous cycle on alarm USV production and acoustics are unknown.

Although several studies have evaluated how perinatal conditions affect USVs, the effects have primarily been evaluated in the pup offspring rather than the pregnant dams (Table 1). However, two studies have demonstrated that stress affects dam USVs [169,170]. In one study postpartum dams that received brief or long-term separation from their litters, produced more 22 kHz USVs in response to a startle stimulus than control dams [169]. In another study where the non-pregnant female cagemate of a pregnant dam underwent two 30-min stress tests for five consecutive days and then was recorded with pregnant female cagemate, the stressed non-pregnant females produced more 22 kHz USVs during interaction while the pregnant bystander produced more 50 kHz USVs than the stressed non-pregnant females [170]. Therefore, although perinatal models have demonstrated effects on pup USVs, perinatal effects can also affect dam USVs, and currently it is unclear if dam USVs influence the USVs of their pups revealing a large gap in the literature.

Adult 50 kHz USVs

In mating contexts, several studies have found that female Long–Evans rats produce more 50 kHz USVs during receptive (proestrus and estrus) stages of the estrous cycle [153,171,172]. Because high rates of 50 kHz USVs are observed at both high hormone (proestrus) and low hormone (estrus) stages, these studies have collectively suggested that 50 kHz USV production rate may serve as a proceptive cue to male rats in mating contexts rather than a hormonal effect [153,171,172].

In a mating context, the USV acoustics of Long–Evans female rats are influenced by the estrous cycle [153]. In general, USV frequency parameters (such as lowest, highest, and median frequencies) are highest during high hormonal states (diestrus II and proestrus) and lowest during low hormonal states (estrus and metestrus) [153]. USV intensity (power) tends to be greatest during low hormonal states (estrus and metestrus), whereas duration and USV complexity (frequency modulation) tend to be greatest during high hormone (diestrus II and proestrus) and receptive states (proestrus and estrus) [153]. Therefore, USV acoustics are influenced by copulation behavior and hormone levels in normal-cycling female rats.

Another study found that hormonal injections influenced USV acoustic parameters of trill and flat-trill 50 kHz USVs during clitoral stimulation [147]. In this study, rats were ovariectomized and treated with estradiol, progesterone, estradiol + progesterone, or a vehicle. The combined estradiol + progesterone treatment significantly increased the rate of USVs as well as the duration and complexity of the USVs [147]. This finding mirrors the previously mentioned study that found USV duration and complexity to be greatest during high hormone/receptive states.

In a mating context, an ovariectomy affects the rate of USV production but has minimal effects on USV acoustics when compared to USVs of normal-cycling rats. The elimination of the cycle via ovariectomy overall reduces the number of USVs produced during mating contexts [153,171,172]. Studies found that ovariectomized Long–Evans female rats produced fewer 50 kHz USVs in a mating context than receptive age-matched females [153,171,172]. Additionally, in ovariectomized rats the USV acoustic parameters of frequency, complexity, intensity, and duration did not differ from control rats when compared across the estrous cycle [153]. Nevertheless, this reduction in USV rate can be counteracted with estradiol + progesterone injections [151,173]; however, estradiol or progesterone alone does not increase the number of 50 kHz USVs in ovariectomized Long–Evans rats in mating contexts [173].

In a non-mating social context, ovariectomized Sprague–Dawley rats receiving estradiol produced fewer 50 kHz USVs than ovariectomized Sprague–Dawley rats without hormone treatment [174]. Although the study's authors predicted a higher 50 kHz USV production rate in rats receiving estradiol treatment, the decreased USV production may be indicative of improved social memory of the estradiol treatment group. Additionally, combined estrogen and progesterone treatments may be required to enhance social USVs. In a similar non-mating social context, ovariectomized Long–Evans female rats produce a similar number of USVs with similar acoustics to age-matched normal-cycling rats [153]. These results indicate that the estrous cycle influences the rate and acoustics of 50 kHz USVs during mating contexts more than non-mating social contexts.

In social isolation, the estrous cycle has less influence on USV parameters [153]. In isolation, female rats in estrus produced the most USVs with the lowest frequency parameters, greatest intensity, and complexity [153]. Female rats in metestrus produced the USVs with the least complexity, and shortest duration [153]. These results indicate that both the behavior and hormonal levels of the ovarian cycle influence USV acoustics of the normal-cycling female rat.

Ovarian Hormone Summary

In summary, the effects of the estrous cycle and ovarian hormones require further study to determine their influence on the female rat USV. To date, in the normal-cycling female rat, the estrous cycle is known to influence the number and acoustics of the 50 kHz USV in social contexts (primarily mating contexts). Additionally, the ovariectomy is known to reduce the number of 50 kHz USVs produced during mating contexts but not the acoustic parameters. Collectively, studies demonstrate an effect of ovarian hormones on 50 kHz USVs.

3. Conclusions

Biological and acoustic sex differences are apparent in the rat laryngeal mechanism. The type and acoustic features of USVs are different between male and female rats and are dependent on age, strain, and experimental models. Additionally, rat vocal folds are also sexually dimorphic which may contribute to the observed USV production and acoustic sex differences. This sexual dimorphism has been partially attributed to sex hormones; however, few studies of the laryngeal mechanism have investigated the role of hormones in influencing USV production and acoustic features. With recent advances in technology (such as DeepSqueak [138]) which simplifies and reduces the time burden of USV analysis, sexual dimorphism of the rat larynx can be further explored.

Author Contributions: Conceptualization, C.L. and M.R.C.; methodology, C.L.; resources, C.L., C.K.B. and N.E.S.-H.; data curation, C.L.; writing—original draft preparation, C.L., C.K.B. and N.E.S.-H.; writing—review and editing, C.L. and M.R.C.; visualization, C.L.; supervision, C.L. and M.R.C.; project administration, C.L.; funding acquisition, M.R.C. All authors have read and agreed to the published version of the manuscript.

Funding: This research was funded by National Institutes of Health, grant numbers: T32DC009401-12 (PI: Thibeault), F31DC018726 (PI: Broadfoot), and R01DC018584 (PI: Ciucci). The APC was funded by R01DC018584(PI: Ciucci).

Acknowledgments: Thank you to the members of Ciucci's lab who provided valuable feedback and editing for this manuscript.

Conflicts of Interest: The authors declare no conflict of interest.

References

1. Wöhr, M.; Schwarting, R.K. Affective communication in rodents: Ultrasonic vocalizations as a tool for research on emotion and motivation. *Cell Tissue Res.* **2013**, *354*, 81–97. [CrossRef] [PubMed]
2. Sewell, G.D. Ultrasonic communication in rodents. *Nature* **1970**, *227*, 410. [CrossRef] [PubMed]
3. Simola, N.; Granon, S. Ultrasonic vocalizations as a tool in studying emotional states in rodent models of social behavior and brain disease. *Neuropharmacology* **2019**, *159*, 107420. [CrossRef] [PubMed]
4. Seffer, D.; Schwarting, R.K.; Wöhr, M. Pro-social ultrasonic communication in rats: Insights from playback studies. *J. Neurosci. Methods* **2014**, *234*, 73–81. [CrossRef]
5. Wöhr, M.; Engelhardt, K.A.; Seffer, D.; Sungur, A.Ö.; Schwarting, R.K. Acoustic Communication in Rats: Effects of Social Experiences on Ultrasonic Vocalizations as Socio-Affective Signals. In *Social Behavior from Rodents to Humans*; Springer: Berlin/Heidelberg, Germany, 2015; pp. 67–89.
6. Brudzynski, S.M. Ethotransmission: Communication of emotional states through ultrasonic vocalization in rats. *Curr. Opin. Neurobiol.* **2013**, *23*, 310–317. [CrossRef]
7. Burgdorf, J.; Kroes, R.A.; Moskal, J.R.; Pfaus, J.G.; Brudzynski, S.M.; Panksepp, J. Ultrasonic vocalizations of rats (*Rattus norvegicus*) during mating, play, and aggression: Behavioral concomitants, relationship to reward, and self-administration of playback. *J. Comp. Psychol.* **2008**, *122*, 357–367. [CrossRef]
8. Portfors, C.V. Types and functions of ultrasonic vocalizations in laboratory rats and mice. *J. Am. Assoc. Lab. Anim. Sci.* **2007**, *46*, 28–34.
9. Schwarting, R.; Wöhr, M. On the relationships between ultrasonic calling and anxiety-related behavior in rats. *Braz. J. Med. Biol. Res.* **2012**, *45*, 337–348. [CrossRef]
10. Brudzynski, S.M. Pharmacology of ultrasonic vocalizations in adult rats: Significance, call classification and neural substrate. *Curr. Neuropharmacol.* **2015**, *13*, 180–192. [CrossRef] [PubMed]
11. Hofer, M.A.; Shair, H.N. Ultrasonic vocalization, laryngeal braking, and thermogenesis in rat pups: A reappraisal. *Behav. Neurosci.* **1993**, *107*, 354–362. [CrossRef]
12. Fernandez-Vargas, M.; Johnston, R.E. Ultrasonic vocalizations in golden hamsters (*Mesocricetus auratus*) reveal modest sex differences and nonlinear signals of sexual motivation. *PLoS ONE* **2015**, *10*, 0116789. [CrossRef]

13. Warren, M.R.; Spurrier, M.S.; Roth, E.D.; Neunuebel, J.P. Sex differences in vocal communication of freely interacting adult mice depend upon behavioral context. *PLoS ONE* **2018**, *13*, 0204527. [CrossRef]
14. Wright, S.L.; Brown, R.E. Sex differences in ultrasonic vocalizations and coordinated movement in the California mouse (*Peromyscus californicus*). *Behav. Processes* **2004**, *65*, 155–162. [CrossRef] [PubMed]
15. Zala, S.M.; Reitschmidt, D.; Noll, A.; Balazs, P.; Penn, D.J. Sex-dependent modulation of ultrasonic vocalizations in house mice (*Mus musculus musculus*). *PLoS ONE* **2017**, *12*, 0188647. [CrossRef] [PubMed]
16. Heckman, J.J.; Proville, R.; Heckman, G.J.; Azarfar, A.; Celikel, T.; Englitz, B. High-precision spatial localization of mouse vocalizations during social interaction. *Sci. Rep.* **2017**, *7*, 3017. [CrossRef] [PubMed]
17. Bowers, J.M.; Perez-Pouchoulen, M.; Edwards, N.S.; McCarthy, M.M. Foxp2 mediates sex differences in ultrasonic vocalization by rat pups and directs order of maternal retrieval. *J. Neurosci.* **2013**, *33*, 3276–3283. [CrossRef]
18. Bowers, J.M.; Perez-Pouchoulen, M.; Roby, C.R.; Ryan, T.E.; McCarthy, M.M. Androgen modulation of Foxp1 and Foxp2 in the developing rat brain: Impact on sex specific vocalization. *Endocrinology* **2014**, *155*, 4881–4894. [CrossRef]
19. Frohlich, H.; Rafiullah, R.; Schmitt, N.; Abele, S.; Rappold, G.A. Foxp1 expression is essential for sex-specific murine neonatal ultrasonic vocalization. *Hum. Mol. Genet.* **2017**, *26*, 1511–1521. [CrossRef]
20. Wöhr, M.; Houx, B.; Schwarting, R.K.; Spruijt, B. Effects of experience and context on 50-kHz vocalizations in rats. *Physiol. Behav.* **2008**, *93*, 766–776. [CrossRef] [PubMed]
21. Himmler, B.T.; Kisko, T.M.; Euston, D.R.; Kolb, B.; Pellis, S.M. Are 50-kHz calls used as play signals in the playful interactions of rats? I. Evidence from the timing and context of their use. *Behav. Processes* **2014**, *106*, 60–66. [CrossRef]
22. Himmler, B.T.; Pellis, V.C.; Pellis, S.M. Peering into the dynamics of social interactions: Measuring play fighting in rats. *JoVE J. Vis. Exp.* **2013**, 4288. [CrossRef]
23. Thomas, D.A.; Takahashi, L.K.; Barfield, R.J. Analysis of ultrasonic vocalizations emitted by intruders during aggressive encounters among rats (*Rattus norvegicus*). *J. Comp. Psychol.* **1983**, *97*, 201. [CrossRef] [PubMed]
24. Ziemka-Nalecz, M.; Jaworska, J.; Sypecka, J.; Polowy, R.; Filipkowski, R.K.; Zalewska, T. Sodium Butyrate, a Histone Deacetylase Inhibitor, Exhibits Neuroprotective/Neurogenic Effects in a Rat Model of Neonatal Hypoxia-Ischemia. *Mol. Neurobiol.* **2017**, *54*, 5300–5318. [CrossRef] [PubMed]
25. Saucier, D.M.; Ehresman, C.A.; Keller, A.J.; Armstrong, E.; Elderkin, A.; Yager, J.Y. Hypoxia ischemia affects ultrasonic vocalization in the neonatal rat. *Behav. Brain Res.* **2008**, *190*, 243–247. [CrossRef] [PubMed]
26. Johnson, R.A.; Kelm-Nelson, C.A.; Ciucci, M.R. Changes to Ventilation, Vocalization, and Thermal Nociception in the Pink1-/- Rat Model of Parkinson's Disease. *J. Parkinsons Dis.* **2020**, *10*, 489–504. [CrossRef]
27. Kelm-Nelson, C.A.; Brauer, A.F.; Ciucci, M.R. Vocal training, levodopa, and environment effects on ultrasonic vocalizations in a rat neurotoxin model of Parkinson disease. *Behav. Brain Res.* **2016**, *307*, 54–64. [CrossRef]
28. Ciucci, M.R.; Vinney, L.; Wahoske, E.J.; Connor, N.P. A translational approach to vocalization deficits and neural recovery after behavioral treatment in Parkinson disease. *J. Commun. Disord.* **2010**, *43*, 319–326. [CrossRef]
29. Ciucci, M.R.; Ahrens, A.M.; Ma, S.T.; Kane, J.R.; Windham, E.B.; Woodlee, M.T.; Schallert, T. Reduction of dopamine synaptic activity: Degradation of 50-kHz ultrasonic vocalization in rats. *Behav. Neurosci.* **2009**, *123*, 328–336. [CrossRef]
30. Johnson, A.M.; Grant, L.M.; Schallert, T.; Ciucci, M.R. Changes in Rat 50-kHz Ultrasonic Vocalizations During Dopamine Denervation and Aging: Relevance to Neurodegeneration. *Curr. Neuropharmacol.* **2015**, *13*, 211–219. [CrossRef]
31. Johnson, A.M.; Ciucci, M.R.; Connor, N.P. Vocal training mitigates age-related changes within the vocal mechanism in old rats. *J. Gerontol. A Biol. Sci. Med. Sci.* **2013**, *68*, 1458–1468. [CrossRef]
32. Johnson, A.M.; Doll, E.J.; Grant, L.M.; Ringel, L.; Shier, J.N.; Ciucci, M.R. Targeted training of ultrasonic vocalizations in aged and Parkinsonian rats. *J. Vis. Exp.* **2011**, *54*, e2835. [CrossRef]
33. Lenell, C.; Kelm-Nelson, C.A.; Ciucci, M.R.; Johnson, A.M. Chapter 36–Changes in Ultrasonic Vocalizations in Senescent Rats. In *Handbook of Behavioral Neuroscience*; Brudzynski, S.M., Ed.; Elsevier: Amsterdam, The Netherlands, 2018; Volume 25, pp. 383–386.
34. Basken, J.N.; Connor, N.P.; Ciucci, M.R. Effect of aging on ultrasonic vocalizations and laryngeal sensorimotor neurons in rats. *Exp. Brain Res.* **2012**, *219*, 351–361. [CrossRef]
35. Peterson, J.R.; Watts, C.R.; Morris, J.A.; Shelton, J.M.; Cooper, B.G. Laryngeal aging and acoustic changes in male rat ultrasonic vocalizations. *Dev. Psychobiol.* **2013**, *55*, 818–828. [CrossRef]
36. Simola, N. Rat Ultrasonic Vocalizations and Behavioral Neuropharmacology: From the Screening of Drugs to the Study of Disease. *Curr. Neuropharmacol.* **2015**, *13*, 164–179. [CrossRef] [PubMed]
37. Wright, J.M.; Gourdon, J.C.; Clarke, P.B. Identification of multiple call categories within the rich repertoire of adult rat 50-kHz ultrasonic vocalizations: Effects of amphetamine and social context. *Psychopharmacology (Berlin)* **2010**, *211*, 1–13. [CrossRef] [PubMed]
38. Barker, D.J.; Simmons, S.J.; West, M.O. Ultrasonic vocalizations as a measure of affect in preclinical models of drug abuse: A review of current findings. *Curr. Neuropharmacol.* **2015**, *13*, 193–210. [CrossRef] [PubMed]
39. Hodgson, R.A.; Guthrie, D.H.; Varty, G.B. Duration of ultrasonic vocalizations in the isolated rat pup as a behavioral measure: Sensitivity to anxiolytic and antidepressant drugs. *Pharmacol. Biochem. Behav.* **2008**, *88*, 341–348. [CrossRef]
40. Mu, P.; Fuchs, T.; Saal, D.B.; Sorg, B.A.; Dong, Y.; Panksepp, J. Repeated cocaine exposure induces sensitization of ultrasonic vocalization in rats. *Neurosci. Lett.* **2009**, *453*, 31–35. [CrossRef] [PubMed]

41. Naito, H.; Okumara, T.; Inoue, M.; Suzuki, Y. Ultrasonic vocalization response elicited in adjuvant-induced arthritic rats as a useful method for evaluating analgesic drugs. *Exp. Anim.* **2006**, *55*, 125–129. [CrossRef]
42. Naito, H.; Nakamura, A.; Inoue, M.; Suzuki, Y. Effect of anxiolytic drugs on air-puff-elicited ultrasonic vocalization in adult rats. *Exp. Anim.* **2003**, *52*, 409–414. [CrossRef]
43. Wintink, A.J.; Brudzynski, S.M. The related roles of dopamine and glutamate in the initiation of 50-kHz ultrasonic calls in adult rats. *Pharmacol. Biochem. Behav.* **2001**, *70*, 317–323. [CrossRef]
44. Simola, N.; Fenu, S.; Costa, G.; Pinna, A.; Plumitallo, A.; Morelli, M. Pharmacological characterization of 50-kHz ultrasonic vocalizations in rats: Comparison of the effects of different psychoactive drugs and relevance in drug-induced reward. *Neuropharmacology* **2012**, *63*, 224–234. [CrossRef]
45. Riede, T. Peripheral Vocal Motor Dynamics and Combinatory Call Complexity of Ultrasonic Vocal Production in Rats. In *Handbook of Behavioral Neuroscience*; Elsevier: Amsterdam, The Netherlands, 2018; Volume 25, pp. 45–60.
46. Riede, T. Stereotypic laryngeal and respiratory motor patterns generate different call types in rat ultrasound vocalization. *J. Exp. Zool. A Ecol. Genet. Physiol.* **2013**, *319*, 213–224. [CrossRef] [PubMed]
47. Riede, T. Subglottal pressure, tracheal airflow, and intrinsic laryngeal muscle activity during rat ultrasound vocalization. *J. Neurophysiol.* **2011**, *106*, 2580–2592. [CrossRef] [PubMed]
48. Håkansson, J.; Jiang, W.; Xue, Q.; Zheng, X.; Ding, M.; Agarwal, A.A.; Elemans, C.P.H. Aerodynamics and motor control of ultrasonic vocalizations for social communication in mice and rats. *bioRxiv* **2021**. [CrossRef]
49. Kobler, J.B.; Datta, S.; Goyal, R.K.; Benecchi, E.J. Innervation of the larynx, pharynx, and upper esophageal sphincter of the rat. *J. Comp. Neurol.* **1994**, *349*, 129–147. [CrossRef] [PubMed]
50. Pascual-Font, A.; Hernández-Morato, I.; McHanwell, S.; Vázquez, T.; Maranillo, E.; Sañudo, J.; Valderrama-Canales, F.J. The central projections of the laryngeal nerves in the rat. *J. Anat.* **2011**, *219*, 217–228. [CrossRef] [PubMed]
51. Van Daele, D.J.; Cassell, M.D. Multiple forebrain systems converge on motor neurons innervating the thyroarytenoid muscle. *Neuroscience* **2009**, *162*, 501–524. [CrossRef] [PubMed]
52. Wetzel, D.M.; Kelley, D.B.; Campbell, B.A. Central control of ultrasonic vocalizations in neonatal rats: I. Brain stem motor nuclei. *J. Comp. Physiol. Psychol.* **1980**, *94*, 596. [CrossRef] [PubMed]
53. Kelm-Nelson, C.A.; Lenell, C.; Johnson, A.M.; Ciucci, M.R. Laryngeal Activity for Production of Ultrasonic Vocalizations in Rats. In *Handbook of Behavioral Neuroscience*; Elsevier: Amsterdam, The Netherlands, 2018; Volume 25, pp. 37–43.
54. Johnson, A.M.; Ciucci, M.R.; Russell, J.A.; Hammer, M.J.; Connor, N.P. Ultrasonic output from the excised rat larynx. *J. Acoust. Soc. Am.* **2010**, *128*, 75–79. [CrossRef]
55. Riede, T. Rat ultrasonic vocalization shows features of a modular behavior. *J. Neurosci.* **2014**, *34*, 6874–6878. [CrossRef] [PubMed]
56. Riede, T.; Borgard, H.L.; Pasch, B. Laryngeal airway reconstruction indicates that rodent ultrasonic vocalizations are produced by an edge-tone mechanism. *R. Soc. Open Sci.* **2017**, *4*, 170976. [CrossRef] [PubMed]
57. Inagi, K.; Schultz, E.; Ford, C.N. An anatomic study of the rat larynx: Establishing the rat model for neuromuscular function. *Otolaryngol. Head Neck Surg.* **1998**, *118*, 74–81. [CrossRef]
58. Toya, Y.; Riabroy, N.; Davis, C.R.; Kishimoto, Y.; Tanumihardjo, S.A.; Bless, D.M.; Welham, N.V. Interspecies comparison of stellate cell-containing macula flavae and vitamin A storage in vocal fold mucosa. *J. Anat.* **2014**, *225*, 298–305. [CrossRef] [PubMed]
59. Brudzynski, S.M.; Fletcher, N.H. Rat Ultrasonic Vocalization: Short-Range Communication. In *Handbook of Behavioral Neuroscience*; Elsevier: Amsterdam, The Netherlands, 2010; Volume 19, pp. 69–76.
60. Kaltwasser, M.-T. Acoustic signaling in the black rat (*Rattus rattus*). *J. Comp. Psychol.* **1990**, *104*, 227. [CrossRef]
61. Lenell, C.; Johnson, A.M. Sexual dimorphism in laryngeal muscle fibers and ultrasonic vocalizations in the adult rat. *Laryngoscope* **2017**, *127*, 270–276. [CrossRef]
62. Feng, X.; Zhang, T.; Ralston, E.; Ludlow, C.L. Differences in neuromuscular junctions of laryngeal and limb muscles in rats. *Laryngoscope* **2012**, *122*, 1093–1098. [CrossRef]
63. Lenell, C.; Johnson, A.M. The Effects of Menopause on Neuromuscular Parameters of the Rat Vocal Folds. *Laryngoscope* **2020**. [CrossRef]
64. Kim, J.M.; Shin, S.C.; Park, G.C.; Lee, J.C.; Jeon, Y.K.; Ahn, S.J.; Thibeault, S.; Lee, B.J. Effect of sex hormones on extracellular matrix of lamina propria in rat vocal fold. *Laryngoscope* **2020**, *130*, 732–740. [CrossRef] [PubMed]
65. Oyarzún, P.; Sepúlveda, A.; Valdivia, M.; Roa, I.; Cantín, M.; Trujillo, G.; Zavando, D.; Galdames, I.S. Variations of the Vocal Fold Epithelium in a Menopause Induced Model. *Int. J. Morphol.* **2011**, *29*, 377–381. [CrossRef]
66. Tatlipinar, A.; Gunes, P.; Ozbeyli, D.; Cimen, B.; Gokceer, T. Effects of ovariectomy and estrogen replacement therapy on laryngeal tissue: A histopathological experimental animal study. *Otolaryngol. Head Neck Surg.* **2011**, *145*, 987–991. [CrossRef] [PubMed]
67. Ulkumen, B.; Ulkumen, B.A.; Batir, M.B.; Pala, H.G.; Vatansever, S.; Cam, S. Impact of pregnancy and glucocorticoid treatment on NF-κB and MUC5AC in mucosa of rat larynx. *J. Voice* **2019**. [CrossRef] [PubMed]
68. Şanal, S.K.; Biçer, Y.Ö.; Kükner, A.; Tezcan, E. Effect of pregnancy on vocal cord histology: An animal experiment. *Balk. Med. J.* **2016**, *33*, 448. [CrossRef] [PubMed]
69. Inagaki, H. Chapter 32–Sex Differences in Ultrasonic Vocal Expression of Negative Emotional States in Rats. In *Handbook of Behavioral Neuroscience*; Brudzynski, S.M., Ed.; Elsevier: Amsterdam, The Netherlands, 2018; Volume 25, pp. 337–344.
70. Blanchard, R.J.; Agullana, R.; McGee, L.; Weiss, S.; Blanchard, D.C. Sex differences in the incidence and sonographic characteristics of antipredator ultrasonic cries in the laboratory rat (*Rattus norvegicus*). *J. Comp. Psychol.* **1992**, *106*, 270.

71. Graham, L.K.; Yoon, T.; Lee, H.J.; Kim, J.J. Strain and sex differences in fear conditioning: 22 kHz ultrasonic vocalizations and freezing in rats. *Psychol. Neurosci.* **2009**, *2*, 219–225. [CrossRef]
72. Willadsen, M.; Uengoer, M.; Schwarting, R.K.W.; Homberg, J.R.; Wöhr, M. Reduced emission of alarm 22-kHz ultrasonic vocalizations during fear conditioning in rats lacking the serotonin transporter. *Prog. Neuropsychopharmacol. Biol. Psychiatry* **2020**. [CrossRef] [PubMed]
73. Mallo, T.; Matrov, D.; Herm, L.; Koiv, K.; Eller, M.; Rinken, A.; Harro, J. Tickling-induced 50-kHz ultrasonic vocalization is individually stable and predicts behaviour in tests of anxiety and depression in rats. *Behav. Brain Res.* **2007**, *184*, 57–71. [CrossRef] [PubMed]
74. Mällo, T.; Matrov, D.; Kõiv, K.; Harro, J. Effect of chronic stress on behavior and cerebral oxidative metabolism in rats with high or low positive affect. *Neuroscience* **2009**, *164*, 963–974. [CrossRef]
75. Kosten, T.A.; Lee, H.J.; Kim, J.J. Early life stress impairs fear conditioning in adult male and female rats. *Brain Res.* **2006**, *1087*, 142–150. [CrossRef]
76. Dimatelis, J.J.; Stein, D.J.; Russell, V.A. Behavioral changes after maternal separation are reversed by chronic constant light treatment. *Brain Res.* **2012**, *1480*, 61–71. [CrossRef]
77. Dimatelis, J.J.; Vermeulen, I.M.; Bugarith, K.; Stein, D.J.; Russell, V.A. Female rats are resistant to developing the depressive phenotype induced by maternal separation stress. *Metab. Brain Dis.* **2016**, *31*, 109–119. [CrossRef] [PubMed]
78. Wöhr, M.; Willadsen, M.; Kisko, T.M.; Schwarting, R.K.; Fendt, M. Sex-dependent effects of Cacna1c haploinsufficiency on behavioral inhibition evoked by conspecific alarm signals in rats. *Prog. Neuropsychopharmacol. Biol. Psychiatry* **2020**, *99*, 109849. [CrossRef] [PubMed]
79. Inagaki, H.; Mori, Y. The emission of stress-induced 22-kHz calls in female rats is independent of testosterone levels. *Horm. Behav.* **2015**, *69*, 116–118. [CrossRef] [PubMed]
80. Inagaki, H.; Mori, Y. Relationship between 22-kHz calls and testosterone in male rats. *Horm. Behav.* **2014**, *65*, 42–46. [CrossRef]
81. Barfield, R.J.; Auerbach, P.; Geyer, L.A.; Mcintosh, T.K. Ultrasonic Vocalizations in Rat Sexual-Behavior. *Am. Zool.* **1979**, *19*, 469–480. [CrossRef]
82. Barfield, R.J.; Geyer, L.A. The ultrasonic postejaculatory vocalization and the postejaculatory refractory period of the male rat. *J. Comp. Physiol. Psychol.* **1975**, *88*, 723–734. [CrossRef]
83. Miczek, K.A.; Van Der Poel, A.M. Long ultrasonic calls in male rats following mating, defeat and aversive stimulation: Frequency modulation and bout structure. *Behaviour* **1991**, *119*, 127–142. [CrossRef]
84. Barfield, R.J.; Geyer, L.A. Sexual behavior: Ultrasonic postejaculatory song of the male rat. *Science* **1972**, *176*, 1349–1350. [CrossRef]
85. Delius, J.D. Irrelevant behaviour, information processing and arousal homeostasis. *Psychol. Forsch.* **1970**, *33*, 165–188. [CrossRef] [PubMed]
86. Brudzynski, S.M.; Kehoe, P.; Callahan, M. Sonographic structure of isolation-induced ultrasonic calls of rat pups. *Dev. Psychobiol.* **1999**, *34*, 195–204. [CrossRef]
87. Stark, R.A.; Harker, A.; Salamanca, S.; Pellis, S.M.; Li, F.; Gibb, R.L. Development of ultrasonic calls in rat pups follows similar patterns regardless of isolation distress. *Dev. Psychobiol.* **2020**, *62*, 617–630. [CrossRef]
88. Boulanger-Bertolus, J.; Rincón-Cortés, M.; Sullivan, R.M.; Mouly, A.-M. Understanding pup affective state through ethologically significant ultrasonic vocalization frequency. *Sci. Rep.* **2017**, *7*, 1–12. [CrossRef]
89. Gardner, C.R. Distress vocalization in rat pups. A simple screening method for anxiolytic drugs. *J. Pharmacol. Methods* **1985**, *14*, 181–187. [CrossRef]
90. Insel, T.R.; Hill, J.L.; Mayor, R.B. Rat pup ultrasonic isolation calls: Possible mediation by the benzodiazepine receptor complex. *Pharm. Biochem. Behav.* **1986**, *24*, 1263–1267. [CrossRef]
91. Branchi, I.; Santucci, D.; Alleva, E. Ultrasonic vocalisation emitted by infant rodents: A tool for assessment of neurobehavioural development. *Behav. Brain Res.* **2001**, *125*, 49–56. [CrossRef]
92. Brunelli, S.A. Selective breeding for an infant phenotype: Rat pup ultrasonic vocalization (USV). *Behav. Genet.* **2005**, *35*, 53–65. [CrossRef] [PubMed]
93. Hofer, M.A.; Shair, H.N.; Masmela, J.R.; Brunelli, S.A. Developmental effects of selective breeding for an infantile trait: The rat pup ultrasonic isolation call. *Dev. Psychobiol.* **2001**, *39*, 231–246. [CrossRef] [PubMed]
94. Naito, H.; Tonoue, T. Sex difference in ultrasound distress call by rat pups. *Behav. Brain Res.* **1987**, *25*, 13–21. [CrossRef]
95. Manduca, A.; Servadio, M.; Melancia, F.; Schiavi, S.; Manzoni, O.J.; Trezza, V. Sex-specific behavioural deficits induced at early life by prenatal exposure to the cannabinoid receptor agonist WIN55, 212-2 depend on mGlu5 receptor signalling. *Br. J. Pharmacol.* **2020**, *177*, 449–463. [CrossRef] [PubMed]
96. Shahrier, M.A.; Wada, H. Effects of prenatal ethanol exposure on acoustic characteristics of play fighting-induced ultrasonic vocalizations in juvenile rats. *Neurotoxicology* **2020**, *79*, 25–39. [CrossRef]
97. Bird, C.W.; Barto, D.; Magcalas, C.M.; Rodriguez, C.I.; Donaldson, T.; Davies, S.; Savage, D.D.; Hamilton, D.A. Ifenprodil infusion in agranular insular cortex alters social behavior and vocalizations in rats exposed to moderate levels of ethanol during prenatal development. *Behav. Brain Res.* **2017**, *320*, 1–11. [CrossRef]
98. Waddell, J.; Yang, T.; Ho, E.; Wellmann, K.A.; Mooney, S.M. Prenatal Ethanol Exposure and Whisker Clipping Disrupt Ultrasonic Vocalizations and Play Behavior in Adolescent Rats. *Brain Sci.* **2016**, *6*, 43. [CrossRef] [PubMed]

99. Wellmann, K.; Lewis, B.; Barron, S. Agmatine reduces ultrasonic vocalization deficits in female rat pups exposed neonatally to ethanol. *Neurotoxicol. Teratol.* **2010**, *32*, 158–163. [CrossRef]
100. Brasser, S.M.; Spear, N.E. Physiological and behavioral effects of acute ethanol hangover in juvenile, adolescent, and adult rats. *Behav. Neurosci.* **2002**, *116*, 305–320. [CrossRef] [PubMed]
101. Mittal, N.; Thakore, N.; Bell, R.L.; Maddox, W.T.; Schallert, T.; Duvauchelle, C.L. Sex-specific ultrasonic vocalization patterns and alcohol consumption in high alcohol-drinking (HAD-1) rats. *Physiol. Behav.* **2019**, *203*, 81–90. [CrossRef] [PubMed]
102. Doncheck, E.M.; Liddiard, G.T.; Konrath, C.D.; Liu, X.; Yu, L.; Urbanik, L.A.; Herbst, M.R.; DeBaker, M.C.; Raddatz, N.; Van Newenhizen, E.C.; et al. Sex, stress, and prefrontal cortex: Influence of biological sex on stress-promoted cocaine seeking. *Neuropsychopharmacology* **2020**, *45*, 1974–1985. [CrossRef]
103. Cox, E.T.; Hodge, C.W.; Sheikh, M.J.; Abramowitz, A.C.; Jones, G.F.; Jamieson-Drake, A.W.; Makam, P.R.; Zeskind, P.S.; Johns, J.M. Delayed developmental changes in neonatal vocalizations correlates with variations in ventral medial hypothalamus and central amygdala development in the rodent infant: Effects of prenatal cocaine. *Behav. Brain Res.* **2012**, *235*, 166–175. [CrossRef] [PubMed]
104. Campbell, J.O.; Fogarty, J.A.; Spear, L.P. Inhibition of nitric oxide synthesis with L-NAME suppresses isolation-induced ultrasounds in rat pups. *Pharmacol. Biochem. Behav.* **1999**, *63*, 45–53. [CrossRef]
105. Shepherd, J.K.; Blanchard, D.C.; Weiss, S.M.; Rodgers, R.J.; Blanchard, R.J. Morphine attenuates antipredator ultrasonic vocalizations in mixed-sex rat colonies. *Pharmacol. Biochem. Behav.* **1992**, *41*, 551–558. [CrossRef]
106. Vassoler, F.M.; Oranges, M.L.; Toorie, A.M.; Byrnes, E.M. Oxycodone self-administration during pregnancy disrupts the maternal-infant dyad and decreases midbrain OPRM1 expression during early postnatal development in rats. *Pharmacol. Biochem. Behav.* **2018**, *173*, 74–83. [CrossRef]
107. Houwing, D.J.; Staal, L.; Swart, J.M.; Ramsteijn, A.S.; Wöhr, M.; de Boer, S.F.; Olivier, J.D.A. Subjecting Dams to Early Life Stress and Perinatal Fluoxetine Treatment Differentially Alters Social Behavior in Young and Adult Rat Offspring. *Front. Neurosci.* **2019**, *13*, 229. [CrossRef]
108. Naito, H.; Inoue, M.; Makino, J. Ultrasonic isolation calls in genetically high- and low-emotional rat pups. *Exp. Anim.* **2000**, *49*, 289–294. [CrossRef]
109. Bekkedal, M.Y.; Rossi, J., 3rd; Panksepp, J. Fetal and neonatal exposure to trimethylolpropane phosphate alters rat social behavior and emotional responsivity. *Neurotoxicol. Teratol.* **1999**, *21*, 435–443. [CrossRef]
110. Berg, E.L.; Copping, N.A.; Rivera, J.K.; Pride, M.C.; Careaga, M.; Bauman, M.D.; Berman, R.F.; Lein, P.J.; Harony-Nicolas, H.; Buxbaum, J.D.; et al. Developmental social communication deficits in the Shank3 rat model of phelan-mcdermid syndrome and autism spectrum disorder. *Autism Res.* **2018**, *11*, 587–601. [CrossRef]
111. Umeda, T.; Takashima, N.; Nakagawa, R.; Maekawa, M.; Ikegami, S.; Yoshikawa, T.; Kobayashi, K.; Okanoya, K.; Inokuchi, K.; Osumi, N. Evaluation of Pax6 mutant rat as a model for autism. *PLoS One* **2010**, *5*, e15500. [CrossRef]
112. Gzielo, K.; Potasiewicz, A.; Holuj, M.; Litwa, E.; Popik, P.; Nikiforuk, A. Valproic acid exposure impairs ultrasonic communication in infant, adolescent and adult rats. *Eur. Neuropsychopharmacol.* **2020**, *41*, 52–62. [CrossRef] [PubMed]
113. Morales-Navas, M.; Castano-Castano, S.; Perez-Fernandez, C.; Sanchez-Gil, A.; Teresa Colomina, M.; Leinekugel, X.; Sanchez-Santed, F. Similarities between the Effects of Prenatal Chlorpyrifos and Valproic Acid on Ultrasonic Vocalization in Infant Wistar Rats. *Int. J. Environ. Res. Public Health* **2020**, *17*, 6376. [CrossRef]
114. Potasiewicz, A.; Gzielo, K.; Popik, P.; Nikiforuk, A. Effects of prenatal exposure to valproic acid or poly(I:C) on ultrasonic vocalizations in rat pups: The role of social cues. *Physiol. Behav.* **2020**, *225*, 113113. [CrossRef] [PubMed]
115. Kisko, T.M.; Schwarting, R.K.W.; Wöhr, M. Sex differences in the acoustic features of social play-induced 50-kHz ultrasonic vocalizations: A detailed spectrographic analysis in wild-type Sprague-Dawley and Cacna1c haploinsufficient rats. *Dev. Psychobiol.* **2021**, *63*, 262–276. [CrossRef]
116. Kisko, T.M.; Braun, M.D.; Michels, S.; Witt, S.H.; Rietschel, M.; Culmsee, C.; Schwarting, R.K.W.; Wöhr, M. Sex-dependent effects of Cacna1c haploinsufficiency on juvenile social play behavior and pro-social 50-kHz ultrasonic communication in rats. *Genes Brain Behav.* **2020**, *19*, 12552. [CrossRef]
117. Potasiewicz, A.; Holuj, M.; Litwa, E.; Gzielo, K.; Socha, L.; Popik, P.; Nikiforuk, A. Social dysfunction in the neurodevelopmental model of schizophrenia in male and female rats: Behavioural and biochemical studies. *Neuropharmacology* **2020**, *170*, 108040. [CrossRef]
118. Potasiewicz, A.; Holuj, M.; Piotrowska, D.; Zajda, K.; Wojcik, M.; Popik, P.; Nikiforuk, A. Evaluation of ultrasonic vocalizations in a neurodevelopmental model of schizophrenia during the early life stages of rats. *Neuropharmacology* **2019**, *146*, 28–38. [CrossRef]
119. Paul, M.J.; Peters, N.V.; Holder, M.K.; Kim, A.M.; Whylings, J.; Terranova, J.I.; de Vries, G.J. Atypical Social Development in Vasopressin-Deficient Brattleboro Rats. *eNeuro* **2016**, *3*. [CrossRef]
120. Schatz, K.C.; Kyne, R.F.; Parmeter, S.L.; Paul, M.J. Investigation of social, affective, and locomotor behavior of adolescent Brattleboro rats reveals a link between vasopressin's actions on arousal and social behavior. *Horm. Behav.* **2018**, *106*, 1–9. [CrossRef]
121. Marquis, J.M.; Lettenberger, S.E.; Kelm-Nelson, C.A. Early-onset Parkinsonian behaviors in female Pink1−/− rats. *Behav. Brain Res.* **2020**, *377*, 112175. [CrossRef]

122. Lopez-Meraz, M.L.; Medel-Matus, J.S.; Morgado-Valle, C.; Beltran-Parrazal, L.; Perez-Estudillo, C.; Manzo, J. Effect of lithium-pilocarpine-induced status epilepticus on ultrasonic vocalizations in the infant rat pup. *Epilepsy Behav.* **2014**, *31*, 263–266. [CrossRef] [PubMed]
123. Kirsten, T.B.; Chaves-Kirsten, G.P.; Chaible, L.M.; Silva, A.C.; Martins, D.O.; Britto, L.R.; Dagli, M.L.; Torrao, A.S.; Palermo-Neto, J.; Bernardi, M.M. Hypoactivity of the central dopaminergic system and autistic-like behavior induced by a single early prenatal exposure to lipopolysaccharide. *J. Neurosci. Res.* **2012**, *90*, 1903–1912. [CrossRef] [PubMed]
124. Krishnan, K.; Mittal, N.; Thompson, L.M.; Rodriguez-Santiago, M.; Duvauchelle, C.L.; Crews, D.; Gore, A.C. Effects of the Endocrine-Disrupting Chemicals, Vinclozolin and Polychlorinated Biphenyls, on Physiological and Sociosexual Phenotypes in F2 Generation Sprague-Dawley Rats. *Environ. Health Perspect.* **2018**, *126*, 97005. [CrossRef] [PubMed]
125. Topper, V.Y.; Reilly, M.P.; Wagner, L.M.; Thompson, L.M.; Gillette, R.; Crews, D.; Gore, A.C. Social and neuromolecular phenotypes are programmed by prenatal exposures to endocrine-disrupting chemicals. *Mol. Cell. Endocrinol.* **2019**, *479*, 133–146. [CrossRef]
126. Bell, M.R.; Thompson, L.M.; Rodriguez, K.; Gore, A.C. Two-hit exposure to polychlorinated biphenyls at gestational and juvenile life stages: 1. Sexually dimorphic effects on social and anxiety-like behaviors. *Horm. Behav.* **2016**, *78*, 168–177. [CrossRef]
127. Abuaish, S.; Tse, E.K.; McGowan, P.O. Perinatal high-fat diet impairs pup retrieval and induces sex-specific changes in ultrasonic vocalization characteristics of rat pups. *Dev. Psychobiol.* **2020**, *62*, 436–445. [CrossRef]
128. Kosten, T.A.; Miserendino, M.J.; Bombace, J.C.; Lee, H.J.; Kim, J.J. Sex-selective effects of neonatal isolation on fear conditioning and foot shock sensitivity. *Behav. Brain Res.* **2005**, *157*, 235–244. [CrossRef]
129. Kalinichev, M.; Easterling, K.W.; Plotsky, P.M.; Holtzman, S.G. Long-lasting changes in stress-induced corticosterone response and anxiety-like behaviors as a consequence of neonatal maternal separation in Long-Evans rats. *Pharmacol. Biochem. Behav.* **2002**, *73*, 131–140. [CrossRef]
130. Keller, A.; Saucier, D.; Sheerin, A.; Yager, J. Febrile convulsions affect ultrasonic vocalizations in the rat pup. *Epilepsy Behav.* **2004**, *5*, 649–654. [CrossRef] [PubMed]
131. Blumberg, M.S.; Stolba, M.A. Thermogenesis, myoclonic twitching, and ultrasonic vocalization in neonatal rats during moderate and extreme cold exposure. *Behav. Neurosci.* **1996**, *110*, 305–314. [CrossRef]
132. Williams, M.T.; Hennessy, M.B.; Davis, H.N. Stress during pregnancy alters rat offspring morphology and ultrasonic vocalizations. *Physiol. Behav.* **1998**, *63*, 337–343. [CrossRef]
133. Williams, M.T.; Hennessy, M.B.; Davis, H.N. CRF administered to pregnant rats alters offspring behavior and morphology. *Pharmacol. Biochem. Behav.* **1995**, *52*, 161–167. [CrossRef]
134. Knutson, B.; Burgdorf, J.; Panksepp, J. Ultrasonic vocalizations as indices of affective states in rats. *Psychol. Bull.* **2002**, *128*, 961. [CrossRef] [PubMed]
135. Brudzynski, S.M.; Holland, G. Acoustic characteristics of air puff-induced 22-kHz alarm calls in direct recordings. *Neurosci. Biobehav. Rev.* **2005**, *29*, 1169–1180. [CrossRef]
136. Brudzynski, S.M. Principles of rat communication: Quantitative parameters of ultrasonic calls in rats. *Behav. Genet.* **2005**, *35*, 85–92. [CrossRef]
137. Simola, N.; Costa, G. Emission of categorized 50-kHz ultrasonic vocalizations in rats repeatedly treated with amphetamine or apomorphine: Possible relevance to drug-induced modifications in the emotional state. *Behav. Brain Res.* **2018**, *347*, 88–98. [CrossRef]
138. Coffey, K.R.; Marx, R.G.; Neumaier, J.F. DeepSqueak: A deep learning-based system for detection and analysis of ultrasonic vocalizations. *Neuropsychopharmacology* **2019**, *44*, 859–868. [CrossRef] [PubMed]
139. Manduca, A.; Campolongo, P.; Palmery, M.; Vanderschuren, L.J.; Cuomo, V.; Trezza, V. Social play behavior, ultrasonic vocalizations and their modulation by morphine and amphetamine in Wistar and Sprague-Dawley rats. *Psychopharmacology (Berlin)* **2014**, *231*, 1661–1673. [CrossRef] [PubMed]
140. Knutson, B.; Burgdorf, J.; Panksepp, J. Anticipation of play elicits high-frequency ultrasonic vocalizations in young rats. *J. Comp. Psychol.* **1998**, *112*, 65–73. [CrossRef] [PubMed]
141. Brunelli, S.A.; Nie, R.; Whipple, C.; Winiger, V.; Hofer, M.A.; Zimmerberg, B. The effects of selective breeding for infant ultrasonic vocalizations on play behavior in juvenile rats. *Physiol. Behav.* **2006**, *87*, 527–536. [CrossRef] [PubMed]
142. Pellis, S.M.; Pellis, V.C. The prejuvenile onset of play fighting in laboratory rats (*Rattus norvegicus*). *Dev. Psychobiol.* **1997**, *31*, 193–205. [CrossRef]
143. Kisko, T.M.; Euston, D.R.; Pellis, S.M. Are 50-khz calls used as play signals in the playful interactions of rats? III. The effects of devocalization on play with unfamiliar partners as juveniles and as adults. *Behav. Processes* **2015**, *113*, 113–121. [CrossRef]
144. Willadsen, M.; Seffer, D.; Schwarting, R.K.; Wöhr, M. Rodent ultrasonic communication: Male prosocial 50-kHz ultrasonic vocalizations elicit social approach behavior in female rats (*Rattus norvegicus*). *J. Comp. Psychol.* **2014**, *128*, 56–64. [CrossRef]
145. Barfield, R.J.; Thomas, D.A. The role of ultrasonic vocalizations in the regulation of reproduction in rats. *Ann. N. Y. Acad. Sci.* **1986**, *474*, 33–43. [CrossRef]
146. Snoeren, E.M.; Agmo, A. The incentive value of males' 50-kHz ultrasonic vocalizations for female rats (*Rattus norvegicus*). *J. Comp. Psychol.* **2014**, *128*, 40–55. [CrossRef]
147. Gerson, C.A.; Mac Cionnaith, C.E.; Quintana, G.R.; Pfaus, J.G. Effects of ovarian hormones on the emission of 50-kHz ultrasonic vocalizations during distributed clitoral stimulation in the rat. *Horm. Behav.* **2019**, *109*, 1–9. [CrossRef]

148. Snoeren, E.M.; Agmo, A. Female ultrasonic vocalizations have no incentive value for male rats. *Behav. Neurosci.* **2013**, *127*, 439–450. [CrossRef] [PubMed]
149. Agmo, A.; Snoeren, E.M. Silent or Vocalizing Rats Copulate in a Similar Manner. *PLoS ONE* **2015**, *10*, 0144164. [CrossRef] [PubMed]
150. Snoeren, E.M.; Helander, L.R.; Iversen, E.E.; Agmo, A. On the role of individual differences in female odor and ultrasonic vocalizations for male's choice of partner. *Physiol. Behav.* **2014**, *132*, 17–23. [CrossRef] [PubMed]
151. McGinnis, M.Y.; Vakulenko, M. Characterization of 50-kHz ultrasonic vocalizations in male and female rats. *Physiol. Behav.* **2003**, *80*, 81–88. [CrossRef]
152. McIntosh, T.K.; Barfield, R.J.; Geyer, L.A. Ultrasonic vocalisations facilitate sexual behaviour of female rats. *Nature* **1978**, *272*, 163–164. [CrossRef]
153. Lenell, C.; Johnson, A.M. The effects of the estrous cycle, menopause, and recording condition on female rat ultrasonic vocalizations. *Physiol. Behav.* **2021**, *229*, 113248. [CrossRef]
154. Börner, A.; Hjemdahl, R.; Götz, T.; Brown, G.R. Ultrasonic vocalizations of female Norway rats (*Rattus norvegicus*) in response to social partners. *J. Comp. Psychol.* **2016**, *130*, 76. [CrossRef]
155. Inagaki, H.; Kuwahara, M.; Tsubone, H.; Mori, Y. The effect of post-weaning individual housing on 50-kHz calls emitted from male rats to sexually receptive female rats. *Physiol. Behav.* **2013**, *110–111*, 30–33. [CrossRef]
156. Johnson, A.M. Social isolation alters ultrasonic vocalizations but not thyroarytenoid neuromuscular junctions in old rats. *Laryngoscope* **2019**, *129*, 9–14. [CrossRef]
157. Shembel, A.C.; Lenell, C.; Chen, S.; Johnson, A.M. Effects of Vocal Training on Thyroarytenoid Muscle Neuromuscular Junctions and Myofibers in Young and Older Rats. *J. Gerontol. A Biol. Sci. Med. Sci.* **2021**, *76*, 244–252. [CrossRef]
158. McMullen, C.A.; Andrade, F.H. Functional and morphological evidence of age-related denervation in rat laryngeal muscles. *J. Gerontol. A Biol. Sci. Med. Sci.* **2009**, *64*, 435–442. [CrossRef] [PubMed]
159. Connor, N.P.; Suzuki, T.; Lee, K.; Sewall, G.K.; Heisey, D.M. Neuromuscular junction changes in aged rat thyroarytenoid muscle. *Ann. Otol. Rhinol. Laryngol.* **2002**, *111*, 579–586. [CrossRef]
160. Yamamoto, Y.; Tanaka, S.; Tsubone, H.; Atoji, Y.; Suzuki, Y. Age-related changes in sensory and secretomotor nerve endings in the larynx of F344/N rat. *Arch. Gerontol. Geriatr.* **2003**, *36*, 173–183. [CrossRef]
161. McMullen, C.A.; Andrade, F.H. Contractile dysfunction and altered metabolic profile of the aging rat thyroarytenoid muscle. *J. Appl. Physiol.* **2006**, *100*, 602–608. [CrossRef] [PubMed]
162. Nishida, N.; Taguchi, A.; Motoyoshi, K.; Hyodo, M.; Gyo, K.; Desaki, J. Age-related changes in rat intrinsic laryngeal muscles: Analysis of muscle fibers, muscle fiber proteins, and subneural apparatuses. *Eur. Arch. Otorhinolaryngol.* **2013**, *270*, 975–984. [CrossRef]
163. Inagaki, H.; Sato, J. Air puff-induced 22-kHz calls in F344 rats. *Physiol. Behav.* **2016**, *155*, 237–241. [CrossRef]
164. Haney, M.; Miczek, K.A. Ultrasounds during agonistic interactions between female rats (*Rattus norvegicus*). *J. Comp. Psychol.* **1993**, *107*, 373–379. [CrossRef] [PubMed]
165. Lu, K.H.; Hoper, B.R.; Vargo, T.M.; Yen, S.S.C. Chronological changes in sex steroid, gonadotropin and prolactin secretion in aging female rats displaying diferent reproductive states. *Biol. Reprod.* **1979**, *21*, 193–203. [CrossRef]
166. Long, J.A.; Evans, H.M. *The Oestrous Cycle in the Rat and Its Associated Phenomena*; University of California Press: Barkeley, CA, USA, 1922; p. 328.
167. Marcondes, F.K.; Bianchi, F.J.; Tanno, A.P. Determination of the estrous cycle phases of rats: Some helpful considerations. *Braz J. Biol.* **2002**, *62*, 609–614. [CrossRef]
168. Berkley, K.J.; McAllister, S.L.; Accius, B.E.; Winnard, K.P. Endometriosis-induced vaginal hyperalgesia in the rat: Effect of estropause, ovariectomy, and estradiol replacement. *Pain* **2007**, *132* (Suppl. 1), 150–159. [CrossRef]
169. Kalinichev, M.; Easterling, K.W.; Holtzman, S.G. Periodic postpartum separation from the offspring results in long-lasting changes in anxiety-related behaviors and sensitivity to morphine in Long-Evans mother rats. *Psychopharmacology (Berlin)* **2000**, *152*, 431–439. [CrossRef] [PubMed]
170. Mychasiuk, R.; Schmold, N.; Ilnytskyy, S.; Kovalchuk, O.; Kolb, B.; Gibb, R. Prenatal bystander stress alters brain, behavior, and the epigenome of developing rat offspring. *Dev. Neurosci.* **2011**, *33*, 159–169. [CrossRef] [PubMed]
171. Matochik, J.A.; White, N.R.; Barfield, R.J. Variations in scent marking and ultrasonic vocalizations by Long-Evans rats across the estrous cycle. *Physiol. Behav.* **1992**, *51*, 783–786. [CrossRef]
172. Thomas, D.A.; Barfield, R.J. Ultrasonic vocalization of the female rat (*Rattus norvegicus*) during mating. *Anim. Behav.* **1985**, *33*, 720–725. [CrossRef]
173. Matochik, J.A.; Barfield, R.J.; Nyby, J. Regulation of sociosexual communication in female Long-Evans rats by ovarian hormones. *Horm. Behav.* **1992**, *26*, 545–555. [CrossRef]
174. Garcia, A.N.; Bezner, K.; Depena, C.; Yin, W.; Gore, A.C. The effects of long-term estradiol treatment on social behavior and gene expression in adult female rats. *Horm. Behav.* **2017**, *87*, 145–154. [CrossRef]

brain sciences

MDPI

Article

Observational Fear Learning in Rats: Role of Trait Anxiety and Ultrasonic Vocalization

Markus Fendt [1,2,*], Claudia Paulina Gonzalez-Guerrero [1,3] and Evelyn Kahl [1]

1 Institute for Pharmacology and Toxicology, Medical Faculty, Otto-von-Guericke University Magdeburg, 39120 Magdeburg, Germany; claudia.gonzalez@st.ovgu.de (C.P.G.-G.); evelyn.kahl@med.ovgu.de (E.K.)
2 Center for Behavioral Brain Sciences, Otto-von-Guericke University Magdeburg, 39120 Magdeburg, Germany
3 Integrative Neuroscience Program, Otto-von-Guericke University Magdeburg, 39120 Magdeburg, Germany
* Correspondence: markus.fendt@med.ovgu.de

Abstract: Rats can acquire fear by observing conspecifics that express fear in the presence of conditioned fear stimuli. This process is called observational fear learning and is based on the social transmission of the demonstrator rat's emotion and the induction of an empathy-like or anxiety state in the observer. The aim of the present study was to investigate the role of trait anxiety and ultrasonic vocalization in observational fear learning. Two experiments with male Wistar rats were performed. In the first experiment, trait anxiety was assessed in a light–dark box test before the rats were submitted to the observational fear learning procedure. In the second experiment, ultrasonic vocalization was recorded throughout the whole observational fear learning procedure, and 22 kHz and 50 kHz calls were analyzed. The results of our study show that trait anxiety differently affects direct fear learning and observational fear learning. Direct fear learning was more pronounced with higher trait anxiety, while observational fear learning was the best with a medium-level of trait anxiety. There were no indications in the present study that ultrasonic vocalization, especially emission of 22 kHz calls, but also 50 kHz calls, are critical for observational fear learning.

Keywords: anxiety; observational fear learning; rat; ultrasonic vocalization

check for updates

Citation: Fendt, M.; Gonzalez-Guerrero, C.P.; Kahl, E. Observational Fear Learning in Rats: Role of Trait Anxiety and Ultrasonic Vocalization. *Brain Sci.* **2021**, *11*, 423. https://doi.org/10.3390/brainsci11040423

Academic Editor: Stefan Brudzynski

Received: 25 February 2021
Accepted: 23 March 2021
Published: 26 March 2021

1. Introduction

In potentially threatening situations, rats—as other animals and humans—express a variety of defensive behavior that ultimately helps to survive such situations [1]. The choice of the defensive behavior is dependent on several variables, such as the intensity and the proximity of the threatening stimulus but also on the local conditions, e.g., whether there is a way to escape or a place to hide [2]. A defensive attack may occur if the threat is too close, whereas freezing, a cessation of movements [3], is expressed if the threat is not too close, but escape is not possible. After the initial defensive response, e.g., hiding or freezing, many rats emit ultrasonic vocalizations [1,4,5]. These calls are characterized by a peak frequency around 22 kHz, a narrow bandwidth, and call durations around one second [6,7]. The function of these 22 kHz calls is not completely understood. It is well recognized that they reflect the negative affective state of the emitting rat [8]. Thereby, they serve as species-specific aversive communication signals, i.e., they are recognized by conspecifics and have the capability to change their emotional state [9] or even to warn them about a threat [4]. However, only modest effects are reported in experiments using playbacks of natural 22 kHz calls [10–13], eventually questioning the direct behavioral effects of 22 kHz calls.

Of note, threatening situations are also associated with very robust and rapid learning processes [14,15]. Both contextual and discrete stimuli that are present during the threatening situation can be associated with the threat, a process called threat or fear-conditioning [16]. Laboratory protocols of fear-conditioning are widely used in neuroscience research to investigate the neural basis of fear/anxiety learning [17–20]. There is a high translational relevance of this research field since fear-conditioning itself, its specificity, extinction, and inhibition are affected in patients with anxiety disorder [21–23]. Some of these changes are not only seen in clinical populations but also in human subjects with high but not yet pathological trait anxiety [24–26]. Of note, also rats with high trait anxiety showed exaggerated fear learning, a less specific fear memory, and impaired fear extinction [27,28].

Several studies have described observational fear learning, also called "social fear learning", "vicarious fear learning", or "fear-conditioning by proxy", in rats [29–32]. In these studies, rats learned fear by observation, i.e., they joined an already fear-conditioned conspecific (the "demonstrator") in a retention session on conditioned fear and could observe its fear responses to the conditioned context and/or discrete stimulus (that are neutral for the observer rat). To test whether fear was acquired by observation, a further retention test was performed one day later in which the observer rat alone was exposed to the conditioned context or stimulus [33]. Observational fear learning has been successful in male and female cage-mates [30,32], was more pronounced in related and submissive rats [31,32] and less pronounced in dominant rats [31]. Furthermore, the amount of social interaction and ultrasonic vocalization during the observational fear learning session was positively correlated with the observers' fear response on the following day [30–32]. Of note, only 22 kHz calls were analyzed in this experiment [31]. Of potential interest would also be the so-called "50 kHz calls" that are typically emitted during social contacts or other appetitive situations [34–36]. Since social interaction was positively correlated with the observers' fear learning [30–32] and 50 kHz calls are typically increased during social contacts, the amount of 50 kHz calls should also positively correlate with the observers' fear learning.

The role of trait anxiety in observational fear learning has not been investigated in rats so far. There are hints for a possible role of trait anxiety from a human study. In this study, participants had to observe mock panic attacks while their fear responses were measured [37]. "Anxiety sensitivity" was correlated with self-reported fear levels but not physiological arousal during this observational fear challenge procedure. However, potential learning processes during the challenge were not investigated in this study.

In the present study, the role of trait anxiety and ultrasonic vocalization on observational fear learning in male Wistar rats was investigated. Two experiments using a published protocol [33] were performed: In the first experiment, trait anxiety was assessed in the light–dark box test before the rats were submitted to the observational fear learning procedure. In the second experiment, ultrasonic vocalization was recorded throughout the whole observational fear learning procedure and 22 kHz and 50 kHz calls were analyzed. Our working hypotheses were that (1) rats with higher trait anxiety show more pronounced observational fear learning and that (2) higher emission of both 22 kHz and 50 kHz calls in the observational fear learning session leads to enhanced fear memory in the observer rats.

2. Materials and Methods

2.1. Animals and Housing

Testing was carried out using 102 experimentally naive male Wistar rats aged 8–12 weeks. They were bred and reared at the local animal facility (original breeding stock: Taconic, Silkeborg, Denmark). They were housed in groups of six animals in standard Macrolon Type IV cages (58 cm × 33 cm × 20 cm) with water and standard lab chow (ssniff, Soest, Germany) available ad libitum. Cages were kept in temperature- and humidity-controlled rooms (22 ± 2 °C, 55 ± 10%) with a 12:12 h light/dark cycle (lights on at 6:00 am). All behavioral tests were conducted during the light phase between 8:00 am and 3:00 pm. Our study was carried out in accordance with international guidelines for the use of animals in experiments (2010/63/EU) and was approved by the local ethical committee (Landesverwaltungsamt Sachsen-Anhalt, Az. 42502-2-1587 UniMD).

2.2. Apparatus

The light–dark box system consisted of four identical boxes (49.5 cm × 49.5 × 41.5 cm; TSE Systems, Bad Homburg, Germany) divided into two compartments of the same size. The light compartment (135–310 lux) had transparent acryl glass walls, while the dark compartment (0.2–1.5 lux) had black walls. The two compartments were connected by an 8 cm × 6 cm opening. Position and movements of the rats were detected by animal detection infrared sensor frames (detector distance: 14 mm) and analyzed by TSE PhenoMaster software (version 4.9.4).

The fear-conditioning system consisted of four identical transparent acryl glass box (46 cm × 46 cm × 32 cm) located in sound-attenuating chambers (70 cm × 80 cm × 70 cm; TSE Systems, Bad Homburg, Germany). The chambers were equipped with loudspeakers for the acoustic stimuli (background noise of 55 dB SPL and the tone stimuli for fear-conditioning), light sources (continuous illumination of ca. 10 lux), ventilation fans, and video cameras (for monitoring and videotaping) mounted in the ceiling of the chambers. The floor of the boxes consisted of removable stainless steel grids (bars: 4 mm diameter, distance: 9 mm), which were connected to a shock unit and able to deliver foot shocks. Delivery of the different stimuli was controlled by TSE fear-conditioning software (version 09.10). This software also analyzed the position and movements of the rats, recorded by infrared animal detection sensor frames (detector distance: 14 mm). Freezing behavior was defined as no infrared beam crosses for more than 1 s. This automated measurement of freezing was previously validated by demonstrating a high correlation with observer scoring of freezing [13,38].

Recording and analyses of ultrasonic vocalization were performed with the UltraSoundGate system (Avisoft Bioacoustics, Berlin, Germany). For recording, an ultrasound condenser microphone (CM16/CMPA, Avisoft Bioacoustics, Berlin, Germany) sensitive to frequencies of 15–180 kHz (flat frequency response between 25 and 140 kHz; ±6 dB) mounted on one corner of the box was used, which was connected to a laptop via a USB audio device (UltraSoundGate 116H). Acoustic data were recorded by Avisoft Recorder USGH software (version 5.2) using a sampling rate of 250,000 Hz in 16-bit format and a recording range of 0–125 kHz.

2.3. Behavioral Procedure

2.3.1. Experiment 1: Role of Trait Anxiety

Forty-eight rats were used in this experiment. On day 1, all rats were tested in the light–dark box (see Figure 1a). Hence, they were placed centrally into the dark compartment and could freely move for 10 min. Then, the rats were returned to their home cages. Based on the results of the light–dark box test, the rats were grouped into sixteen triads, in which one rat served later as the "demonstrator" (DEM), one as the "observer" (OBS), and one as the "naive" (NAIVE). The grouping was done in a way that there were always two triads per cage and that the DEM, OBS and NAIVE rats had similar mean anxiety scores in the light–dark box test (based on time spent in the light compartment).

Three days later, the observational fear learning protocol started. On day 4, the DEM rats were directly fear-conditioned. Therefore, these rats were placed into the fear-conditioning boxes, and after an acclimation time of 10 min, they received three presentations of an auditory stimulus (CS; frequency: 10 kHz, duration: 20 s; mean inter-stimulus interval: 180 s) each co-terminated with an electric stimulus through the floor grids (intensity: 0.7 mA; duration: 0.5 s). Three minutes after the last electric stimulus, the rats were returned to their home cages. One day later (day 5), the observational fear learning session was performed. The DEM rats were returned to the conditioning boxes, each one accompanied by the OBS rat from the same triad. Both rats could freely interact with each other during the following session. After three minutes, the CS was presented three times with a mean inter-stimulus interval of 180 s. There were no electric stimuli during this session. Three minutes after the last CS presentation, the rats were returned to their home cages. Again one day later (day 6), all rats (DEM, OBS, NAIVE) were submitted to a fear retention test. The rats were placed alone into the conditioning boxes and tested with the same protocol as the day before.

2.3.2. Experiment 2: Role of Ultrasonic Vocalization

Fifty-four rats were used in this experiment, i.e., rats were grouped into eighteen triads. The experiment was identical to experiment 1 despite that the rats were not tested in the light–dark box. Furthermore, the box for the observational fear learning procedure was equipped with the ultrasound condenser microphone, and ultrasonic vocalization was recorded throughout the experiment.

2.4. *Offline Analyses of Behavior and Ultrasonic Vocalization*

Automated scoring of behavior via the infrared detection sensor frames could not be used when two rats were together in the box (DEM and OBS in the observational fear learning session). In these cases, the videotapes were used for offline scoring of behavior. The manual scoring was performed by an experienced observer using the Solomon Coder software (version beta 19.08; 16 August 2019; https://solomon.andraspeter.com). Freezing behavior (i.e., cessation of all body movements except those for breathing) of the DEM and OBS rats were scored by two focal sampling sessions per recording.

For the offline analysis of the acoustic data, SASLab Pro software (version 5.2; Avisoft Bioacoustics, Berlin, Germany) was used. After a fast Fourier transformation (512 FFT length, 100% frame, Hamming window and 75% time window overlap), high-resolution spectrograms were produced with a frequency resolution of 488 Hz and a time resolution of 0.512 ms. Onset and offset of the recorded 22 kHz calls were manually marked and 50 kHz calls (all calls > 30 kHz) were counted and categorized by a person, who was not aware of the experimental condition. The following parameters were determined and calculated for each single session: latency of the first 22 kHz call from start of the session, number of 22 kHz calls, mean 22 kHz call duration, total 22 kHz call duration/session, mean peak frequency of 22 kHz calls, and number of 50 kHz calls (peak frequency greater than 30 kHz). Furthermore, the call profiles were determined by visually sub-categorizing the 50 kHz calls into the following categories originally described by Wright and colleagues [39]: (i) downward ramp: monotonically decreasing frequency; (ii) upward ramp: monotonically increasing frequency; (iii) flat: near-constant frequency; (iv) complex: two or more directional changes in frequency; (v) trills: oscillation with a period of ca. 15 ms, can be flanked by a monotonic portion or can contain higher-frequency components; (vi) short: duration less than 12 ms; (vii) steps: one or more instantaneous frequency change to a higher or lower frequency; (viii) inverted-U: monotonic frequency increase followed by a monotonic decrease; (ix) composite: calls that compromise more than one category.

2.5. *Descriptive and Analytical Statistics*

Behavioral data are expressed as means ± standard errors of the mean (SEM), whereas acoustic data are shown as whisker box plots. Statistical analyses were performed with GraphPad Prism (version 8.00, GraphPad Software Inc., La Jolla, CA, USA). Data were checked for normal distribution (D'Agostino and Pearson's omnibus normality test) and were analyzed by analysis of variance (ANOVA) or Kruskal–Wallis test followed by appropriate post hoc comparisons (Holm–Sidak's or Dunn's multiple comparison tests) or Mann–Whitney tests. A $p < 0.05$ was considered statistically significant.

3. Results

3.1. *Experiment 1: Role of Trait Anxiety*

In this experiment, the rats were first submitted to a light–dark box test (Figure 1a). Then, the rats were grouped with respect to their prospective roles (DEM, OBS, NAIVE; $n = 16$/group) in the upcoming observational fear learning experiment with the restriction that there must be two triads with DEM OBS and NAIVE rats per cage. Figure 1b shows the mean percent time the rats of the different groups spent in the light compartment of the light–dark box (lower values indicate higher anxiety). An ANOVA revealed no differences between the groups ($F_{2,45} = 0.06$, $p = 0.95$) and a Bartlett's test no differences in variances ($\chi^2 = 2.78$, $p = 0.25$).

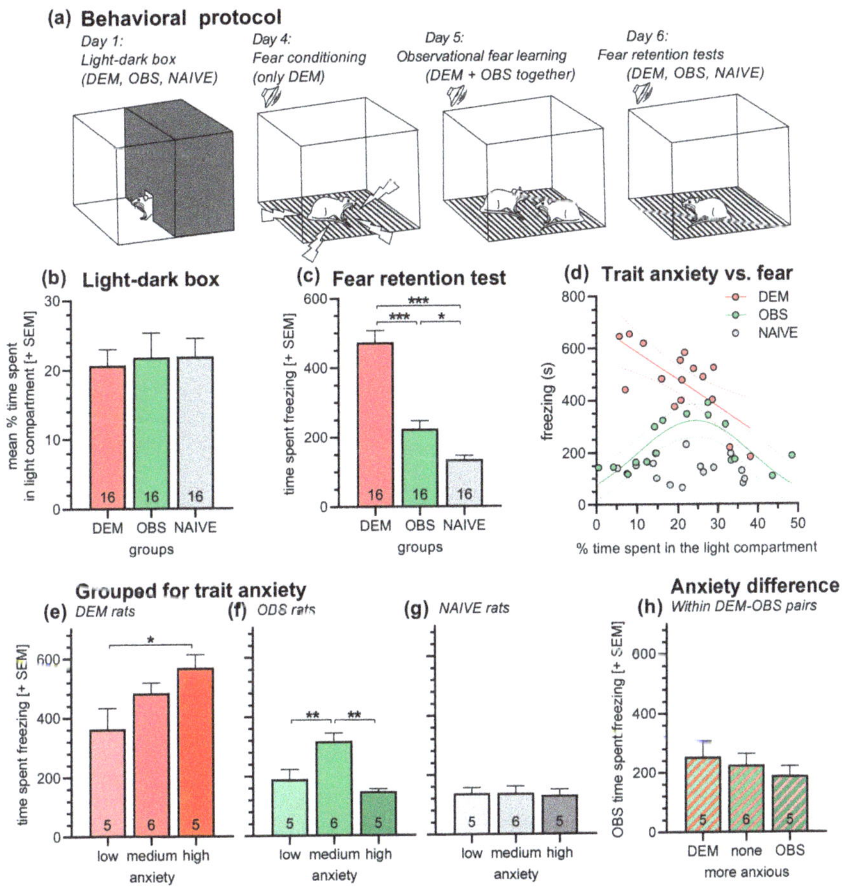

Figure 1. Role of trait anxiety in observational fear learning. (**a**) Behavioral protocol (for more details, see Materials and Methods): After a light–dark box test, rats were divided into three groups ("demonstrator" (DEM), "observer" (OBS), "naive" (NAIVE)) with similar anxiety levels. DEM rats were fear-conditioned. One day later, the DEM and OBS pairs were exposed to the conditioning boxes. Last, all rats were individually tested for conditioned fear. (**b**) Mean percent time spent in the light compartment of the light–dark box. DEM, OBS, and NAIVE rats had similar anxiety levels. (**c**) DEM rats expressed more freezing in the retention test than rats of the two other groups. Importantly, OBS rats had significantly higher freezing scores than NAIVE rats, indicating observational fear learning in OBS rats. (**d**) Freezing scores as a function of trait anxiety (more% time spent in the light compartment indicate less anxiety). In DEM rats, higher freezing scores could be observed with higher trait anxiety. In OBS rats, a bell-shaped correlation was found, freezing was highest with a medium levels of trait anxiety. These correlations were confirmed by grouping the animals into low, medium, and high anxiety. (**e**) High anxious DEM rats had significantly higher freezing scores than low-anxiety DEM rats. (**f**) Medium anxious OBS rats had significantly higher freezing scores than low- and high-anxiety OBS rats. (**g**) Trait anxiety did not affect freezing scores in NAIVE rats. (**h**) The difference of trait anxiety within a DEM-OBS pair did not influence freezing scores in the retention test. In this analysis, the DEM-OBS pairs were separated into subgroups in which DEM rats being more anxious than the OBS rat ("DEM more anxious"), DEM and OBS rats that were similarly anxious ("none more anxious"), and OBS rats that were more anxious than the DEM rats ("OBS more anxious"). Abbreviations: DEM, demonstrator rats; OBS, observer rats; NAIVE, naive rats. Numbers in the bars indicate group and subgroup sizes. *** $p < 0.001$, ** $p < 0.01$, * $p < 0.05$; comparisons as indicated, after significant main effects in ANOVA.

As expected, there were strong differences in the time spent freezing during the retention test on conditioned fear (Figure 1c; ANOVA: $F_{2,45} = 50.08$, $p < 0.0001$). Post hoc comparisons revealed higher freezing scores in the DEM rats compared to the OBS and NAIVE rats (Holm–Sidak's tests: $t_{16} = 7.12$, $p < 0.0001$ and $t_{16} = 9.65$, $p < 0.0001$, respectively). Importantly, the OBS rats had higher freezing levels than the NAIVE rats ($t_{16} = 2.54$; $p = 0.015$), indicating observational fear learning in the OBS rats.

Next, we were interested if anxiety-like behavior in the light–dark box was associated with the freezing behavior in the retention test on conditioned fear. Figure 1d depicts the freezing behavior of the individual rats as a function of the percent time the rats spent in the light compartment of the light–dark box. We tested several models for linear and nonlinear regressions and found best-fits for a simple linear regression in the DEM group ($R^2 = 0.52$, $p = 0.002$) and a Gaussian curve OBS group ($R^2 = 0.58$). For the NAIVE group, we did not find a model with $R^2 > 0.20$. This indicates that in DEM rats conditioned fear is higher with higher anxiety scores (i.e., less time spent in the light compartment), while in OBS rats conditioned fear is highest in medium-anxious rats. This is supported by further analyses in which we sorted the rats within the DEM, OBS and NAIVE groups regarding their anxiety scores in the light–dark box test. The thirds of rats with most percent time spent in the light compartment formed the low anxiety subgroups, the thirds of rats with the lowest time spent in the light compartment formed the high anxiety subgroups, and the remaining rats formed the medium anxiety subgroups (Figure 1e–g). Separated ANOVAs for DEM, OBS, and NAIVE rats revealed effects of trait anxiety in DEM ($F_{2,13} = 4.04$, $p = 0.04$) and OBS ($F_{2,13} = 12.87$, $p = 0.0008$) rats, but not in NAIVE rats ($F_{2,13} = 0.04$, $p = 0.96$). Post hoc comparisons showed significantly more freezing in high-anxiety DEM rats compared with low-anxiety DEM rats (Figure 1e; $t_9 = 2.83$, $p = 0.04$), and more freezing in medium anxious OBS rats compared to low- and high-anxiety OBS rats (Figure 1f; $t_9 = 3.61$; $p = 0.006$ and $t_9 = 4.12$, $p = 0.001$, respectively).

Last, to evaluate whether the differences in trait anxiety levels between the DEM and OBS rats affect observational fear learning, we also calculated the differences between the anxiety scores (% time spent in the light compartment) of the DEM-OBS pairs that had the observational fear learning session together. Using these differences, we built DEM-OBS subgroups, in which the DEM rat was clearly more anxious than the OBS rat ("DEM more anxious", difference > 10%), in which DEM and OBS rats were similarly anxious ("none more anxious", difference < 10%), and in which the OBS rat was more anxious than the DEM rat ("OBS more anxious", difference > 10%). As indicated in Figure 1h, the trait anxiety difference within the DEM-OBS pairs did not affect the freezing of the OBS rat in the retention test (ANOVA: $F_{2,13} = 0.58$; $p = 0.57$).

3.2. *Experiment 2: Role of Ultrasonic Vocalization*

In this experiment, the rats ($n = 18$/group) were not tested in the light–dark box. The observation fear learning protocol was identical to the one used in experiment 1. However, ultrasonic vocalization was recorded throughout the experiment (Figure 2a). As expected and already observed in experiment 1, there were significant group differences in the fear retention test (Figure 2b; ANOVA: $F_{2,51} = 100.90$; $p < 0.0001$). DEM rats expressed more freezing behavior than OBS and NAIVE rats (Holm–Sidak's tests: $t_{18} = 11.03$, $p < 0.0001$ and $t_{18} = 13.27$, $p < 0.0001$, respectively). Of note, OBS rats had significantly higher freezing scores than NAIVE rats ($t_{18} = 2.24$, $p = 0.03$), indicating successful observational fear learning in the OBS rats.

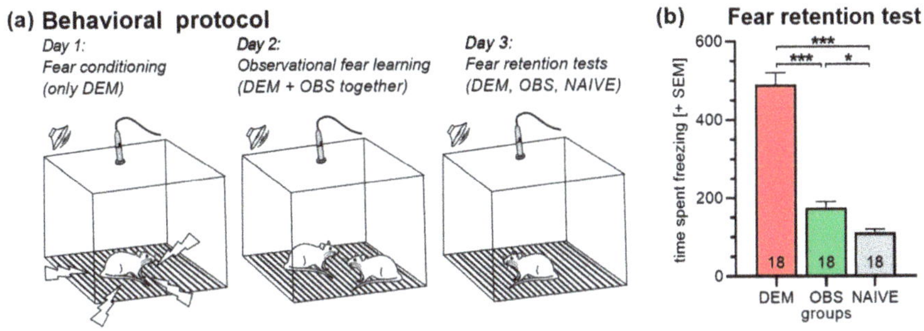

Figure 2. (**a**) Behavioral protocol of experiment 2 was identical to experiment 1, but no light–dark test was performed, and ultrasonic vocalization was recorded throughout the experiment. (**b**) In the retention test, DEM rats showed more freezing than OBS and NAIVE rats. OBS rats had significantly higher freezing scores than NAIVE rats, indicating observational fear learning in OBS rats. Abbreviations: DEM, demonstrator rats; OBS, observer rats; NAIVE, naive rats. Numbers in the bars indicate group sizes. *** $p < 0.001$, * $p < 0.05$; comparisons as indicated, after significant main effects in ANOVA.

During all phases of this experiment, ultrasonic vocalization was recorded and analyzed. Our first analysis was focused on 22 kHz calls (Figures 3 and 4). 22 kHz calls were exclusively emitted by DEM rats with the exception of one NAIVE rat (four 22 kHz calls in the retention test). The number of DEM rats emitting 22 kHz calls in the different phases of the experiment was not different (Figure 3a; chi-squared test: $\chi^2 = 1.99$, $p = 0.37$). From the fourteen DEM rats that emitted at least once during the experiment 22 kHz calls, seven rats emitted them in all three phases. The other seven rats emitted 22 kHz calls mainly in the observational fear learning session (75%), i.e., when a conspecific was present, and less during the sessions without conspecific (50%). It should be emphasized that the recordings from the observational fear learning phase potentially include vocalizations from two rats, the DEM and the OBS rats. However, since we never detected simultaneously emitted 22 kHz calls in the sonograms and the OBS rats never emitted 22 kHz calls in the retention test, we are confident that all recorded 22 kHz calls in the observational fear learning phase were emitted by the DEM rats.

The latency of the first 22 kHz call was very different in the three sessions (Figure 3b). Latencies were longest during the fear-conditioning session. The 22 kHz calls were never emitted before the first electric stimulus but usually appeared first after the 2nd electric stimulus (in fewer cases also after the 1st or third stimulus). The observational fear learning session and the retention test had identical protocols (including no electric stimuli); therefore, these sessions can be directly compared. The latencies of the first 22 kHz calls in the observational fear learning session were significantly longer than those in the retention tests (Mann–Whitney test: $U = 29$, $p = 0.03$). In the observational fear learning session, most rats started to emit 22 kHz calls after the first tone presentation, while in the retention test, most rats started before or during this tone.

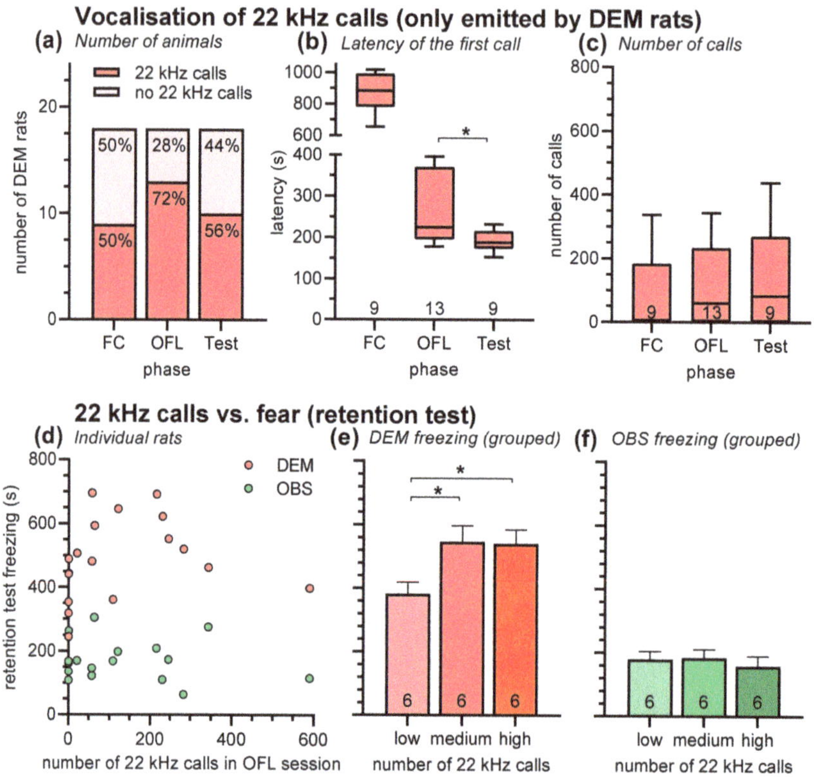

Figure 3. (**a**) Number of rats emitting 22 kHz calls, (**b**) latency of the first 22 kHz call, and (**c**) number of 22 kHz calls in the different phases of the experiment. 22 kHz calls were almost exclusively emitted by DEM rats. There was no difference in the number of rats emitting calls and in the number of calls. However, latency was longest in the fear-conditioning session and significantly longer during the observational fear learning session than during the retention tests. (**d**) Freezing behavior of DEM and OBS rats in the retention session as a function of the number of 22 kHz calls in the observational fear learning session. No correlations were found. (**e**) Next, the observational fear learning session were grouped into sessions with low, medium, and high number of 22 kHz calls. DEM rats of OFL sessions with low number of 22 kHz calls showed lower freezing behavior than the DEM rats from the other session. (**f**) Freezing behavior of OBS rats was not affected by the number of 22 kHz calls in the observational fear learning session. *Y*-axis scale and units in panels (**e**,**f**) are the same as in panel (**d**). Abbreviations: DEM, demonstrator rats; FC, fear-conditioning session; OBS, observer rats; OFL, observational fear learning session; Test, retention test. Numbers in or below the bars or boxes indicate group and subgroup sizes. * $p < 0.05$; comparisons as indicated, after significant main effects in Kruskal–Wallis test or ANOVA, respectively.

Despite the different latencies, the number of 22 kHz calls during the different phases of the experiments was not different (Figure 3c; Kruskal–Wallis test: $H = 1.00$, $p = 0.61$). When we compared the number of 22 kHz calls in the observational fear learning session with the freezing behavior of DEM or OBS rats in the retention tests (Figure 3d), we did not find any correlations with acceptable R^2-values (for example, linear regression: both $R^2 = 0.03$). However, after grouping the observational fear learning sessions-in-sessions with low (0–1), medium (21–121), and high number (215–590) of 22 kHz calls (lower, medium and higher third, i.e., $n = 6$/group), we found that the number of 22 kHz calls in the observational fear learning sessions affected the freezing behavior of DEM rats in the retention test (Figure 3e; ANOVA: $F_{2,15} = 4.61$, $p = 0.03$). Post hoc comparisons revealed that DEM rats from observational fear learning sessions with medium or high numbers of 22 kHz calls showed higher freezing scores in the fear retention test than those from a session with low numbers (Holm–Sidak's tests: t values > 2.58, p values = 0.05). Against that, the freezing of OBS rats in the retention test was not affected by the number of 22 kHz calls in the observational fear learning sessions (Figure 3f; ANOVA: $F_{2,15} = 0.26$, $p = 0.77$).

We further analyzed several parameters of the 22 kHz calls (total number: 6725 calls). The mean call duration was approximately 0.90–1.00 s (Figure 4a), and the peak frequency of the calls approximately 21.8–22.2 kHz (Figure 4b). These parameters were not affected by the phase of the experiment (ANOVAs: $F_{2,30} = 0.10$, $p = 0.91$ and $F_{2,30} = 0.31$, $p = 0.73$, respectively). Regularly, calls with very short durations were recorded (2.3% of the calls were <300 ms and 0.8% < 150 ms; Figure 4c–e), and about 0.2% of the calls had an exceptionally long duration (>3000 ms). The maximal call duration in our experiment was 4.341 s (Figure 4f).

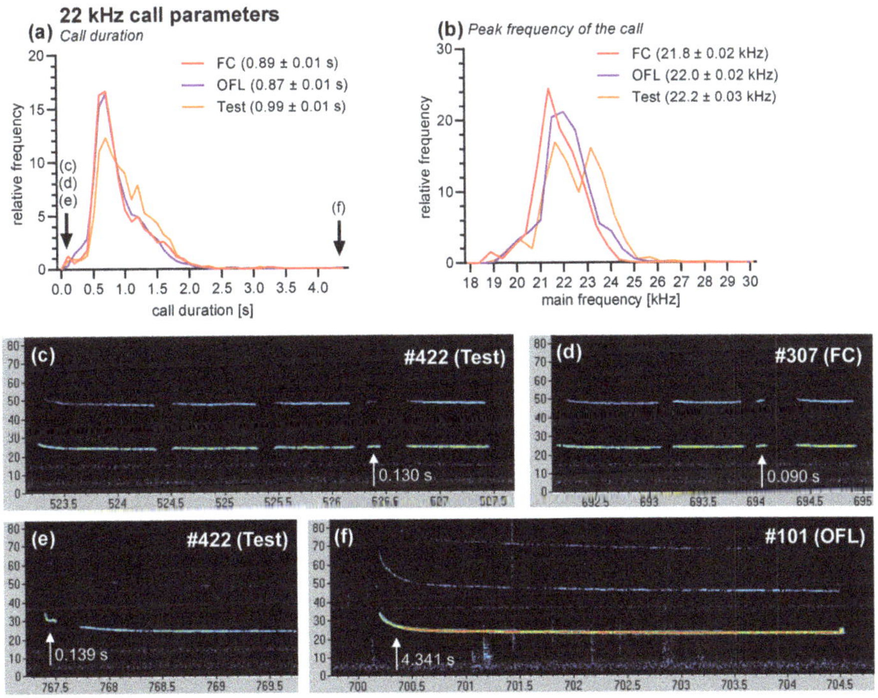

Figure 4. Parameters of the 22 kHz calls. Histogram depicting (**a**) the distribution of the call duration and (**b**) peak frequencies of the calls in the different phases (mean ± SEM). (**c**–**f**) Examples of calls with very short (**c**–**e**) or exceptionally long (**f**) call duration. Abbreviations: FC, fear-conditioning session; OFL, observational fear learning session; Test, retention test.

In contrast to the 22 kHz calls, the 50 kHz calls were emitted by all rats (DEM, OBS, NAIVE) in all phases of the experiment (Figure 5a,b shows some representative examples of 50 kHz calls). Clearly, the highest number of 50 kHz calls were emitted in the observational fear learning phase. Of note, two rats were present during this phase, and very likely, both of them emitted 50 kHz calls. A comparison of this phase with the other phases, in which always only one rat was present, is, therefore, difficult, if not impossible. However, such an analysis is possible for the retention test, and this analysis revealed different numbers of 50 kHz calls in DEM, OBS, and NAIVE rats (Kruskal–Wallis test: $H = 14.21$ $p = 0.0008$). Pairwise comparisons show less calls in DEM and OBS rats than in NAIVE rats ($p = 0.002$ and $p = 0.02$, respectively), but no difference between DEM and OBS rats ($p = 0.12$). Figure 5b depicts the 50 kHz call category profile (Figure 5b) using the categories described by Wright and colleagues [39], see examples in Figure 5c. The call profile significantly changed across the different sessions of the experiment. Of note, downward and upward ramp calls were emitted less often in the observation fear learning session and in the retention test of the DEM rats than in the other phases (χ^2 tests: χ^2 values > 67; p values < 0.0001). In contrast, the proportion of complex calls and trills were higher in these two phases (χ^2 test: χ^2 values > 125; p values < 0.0001).

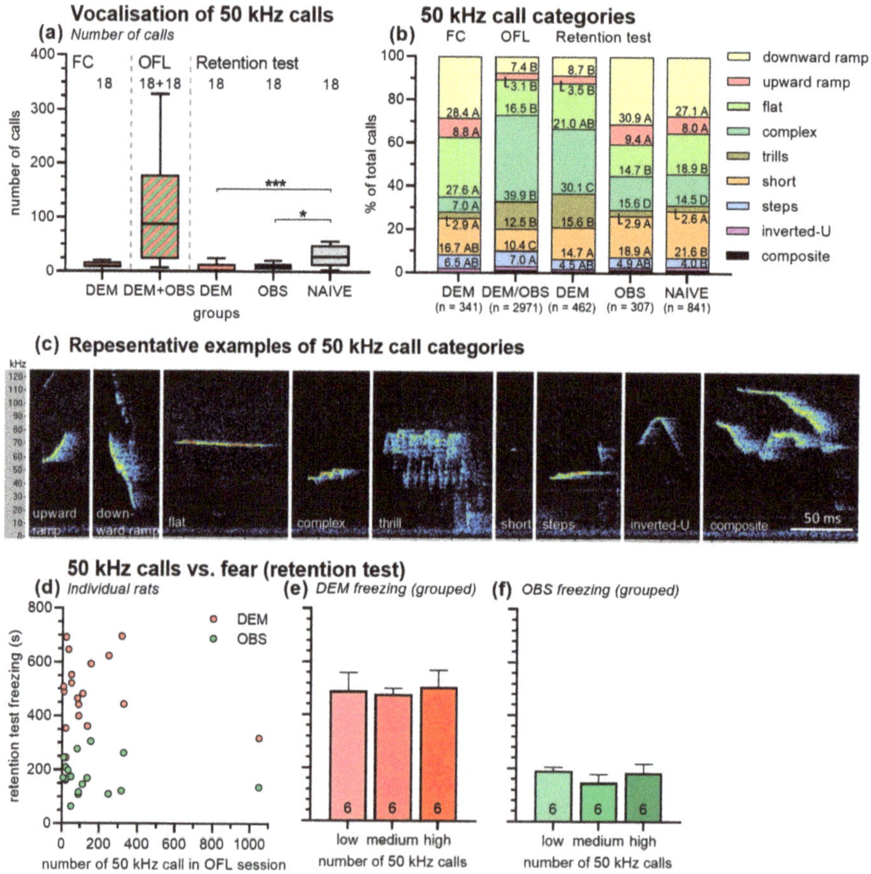

Figure 5. (**a**) Number of 50 kHz calls in the different phases of the experiment. Most 50 kHz calls were emitted in the observational fear learning session in which two rats were present. In the retention test, NAIVE rats emitted more 50 kHz calls than DEM and OBS rats. (**b**) 50 kHz call category profiles in the different sessions. The different colors of the bar sections indicate the different 50 kHz call categories. The numbers in the sectors represent the percent number of calls in a given category, related to the total number of calls in the respective session (indicated on the bottom). The letters after the number indicate whether the proportion of a call category differs across the sessions (χ^2 test). Different letters symbolize significantly different proportions; equal letters stand for non-significant proportions. (**c**) Representative examples of the different 50 kHz call categories. (**d**) Freezing behavior of DEM and OBS rats in the retention session as a function of the number of 50 kHz calls in the observational fear learning session. No correlations were found. (**e**,**f**) Therefore, the observational fear learning sessions were grouped into sessions with low, medium, and high number of 50 kHz calls. No effects of 50 kHz calls on the freezing behavior of DEM and OBS rats in the retention test were found. *Y*-axis scale and units in panels (**e**,**f**) are the same as in panel (**d**). Abbreviations: DEM, demonstrator rats; FC, fear conditioning session; OBS, observer rats; OFL, observational fear learning session; NAIVE, naive rats. Numbers in the bars or above the boxes indicate group and subgroup sizes. *** p < 0.001, * p < 0.05; comparisons as indicated, after significant main effects in Kruskal–Wallis test.

We further analyzed whether the number of 50 kHz calls in the observational fear learning session was correlated with the freezing of DEM or OBS rats in the retention test (Figure 5d). Again, no correlations with acceptable R^2-values were found (for example,: linear regression: $R^2 = 0.04$ or 0.02, respectively). In addition, grouping the observational fear learning sessions-in-sessions with low (7–34), medium (51–113), and high number (137–1047) of 50 kHz calls ($n = 6$/group) did not reveal effects of 50 kHz calls on the freezing behavior of DEM rats (Figure 5e; ANOVA: $F_{2,15} = 0.07$, $p = 0.93$) or OBS rats (Figure 5f; ANOVA: $F_{2,15} = 0.77$, $p = 0.48$) in the fear retention session.

4. Discussion

The present study investigated the role of trait anxiety and ultrasonic vocalization on observational fear learning in rats. For our experiments, we used a published protocol [33] in which rats from the same cage were grouped in triads (DEM, OBS, NAIVE). The DEM rat was directly fear-conditioned, the OBS rat had to learn fear by observing the DEM rat, and the NAIVE rats stayed naive until the fear retention test. In both retention tests of the present study, DEM rats expressed a freezing behavior, followed by the OBS rats and the NAIVE rats (Figures 1c and 2b). Importantly, OBS rats had significantly higher freezing scores than the NAIVE rats indicating successful observational fear learning in the present experiments. Very similar differences were observed when freezing behavior during context exposure only or during the tone presentations was separately analyzed (data not shown), supporting previous data showing that both cued and contextual fear can be learned by observation [32]. It is important to note that in the used observational fear learning protocol, the OBS rats learned by observing a learned fear response that the DEM rats expressed during the observational fear learning session in which no unconditioned stimuli (the electric stimuli) are presented. This is different in other protocols in which the OBS rats observed the DEM rats directly during fear-conditioning, i.e., during the application of aversive stimuli that usually induce strong unconditioned responses [40,41] or in which observing fear responses affects later fear-conditioning in the OBS rats [42]. Importantly, all of these protocols are based on the concept that the recognition of the DEM rats' fear triggers a similar feeling in the OBS rats, i.e., an empathy-like or anxiety state, which then leads to a learning experience or influences a following learning process [43,44]. However, the different protocols may lead to different subtypes of observational learning, and trait anxiety and ultrasonic vocalization may play different roles in these different learning subtypes.

Our first experiment was focused on the role of trait anxiety. The present data show that trait anxiety differently affected the fear response of DEM and OBS rats. In DEM rats, we observed a linear correlation of trait anxiety and fear learning that was also obvious after subgrouping the DEM rats in low, medium and high anxiety. The more anxious the subgroup was the more freezing in the retention test they expressed (Figure 1d,e). This finding is in line with previous publications showing that trait anxiety in laboratory rodents is positively correlated with the strength of fear learning [27] and that breeding lines or strains with higher trait anxiety show enhanced fear expression [28,45,46]. For the OBS rats, we expected a similar relationship between trait anxiety and observational fear learning. However, OBS rats with a medium level of trait anxiety expressed highest freezing responses in the retention test, while low- and high-anxiety OBS rats expressed very low freezing responses (Figure 1d,f). Of note, mean trait anxiety was not different in DEM, OBS, and NAIVE rats (Figure 1b) and played no role in the freezing behavior of NAIVE rats (Figure 1g).

Our study strongly indicates that observational fear learning is best in medium anxious OBS rats and relatively poor or absent in low- and high-anxiety OBS rats. To the best of our knowledge, the present study is the first one investigating the role of trait anxiety in observational fear learning. We also did not find such studies in other species, including humans or on other observational learning processes. Actually, we only find one human study showing that trait anxiety of parents, who served as demonstrators in this study, was positively correlated with observational fear learning of the parents' kids [47]. This effect of demonstrators' trait anxiety on observers' observational fear learning was not seen in the present study after grouping the DEM rats into low, medium and high-anxiety subgroups (data not shown).

One explanation of the present data could be that both too low and too high anxiety levels interfere with several processes that are important for observational fear learning. These processes include recognizing the DEM rats' emotional state, experiencing a similar emotional state (an empathy-like or anxiety process), associating this state with the stimuli that triggered the DEM rats' state, and consolidating such associations. For example, OBS rats with too low trait anxiety might simply not develop this empathy-like or anxiety process, which is necessary to learn by observation. In contrast, OBS rats with high trait anxiety might develop too much fear/anxiety during the observational fear learning session, which then interferes with learning abilities and/or lead to very unspecific learning. The role of trait anxiety in these processes is poorly investigated in laboratory rodents and even in humans poorly understood. For example, one study shows that high-anxiety participants recognize fear faces better than low anxious participants [48], while another study did not find this effect of trait anxiety [49]. Unfortunately, very often, only low- and high-anxiety participants are compared in these studies, and a medium level of anxiety is neglected.

We also had the suspicion that the relationship between the DEM rats' and OBS rats' trait anxiety plays a role in observational fear learning, i.e., that OBS rats maybe learn better from DEM rats that are more or less anxious than they are or that similar levels of trait anxiety in DEM and OBS rates are optimal. However, we did not find that any of these relationships lead to better observational fear learning (Figure 1h), even though it may be beneficial for observational fear learning if the demonstrator is more anxious than the observer.

The aim of our second experiment was to get some insight into the role of ultrasonic vocalization during observational fear learning. Our hypothesis was that both 22 kHz calls and 50 kHz calls should play a role in observational fear learning. The 22 kHz calls should be of impact since they are discussed as species-specific aversive communication signals [4,9,50], while 50 kHz calls should matter since they are typically emitted during positive social interaction [34–36], and the amount of social interaction was found to be correlated with observational fear learning [30–32]. Actually, Jones and colleagues showed a correlation of the total duration of 22 kHz calls in the observational fear learning session with the OBS rats' freezing during the retention test [31]. As in the aforementioned study, we recorded 22 kHz calls only in DEM rats—with the exception of one NAIVE rat emitting four 22 kHz calls during the retention test. Of note, 22 kHz calls during the observational fear learning session could potentially also be emitted by the OBS rats. However, analyses of the video recordings of this session revealed that there were always the DEM rats and rarely the OBS rats showing the typical body posture and flank movements that are associated with the emission of 22 kHz calls. Furthermore, no simultaneously emitted 22 kHz calls were identified in the sonograms. Therefore, we are confident that all recorded 22 kHz calls in the observational fear learning session were emitted by the DEM rats.

During the fear-conditioning session and the retention test, about half of the DEM rats emitted 22 kHz calls, while more of them (72%) emitted these calls during the observational fear learning session, i.e., when a conspecific was present (Figure 3a). This difference was not statistically confirmed but could indicate a so-called "audience effect". The audience effect refers to the idea that 22 kHz calls serve as alarm calls to conspecifics and, therefore, require their presence [4,9]. While the audience effect was described in one of the seminal studies on rats' ultrasonic vocalization [4], this effect was not observed in other studies [51]. In addition to the potential audience effect, we found that the latency to emit 22 kHz calls was longer in the observational fear learning session than in the retention test (Figure 3b). This effect was most probably caused by the social interaction with the OBS rats and could, therefore, reflect social buffering of fear [52,53]. However, the number of 22 kHz calls was not different in the different phases of the experiment (Figure 3c). Of note, there was also no association of 22 kHz call emission during the observational fear learning session with the OBS rats' fear in the retention test (Figure 3d,f), which stands in contrast to a previous report that found such a correlation [31]. However, we found that DEM rats that emitted medium and high number of 22 kHz calls in the observational fear learning session express more fear in the retention test than the DEM rats from the session with a low number of 22 kHz calls (Figure 3e). Very similar results were obtained when the total duration of 22 kHz calls was analyzed (data not shown).

The analyses of the call parameters of the recorded 22 kHz calls showed no differences to the 22 kHz calls described in the literature [6,7,54], i.e., the mean duration was between 0.87 and 0.99 s and the peak frequency between 21.8 and 22.2 kHz calls in the different phases of the experiment. These parameters were not affected by the phase of the experiment. As already previously reported [6], we observed a high variation in the duration of the 22 kHz calls. Approximately 2.3% of the analyzed 22 kHz calls were shorter than 300 ms, and about a third of them (0.8%) were even shorter than 150 ms, two thresholds that are sometimes used [6,9]. In our opinion, these thresholds are arbitrary and not justified since such short calls occurred regularly within sequences of longer calls, without any "external disturbances". However, some of the short 22 kHz calls were obviously caused by such external disturbances, such as interactions with the OBS rats or the onset and offset of the tone stimulus or the electric stimulus. Rarely (0.2%), calls with call durations longer than 3 s were recorded. Of note, the maximal duration in our study was 4.341 s (Figure 4f), which is ca. 400 ms longer than the published maximal call duration [6].

In contrast to the 22 kHz calls, 50 kHz calls were emitted by DEM, OBS, and NAIVE rats and recorded in all phases of the experiment. The number of calls varied strongly in the different phases (Figure 5a). Most calls were recorded in the observational fear learning session, which is not surprising since two rats were present in this phase. Of note, we cannot distinguish which rat emitted the 50 kHz calls in this session. Only a little amount of 50 kHz calls were emitted in the fear-conditioning session. Usually, these calls were present until the first aversive electric stimulus, then no calls were emitted for a while and then, often 22 kHz calls were emitted (see above). However, 50 kHz calls were also occasionally recorded a few seconds after the electric stimuli. In the retention test, the number of 50 kHz calls was similar in DEM and OBS rats, while NAIVE rats emitted higher amounts of 50 kHz calls. Since 50 kHz calls are associated with positive emotional states [34,36,55,56], this difference between OBS and NAIVE rats may indicate observational fear learning in the OBS rats. However, 50 kHz calls are also emitted during exploratory behavior [35]; the high numbers of 50 kHz calls in NAIVE rats could also be explained by the fact that the retention test was the first exposure of the naive rats to the experimental setup.

Based on the idea that the number of 50 kHz calls during the observational fear learning session may indicate the amount of social interaction between DEM and OBS rats, we also analyzed whether the number of 50 kHz calls in this session is correlated with the freezing behavior in the retention tests. This was neither the case for DEM nor for OBS rats (Figure 5d–f). We also analyzed the 50 kHz call profiles in the different sessions using the categories published by Wright and colleagues [39], Figure 5c. Actually, the call profile significantly changed across the different sessions (Figure 5b). Of note, the profiles of the observational fear learning session and of the retention test of the DEM rats were similar by consisting of more complex calls and trills and fewer downward and upward ramp calls than the other sessions. So far, the meaning of the different 50 kHz call categories is poorly understood. Some categories seem to be associated with certain behaviors, e.g., split and composite calls with running and jumping and trills with slower movements [57] but also with playful attacks in juvenile rats [58]. We detected trills and complex calls more in the observational fear learning session, i.e., in a social situation, but surprisingly also in the retention test of the demonstrators. The presence of these call types in the latter session maybe indicate the search for the social partner the DEM rats had in the session before. Upward and downward ramp calls were most emitted in the fear-conditioning session (before the aversive electric stimuli) and during the retention tests of OBS and NAIVE rats. During these sessions, mainly exploratory behavior was expressed, which suggests that these two call types might be associated with exploration.

Together, these findings suggest no important roles of 22 kHz calls and 50 kHz calls in observational fear learning. We think that the number, total duration and latency of 22 kHz calls potentially signal the strength of conditioned fear of the DEM rats in the different phases of the experiment. Furthermore, there were minor hints for audience effects as well as for social buffering with respect to 22 kHz calls in the observational fear learning phase. In this phase, 50 kHz calls were emitted in a high amount, but this is expected in a situation in which two rats can socially interact. The lower number of 50 kHz calls in DEM and OBS rats in the retention test may indicate a less appetitive state compared with the naive rats and thereby support that OBS rats learned fear by observation.

Based on the present findings, the question arises: Which information source is used by the OBS rats during observational fear learning if 22 kHz calls and 50 kHz calls are not critically involved? The DEM rats express several species-specific behavioral signs of fear, including freezing (in mice, even facial expression of fear was described [59]), which may serve as visual stimuli for observational fear learning [43]. Of note, rats also release an alarm pheromone during aversive emotional states that can induce defensive behavior in conspecifics [60–62]. During social interaction, this alarm pheromone might be best perceived by the OBS rats, which could be the reason why social interaction during the observational fear learning session correlates with the learned fear in OBS rats [30,31]. Ultimately, these visual and/or olfactory stimuli are most likely sufficient for the OBS rats to learn fear by observation.

5. Conclusions

The present study showed that trait anxiety affects direct fear learning and observational fear learning slightly differently. Direct fear learning was more pronounced with higher trait anxiety (DEM rats), while observational fear learning was the best with a medium-level of trait anxiety (OBS rats). There were no indications in the present study that ultrasonic vocalization, especially emission of 22 kHz calls, is critical for observational fear learning. We believe that both 22 kHz calls and 50 kHz calls are indicative of the rats' individual emotional state but play no major role in observational fear learning.

Author Contributions: Conceptualization, M.F.; formal analysis, M.F. and C.P.G.-G.; investigation, M.F., C.P.G.-G. and E.K.; resources, M.F.; writing—original draft preparation, M.F.; writing—review and editing, M.F. and C.P.G.-G.; visualization, M.F.; supervision, M.F. All authors have read and agreed to the published version of the manuscript.

Funding: This research received no external funding.

Institutional Review Board Statement: This study was conducted according to the European guidelines for the use of animals in experiments (2010/63/EU) and was approved by the Ethics Committee and the Landesverwaltungsamt of the state of Sachsen-Anhalt (Az. 42502-2-1587 UniMD; approved 14 October 2019).

Informed Consent Statement: Not applicable.

Data Availability Statement: The data presented in this study are available on request from the corresponding author.

Acknowledgments: We are grateful to Iris Müller for her valuable comments on the manuscript.

Conflicts of Interest: The authors declare no conflict of interest.

References

1. Blanchard, R.J.; Blanchard, D.C.; Rodgers, J.; Weiss, S.M. The characterization and modelling of antipredator defensive behavior. *Neurosci. Biobehav. Rev.* **1990**, *14*, 463–472. [CrossRef]
2. Blanchard, R.J.; Blanchard, D.C. An ethoexperimental analysis of defense, fear, and anxiety. In *Anxiety*; McNaughton, N., Andrews, G., Eds.; University of Otago Press: Dunedin, New Zealand, 1990; pp. 124–133.

3. Blanchard, R.J.; Blanchard, D.C. Crouching as an index of fear. *J. Comp. Physiol. Psychol.* **1969**, *67*, 370–375. [CrossRef]
4. Blanchard, R.J.; Blanchard, D.C.; Agullana, R.; Weiss, S.M. Twenty-two kHz alarm cries to presentation of a predator, by laboratory rats living in visible burrow systems. *Physiol. Behav.* **1991**, *50*, 967–972. [CrossRef]
5. Brudzynski, S.M. Pharmacological and behavioral characteristics of 22 kHz alarm calls in rats. *Neurosci. Biobehav. Rev.* **2001**, *25*, 611–617. [CrossRef]
6. Brudzynski, S.M.; Bihari, F.; Ociepa, D.; Fu, X.W. Analysis of 22 kHz ultrasonic vocalization in laboratory rats: Long and short calls. *Physiol. Behav.* **1993**, *54*, 215–221. [CrossRef]
7. Wöhr, M.; Borta, A.; Schwarting, R.K.W. Overt behavior and ultrasonic vocalization in a fear conditioning paradigm: A dose-response study in the rat. *Neurobiol. Learn. Mem.* **2005**, *84*, 228–240. [CrossRef] [PubMed]
8. Wöhr, M.; Schwarting, R.K.W. Affective communication in rodents: Ultrasonic vocalizations as a tool for research on emotion and motivation. *Cell Tiss. Res.* **2013**, *354*, 81–97. [CrossRef]
9. Brudzynski, S.M. Ethotransmission: Communication of emotional states through ultrasonic vocalization in rats. *Curr. Opin. Neurobiol.* **2013**, *23*, 310–317. [CrossRef]
10. Brudzynski, S.M.; Chiu, E.M. Behavioural responses of laboratory rats to playback of 22 kHz ultrasonic calls. *Physiol. Behav.* **1995**, *57*, 1039–1044. [CrossRef]
11. Wöhr, M.; Schwarting, R.K.W. Ultrasonic communication in rats: Can playback of 50-kHz calls induce approach behavior? *PLoS ONE* **2007**, *2*, e1365. [CrossRef]
12. Sales, G.D. The effect of 22 kHz calls and artificial 38 kHz signals on activity in rats. *Behav. Proc.* **1991**, *24*, 83–93. [CrossRef]
13. Endres, T.; Widmann, K.; Fendt, M. Are rats predisposed to learn 22 kHz calls as danger-predicting signals? *Behav. Brain Res.* **2007**, *185*, 69–75. [CrossRef] [PubMed]
14. Blanchard, D.C.; Blanchard, R.J. Defensive behaviors, fear, and anxiety. In *Handbook of Behavioral Neuroscience*; Blanchard, R.J., Blanchard, D.C., Griebel, G., Nutt, D.J., Eds.; Elsevier: Amsterdam, The Netherlands, 2008; Volume 17, pp. 63–79.
15. Fanselow, M.S. The role of learning in threat imminence and defensive behaviors. *Curr. Opin. Behav. Sci.* **2018**, *24*, 44–49. [CrossRef]
16. LeDoux, J.E. Coming to terms with fear. *Proc. Natl. Acad. Sci. USA* **2014**, *111*, 2871–2878. [CrossRef]
17. LeDoux, J.E. Emotion circuits in the brain. *Annu. Rev. Neurosci.* **2000**, *23*, 155–184. [CrossRef]
18. Fendt, M.; Fanselow, M.S. The neuroanatomical and neurochemical basis of conditioned fear. *Neurosci. Biobehav. Rev.* **1999**, *23*, 743–760. [CrossRef]
19. Maren, S. Neurobiology of Pavlovian fear conditioning. *Annu. Rev. Neurosci.* **2001**, *24*, 897–931. [CrossRef]
20. Pape, H.C.; Pare, D. Plastic synaptic networks of the amygdala for the acquisition, expression, and extinction of conditioned fear. *Physiol. Rev.* **2010**, *90*, 419–463. [CrossRef]
21. Lissek, S.; Rabin, S.; Heller, R.E.; Lukenbaugh, D.; Geraci, M.; Pine, D.S.; Grillon, C. Overgeneralization of conditioned fear as a pathogenic marker of panic disorder. *Am. J. Psychiatry* **2010**, *167*, 47–55. [CrossRef]
22. Lissek, S.; Kaczkurkin, A.N.; Rabin, S.; Geraci, M.; Pine, D.S.; Grillon, C. Generalized anxiety disorder is associated with overgeneralization of classically conditioned fear. *Biol. Psychiatry* **2014**, *75*, 909–915. [CrossRef] [PubMed]
23. Rabinak, C.A.; Mori, S.; Lyons, M.; Milad, M.R.; Phan, K.L. Acquisition of CS-US contingencies during Pavlovian fear conditioning and extinction in social anxiety disorder and posttraumatic stress disorder. *J. Affect. Dis.* **2017**, *207*, 76–85. [CrossRef]
24. Grillon, C.; Ameli, R.; Foot, M.; Davis, M. Fear-potentiated startle: Relationship to the level of state/trait anxiety in healthy subjects. *Biol. Psychiatry* **1993**, *33*, 566–574. [CrossRef]
25. Gazendam, F.J.; Kamphuis, J.H.; Kindt, M. Deficient safety learning characterizes high trait anxious individuals. *Biol. Psychol.* **2013**, *92*, 342–352. [CrossRef] [PubMed]
26. Haaker, J.; Lonsdorf, T.B.; Schumann, D.; Menz, M.; Brassen, S.; Bunzeck, N.; Gamer, M.; Kalisch, R. Deficient inhibitory processing in trait anxiety: Evidence from context-dependent fear learning, extinction recall and renewal. *Biol. Psychol.* **2015**, *111*, 65–72. [CrossRef] [PubMed]
27. Ilse, A.; Prameswari, V.; Kahl, E.; Fendt, M. The role of trait anxiety in associative learning during and after an aversive event. *Learn. Mem.* **2019**, *26*, 56–59. [CrossRef] [PubMed]
28. Muigg, P.; Hetzenauer, A.; Hauer, G.; Hauschild, M.; Gaburro, S.; Frank, E.; Landgraf, R.; Singewald, N. Impaired extinction of learned fear in rats selectively bred for high anxiety—Evidence of altered neuronal processing in prefrontal-amygdala pathways. *Eur. J. Neurosci.* **2008**, *28*, 2299–2309. [CrossRef]
29. Atsak, P.; Orre, M.; Bakker, P.; Cerliani, L.; Roozendaal, B.; Gazzola, V.; Moita, M.; Keysers, C. Experience modulates vicarious freezing in rats: A model for empathy. *PLoS ONE* **2011**, *6*, e21855. [CrossRef]
30. Bruchey, A.K.; Jones, C.E.; Monfils, M.H. Fear conditioning by-proxy: Social transmission of fear during memory retrieval. *Behav. Brain Res.* **2010**, *214*, 80–84. [CrossRef]
31. Jones, C.E.; Monfils, M.H. Dominance status predicts social fear transmission in laboratory rats. *Anim. Cogn.* **2016**, *19*, 1051–1069. [CrossRef]
32. Jones, C.E.; Riha, P.D.; Gore, A.C.; Monfils, M.H. Social transmission of Pavlovian fear: Fear-conditioning by-proxy in related female rats. *Anim. Cogn.* **2014**, *17*, 827–834. [CrossRef]
33. Jones, C.E.; Agee, L.; Monfils, M.H. Fear conditioning by proxy: Social transmission of fear between interacting conspecifics. *Curr. Protoc. Neurosci.* **2018**, *83*, e43. [CrossRef]

34. Wöhr, M. Ultrasonic communication in rats: Appetitive 50-kHz ultrasonic vocalizations as social contact calls. *Behav. Ecol. Sociobiol.* **2017**, *72*, 14. [CrossRef]
35. Brudzynski, S.M.; Pniak, A. Social contacts and production of 50-kHz short ultrasonic calls in adult rats. *J. Comp. Psychol.* **2002**, *116*, 73–82. [CrossRef]
36. Mulvihill, K.G.; Brudzynski, S.M. Non-pharmacological induction of rat 50 kHz ultrasonic vocalization: Social and non-social contexts differentially induce 50 kHz call subtypes. *Physiol. Behav.* **2018**, *196*, 200–207. [CrossRef] [PubMed]
37. Kelly, M.M.; Forsyth, J.P. Associations between emotional avoidance, anxiety sensitivity, and reactions to an observational fear challenge procedure. *Behav. Res. Ther.* **2009**, *47*, 331–338. [CrossRef] [PubMed]
38. Misane, I.; Tovote, P.; Meyer, M.; Spiess, J.; Ogren, S.O.; Stiedl, O. Time-dependent involvement of the dorsal hippocampus in trace fear conditioning in mice. *Hippocampus* **2005**, *15*, 418–426. [CrossRef]
39. Wright, J.M.; Gourdon, J.C.; Clarke, P.B. Identification of multiple call categories within the rich repertoire of adult rat 50-kHz ultrasonic vocalizations: Effects of amphetamine and social context. *Psychopharmacology* **2010**, *211*, 1–13. [CrossRef]
40. Jeon, D.; Kim, S.; Chetana, M.; Jo, D.; Ruley, H.E.; Lin, S.Y.; Rabah, D.; Kinet, J.P.; Shin, H.S. Observational fear learning involves affective pain system and Ca v 1.2 Ca 2+ channels in ACC. *Nat. Neurosci.* **2010**, *13*, 482–488. [CrossRef]
41. Kim, S.; Mátyás, F.; Lee, S.; Acsády, L.; Shin, H.S. Lateralization of observational fear learning at the cortical but not thalamic level in mice. *Proc. Natl. Acad. Sci. USA* **2012**, *109*, 15497–15501. [CrossRef]
42. Knapska, E.; Mikosz, M.; Werka, T.; Maren, S. Social modulation of learning in rats. *Learn. Mem.* **2010**, *17*, 35–42. [CrossRef]
43. Keum, S.; Shin, H.S. Neural basis of observational fear learning: A potential model of affective empathy. *Neuron* **2019**, *104*, 78–86. [CrossRef]
44. Debiec, J.; Olsson, A. Social fear learning: From animal models to human function. *Trends Cogn. Sci.* **2017**, *21*, 546–555. [CrossRef]
45. Sartori, S.B.; Hauschild, M.; Bunck, M.; Gaburro, S.; Landgraf, R.; Singewald, N. Enhanced fear expression in a psychopathological mouse model of trait anxiety: Pharmacological interventions. *PLoS ONE* **2011**, *6*, e16849. [CrossRef] [PubMed]
46. Nobre, M.J.; Cabral, A.; Brandão, M.L. GABAergic regulation of auditory sensory gating in low- and high-anxiety rats submitted to a fear conditioning procedure. *Neuroscience* **2010**, *171*, 1152–1163. [CrossRef]
47. Silvers, J.A.; Callaghan, B.L.; VanTieghem, M.; Choy, T.; O'Sullivan, K.; Tottenham, N. An exploration of amygdala-prefrontal mechanisms in the intergenerational transmission of learned fear. *Dev. Sci.* **2020**, e13056. [CrossRef]
48. Surcinelli, P.; Codispoti, M.; Montebarocci, O.; Rossi, N.; Baldaro, B. Facial emotion recognition in trait anxiety. *J. Anxiety Dis.* **2006**, *20*, 110–117. [CrossRef] [PubMed]
49. Cooper, R.M.; Rowe, A.C.; Penton-Voak, I.S. The role of trait anxiety in the recognition of emotional facial expressions. *J. Anxiety Dis.* **2008**, *22*, 1120–1127. [CrossRef]
50. Schwarting, R.K.W.; Wöhr, M. On the relationships between ultrasonic calling and anxiety-related behavior in rats. *Braz. J. Med. Biol. Res.* **2012**, *45*, 337–348. [CrossRef]
51. Wöhr, M.; Schwarting, R.K.W. Ultrasonic calling during fear conditioning in the rat: No evidence for an audience effect. *Anim. Behav.* **2008**, *76*, 749–760. [CrossRef]
52. Davitz, J.R.; Mason, D.J. Socially facilitated reduction of a fear response in rats. *J. Comp. Physiol. Psychol.* **1955**, *48*, 149–151. [CrossRef]
53. Kiyokawa, Y.; Hennessy, M.B. Comparative studies of social buffering: A consideration of approaches, terminology, and pitfalls. *Neurosci. Biobehav. Rev.* **2018**, *86*, 131–141. [CrossRef] [PubMed]
54. Fendt, M.; Brosch, M.; Wernecke, K.E.A.; Willadsen, M.; Wöhr, M. Predator odour but not TMT induces 22-kHz ultrasonic vocalizations in rats that lead to defensive behaviours in conspecifics upon replay. *Sci. Rep.* **2018**, *8*, 11041. [CrossRef] [PubMed]
55. Knutson, B.; Burgdorf, J.; Panksepp, J. Ultrasonic Vocalizations as indices of affective states in rats. *Psychol. Bull.* **2002**, *128*, 961–977. [CrossRef] [PubMed]
56. Burgdorf, J.; Kroes, R.A.; Moskal, J.R.; Pfaus, J.G.; Brudzynski, S.M.; Panksepp, J. Ultrasonic vocalizations of rats (*Rattus norvegicus*) during mating, play, and aggression: Behavioral concomitants, relationship to reward, and self-administration of playback. *J. Comp. Psychol.* **2008**, *122*, 357–367. [CrossRef] [PubMed]
57. Burke, C.J.; Kisko, T.M.; Swiftwolfe, H.; Pellis, S.M.; Euston, D.R. Specific 50-kHz vocalizations are tightly linked to particular types of behavior in juvenile rats anticipating play. *PLoS ONE* **2017**, *12*, e0175841. [CrossRef] [PubMed]
58. Himmler, B.T.; Kisko, T.M.; Euston, D.R.; Kolb, B.; Pellis, S.M. Are 50-kHz calls used as play signals in the playful interactions of rats? I. Evidence from the timing and context of their use. *Behav. Proc.* **2014**, *106*, 60–66. [CrossRef] [PubMed]
59. Defensor, E.B.; Corley, M.J.; Blanchard, R.J.; Blanchard, D.C. Facial expressions of mice in aggressive and fearful contexts. *Physiol. Behav.* **2012**, *107*, 680–685. [CrossRef]
60. Inagaki, H.; Kiyokawa, Y.; Tamogami, S.; Watanabe, H.; Takeuchi, Y.; Mori, Y. Identification of a pheromone that increases anxiety in rats. *Proc. Natl. Acad. Sci. USA* **2014**, *111*, 18751–18756. [CrossRef]
61. Breitfeld, T.; Bruning, J.E.A.; Inagaki, H.; Takeuchi, Y.; Kiyokawa, Y.; Fendt, M. Temporary inactivation of the anterior part of the bed nucleus of the stria terminalis blocks alarm pheromone-induced defensive behavior in rats. *Front. Neurosci.* **2015**, *9*, 321. [CrossRef]
62. Kiyokawa, Y.; Shimozuru, M.; Kikusui, T.; Takeuchi, Y.; Mori, Y. Alarm pheromone increases defensive and risk assessment behaviors in male rats. *Physiol. Behav.* **2006**, *87*, 383–387. [CrossRef]

brain sciences

MDPI

Article

Relaying Aversive Ultrasonic Alarm Calls Depends on Previous Experience. Empathy, Social Buffering, or Panic?

Wiktoria Karwicka [1], Marta Wiatrowska [2], Kacper Kondrakiewicz [2], Ewelina Knapska [2], Miron Bartosz Kursa [3] and Adam Hamed [1,*]

[1] Laboratory of Spatial Memory, Nencki Institute of Experimental Biology, Polish Academy of Sciences, 3 Pasteur Street, 02-093 Warsaw, Poland; w.karwicka@nencki.edu.pl

[2] Laboratory of Emotions Neurobiology, BRAINCITY—Centre of Excellence for Neural Plasticity and Brain Disorders, Nencki Institute of Experimental Biology, Polish Academy of Sciences, 3 Pasteur Street, 02-093 Warsaw, Poland; m.wiatrowska@nencki.edu.pl (M.W.); k.kondrakiewicz@nencki.edu.pl (K.K.); e.knapska@nencki.edu.pl (E.K.)

[3] Interdisciplinary Centre for Mathematical and Computational Modelling, University of Warsaw, Pawinskiego 5A, 02-106 Warsaw, Poland; m.kursa@icm.edu.pl

* Correspondence: a.hamed@nencki.edu.pl

Citation: Karwicka, W.; Wiatrowska, M.; Kondrakiewicz, K.; Knapska, E.; Kursa, M.B.; Hamed, A. Relaying Aversive Ultrasonic Alarm Calls Depends on Previous Experience. Empathy, Social Buffering, or Panic? *Brain Sci.* **2021**, *11*, 759. https://doi.org/10.3390/brainsci11060759

Academic Editors: Stefan M. Brudzynski and Anne-Marie Mouly

Received: 30 April 2021
Accepted: 1 June 2021
Published: 8 June 2021

Abstract: Ultrasonic vocalizations are among the oldest evolutionarily forms of animal communication. In order to study the communication patterns in an aversive social situation, we used a behavioral model in which one animal, the observer, is witnessing as his cagemate, the demonstrator, is experiencing a series of mild electrical foot shocks. We studied the effect of the foot shock experience on the observer and the influence of a warning sound (emitted shortly before the shock) on USV communication. These experiments revealed that such a warning seems to increase the arousal level, which differentiates the responses depending on previous experience. This can be identified by the emission of characteristic, short 22 kHz calls of a duration below 100 ms. Two rats emitted calls that overlapped in time. Analysis of these overlaps revealed that in 'warned' pairs with a naive observer, 22 kHz calls were mixed with 50 kHz calls. This fact, combined with a high fraction of very high-pitched 50 kHz calls (over 75 kHz), suggests the presence of the phenomenon of social buffering. Pure 22 kHz overlaps were mostly found in 'warned' pairs with an experienced observer, suggesting a possible fear contagion with distress sharing. The results show the importance of dividing 22 kHz calls into long and short categories.

Keywords: ultrasonic vocalization; social buffering; 50 kHz calls; 22 kHz calls; distress; emotional contagion; fear contagion; aversive state; communication

1. Introduction

The ultrasonic vocalizations (USVs) of animals are among the oldest evolutionarily forms of communication [1]. Rats emit sounds varying in frequency that are inaudible to humans and in a band above 20 kHz, reaching a frequency of up to 125 kHz. Most rodent vocal communication research focuses on appetitive sounds—in the so-called "50 kHz" class—that derive from a positive emotional state [2,3]. Ultrasonic vocalizations ranging from 30 to 125 kHz can be induced by addictive substances [4–8], positive social interaction [9–15], the anticipation of reward [16,17], and in response to a context associated with appetitive conditioning [4,17–20].

However, rodents, like humans, not only experience and show their positive states but also express negative emotional states [21–23]. Jaak Panksepp, an author of the concept of affective neuroscience, classified seven basic emotions as a result of artificial stimulation of the mammalian brain: seeking, rage, fear, lust, care, panic/grief, and play [24].

22 kHz USV calls, with a frequency bandwidth of 18 to 28 kHz, serve as indicators of rats' negative emotional states, such as distress, discomfort, or anxiety [25–28].

These aversive types of emissions are usually observed in stressful or endangering situations, such as the presence of a predator [29–32], the aggressive behavior of other conspecific(s) [26,33–35], or when exposed to negative stimuli [36–38]. Experiments using playback techniques have shown that exposure to 22 kHz calls elicits freezing and avoidance responses [39–41]. Several reports [21,22,35,42] state that 22 kHz calls have a communicative function in a group of rats and are not merely an expression of negative emotions. Litvin et al. (2007) proposed a subdivision of calls based on the situation and purpose of their emission: ultrasonic "warning calls" that are meant to warn conspecifics about danger [39,43] and sonic "alarm calls" that are supposed to discourage predators [29,31,32,43]. In both cases, the calls are risk assessment dependent and appear only if potential benefits outweigh the costs [43]. Interestingly, the documented functions of ultrasound emissions in the 22 kHz class include producing alarm signals to protect the social group [44,45]. Thanks to the modern technology of recording and analyzing USVs, we can determine the animal's emotional state with increasing precision. Moreover, we can register sounds lasting even a few milliseconds with greater accuracy and resolution. With such tools, we can determine the course of emotional states during behaviorally modulated social interaction.

The neurobiological foundations of ultrasonic vocalizations and their association with the emotional states of rodents are still under investigation. It has been demonstrated that stimulation of cholinergic neurons in the laterodorsal tegmental nucleus triggers 22 kHz calls [46]. Moreover, previous research has shown that the mPFC (medial prefrontal cortex) plays an important role in modulation of 22 kHz calls and that lesion of this structure considerably or completely reduces the number of this type of call [12,47,48]. Dupin et al. (2019) investigated the relationship between electrophysiological data, respiration, and the emission of USVs, suggesting that sequences of USVs could result in a differential gating of information within the network of structures sustaining fear or anxiety behavior. The emission of 22 kHz ultrasonic vocalization calls converges with decreased theta power and increased delta and gamma power in the BLA (basolateral amygdala), the mPFC, and the olfactory piriform cortex (PIR) [49].

In our previous research, we noted occasional 22 kHz signal amplification in pairs of rats. Based on that observation, we aimed to verify the hypothesis that the experience of an electric shock in the past could change the organization of communication between two familiar rats. An in-depth examination of the communication patterns during stressful situations would explain why an observer might behave differently, depending on previous experience.

To study the communicative function of 22 kHz USVs, we used a behavioral model in which one animal, the observer, is witnessing as his cagemate, the demonstrator, is experiencing a series of mild electrical foot shocks [50]. Aversive stimuli in this model elicited 22 kHz vocalizations [50,51]. Previous works have shown that the experience of aversive stimuli may modulate the future empathetic responses of animals [52–54]. For instance, previous studies that used the same behavioral model showed that prior experience with foot shocks increases the freezing of the observers [55]. Thus, we compared the vocalizations of the pairs of rats with experienced and inexperienced observers. In addition, in half of the groups, we added the protocol of a 19 s, 1.75 kHz audio signal, which was the foot shock introduction (warning signal). This signal was intended to associate the external stimulus with the demonstrator's foot shock. Simultaneously, this allowed us to investigate how the prediction of the stimulus affects the behavior of the demonstrator and of the observer.

An essential part of the study design was to exclude the animals' aversive conditioning response. The electric shocks, intended to familiarize experienced observers with aversive stimuli, were administered in a different spatial context. The training cage differed in shape, lighting, smell, and sound from the one in which we performed the test.

In our study, we aimed to (1) compare USV emissions in pairs of rats between foot-shock-experienced and -inexperienced (naive) observers, (2) examine whether experienced observers amplify the 22 kHz aversive signal emitted by demonstrators, and (3) test whether the warning sound signaling foot shocks changes communication between rats.

2. Materials and Methods

2.1. Animals

Forty experimentally naive male Wistar rats (250–300 g at the beginning of the experiment) were used in the experiment. The animals were supplied by the Center of Experimental Medicine in Bialystok, Poland. Subjects were randomly paired and housed together in standard home cages (43.0 × 25.0 × 18.5 cm). They were kept in standard laboratory conditions under a 12/12 light–dark cycle and were provided with free access to food and water. All experiments were carried out in accordance with the Polish Act on Animal Welfare, after obtaining permission (126/2016) from the First Warsaw Ethical Committee on Animal Research.

2.2. Experimental Procedure

2.2.1. Habituation

After initial acclimatization to a home cage (4 days), the rats were habituated for 10 days to the experimenter's hand (4–5 min/pair/day) and to the experimental room (20 min for 3 consecutive days in dimmed light), and they were transported between rooms (Figure 1). Pairs of rats were randomly assigned to three experimental groups: control, naive, and experienced. Within each pair, rats were additionally marked either as a demonstrator or as an observer; this division was not necessary for pairs in the control group, as both of the rats underwent the same procedure. The experiment was carried out in two groups—'warned' and 'unwarned,' with the 3 subgroups that underwent each procedure consisting of the following number of animals:

1. Naive—4 demonstrators and 4 observers;
2. Experienced—4 demonstrators and 4 observers;
3. Control—4 animals, undivided into demonstrators and observers.

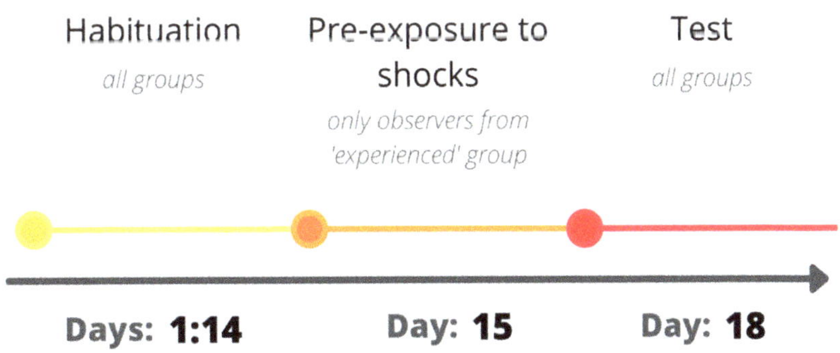

Figure 1. A graphic representation of the experimental design.

2.2.2. Pre-Exposure to Shocks

Three days prior to the experiment, observers from the experienced subgroup were placed into the aversive conditioning cage (Panlab). After 1 min habituation, the rats received 3 electric shocks (with an intensity of 0.7 mA and a duration of 1 s), with a 1 min time interval between each shock. To avoid aversive conditioning to the specific context, the following measures were taken:

- The exposure was performed in a different behavioral room than the following test;
- The cage was illuminated with a bright, white light;
- The interior was sprayed with 1% acetic acid which left a strong smell;
- A plastic rooftop was installed to obtain a triangular shape for different spatial cues.

During conditioning of the unwarned group, a 1.75 kHz sound (19 s duration) was also emitted prior to each shock (Figures 1 and 2).

Figure 2. Illustration of the training: pre-exposure to the electric foot shocks. To exclude the rats' aversive conditioning response, the training cage differed from the test cage in terms of smell, light intensity, and shape. Depending on the experimental group, 'unwarned' or 'warned,' (the group names are according to the test day procedure), a 1.75 kHz sound signaling a foot shock was emitted or not emitted, respectively.

2.2.3. Test Day

On the 18th day of the experiment, the test was carried out in a specially constructed cage (62 × 48 × 25 cm) made of transparent plastic (details in [50]). A perforated, transparent plexiglass divided the interior into two halves—one intended for the observer and the other for the demonstrator. Metal rods installed in the demonstrators' section were connected to the current generator (MedAssociates), which allowed for the administration of mild electrical shocks. The rats could see, hear, and smell each other throughout the test. During the test, the animals were recorded with a digital camera and with 4 microphones (UltraSoundGate, Avisoft). The test session and the pre-exposure to shocks were conducted in separate rooms. To provide a different context, the lights were dimmed, and the cage was cleaned with a 1% acetone solution after testing each pair. Two minutes after inserting the rats into the appropriate chambers, the demonstrator was given a series of 10 foot shocks (1 s, 1.0 mA). In the 'warned' group, each electric shock was signaled by a 19 s, 1.75 kHz sound, equaling a total of 80 s between 2 shocks, whereas in the 'unwarned' group, there was no sound indication and the shocks were administered in a 1 min interval (Figures 1 and 3).

2.3. Apparatus and USV Recordings

USVs were recorded using an UltraSoundGate Condenser Microphone CM16 (Avisoft Bioacoustics, Berlin, Germany) that was positioned 25–30 cm above the floor of the cage. The microphone was sensitive to frequencies of 15–180 kHz with a flat frequency response (±6 dB) between 25 and 140 kHz. The microphone was connected to an amplifier (custom-made, Warsaw) that had the following parameters: a voltage gain of 16 v/v (12 dB), a frequency response of ±0.1 dB, a range of 30 Hz to 120 kHz, and an input impedance of 600 Ω. The signal was then transferred through a 120 kHz anti-aliasing filter (custom-made, Warsaw). The filtered sounds were sent to a PCI-703-16A data acquisition board (Eagle Technology, Chicago, IL, USA). This board was a 14 bit, 400 kHz analogue input and output board for PCI-based systems. The recorded data were processed using the RAT-REC PRO 7.3 software (custom-made, Warsaw, Poland). The signals were processed through a fast Fourier transformation (1024, Hamming or Hann window) and displayed as color spectrograms. Each signal was manually marked with the section label included in the automated parameter measurement.

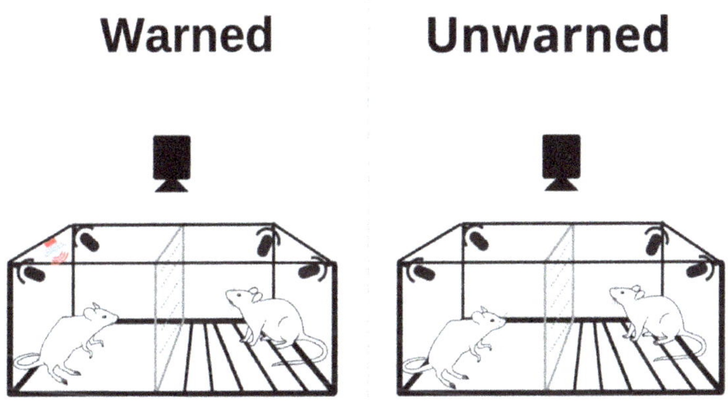

Figure 3. Illustration of the test session. To exclude the rats' aversive conditioning response, the test cage differed from the training cage in terms of smell, light intensity, and shape. Depending on the experimental group, 'warned' or 'unwarned,' a 1.75 kHz sound signaling a foot shock was delivered or not delivered, respectively.

Two rats from each pair were recorded simultaneously. All vocalizations from calls that overlapped in time (occurrence of the two vocalizations at the same time) were marked as separate episodes and analyzed individually. All harmonic sounds were excluded; only fundamental frequencies, which indicate at most two actual calls in every case, were analyzed. Various parameters were determined automatically, including the number of USV calls, the total calling time (s), the mean call duration (s), the frequency bandwidth (kHz), the number of gaps, the mean gap duration (s), and the mean peak frequency (kHz). The signal from the microphone was sent to another room where the computer and research observer were situated (more in [4,10,11]).

2.4. Statistics

To assess the relationship between continuous and categorical variables, we used the Kruskal–Wallis test, followed by the Conover–Iman post-hoc test for the identification of precise pairwise differences, but only when the result of the Kruskal–Wallis test was significant. A significance level of 0.05 and two-sided testing were employed. All statistical analyses were performed in R, version 4.0.5, using the conover.test package version 1.1.5.

3. Results

3.1. The Effects of the Warning Signal. Fractions of the Ultrasonic Vocalizations during Social Communication in the Social Transfer of Fear Paradigm

Figure 4 presents the joint distribution of the duration and frequency of every call considered in the paper, split into experimental groups. The various fractions of calls are clearly visible, so there are quantitative and qualitative differences between them across the groups. The 22 kHz fraction is strongly present in all groups, except in the control group. In 'unwarned' groups, however, its duration span is substantially limited, which corresponds to a lack of episodes, which we later refer to as "short 22 kHz" calls.

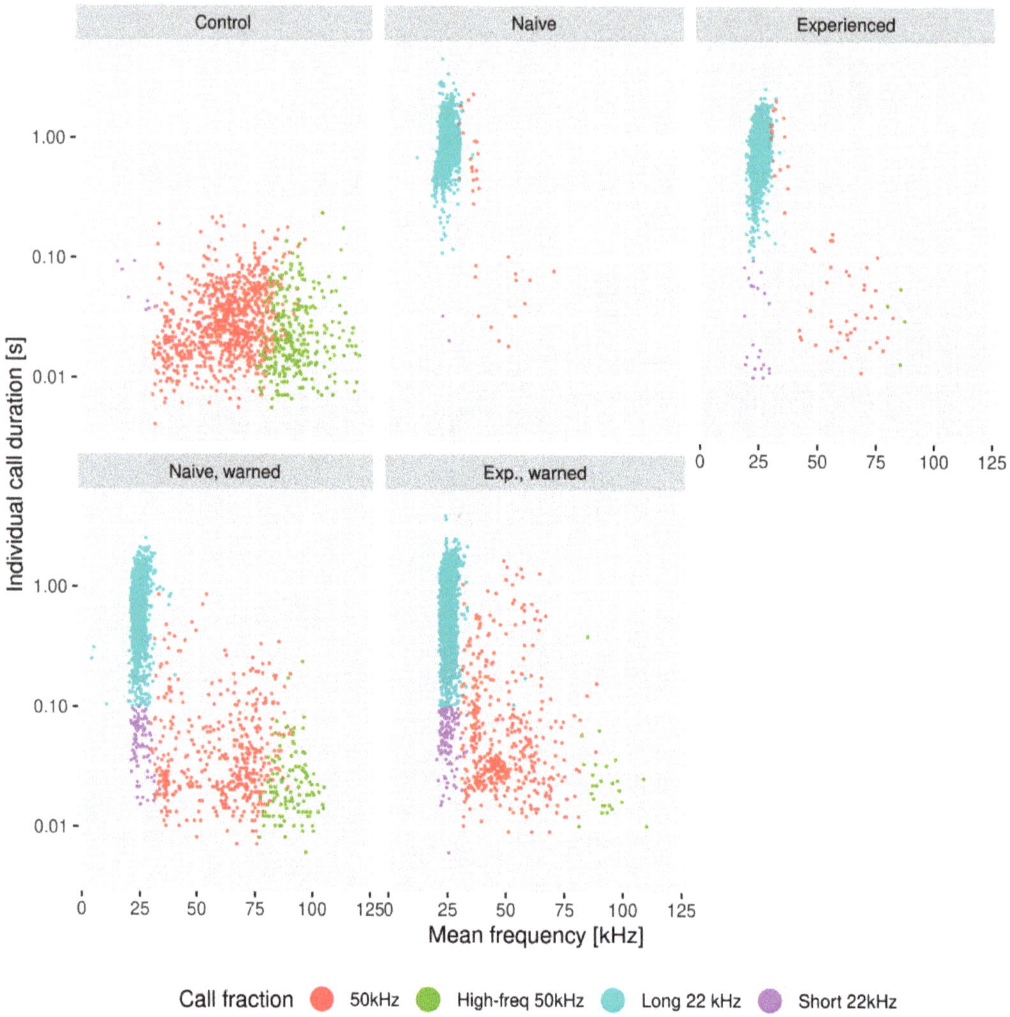

Figure 4. Scatterplot of the basic properties of the recorded USV calls, illustrating their mean frequency and duration. Each panel collects all vocalizations recorded for pairs of a certain class. Each of the presented groups were represented by four pairs of rats (eight animals per group). Color denotes the call fraction. Note that the duration is shown on a logarithmic scale.

Figure 5 shows the quantitative differences between short 22 kHz calls in the different groups together with the results of the Conover–Iman test. One can see that short 22 kHz calls are especially abundant in the 'experienced, warned' class, and significantly more numerous than in either of the 'unwarned' groups; similarly, the 'naive, warned' group emitted a significantly increased number of short 22 kHz calls than the aforementioned 'unwarned' groups.

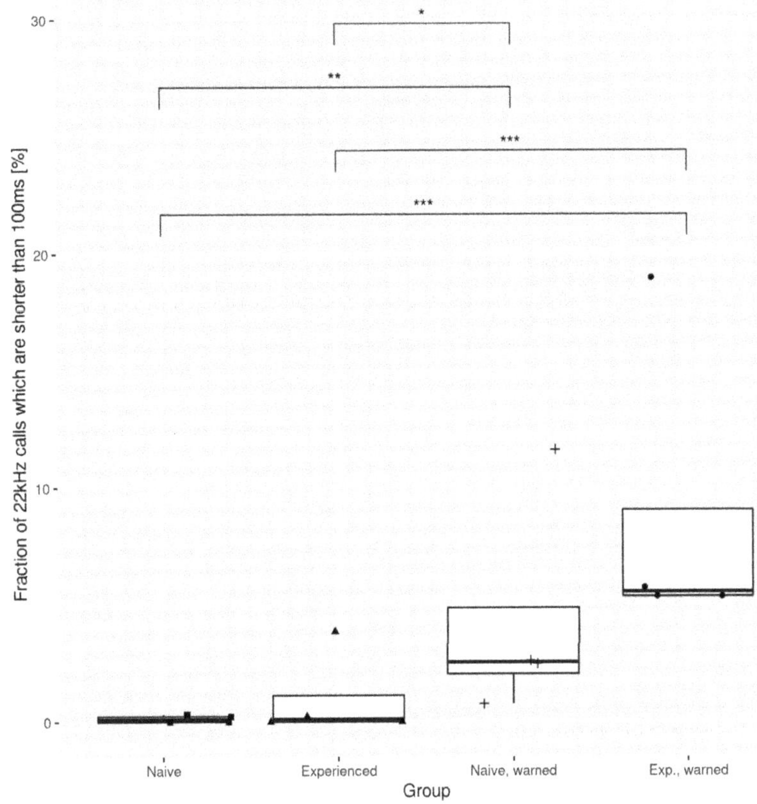

Figure 5. Fraction of short (<100 ms) 22 kHz calls in the different experimental groups. Brackets mark significant differences identified by the Conover–Iman test. * $p < 0.05$; ** $p < 0.01$; *** $p < 0.001$.

Returning to Figure 4, one can note that the 50 kHz calls build a sparse cluster with a wide span in both parameters. Again, it is less evident in the 'unwarned' groups. The center of the 50 kHz cluster also varies between groups, which is reflected in the average frequency, as well as in the abundance of 50 kHz calls above the 75 kHz boundary, which we later refer to as "high-frequency 50 kHz" calls (Figure 6).

Figure 6 shows the comparison of high-frequency 50 kHz calls between groups. As with the fraction of short 22 kHz calls, each of the 'warned' groups exhibited a significantly higher abundance of these 50 kHz calls than either of the 'unwarned' groups. The highest fraction of high-frequency calls was found in the 'naive, warned' group, which formed over 10% of all calls. The same conclusions can be drawn from Figure 7, which presents the mean frequencies of 50 kHz calls in each pair. Here, the 'naive, warned' group generated the highest pitched calls of all the experimental groups, with a median over 75 kHz. This result was similar to that achieved by the control pairs. Beyond those findings, we can also see that among 'unwarned' groups, 'experienced' rats produced higher USVs than 'naive' ones.

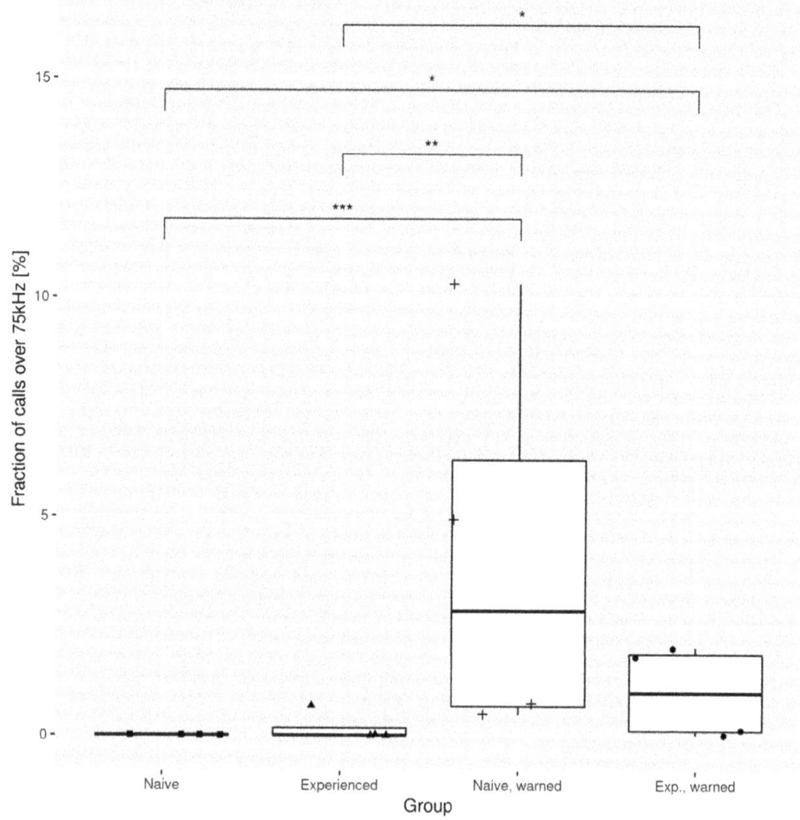

Figure 6. Fraction of 50 kHz calls with a frequency above 75 kHz in different experimental groups. Brackets mark significant differences identified by the Conover–Iman test. * $p < 0.05$; ** $p < 0.01$; *** $p < 0.001$.

3.2. The Effects of the Warning Signal. Overlapping of the Ultrasonic Vocalizations during Social Communication in the Social Transfer of the Fear Paradigm

Some of the reported USV calls were temporally overlapping (these overlapping calls originated from the vocalizations of two rats at the same time); that is, one coherent signal could be seen as superimposed on another, as can be seen in Figure 8. We extracted all such events and denoted the frequencies of each of the two involved vocalizations. Using this key, overlaps can be divided into three types: a pair of 22 kHz calls, a pair of two 50 kHz calls, and an overlap of 22 and 50 kHz calls, which we later refer to as a mixed overlap. Figure 9 reports the distribution of said overlap types, both in relation to all recorded calls and to the overall overlapping USVs.

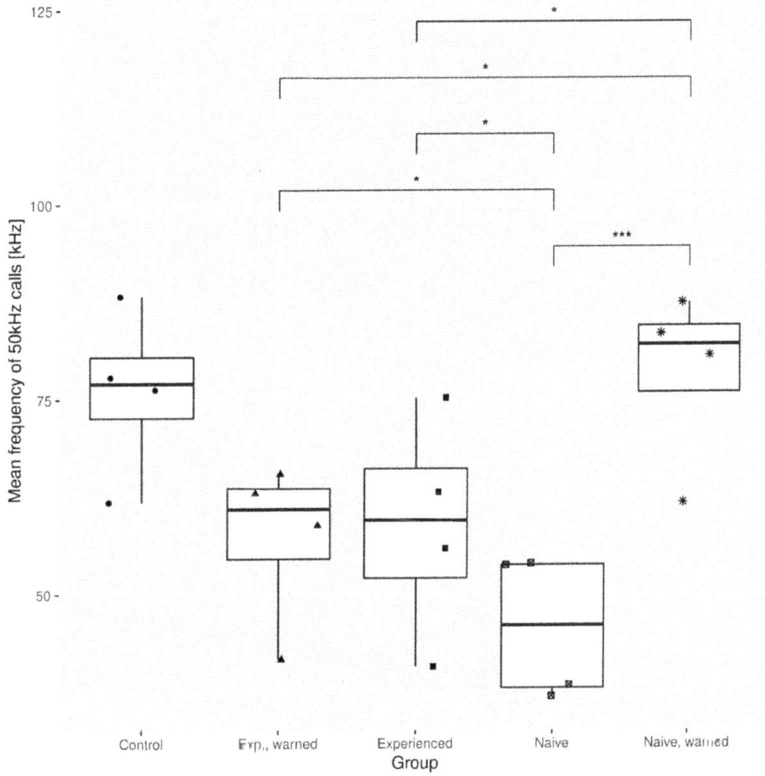

Figure 7. Mean frequency of 50 kHz calls in different experimental groups. Brackets mark significant differences identified by the Conover–Iman test. * $p < 0.05$; *** $p < 0.001$.

One can see that 'experienced, warned' rats had a substantial number of pure 22 kHz overlaps, up to almost 3 per 10 episodes in the case of 1 pair. This is significantly higher than in any other experimental group.

In the 'naive, warned' rats, the overlap rate was smaller than in the 'experienced, warned' rats, but larger than in either of the 'unwarned' experimental groups. In relative terms, the mixed overlaps constituted from approximately 50% to 100% of all overlaps identified in this group.

Finally, in the control group, the most substantial type of overlaps were pure 50 kHz calls.

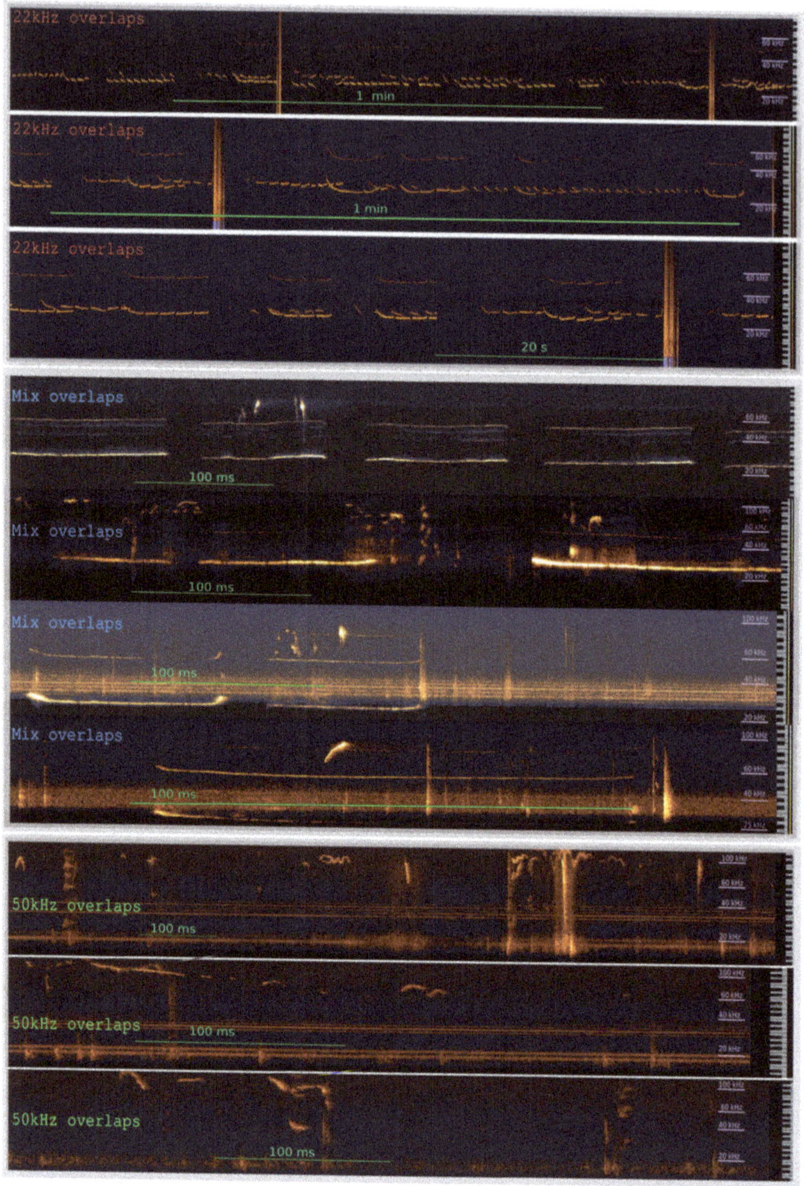

Figure 8. An example of overlapping ultrasonic vocalization presented on a spectrogram.

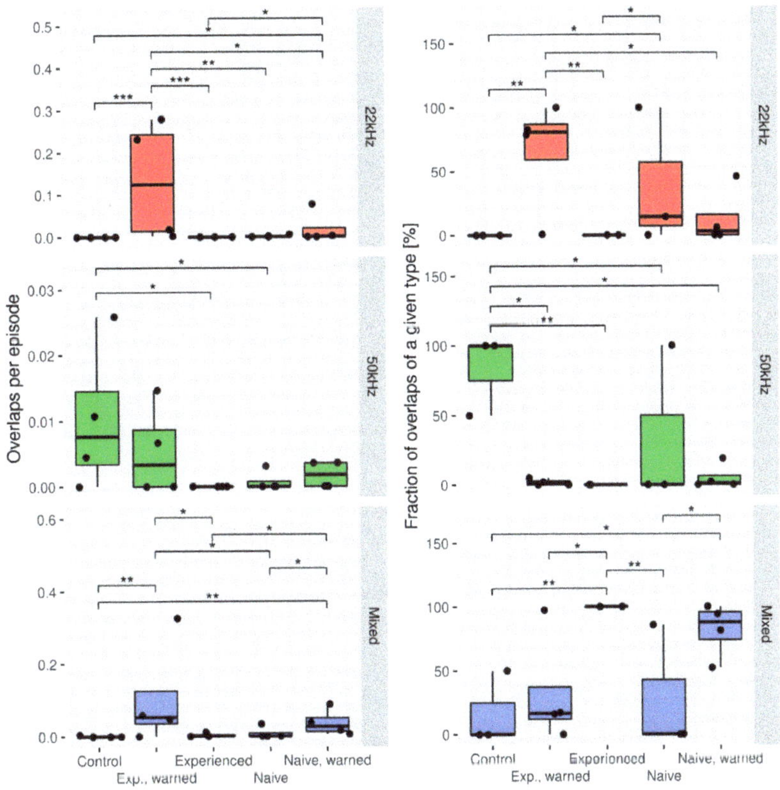

Figure 9. Fractions of temporal USV overlaps which belong to a certain class: 22 kHz—two 22 kHz calls overlapping, 50 kHz—two 50 kHz calls overlapping, mixed—22 and 50 kHz calls overlapping in different experimental groups. The left panels show the count normalized by the episode count, while the right panels show the count normalized by the total count of temporally overlapping USVs. Brackets mark significant differences identified by the Conover–Iman test. * $p < 0.05$; ** $p < 0.01$; *** $p < 0.001$.

4. Discussion

In this study, we conducted a detailed analysis of the ultrasonic vocalizations emitted by pairs of rats during two slightly different aversive situations. In one of the groups, the electrical shocks were signaled by the emission of a 19-s audible (warning) sound ('warned' group), while the other experimental group did not receive any external acoustic signals before the electrical stimulus ('unwarned' group). The tested animals either had previous experience with foot shocks ('experienced' group) or did not have such experience ('naive' group). We were interested in whether there was a difference in communication patterns between 'warned' and 'unwarned' animals and whether the experience of the electrical stimulation would change the animals' behavior in an aversive situation. The warning sound emitted before the electrical stimulus seemed to increase arousal level, which differentiated the responses, depending on previous experience.

In the animals from the 'warned' group, we identified a fraction of short calls in the "22 kHz" class, which was practically absent in the 'unwarned' animals. It has been previously suggested [26,56] that such calls are a sign of distress (internal negative emotional state) rather than a response to external, aversive stimuli triggering the common, longer 22 kHz calls. Our observations provide additional, substantial evidence in favor of this hypothesis. This could mean that the anticipation of electric shock, induced by the warning signal, leads to a higher stress level in contrast to unpredictable shocks, which is in line with the results of fear conditioning studies [57].

Many studies have shown that USV playback or synthetic sound presentation in the 22 kHz class activates the perirhinal cortex, periaqueductal grey matter, the amygdala, or the hypothalamus—the structures involved in defensive behavior [41,58,59]. The ultrasonic vocalizations produced by rats in a stressful or threatening situation may be crucial for their conspecifics' survival or wellbeing [43]. Both 22 kHz calls [56] and freezing behavior [60] can be interpreted as a warning and used by the observers to learn about danger. Importantly, the reception of the 22 kHz calls changes the function of the brain fear circuit, which probably helps the recipients to quickly adapt to the threatening situation. For example, Dupin et al. (2019), who investigated the relationship between the emission of USVs, respiration, and the electrophysiological activity of brain structures, showed that 22 kHz calls result in a decrease in theta power and an increase in delta and gamma power in the BLA, the mPFC, and the piriform olfactory cortex (PIR)—the structures involved in fear responses [49]. Intense fear, observed as a panic attack, is associated with hyperventilation and breathlessness [61], which can be reduced by paroxetine, a selective serotonin reuptake inhibitor [62]. Interestingly, Willadsen et al. (2020) documented that animals lacking the serotonin transporter emitted a lower number of 22 kHz USVs [63]. These results, together with the correlational studies demonstrated by Dupin et al. (2019), indicate that the animals warned by the audible sound in our research (the ones emitting short 22 kHz signals) were in intense distress. This indicates the crucial role of 22 kHz USVs in emotional contagion within a social group.

We detected a difference between the 'experienced' and 'naive' groups in the distribution of the frequency of the emitted ultrasonic vocalizations.

The simplest form of empathy, which can be observed in animals, is defined as the ability to understand or to share another individual's emotional state [64]. It plays an essential role in regulating social behavior and can be modulated by prior experience [65]. An interesting phenomenon observed in our study is that short 22 kHz episodes that reflect distress are emitted both by the demonstrator and by the observer, as reflected in overlapping 22 kHz calls. Additional intensification of the ultrasound signal in the 22 kHz class appears shortly after the initial short 22 kHz calls, thus indicating emotional contagion; the observer is being "infected" with the demonstrator's emotional state [51]. Presence of the short 22 kHz USVs indicates accelerated breathing (tachypnea) that is typically related to hyperventilation, which is characteristic of panic states. Henceforth, the occurrence of the same 22 kHz calls in the demonstrator and in the observer indicates that distress may be shared by the pair of rats.

On the other hand, we noted an interesting fraction of very high-frequency 50 kHz calls (over 75 kHz) that were present almost exclusively in the 'naive, warned' experimental group. We believe this indicates that naive rats are more inclined to emit soothing calls to reduce the distress of their partners, as explained by the social buffering phenomenon [64,66]. It is known that parallel to fear transfer from the demonstrators to the observers, the observers may provide social support and a moderate stress response to demonstrators, a phenomenon known as social buffering [64]. It has been shown that social buffering is more effective among familiar animals [67]. Furthermore, naive animals are more effective in social buffering than animals subjected to fear conditioning [66,67].

While we were not able to unambiguously attribute each call to the individual rat, we relied on a phenomenon of temporally overlapping calls, which we assumed came from either rat at a particular time. This allowed us to elaborate on their interactions. In particular, we observed all of the possible overlap classes: two 22 kHz calls, two 50 kHz calls, and a mixture of both.

5. Conclusions

Analysis of USV communication patterns reveals that in the presence of an experienced observer, most of the overlapping ultrasonic vocalizations covered a 22 kHz class (Figure 7).

In the case of the 'naive, warned' observer, we noticed that USV overlaps are mixed signals, including 22 and 50 kHz calls (Figure 7). Moreover, the ultrasounds from the 50 kHz class in the 'naive, warned' group mostly consist of calls over 75 kHz (Figures 4, 6 and 8).

Detailed analysis of the USV communication patterns has shown that it is critical to divide the 22 kHz calls into long and short classes in order to precisely quantify emotional processing in rats. In addition, for future studies, we suggest extracting the 50 kHz fraction and dividing it into low- and high-frequency 50 kHz using a 75 kHz threshold, or analyzing their dominant frequency distributions. Currently, the typical 50 kHz analysis approach distinguishes a whole range of subtypes based on their spectrographic shape but without simultaneously considering their frequency. We strongly believe that ultrasonic vocalizations have a prosodic character, which carries the most essential information in social communication, and that the division into specific sub-episodes, distinguished by the different shapes reflected in the FFT, should be taken into account as a second factor.

Author Contributions: Conceptualization, W.K., M.W., K.K., E.K., M.B.K. and A.H.; formal analysis, W.K., M.W., K.K., M.B.K. and A.H.; investigation, M.W., K.K., and A.H.; writing—original draft preparation, W.K., M.B.K. and A.H.; writing—review and editing, W.K., E.K., M.B.K. and A.H.; visualization, W.K., M.B.K. and A.H.; funding acquisition, E.K. and A.H. All authors have read and agreed to the published version of the manuscript.

Funding: K.K., M.W., and E.K. were supported by the European Research Council Starting Grant (H 415148 to EK). A.H. and W.K. were supported by grant UMO-2018/29/B/NZ7/02021 from the National Science Centre Poland.

Institutional Review Board Statement: All applicable international, national, and/or institutional guidelines for the care and use of animals were followed. All experimental procedures using animal subjects were approved by the 1st Local Committee for Animal Care in Warsaw in compliance with Polish Law (21 January 2005). All procedures performed in studies involving animals were in accordance with the ethical standards of the institution or practice at which the studies were conducted.

Informed Consent Statement: Not applicable.

Data Availability Statement: Data available upon request.

Acknowledgments: Special thanks to Laura Karwicka for providing the illustrations of experimental procedures. We would like to thank Tomasz Jaroszewski for co-creating RatRec software for the recording and analysis of ultrasonic vocalizations.

Conflicts of Interest: The authors declare no conflict of interest. National Science Centre, Poland had no role in study design; in the design of the study; in the collection, analyses, or interpretation of data; in the writing of the manuscript; or in the decision to publish the results.

References

1. Brudzynski, S.M. Ethotransmission: Communication of Emotional States through Ultrasonic Vocalization in Rats. *Curr. Opin. Neurobiol.* **2013**, *23*, 310–317. [CrossRef] [PubMed]
2. Panksepp, J.; Burgdorf, J. "Laughing" Rats and the Evolutionary Antecedents of Human Joy? *Physiol. Behav.* **2003**, *79*, 533–547. [CrossRef]
3. Knutson, B.; Burgdorf, J.; Panksepp, J. Ultrasonic Vocalizations as Indices of Affective States in Rats. *Psychol. Bull.* **2002**, *128*, 961–977. [CrossRef] [PubMed]
4. Hamed, A.; Taracha, E.; Szyndler, J.; Krząścik, P.; Lehner, M.; Maciejak, P.; Skórzewska, A.; Płaźnik, A. The Effects of Morphine and Morphine Conditioned Context on 50kHz Ultrasonic Vocalisation in Rats. *Behav. Brain Res.* **2012**, *229*, 447–450. [CrossRef] [PubMed]
5. Simola, N.; Fenu, S.; Costa, G.; Pinna, A.; Plumitallo, A.; Morelli, M. Pharmacological Characterization of 50-kHz Ultrasonic Vocalizations in Rats: Comparison of the Effects of Different Psychoactive Drugs and Relevance in Drug-Induced Reward. *Neuropharmacology* **2012**, *63*, 224–234. [CrossRef]
6. Kuchniak, K.; Wyszogrodzka, E.; Chrapusta, S.J.; Czarna, M.; Michalak, M.; Płaźnik, A.; Krząścik, P.; Mierzejewski, P.; Taracha, E. Using Anticipatory and Drug-Evoked Appetitive Ultrasonic Vocalization for Monitoring the Rewarding Effect of Amphetamine in a Rat Model of Drug Self-Administration. *Behav. Brain Res.* **2019**, *376*, 112187. [CrossRef]
7. Costa, G.; Serra, M.; Marongiu, J.; Morelli, M.; Simola, N. Influence of Dopamine Transmission in the Medial Prefrontal Cortex and Dorsal Striatum on the Emission of 50-kHz Ultrasonic Vocalizations in Rats Treated with Amphetamine: Effects on Drug-Stimulated and Conditioned Calls. *Prog. Neuro-Psychopharmacol. Biol. Psychiatry* **2020**, *79*, 109797. [CrossRef]

8. Simola, N.; Serra, M.; Marongiu, J.; Costa, G.; Morelli, M. Increased Emissions of 50-KHz Ultrasonic Vocalizations in Hemiparkinsonian Rats Repeatedly Treated with Dopaminomimetic Drugs: A Potential Preclinical Model for Studying the Affective Properties of Dopamine Replacement Therapy in Parkinson's Disease. *Prog. Neuro-Psychopharmacol. Biol. Psychiatry* **2021**, *108*, 110184. [CrossRef] [PubMed]
9. Mulvihill, K.G.; Brudzynski, S.M. Non-Pharmacological Induction of Rat 50 kHz Ultrasonic Vocalization: Social and Non-Social Contexts Differentially Induce 50 kHz Call Subtypes. *Physiol. Behav.* **2018**, *196*, 200–207. [CrossRef] [PubMed]
10. Hamed, A.; Szyndler, J.; Taracha, E.; Turzyńska, D.; Sobolewska, A.; Lehner, M.; Krząścik, P.; Daszczuk, P. κ-Opioid Receptor as a Key Mediator in the Regulation of Appetitive 50-kHz Ultrasonic Vocalizations. *Psychopharmacology* **2015**, *232*, 1941–1955. [CrossRef]
11. Hamed, A.; Jaroszewski, T.; Maciejak, P.; Szyndler, J.; Lehner, M.; Kamecka, I.; Olczak, M.; Kuzinska, U.; Taracha, E.; Płaźnik, A. The Effects of Buspirone and Diazepam on Aversive Context- and Social Isolation-Induced Ultrasonic Vocalisation. *Physiol. Behav.* **2009**, *98*, 474–480. [CrossRef]
12. Burgdorf, J.S.; Brudzynski, S.M.; Moskal, J.R. Using Rat Ultrasonic Vocalization to Study the Neurobiology of Emotion: From Basic Science to the Development of Novel Therapeutics for Affective Disorders. *Curr. Opin. Neurobiol.* **2020**, *60*, 192–200. [CrossRef]
13. Lefebvre, E.; Granon, S.; Chauveau, F. Social Context Increases Ultrasonic Vocalizations during Restraint in Adult Mice. *Anim. Cogn.* **2020**, *23*, 351–359. [CrossRef]
14. Brudzynski, S.M.; Pniak, A. Social Contacts and Production of 50-kHz Short Ultrasonic Calls in Adult Rats. *J. Comp. Psychol.* **2002**, *116*, 73–82. [CrossRef] [PubMed]
15. Wöhr, M. Ultrasonic Communication in Rats: Appetitive 50-kHz Ultrasonic Vocalizations as Social Contact Calls. *Behav. Ecol. Sociobiol.* **2017**, *72*. [CrossRef]
16. Burgdorf, J.; Knutson, B.; Panksepp, J. Anticipation of Rewarding Electrical Brain Stimulation Evokes Ultrasonic Vocalization in Rats. *Behav. Neurosci.* **2000**, *114*, 320–327. [CrossRef] [PubMed]
17. Brenes, J.C.; Schwarting, R.K.W. Individual Differences in Anticipatory Activity to Food Rewards Predict Cue-Induced Appetitive 50-kHz Calls in Rats. *Physiol. Behav.* **2015**, *149*, 107–118. [CrossRef] [PubMed]
18. Hamed, A.; Kursa, M.B. Inter-Individual Differences in Serotonin and Glutamate Co-Transmission Reflect Differentiation in Context-Induced Conditioned 50-kHz USVs Response after Morphine Withdrawal. *Brain Struct. Funct.* **2018**, *223*, 3149–3167. [CrossRef]
19. Meyer, P.J.; Ma, S.T.; Robinson, T.E. A Cocaine Cue Is More Preferred and Evokes More Frequency-Modulated 50-kHz Ultrasonic Vocalizations in Rats Prone to Attribute Incentive Salience to a Food Cue. *Psychopharmacology* **2012**, *219*, 999–1009. [CrossRef] [PubMed]
20. Meyer, P.J.; Ma, S.T.; Robinson, T.E. Quantifying Individual Variation in the Propensity to Attribute Incentive Salience to Reward Cues. *Psychopharmacology* **2012**, *7*, 338987. [CrossRef] [PubMed]
21. Brudzynski, S.M.; Ociepa, D. Ultrasonic Vocalization of Laboratory Rats in Response to Handling and Touch. *Physiol. Behav.* **1992**, *52*, 655–660. [CrossRef]
22. Takahashi, L.K.; Thomas, D.A.; Barfield, R.J. Analysis of Ultrasonic Vocalizations Emitted by Residents during Aggressive Encounters among Rats (*Rattus norvegicus*). *J. Comp. Psychol.* **1983**, *97*, 207–212. [CrossRef] [PubMed]
23. Sales, G.D. Ultrasound and Aggressive Behaviour in Rats and Other Small Mammals. *Anim. Behav.* **1972**, *20*, 88–100. [CrossRef]
24. Panksepp, J. Toward a General Psychobiological Theory of Emotions. *Behav. Brain Sci.* **1982**, *5*, 407–422. [CrossRef]
25. Portavella, M.; Depaulis, A.; Vergnes, M. 22–28 kHz Ultrasonic Vocalizations Associated with Defensive Reactions in Male Rats Do Not Result from Fear or Aversion. *Psychopharmacology* **1993**, *111*, 190–194. [CrossRef] [PubMed]
26. Brudzynski, S.M.; Bihari, F.; Ociepa, D.; Fu, X.W. Analysis of 22 kHz Ultrasonic Vocalization in Laboratory Rats: Long and Short Calls. *Physiol. Behav.* **1993**, *54*, 215–221. [CrossRef]
27. Kim, E.J.; Kim, E.S.; Covey, E.; Kim, J.J. Social Transmission of Fear in Rats: The Role of 22-kHz Ultrasonic Distress Vocalization. *PLoS ONE* **2010**, *5*, e15077. [CrossRef]
28. Sewell, G.D. Ultrasound in Adult Rodents. *Nature* **1967**, *215*, 512. [CrossRef]
29. Blanchard, R.J.; Blanchard, D.C.; Agullana, R.; Weiss, S.M. Twenty-Two kHz Alarm Cries to Presentation of a Predator, by Laboratory Rats Living in Visible Burrow Systems. *Physiol. Behav.* **1991**, *50*, 967–972. [CrossRef]
30. Fendt, M.; Brosch, M.; Wernecke, K.E.A.; Willadsen, M.; Wöhr, M. Predator Odour but Not TMT Induces 22-kHz Ultrasonic Vocalizations in Rats That Lead to Defensive Behaviours in Conspecifics upon Replay. *Sci. Rep.* **2018**, *8*, 11041. [CrossRef]
31. Blanchard, R.J.; Agullana, R.; McGee, L.; Weiss, S.; Blanchard, D.C. Sex Differences in the Incidence and Sonographic Characteristics of Antipredator Ultrasonic Cries in the Laboratory Rat (*Rattus norvegicus*). *J. Comp. Psychol.* **1992**, *106*, 270–277. [CrossRef]
32. Blanchard, R.J.; Blanchard, D.C. Defensive Reactions in the Albino Rat. *Learn. Motiv.* **1971**, *2*, 351–362. [CrossRef]
33. Sales, G.; Pye, D. *Ultrasonic Communication by Animals*; Springer: Berlin/Heidelberg, Germany, 1974. [CrossRef]
34. Fu, X.W.; Brudzynski, S.M. High-Frequency Ultrasonic Vocalization Induced by Intracerebral Glutamate in Rats. *Pharmacol. Biochem. Behav.* **1994**, *49*, 835–841. [CrossRef]
35. Van der Poel, A.M.; Miczek, K.A. Long Ultrasonic Calls in Male Rats Following Mating, Defeat and Aversive Stimulation: Frequency Modulation and Bout Structure. *Behaviour* **1991**, *119*, 127–142. [CrossRef]

36. De Vry, J.; Benz, U.; Schreiber, R.; Traber, J. Shock-Induced Ultrasonic Vocalization in Young Adult Rats: A Model for Testing Putative Anti-Anxiety Drugs. *Eur. J. Pharmacol.* **1993**, *249*, 331–339. [CrossRef]
37. Kaltwasser, M.T. Acoustic Startle Induced Ultrasonic Vocalization in the Rat: A Novel Animal Model of Anxiety? *Behav. Brain Res.* **1991**, *43*, 133–137. [CrossRef]
38. Knapp, D.J.; Pohorecky, L.A. An Air-Puff Stimulus Method for Elicitation of Ultrasonic Vocalizations in Rats. *J. Neurosci. Methods* **1995**, *62*, 1–5. [CrossRef]
39. Brudzynski, S.M.; Chiu, E.M.C. Behavioural Responses of Laboratory Rats to Playback of 22 kHz Ultrasonic Calls. *Physiol. Behav.* **1995**, *57*, 1039–1044. [CrossRef]
40. Burgdorf, J.; Kroes, R.A.; Moskal, J.R.; Pfaus, J.G.; Brudzynski, S.M.; Panksepp, J. Ultrasonic Vocalizations of Rats (*Rattus norvegicus*) During Mating, Play, and Aggression: Behavioral Concomitants, Relationship to Reward, and Self-Administration of Playback. *J. Comp. Psychol.* **2008**, *122*, 357–367. [CrossRef] [PubMed]
41. Sadananda, M.; Wöhr, M.; Schwarting, R.K.W. Playback of 22-kHz and 50-kHz Ultrasonic Vocalizations Induces Differential c-Fos Expression in Rat Brain. *Neurosci. Lett.* **2008**, *435*, 17–23. [CrossRef]
42. Panksepp, J.; Burgdorf, J.; Beinfeld, M.C.; Kroes, R.A.; Moskal, J.R. Regional Brain Cholecystokinin Changes as a Function of Friendly and Aggressive Social Interactions in Rats. *Brain Res.* **2004**, *1025*, 75–84. [CrossRef]
43. Litvin, Y.; Blanchard, D.C.; Blanchard, R.J. Rat 22 kHz Ultrasonic Vocalizations as Alarm Cries. *Behav. Brain Res.* **2007**, *182*, 166–172. [CrossRef] [PubMed]
44. Inagaki, H.; Mori, Y. The Emission of Stress-Induced 22-kHz Calls in Female Rats Is Independent of Testosterone Levels. *Horm. Behav.* **2015**, *69*, 116–118. [CrossRef]
45. Brudzynski, S.M.; Holland, G. Acoustic Characteristics of Air Puff-Induced 22-kHz Alarm Calls in Direct Recordings. *Neurosci. Biobehav. Rev.* **2005**, *29*, 1169–1180. [CrossRef]
46. Brudzynski, S.M.; Barnabi, F. Contribution of the Ascending Cholinergic Pathways in the Production of Ultrasonic Vocalization in the Rat. *Behav. Brain Res.* **1996**, *80*, 145–152. [CrossRef]
47. Schwarting, R.K.W.; Wöhr, M. On the Relationships between Ultrasonic Calling and Anxiety-Related Behavior in Rats. *Braz. J. Med. Biol. Res.* **2012**, *45*, 337–348. [CrossRef] [PubMed]
48. Bennett, P.J.G.; Maier, E.; Brecht, M. Involvement of Rat Posterior Prelimbic and Cingulate Area 2 in Vocalization Control. *Eur. J. Neurosci.* **2019**, *50*, 3164–3180. [CrossRef]
49. Dupin, M.; Garcia, S.; Boulanger-Bertolus, J.; Buonviso, N.; Mouly, A.M. New Insights from 22-kHz Ultrasonic Vocalizations to Characterize Fear Responses: Relationship with Respiration and Brain Oscillatory Dynamics. *eNeuro* **2019**, *6*. [CrossRef]
50. Kondrakiewicz, K.; Rokosz-Andraka, K.; Nikolaev, T.; Górkiewicz, T.; Danielewski, K.; Gruszczyńska, A.; Meyza, K.; Knapska, E. Social Transfer of Fear in Rodents. *Curr. Protoc. Neurosci.* **2019**, *90*, e85. [CrossRef] [PubMed]
51. Andraka, K.; Kondrakiewicz, K.; Rojek-Sito, K.; Ziegart-Sadowska, K.; Meyza, K.; Nikolaev, T.; Hamed, A.; Kursa, M.; Wójcik, M.; Danielewski, K.; et al. Distinct Circuits in Rat Central Amygdala for Defensive Behaviors Evoked by Socially Signaled Imminent versus Remote Danger. *Curr. Biol.* **2021**. [CrossRef]
52. Langford, D.J.; Crager, S.E.; Shehzad, Z.; Smith, S.B.; Sotocinal, S.G.; Levenstadt, J.S.; Chanda, M.L.; Levitin, D.J.; Mogil, J.S. Social Modulation of Pain as Evidence for Empathy in Mice. *Science* **2006**, *312*, 1967–1970. [CrossRef]
53. Panksepp, J.B.; Lahvis, G.P. Rodent Empathy and Affective Neuroscience. *Neurosci. Biobehav. Rev.* **2011**, *35*, 1864–1875. [CrossRef]
54. De Waal, F.B.M. Putting the Altruism Back into Altruism: The Evolution of Empathy. *Annu. Rev. Psychol.* **2008**, *59*, 279–300. [CrossRef]
55. Atsak, P.; Orre, M.; Bakker, P.; Cerliani, L.; Roozendaal, B.; Gazzola, V.; Moita, M.; Keysers, C. Experience Modulates Vicarious Freezing in Rats: A Model for Empathy. *PLoS ONE* **2011**, *6*, e21855. [CrossRef] [PubMed]
56. Brudzynski, S. Pharmacology of Ultrasonic Vocalizations in Adult Rats: Significance, Call Classification and Neural Substrate. *Curr. Neuropharmacol.* **2015**, *13*. [CrossRef] [PubMed]
57. Arthur, A.Z. Stress of Predictable and Unpredictable Shock. *Psychol. Bull.* **1986**, *100*, 379–383. [CrossRef] [PubMed]
58. Beckett, S.R.G.; Duxon, M.S.; Aspley, S.; Marsden, C.A. Central C-Fos Expression Following 20kHz Ultrasound Induced Defence Behaviour in the Rat. *Brain Res. Bull.* **1997**, *42*, 421–426. [CrossRef]
59. Allen, T.A.; Furtak, S.C.; Brown, T.H. Single-Unit Responses to 22 kHz Ultrasonic Vocalizations in Rat Perirhinal Cortex. *Behav. Brain Res.* **2007**, *182*, 327–336. [CrossRef]
60. Blanchard, R.J.; Blanchard, D.C. Crouching as an Index of Fear. *J. Comp. Physiol. Psychol.* **1969**, *67*, 370–375. [CrossRef]
61. Gorman, J.M.; Askanazi, J.; Liebowitz, M.R.; Fyer, A.J.; Stein, J.; Kinney, J.M.; Klein, D.F. Response to Hyperventilation in a Group of Patients with Panic Disorder. *Am. J. Psychiatry* **1984**, *141*, 857–861. [CrossRef]
62. Olsson, M.; Annerbrink, K.; Westberg, L.; Melke, J.; Baghaei, F.; Rosmond, R.; Holm, G.; Andersch, S.; Allgulander, C.; Eriksson, E. Angiotensin-Related Genes in Patients with Panic Disorder. *Am. J. Med. Genet. Part B Neuropsychiatr. Genet.* **2004**, *127B*, 81–84. [CrossRef] [PubMed]
63. Willadsen, M.; Uengoer, M.; Schwarting, R.; Homberg, J.; Wöhr, M. Reduced Emission of Alarm 22-kHz Ultrasonic Vocalizations during Fear Conditioning in Rats Lacking the Serotonin Transporter. *Prog. Neuro-Psychopharmacol. Biol. Psychiatry* **2020**, *108*, 110072. [CrossRef]
64. Kikusui, T.; Winslow, J.; Mori, Y. Social Buffering: Relief from Stress and Anxiety. *Philos. Trans. R. Soc. Lond. Ser. B Biol. Sci.* **2007**, *361*, 2215–2228. [CrossRef] [PubMed]

65. Saito, Y.; Yuki, S.; Seki, Y.; Kagawa, H.; Okanoya, K. Cognitive Bias in Rats Evoked by Ultrasonic Vocalizations Suggests Emotional Contagion. *Behav. Process.* **2016**, *132*, 5–11. [CrossRef] [PubMed]
66. Kiyokawa, Y.; Hennessy, M.B. Comparative Studies of Social Buffering: A Consideration of Approaches, Terminology, and Pitfalls. *Neurosci. Biobehav. Rev.* **2018**, *86*, 131–141. [CrossRef] [PubMed]
67. Kiyokawa, Y.; Honda, A.; Takeuchi, Y.; Mori, Y. A Familiar Conspecific Is More Effective than an Unfamiliar Conspecific for Social Buffering of Conditioned Fear Responses in Male Rats. *Behav. Brain Res.* **2014**, *267*, 189–193. [CrossRef] [PubMed]

brain sciences

MDPI

Article

Increased Vocalization of Rats in Response to Ultrasonic Playback as a Sign of Hypervigilance Following Fear Conditioning

Krzysztof H. Olszyński, Rafał Polowy, Agnieszka D. Wardak, Aneta W. Grymanowska and Robert K. Filipkowski *

Behavior and Metabolism Research Laboratory, Mossakowski Medical Research Institute, Polish Academy of Sciences, 5 Pawinskiego St, PL02-106 Warsaw, Poland; kholszynski@imdik.pan.pl (K.H.O.); rpolowy@imdik.pan.pl (R.P.); awardak@imdik.pan.pl (A.D.W.); agrymanowska@imdik.pan.pl (A.W.G.)

* Correspondence: rfilipkowski@imdik.pan.pl

Abstract: We investigated the effects of prior stress on rats' responses to 50-kHz (appetitive) and 22-kHz (aversive) ultrasonic playback. Rats were treated with 0, 1, 6 or 10 shocks (1 s, 1.0 mA each) and were exposed to playbacks the following day. Previous findings were confirmed: (i) rats moved faster during 50-kHz playback and slowed down after 22-kHz playback; (ii) they all approached the speaker, which was more pronounced during and following 50-kHz playback than 22-kHz playback; (iii) 50-kHz playback caused heart rate (HR) increase; 22-kHz playback caused HR decrease; (iv) the rats vocalized more often during and following 50-kHz playback than 22-kHz playback. The previous shock affected the rats such that singly-shocked rats showed lower HR throughout the experiment and a smaller HR response to 50-kHz playback compared to controls and other shocked groups. Interestingly, all pre-shocked rats showed higher locomotor activity during 50-kHz playback and a more significant decrease in activity following 22-kHz playback; they vocalized more often, their ultrasonic vocalizations (USV) were longer and at a higher frequency than those of the control animals. These last two observations could point to hypervigilance, a symptom of post-traumatic stress disorder (PTSD) in human patients. Increased vocalization may be a valuable measure of hypervigilance used for PTSD modeling.

Keywords: hypervigilance; hyperreactivity; exaggerated reactivity; generalization; PTSD; anxiety; depression; heart rate; ultrasonic vocalizations; Wistar

check for updates

Citation: Olszyński, K.H.; Polowy, R.; Wardak, A.D.; Grymanowska, A.W.; Filipkowski, R.K. Increased Vocalization of Rats in Response to Ultrasonic Playback as a Sign of Hypervigilance Following Fear Conditioning. *Brain Sci.* **2021**, *11*, 970. https://doi.org/10.3390/brainsci11080970

Academic Editors: Stefan M. Brudzynski and Jeffrey Burgdorf

Received: 30 April 2021
Accepted: 3 July 2021
Published: 23 July 2021

1. Introduction

Prior stress and its effects on rodents' subsequent behavior have been extensively studied and reflect human symptoms of neuropsychiatric disorders. For example, foot shocks of varying intensity produce behavioral and neurochemical changes which model depression, anxiety, and post-traumatic stress disorder (PTSD) [1,2]. Specifically, electric shocks delivered during fear conditioning in both learning- and trauma-imitating protocols (compared in [3]) were shown to result in increased avoidance, cognitive and mood alterations, increased arousal, social avoidance and sleep disturbance (summarized in [4])

Conditioned fear in rodents is measured by freezing levels to cues or contexts previously paired with the shock. Conditioned fear may be observed in different or partially altered contexts due to fear generalization [5–8]. However, acute exposure to non-traumatic stress (e.g., sound and light) increased locomotor activity, while chronic stress reduced basal activity and eliminated the activation response to acute stress [9].

Heart rate (HR) is one of the most studied psychophysiological parameters in anxiety disorders. It was demonstrated that successful psychological treatment reduces HR physiological reactivity in patients with PTSD and possibly other anxiety disorders [10,11]. However, results concerning HR changes in fear conditioning paradigms are equivocal. Nijsen et al. reported less pronounced tachycardia in the cage where rats experienced fear conditioning compared with control rats [12], while Carnvali et al. showed a reduction of HR lasting for five days after fear conditioning [13]. However, re-exposure of rats to the context where they had received electrical foot shocks evoked an immediate increase in HR [14,15] or a delayed rise in HR [16,17], also in a latent inhibition protocol [18]. Additionally, a decrease in the daily HR rhythm amplitude was observed in a social defeat protocol [19].

Rats communicate via several sensory channels including the emission of ultrasonic vocalizations (USV). USV of 22-kHz type signal aversive states, while 50-kHz USV signal appetitive states [20–23]. Rats emitted 22-kHz USV during fear conditioning and re-exposure to the test apparatus, which corresponded with the number of shocks received [24]. Aversive and appetitive USV emitted by rats or played from speakers evoke physiological and emotional changes in conspecifics [25].

We recently discovered changes in locomotion, USV emission, and HR in Wistar rats exposed to ultrasonic playback from a speaker [26]. In particular, 50-kHz playback elicited abundant appetitive vocalization, increased HR, and locomotor activity. In contrast, 22-kHz playback led to an abrupt decrease in HR and locomotor activity. We wanted to establish whether prior stress affects rats' reactions to ultrasonic playback and whether the response is dependent on the number of shocks previously received (dose-dependent). Towards this end, we fear-conditioned (FC) rats with varying numbers of shocks, encompassing the usually used range [3], and investigated changes in locomotion, USV emission and HR in rats exposed to 50-kHz vs. 22-kHz USV playback.

2. Materials and Methods

2.1. Animals and Housing

All experiments were approved by the Second Local Ethical Committee in Warsaw. Naïve adult male Wistar rats (7 weeks of age, from The Center for Experimental Medicine of the Medical University of Bialystok, Poland), control animals (no electric shock, 0-Trial), and conditioned rats (receiving 1, 6 or 10 conditioning stimuli; 1-Trial, 6-Trial and 10-Trial, respectively) were kept in pairs in two separate rooms. Standard chow and water were provided ad libitum. Both housing rooms had a 12 h light–dark cycle and an ambient temperature of 22–25 °C. Fear conditioning was conducted between 15:00 and 24:00 h when the overall noise level in the animal house is low. All playback experiments were conducted during the light cycle (9:00–21:00 h) on the weekend. There were four weeks between rat arrival and the start of the experiment. In the first week, the rats were habituated to new facilities. All animals were handled once for 2 min per day for 12 days before the experiment. The three experimenters each had at least four handling sessions with the rats. Surgery was performed on the third week.

2.2. Surgical Procedures

A radiotelemetric transmitter (HD-S10, Data Sciences International, St. Paul, MN, USA) was implanted into the abdominal aorta as previously described [26]. The radiotelemetric transmitter (HD-S10, Data Sciences International, St. Paul, MN, USA) for cardiovascular studies was disinfected using Cidex® (Johnson&Johnson, New Brunswick, NJ, USA) and was implanted under ketamine-xylazine anesthesia. The abdominal region was shaved and disinfected (Octenisept, Schulke, Norderstedt, Germany). A midline incision was performed, and the transmitter sensor was implanted into the abdominal aorta by direct puncturing of the vessel (21 g needle) and fixed with tissue glue (Histoacryl®, B. Braun, Melsungen, Germany). The transmitter body was placed in the peritoneal cavity and fixed to the abdominal muscle wall. After surgery, the animal was subcutaneously injected with Metacam (0.4 mg/kg; Boehringer Ingelheim, Ingelheim am Rhein, Germany) for analgesia. An illustrative image with the surgery details can be found elsewhere; Figure 5 in [27]; please note, tissue glue was used instead of cellulose patches and silk sutures. Rats were given at least seven days post-surgery for recovery before the start of the experiment. During recovery, the animals were handled and habituated to the conditions of playback experiment four times. The animals were all 12 weeks of age at the start of the experiment.

2.3. Fear Conditioning

The animals were transported individually to the fear-conditioning room and placed in a sound-attenuated fear conditioning apparatus (MED-VFC2-USB-R, Med Associates, Fairfax, VT, USA; with insides of 54.64 × 64.04 × 29.21 cm). Each animal was habituated to the cage for 10 min with no light inside; rats' freezing, defined as the absence of movement for at least 1 s, was scored automatically by Med Associates Video Freeze software during the first 5 min. The cage was cleaned between animals using detergent, wiped using 10% ethanol and was allowed to dry. The next day, rats were placed in the conditioning cage with no light inside. After 5-min habituation, they received 1, 6 or 10 conditioning stimuli which consisted of a 20 s long white light co-terminating with an electric foot-shock (1 s, 1 mA). The inter-trial interval (ITI) ranged from 180 to 300 s (mean, 240 s) (comp. [28]). Therefore, the conditioning procedure differed in length between groups: 9 min 20 s for 1-Trial, 31 min for 6-Trial, and 48 min 20 s for 10-Trial groups. An equal-time-length control group (no shock) was used for each group. A playback experiment was given one day later (see the next paragraph); the following day (two days after conditioning), rats were returned to the same fear-conditioning context to measure freezing levels (Test). After 5-min habituation, rats were exposed to three blocks of 20 s of white light (CS) followed by 5 min of silence. Freezing was evaluated during the habituation and exposure to CS. The conditioning procedure was executed by an investigator not involved in the playback experiment.

2.4. Playback Experiment

One day after the conditioning, the rats were transferred into individual experimental cages, identical to home cages (plastic; 37 × 25 × 16 cm), and transported to the experimental room, where under white light, in the absence of the experimenter and other rats in the room, acoustic stimuli were presented through an ultrasonic speaker (Vifa, Avisoft Bioacoustics, Berlin, Germany), placed just above the shorter side of the cage, connected to an UltraSoundGate Player 116 (Avisoft Bioacoustics). USV emitted by the rat were recorded by a CM16/CMPA condenser microphone (UltraSoundGate, Avisoft Bioacoustics) placed 33 cm above the center of the cage floor, 20 cm away from the speaker. In this configuration, calls from the speaker were still visible in the recording (monitoring of playback), but they were distinctively weaker than USV emitted from the cage. Both playback and recording were performed using Avisoft Recorder USGH software (Avisoft Bioacoustics). The locomotor activity of the animal was recorded with a camera (acA1300-60gc, Basler AG, Ahrensburg, Germany) mounted above the cage and EthoVision XT software (version 10, Noldus, Wageningen, The Netherlands). Signals from radiotelemetric transmitters were collected by receivers located under the cage floor and then recorded by Ponemah software (version 6.32, Data Sciences International, St. Paul, MN, USA).

2.5. Ultrasonic Playback Presentation

Upon placing a rat into the experimental cage, 10 min of silence with a turned-on speaker, that is, background noise of 20.6 ± 0.2 dB, was followed by four 10-s-long sets of signals, separated by 5-min-long silence intervals; see Figure 1 in [26]. Four sets of signals (playbacks) were presented in counterbalanced order to each rat: (i) 50-kHz natural calls (referred to as "50-kHz USV"), 84 calls in 3 repeats, of 49.2 to 73.4 kHz frequency and 58.6 ± 0.7 kHz mean peak frequency, 28.4 ± 1.6 ms duration, 31.9 ± 0.6 dB sound pressure, recorded during rats' social interactions; (ii) 50-kHz software-generated tones ("50-kHz tones"), 32.6 ± 0.7 dB; (iii) 22-kHz natural calls ("22-kHz USV"), 24 calls in 8 repeats, 21.4–23.0 kHz, 22.1 ± 0.1 kHz, 375.3 ± 21.6 ms, 38.3 ± 1.2 dB, recorded during fear conditioning (Avisoft Bioacoustics [Internet]; c2020, Examples of rat ultrasonic vocalizations (USV), Norwegian rat (*Rattus norvegicus*), Wistar albino strain, males); and

(iv) 22-kHz software-generated tones ("22-kHz tones"), 43.3 ± 3.0 dB, although the sound playbacks of the same frequency range, for example, 50-kHz USV and 50-kHz tones, always followed each other. Artificial tones were generated based on natural tones (mean peak frequency, duration, pauses between tones in the set, but with no frequency modulations) using Avisoft SASLab Pro (Avisoft Bioacoustic). Calls were presented with a sampling rate of 200 kHz in 16-bit format. The sound pressure levels of the background noise and playback signals were assessed in the middle of the test cage's floor, at the height of the animals' typical head position, facing the speaker.

2.6. Analysis of USV and Locomotor Activity

Recordings were transferred to SASLab Pro (Avisoft Bioacoustics), and a fast Fourier transform was conducted (512 FFT-length, 100% frame, Hamming window and 75% time window overlap), resulting in high resolution spectrograms (frequency resolution: 391 Hz; time resolution: 0.64 ms). USV recordings were analyzed using SASLab Pro 5.2.xx. Spectrograms were generated from the.wav files with the following parameters: window type: FlatTop, 512 FFT length, 100% frame size and 75% temporal resolution overlap. An experienced user scored USV on the spectrogram. For analysis, mean peak frequency and element duration were taken via values measured by the software. Automated video tracking system (Ethovision, Noldus, Wageningen, The Netherlands) was used to measure the total distance traveled (cm), a measure of general locomotor activity, and proximity to speaker, that is, time spent (%) in the half of the cage closer to the speaker. The center-point of each animal's shape was used as a reference point for measurements of locomotor activity, thus registering only full-body movements, that is, distance traveled by a given rat.

2.7. Statistical Analysis

All data were analyzed using non-parametric Friedman, Wilcoxon, and Mann-Whitney tests with Statistica 7.1 (Stat-Soft); the p values are given, with a minimal level of significance of $p < 0.05$. Figures were prepared using GraphPad Prism 7 software and depict average values with a standard error of the mean (SEM). Based on preconditioning evaluation (Figure S1), two rats from the 6-Trial group were excluded as outliers (i.e., emitting exceptionally many USV, >3 × Standard Deviation) and subsequently removed from the analysis. However, every reported significant p value was verified to be present with the two rats included (apart from a few exceptions within the supplementary tables marked appropriately).

3. Results

3.1. Rats Showed Freezing after Fear Conditioning

Before conditioning, the rats from all four groups (0, 1, 6, 10-Trial) emitted similar numbers of USV (Figure S1A,D) and showed similar HR (Figure S1B,D) and freezing levels (Figure S1C,D). During the test (two days after conditioning), that is, one day after the playback experiment, the rats from 1-Trial, 6-Trial, and 10-Trial groups showed increased freezing (Figure S1C–E).

3.2. Except for the Periods of Ultrasonic Playback, Rats' Behavior Remained Relatively Constant

Locomotor activity, measured as distance traveled (please note, graphs contain distances travelled in cm per 10 s; speed, when mentioned, is reflected by the distances divided by 10, that is, in cm/s), was the same during the 10-min-silence period at an average speed of 1.70 cm/s (Figure S2A,B), which declined during the playback session to 1.30 cm/s ($p = 0.0000$, Figure 1), for example, to 1.31 cm/s ($p = 0.0000$) and to 1.14 cm/s ($p = 0.0000$) during our control time-intervals (comp. [26], that is, from −120 s to −100 s and −30 s to −10 s, respectively, all Wilcoxon). Within these periods, however, the distance travelled remained relatively constant (Figures 1–3, Figures S3 and S4; Table S1). Please note that Figure 3 has a guiding explanation regarding figures content and some take-home messages.

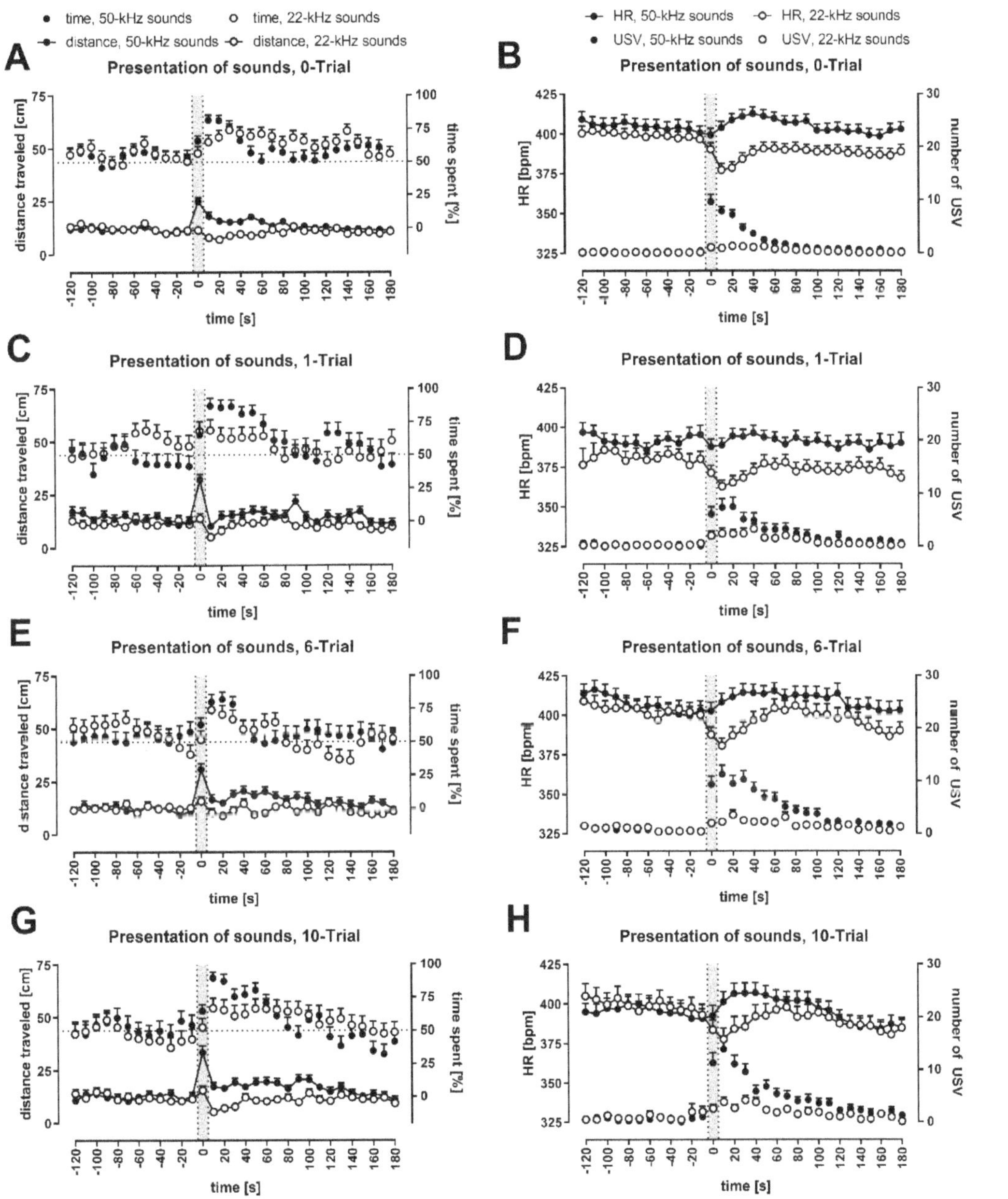

Figure 1. Analysis of changes in distance traveled, time spent in the speaker's half of the cage, heart rate (HR), and the number of emitted vocalizations during playback session in rats exposed to both 50-kHz- vs. 22-kHz playback, with averaged effects of natural and artificial sounds. Gray sections correspond to the 10-s-long ultrasonic presentation. Graphs depict responses after previous exposure to: no shock (**A**,**B**), one shock (**C**,**D**), six shocks (**E**,**F**), and ten shocks (**G**,**H**). In the left column (**A**,**C**,**E**,**G**), distance traveled is presented as connected dots (cm, left Y axis), percentage of time spent in the speaker's cage half—as not connected dots (%, right Y axis). In the right column (**B**,**D**,**F**,**H**), HR is presented as connected.dots (bpm; beats per minute, left Y axis); the number of USV is presented as not connected dots (right Y axis).

Each point is a mean for a 10-s-long time-interval with SEM. The dotted horizontal line marks a 50% chance value for time in a side of the cage. Playback of 50-kHz sounds results in a rise of locomotor activity (the weakest in control rats), copious USV emissions and HR increase (the weakest in 1-Trial rats). Playback of 22-kHz sounds is followed by decrease in locomotor activity (the smallest in 0-Trial rats) and HR as well as modest increase in vocalization; groups: 0-Trial, n = 37; 1-Trial, n = 16; 6-Trial, n = 20; 10-Trial, n = 19.

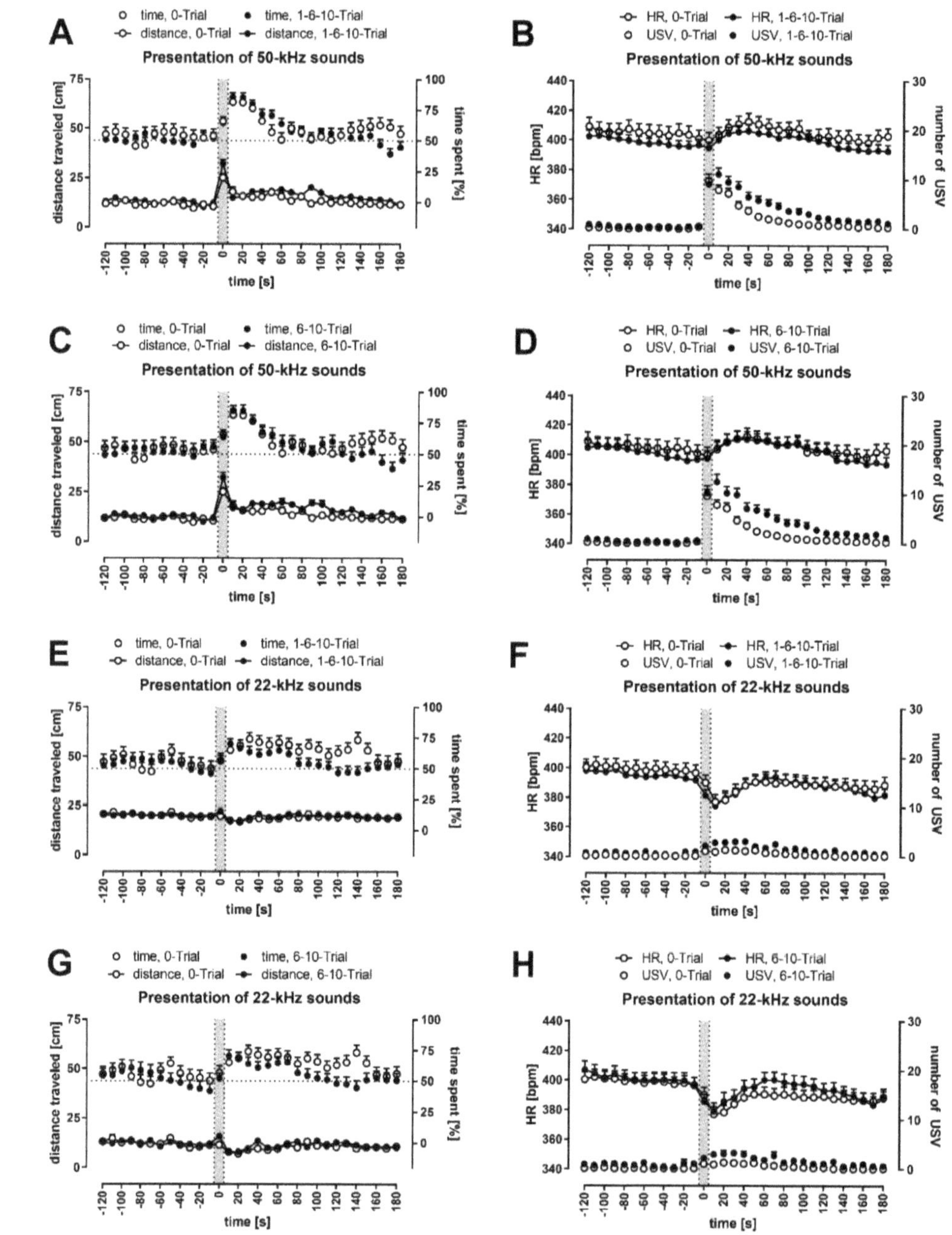

Figure 2. Impact of prior fear conditioning on distance traveled, time spent in the speaker's half of the cage, heart rate (HR) and USV emission. Gray sections correspond to the 10-s-long ultrasonic presentation. Graphs depict responses to

50-kHz sounds (**A–D**) and 22-kHz sounds (**E–H**), that is, with averaged effects of natural and artificial sounds in control animals (0-Trial) vs. FC rats (1-6-10-Trial combined, (**A**,**B**,**E**,**F**) 6-10-Trial combined, (**C**,**D**,**G**,**H**)). In the left column (**A**,**C**,**E**,**G**), distance traveled is presented as connected dots (cm, left Y axis), percentage of time spent in the speaker's cage half—as not connected dots (%, right Y axis). In the right column (**B**,**D**,**F**,**H**), HR is presented as connected dots (bpm; beats per minute, left Y axis); the number of USV is presented as not connected dots (right Y axis). Each point is a mean for a 10-s-long time-interval with SEM. The dotted horizontal line marks a 50% chance value for time in a side of the cage. FC rats had higher locomotor activity during 50-kHz playback, a more significant decrease in activity following 22-kHz playback and more USV in response to playback; groups: 0-Trial, $n = 37$; 1-Trial, $n = 16$; 6-Trial, $n = 20$; 10-Trial, $n = 19$; 1-6-10-Trial, $n = 55$; 6-10-Trial, $n = 39$.

We did not observe a strong preference for either side of the cage during the initial 10 min (Figure S2C,D). Importantly, side-preference was not observed in the playback session, during either of the two control time intervals, before any of the four kinds of playbacks, or in the five groups (including the all-rats group) (Table S2; $p > 0.05$ in all 40 cases, all Wilcoxon). The rats had no cage-side preference before playback presentation, as noted by values around 50% (dotted line) before each ultrasonic playback (Figures 1–3, Figures S3 and S4), as well as a relative lack of changes in preference within the control intervals (Table S3).

3.3. *Animals Moved Faster during 50-kHz Ultrasonic Presentations and Slowed Down after 22-kHz Ultrasonic Presentations*

All rats traveled significantly longer distances during the presentation of 50-kHz signals (Figure 1, Figure 2 and Figure S3; note the p values for −10–10 s time-intervals in Table S1), that is, at 0 s time-interval vs. neighboring −10 s and 10 s time-intervals. These pair-comparisons were significant for both USV and tone playbacks in all rats ($p = 0.0000$ in all four cases, Wilcoxon). The differences were also significant in all four groups, for 50-kHz USV playback (all eight cases, Wilcoxon, with 0.0000–0.0026 p levels).

In the case of 22-kHz playback (Figure 1, Figure 2 and Figure S4; note the p values for 0–30 s time-intervals in Table S1), a reduction in distance traveled appeared immediately after signal presentation, that is, at 0 vs. 10 s time-intervals after both USV and tone playbacks in all rats ($p = 0.0000$ for both conditions, Wilcoxon). These differences were also significant in all groups, for 22-kHz USV playback (all four cases, Wilcoxon, with 0.0003–0.0438 p levels).

3.4. *Conditioned Rats Showed Higher Locomotor Activity during 50-kHz Playback*

The peak of locomotor activity observed during 50-kHz playback seemed to be higher in FC rats (Figure 1, Figure 2 and Figure S3). Indeed, when distances traveled during 50-kHz USV and tone playbacks, that is, at 0 s time-interval, were averaged for analysis, the group comparisons revealed the difference between the 0-Trial rats (2.51 ± 0.20 cm/s) and all fear-conditioned rats combined (3.24 ± 0.19 cm/s, $p = 0.0247$); this was observed for 50-kHz USV playback as well, that is, between 0-Trial rats (2.62 ± 0.24 cm/s) and all fear-conditioned rats combined (3.45 ± 0.22 cm/s, $p = 0.0125/.0095$), as well as 6-Trial and 10-Trial rats combined (3.54 ± 0.29 cm/s, $p = 0.0210$, all Mann-Whitney; Figure 2A,C).

3.5. *Conditioned Rats Showed a More Significant Decrease in Activity Following 22-kHz Playback*

In the case of 22-kHz playback (Figure 1, Figure 2 and Figure S4), a reduction in distance traveled was observed immediately after signal presentation, that is, the difference in locomotor activity at 10 s vs. 0 s time-intervals appeared to be higher in conditioned animals, for example, for 22-kHz USV playback, it was 1.02 ± 0.21 cm/s vs. 0.39 ± 0.21 cm/s in the control group ($p = 0.0525$).

This effect was more pronounced when data following USV and tone playbacks were averaged for analysis; with a 0.36 ± 0.12 cm/s reduction in 0-Trial vs. a 0.82 ± 0.21 cm/s reduction in 6-Trial and 10-Trial rats combined ($p = 0.0188$), and vs. a 0.84 ± 0.17 cm/s reduction in all fear-conditioned rats ($p = 0.0175$; Figure 2E,G).

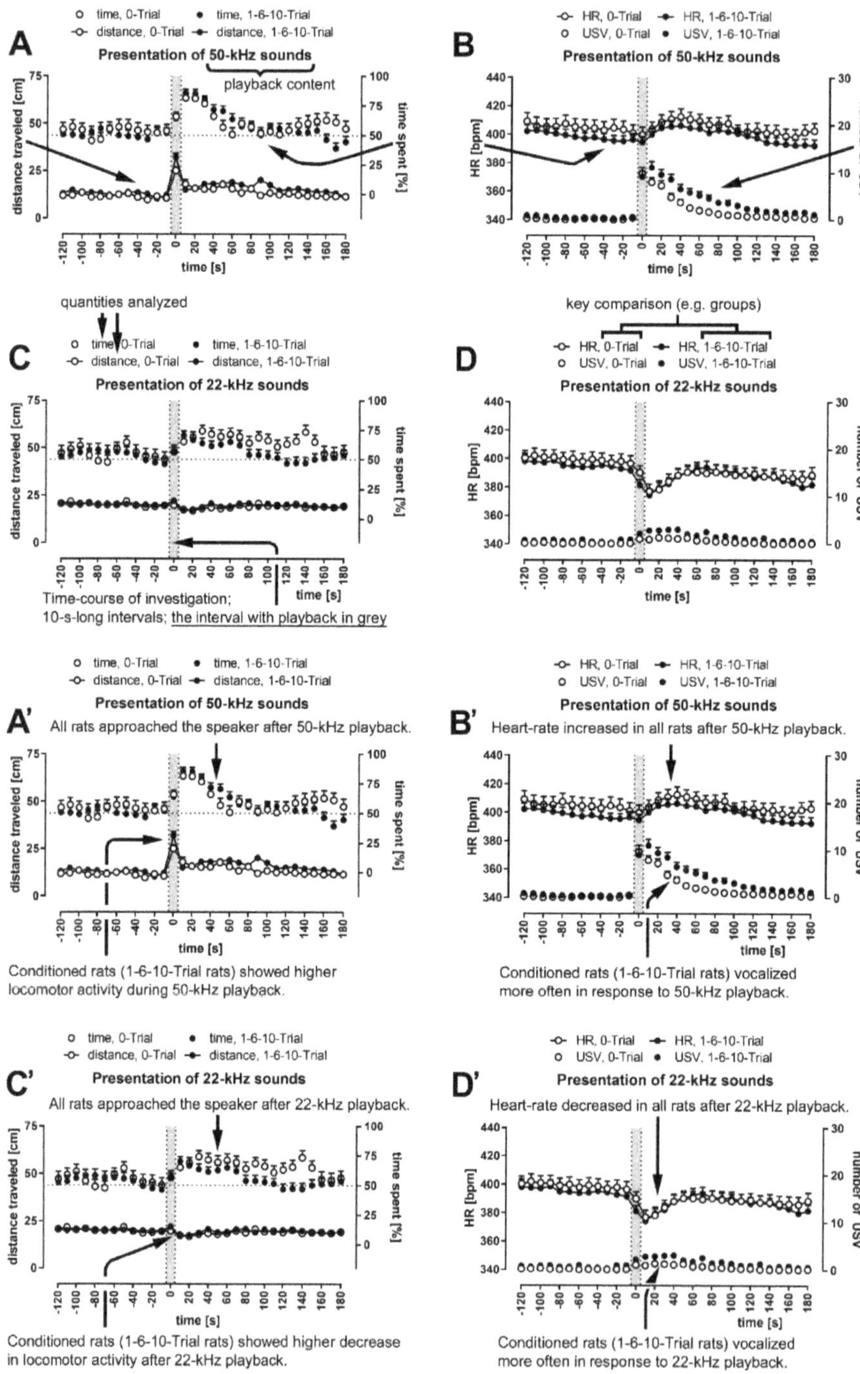

Figure 3. (**A–D**) Main results from Figure 2A,B,E,F with guiding explanations (arrows) pointing to measured quantities and crucial graph elements. (**A′–D′**) Main take-home messages in the form of short titles connected (arrows) with important data points.

3.6. All Rats Approached the Speaker; It Was More Pronounced during and Following 50-kHz Playback

Both during and following playbacks, the rats from all groups approached the speaker (Figures 1–3, Figures S3 and S4). Interestingly, it was observed for both 50 and 22-kHz playbacks; note the p values for 10–30 and 10–60 s time-intervals in Table S2 compared with 50% chance levels as well as at all time intervals, which included 0 s time-interval vs. control intervals in Table S3. An increase in time spent in the speaker's half of the cage between before vs. after/during the playbacks was observed in all groups (Table S4).

However, the rats spent more time in the speaker's half of the cage when presented with 50-kHz playback than when exposed to 22-kHz sounds (Figure 1A,C,E,G). For USV playback, the difference was significant during playback ($p = 0.0477$) and during the 10–30 s time-interval ($p = 0.0088$), while for the tone playback, the difference was significant only during 10–30 s time-interval ($p = 0.0145$). When results from USV and tone playbacks were averaged for analysis, the difference was even more significant during playback ($p = 0.0041$) and the 10–30 s time-interval ($p = 0.0003$, all Wilcoxon). There was no significant difference in control time-intervals for these comparisons, that is, from −120 s to −100 s and −30 s to −10 s ($p > 0.05$, Wilcoxon).

3.7. HR Levels Declined during the Whole Experimental Session

When average levels of HR from the first 5 min of the 10-min-silence period were compared with those from the last 5 min of the playback session, there was a significant decline in the HR of the 0-Trial (476.6 ± 5.5 vs. 383.2 ± 4.9, $p = 0.0000$), 1-Trial (461.0 ± 7.3 vs. 367.8 ± 6.2, $p = 0.0005$), 6-Trial (482.7 ± 7.3 vs. 379.2 ± 7.0, $p = 0.0001$), 10-Trial rats (479.3 ± 10.2 vs. 374.1 ± 6.7, $p = 0.0001$), and all rats (475.7 ± 3.7 vs. 377.9 ± 3.1, $p = 0.0000$; all Wilcoxon). The decline began at 180–200 s and continued for the rest of the 10 min in all groups (Figure S2E,F). It was also observed during the playback sessions (see e.g., Figure 2B,D,F,H) but was never observed during control intervals (Table S5).

3.8. 50-kHz Sounds Caused HR to Increase; 22-kHz Sounds Caused HR to Decrease

Ultrasonic playback affected HR values in all rats (Figures 1–3, Figures S3 and S4, Table S5), that is, the analysis of repeated measures revealed crucial differences from 0 s to 30 s time-intervals following 50-kHz playbacks, as well as from −10 s to 10 s time-intervals following 22-kHz playbacks ($p = 0.0000$ in all cases for all rats analyzed together with the exception of 50-kHz tone with $p = 0.0011$, Wilcoxon). These effects were also present in most of the groups analyzed separately, while no effects ($p > 0.05$) were observed during control time-intervals, that is, from −30 s to −10 s or −120 s to −100 s (Table S5).

The changes in HR around the signal onset, that is, from −10 s to 30 s, were most striking, especially in rats exposed to 50-kHz vs. 22-kHz playback. The former resulted in a significant increase in HR values between 0 s time-interval and following time-intervals (Figure S3B,D,F,H; Tables S5 and S6). Whereas after 22-kHz sounds presentation, the most striking feature was a drop in HR levels from the intervals before the playback vs. subsequent 10–60 s time-intervals (Figure S4B,D,F,H; Tables S5 and S6). Moreover, when results following USV and tone playbacks were averaged, that is, when comparing both 50-kHz and both 22-kHz groups of results (Figure 1B,D,F,H; Tables S5 and S6), the tendencies of HR levels to increase or decrease, and their significances, intensified.

As a consequence, 50-kHz playback resulted in higher HR following 50-kHz USV- vs. 22-kHz USV playback, 50-kHz tone vs. 22-kHz tone, and especially following averaged 50-kHz sounds vs. 22-kHz sounds in all analyzed groups (Figure 1B,D,F,H). Please note that HR values in response to the 50-kHz vs. 22-kHz sounds differed throughout 0–180 s time-intervals in all the animals combined (Table S7).

3.9. One-Trial Rats Showed Lower HR Levels and a Smaller Response to 50-kHz Playback

The HR of 1-Trial rats was lower than the HR in other groups throughout the experiment. There was a group HR effect for averaged values in the whole initial 10 min interval

(Figure S2E,F, p = 0.0312, Kruskal-Wallis); 1-Trial rats displayed lower HR (461.0 ± 7.3) when compared not only to control rats (476.6 ± 5.5, p = 0.0441) but also to 6-Trial (482.7 ± 7.3, p = 0.0135) and 10-Trial groups (479.3 ± 10.2, p = 0.0373; all Mann-Whitney).

Similarly, during the playback session (Figure 1B,D,F,H), there was a group effect at the −120 s to −100 s time interval and the entire −120 s to −10 s time interval (p = 0.0304, p = 0.0436, respectively, Kruskal-Wallis) when data from all presentations, that is, USV, tone, 50-kHz, and 22-kHz, were averaged for analysis. At this latter interval, 1-Trial rats displayed lower HR (386.0 ± 3.5) when compared not only to control rats (402.0 ± 4.6, p = 0.0268) but also 6-Trial (405.3 ± 5.3, p = 0.0039) and 10-Trial groups (397.3 ± 6.3, not significant p = 0.0608; all Mann-Whitney).

In 1-Trial rats, HR response to 50-kHz playback was smaller than in other groups (Figure 1 and Figure S3); this effect was observed especially following 50-kHz USV playback and averaged sound playback (Figure 1D vs. Figure 1B,F,H; Figure S3D vs. S3B,F,H; Table S8). For example, the increase in HR between the −10 s time-interval and post playback averaged 10–60 s interval was smaller and even negative in 1-Trial rats following USV playback (−0.4 ± 4.0) and sound playback (−2.0 ± 3.9) vs. HR increase in, for example, 0-Trial rats following USV playback (12.1 ± 3.8, p = 0.0254) and 10-Trial rats following sound playback (14.2 ± 5.4, p = 0.0341; Mann-Whitney; Table S8, last row).

3.10. Rats Vocalized More Often during and Following 50-kHz Playback Than 22-kHz Playback

The presentation of 50-kHz playback resulted in a dramatic increase in the number of USV emitted (Figure 1, Figure 2 and Figure S3, Tables S9 and S10). In contrast, the increase was modest during and after the presentation of the 22-kHz sounds (Figure 1, Figure 2 and Figure S4; Tables S9 and S10). When the values of USV emissions following 50- vs. 22-kHz playback were compared (Figure 1B,D,F,H), there was a clear and prolonged difference across all analyzed groups, that is, there were more USV following 50-kHz playback throughout 0–180 s time-intervals (Table S11).

Moreover, following the analysis of the number and parameters of USV produced by rats from the different groups (Table 1), we observed higher numbers of USV during and following 50-kHz playback, vs. 22-kHz playback, not only regarding all emitted USV (p = 0.0000), but also regarding 50-kHz USV (p = 0.0000) and short 22-kHz USV (p = 0.0017) in particular (see Table S12 for 50- vs. 22-kHz playbacks comparisons). Additionally, 50-kHz USV emitted in response to 50-kHz playback were both longer (27.8 ± 1.0 ms) and of higher frequency (60.5 ± 0.5 kHz) than 50-kHz calls emitted in response to the 22-kHz playback (23.2 ± 1.5 ms, p = 0.0000; 57.6 ± 0.9 kHz, p = 0.0000; respectively; Table S12).

Table 1. Comparison of number of USV of different types and selected characteristics of 50-kHz USV emitted during the whole experiment, first 10 min of silence and during playback sessions, that is, during the 10-s-long playback and 110 s afterwards in response to 50- vs. 22-kHz USV, tone and sounds (averaged results) in control (0-Trial, n = 37) and fear-conditioned rats (1-Trial, n = 16; 6-Trial, n = 20; 10-Trial, n = 19; all fear-conditioned/FC'ed). MPF—mean peak frequency; categories of USV: 50-kHz (MPF >32 kHz), short 22-kHz (MPF of 18–32 kHz, duration <0.3 s), long 22-kHz (18–32 kHz, >0.3 s); underlined: effect of group showed by the Kruskal–Wallis test; * $p < 0.05$; ** $p < 0.01$, differences between 0-Trial and other groups; # $p < 0.05$, the difference between 1-Trial vs. 10-Trial groups, all Mann–Whitney test. Please note, USV emitted by FC rats are more numerous, prolonged and, usually, of higher frequency.

Groups	Number of USV				Parameters of 50-kHz USV	
	Total USV	50-kHz USV	Short 22-kHz	Long 22-kHz	Duration [ms]	MPF [kHz]
			USV emitted during the whole experiment			
0-Trial	152.7 ± 24.2	141.6 ± 23.8	10.7 ± 2.6	0.4 ± 0.4	25.8 ± 1.2	58.0 ± 0.9
1-Trial	230.0 ± 68.0	205.7 ± 66.7	22.2 ± 8.3	2.0 ± 2.0	26.3 ± 2.0	58.3 ± 1.5
6-Trial	366.1 ± 84.9	346.7 ± 85.2	9.1 ± 1.9	10.4 ± 7.1	25.3 ± 2.1	61.7 ± 1.1
10-Trial	354.2 ± 84.7 **	336.1 ± 84.2 **	10.7 ± 2.2	7.3 ± 7.3	29.0 ± 1.8	60.1 ± 0.4
all FC'ed	322.4 ± 46.8 *	302.0 ± 46.7	13.5 ± 2.7	6.9 ± 3.6	26.9 ± 1.1	60.2 ± 0.6

Table 1. *Cont.*

Groups	Number of USV				Parameters of 50-kHz USV	
	Total USV	50-kHz USV	Short 22-kHz	Long 22-kHz	Duration [ms]	MPF [kHz]
			USV emitted during the first 10 min of silence			
0-Trial	23.9 ± 7.1	19.4 ± 7.0	4.5 ± 1.0	0.0 ± 0.0	15.7 ± 1.7	52.1 ± 1.9
1-Trial	53.6 ± 18.0 *	38.4 ± 15.6	13.1 ± 6.6	2.0 ± 2.0	22.2 ± 3.4	52.7 ± 2.9
6-Trial	72.8 ± 26.6	57.8 ± 26.6	4.6 ± 1.2	10.4 ± 7.1	21.7 ± 2.1 *	55.5 ± 1.1
10-Trial	49.7 ± 23.1	38.1 ± 22.3	4.4 ± 1.3	7.3 ± 7.3	21.0 ± 2.0 *	56.6 ± 1.7
all FC'ed	59.2 ± 13.4	45.3 ± 13.0	7.0 ± 2.0	6.9 ± 3.6	21.6 ± 1.4 **	55.0 ± 1.2
			USV emitted to 50-kHz USV playback (0–120 s time-intervals)			
0-Trial	41.2 ± 7.1	39.9 ± 6.9	0.9 ± 0.2	0.3 ± 0.3	27.6 ± 1.7	60.5 ± 1.5
1-Trial	59.1 ± 19.5	57.8 ± 19.3	1.3 ± 0.6	0.0 ± 0.0	24.8 ± 2.2	60.4 ± 1.1
6-Trial	86.4 ± 16.5	85.7 ± 16.5	0.8 ± 0.3	0.0 ± 0.0	27.3 ± 2.6	62.8 ± 1.4
10-Trial	97.5 ± 17.5 **	95.8 ± 17.2 **	1.7 ± 0.7	0.0 ± 0.0	30.6 ± 1.8	60.9 ± 0.7
all FC'ed	82.3 ± 10.3 *	81.0 ± 10.2 *	1.2 ± 0.3	0.0 ± 0.0	27.8 ± 1.3	61.4 ± 0.6
			USV emitted to 22-kHz USV playback (0–120 s time-intervals)			
0-Trial	13.5 ± 3.6	12.9 ± 3.6	0.6 ± 0.3	0.0 ± 0.0	21.5 ± 1.9	58.7 ± 1.8
1-Trial	14.6 ± 5.4	13.4 ± 5.3	1.1 ± 0.6	0.0 ± 0.0	31.3 ± 13.1	55.9 ± 3.3
6-Trial	32.2 ± 7.6 *	31.9 ± 7.6	0.3 ± 0.1	0.0 ± 0.0	29.4 ± 2.4 *	60.1 ± 0.7
10-Trial	30.9 ± 9.8	29.9 ± 9.8	1.1 ± 0.5	0.0 ± 0.0	23.8 ± 2.4	58.9 ± 1.8
all FC'ed	26.6 ± 4.7	25.8 ± 4.7	0.8 ± 0.3	0.0 ± 0.0	27.7 ± 3.5	58.5 ± 1.1
			USV emitted to 50-kHz tone playback (0–120 s time-intervals)			
0-Trial	41.4 ± 6.6	40.4 ± 6.5	1.0 ± 0.4	0.0 ± 0.0	27.5 ± 1.3	59.1 ± 0.7
1-Trial	49.3 ± 16.2	48.6 ± 16.1	0.8 ± 0.2	0.0 ± 0.0	25.4 ± 2.2	60.6 ± 1.4
6-Trial	76.0 ± 18.5	74.6 ± 18.4	1.4 ± 0.5	0.0 ± 0.0	30.1 ± 2.9	60.1 ± 1.1
10-Trial	76.3 ± 15.3	75.4 ± 15.2	0.9 ± 0.4	0.0 ± 0.0	31.7 ± 2.2	60.6 ± 0.6
all FC'ed	68.3 ± 9.7	67.3 ± 9.7	1.0 ± 0.2	0.0 ± 0.0	29.3 ± 1.4	60.5 ± 0.6 *
			USV emitted to 22-kHz tone playback (0–120 s time-intervals)			
0-Trial	6.6 ± 2.1	6.1 ± 2.0	0.5 ± 0.3	0.0 ± 0.0	17.5 ± 2.1	55.5 ± 1.9
1-Trial	16.0 ± 6.9	15.3 ± 6.9	0.7 ± 0.5	0.0 ± 0.0	27.8 ± 7.1	54.9 ± 2.7
6-Trial	18.1 ± 5.5 *	17.8 ± 5.5	0.4 ± 0.2	0.0 ± 0.0	26.9 ± 3.3 *	60.1 ± 0.9
10-Trial	23.3 ± 8.1	23.2 ± 8.1 *	0.1 ± 0.1	0.0 ± 0.0	25.1 ± 2.2 *	58.3 ± 1.4
all FC'ed	19.3 ± 3.9 *	18.9 ± 3.9 *	0.4 ± 0.2	0.0 ± 0.0	26.5 ± 2.4 **	57.9 ± 1.0
			USV emitted to 50-kHz sound playback (0–120 s time-intervals)			
0-Trial	41.3 ± 6.0	40.1 ± 5.9	1.0 ± 0.3	0.1 ± 0.1	28.0 ± 1.4	59.9 ± 0.8
1-Trial	54.2 ± 17.4	53.2 ± 17.2	1.0 ± 0.4	0.0 ± 0.0	25.1 ± 1.8	60.5 ± 1.1
6-Trial	81.2 ± 16.4	80.1 ± 16.3	1.1 ± 0.3	0.0 ± 0.0	27.6 ± 2.7	62.4 ± 1.4
10-Trial	86.9 ± 15.1 *	85.6 ± 15.0 **	1.3 ± 0.4	0.0 ± 0.0	31.2 ± 1.9 #	60.8 ± 0.6
all FC'ed	75.3 ± 9.4	74.2 ± 9.3	1.1 ± 0.2	0.0 ± 0.0	28.2 ± 1.3	61.2 ± 0.6
			USV emitted to 22-kHz sound playback (0–120 s time-intervals)			
0-Trial	10.1 ± 2.7	9.5 ± 2.6	0.6 ± 0.3	0.0 ± 0.0	18.4 ± 1.6	58.3 ± 1.6
1-Trial	15.3 ± 4.7	14.4 ± 4.7	0.9 ± 0.4	0.0 ± 0.0	36.5 ± 11.6	53.7 ± 3.1
6-Trial	25.2 ± 5.9 *	24.8 ± 5.9 *	0.3 ± 0.1	0.0 ± 0.0	26.8 ± 2.4 **	60.4 ± 0.7
10-Trial	27.1 ± 8.3 *	26.6 ± 8.3 *	0.6 ± 0.3	0.0 ± 0.0	23.2 ± 2.2	58.0 ± 1.6
all FC'ed	23.0 ± 3.8 *	22.4 ± 3.8 *	0.6 ± 0.2	0.0 ± 0.0	28.2 ± 3.4 **	57.6 ± 1.1

3.11. Natural and Artificial Ultrasounds Produced Similar Results, but Still, Some Differences Stood Out

In particular, natural 50-kHz playback produced more USV and more 50-kHz USV responses in FC rats (1, 6, and 10-Trial groups), analyzed together (82.3 ± 10.3, 81.0 ± 10.2, respectively), than the 50-kHz tone playback (68.3 ± 9.7, $p = 0.0283$; 67.3 ± 9.7, $p = 0.0251$; respectively, Wilcoxon; Figure S3B,D,F,H, Table S13). This effect was only observed in each FC group but was not significant; it was not observed in the 0-Trial group.

Similarly, natural 22-kHz playback produced more USV and more 50-kHz USV responses in all rats analyzed together (21.4 ± 3.2, 20.6 ± 3.2, respectively) than 22-kHz tone playback (14.2 ± 2.6, p = 0.0196; 13.8 ± 2.6, p = 0.0238; respectively, Wilcoxon; Figure S4B,D,F,H, Table S13). This effect was observed in most groups but was significant in the 0-Trial group only (p = 0.0025, p = 0.0018, respectively, Wilcoxon). Additionally, 50- and 22-kHz tone playback resulted in a lower frequency of emitted 50-kHz USV than in case of USV playbacks (Table S13).

Natural playback evoked a more pronounced approach to the speaker (Figures S3 and S4), which was mainly observed at 10–60 s time intervals when results from relevant groups were analyzed together. For all rats, there was a difference in time spent in the speaker's half of the cage following 50-kHz USV playback (78.0 ± 2.8%) vs. 50-kHz tone playback (67.6 ± 2.9%, p = 0.0004), as well as following 22-kHz USV playback (73.1 ± 3.7%) vs. 22-kHz tone playback (60.8 ± 4.0%, p = 0.0236; both Wilcoxon; Table S14).

3.12. Previously-Shocked Rats Vocalized More Often, with Longer and Higher Frequency USV

USV of rats from different groups differed in number, duration and frequency (Table 1). There was a group effect for the total number of USV (p = 0.0407) and the total number of 50-kHz USV (p = 0.0419) emitted to the 50-kHz USV playback. There was also a group effect for the duration of 50-kHz USV emitted during the first 10 min of silence (p = 0.0346). The latter effect was also present for averaged responses following 22-kHz playbacks (p = 0.0383; Table 1, all Kruskal–Wallis).

The analysis of between-group differences revealed that previously FC rats vocalized more than controls during the whole experiment, 10 min of introductory silence, and following both 50- and 22-kHz playbacks. For example, control rats emitted on average 41.2 ± 7.1 USV during and after 50-kHz USV playback, while all FC rats emitted 82.3 ± 10.3 USV (p = 0.0275, Mann–Whitney, Table 1).

Moreover, 50-kHz USV emitted by FC rats were also longer, for example, during the 10 min baseline period (21.6 s ± 1.4 ms vs. 15.7 ± 1.7 ms in control rats; p = 0.0025) and following 22-kHz sound playback (28.2 ± 3.4 ms vs. 18.4 ± 1.6 ms in control rats; p = 0.0066) and they were also of higher frequency. Mean peak frequency values were higher in five of the six investigated cases of not averaged data (see Table 1), which was significant in the case of USV emitted to the 50-kHz tone playback (p = 0.0471).

4. Discussion

We used our formerly published model based on rats' exposure to pre-recorded playbacks in home-cage-like conditions. Several observations were the same as described in our previous publication [26]:

- Rats' overt behavior remained relatively constant except during ultrasonic playback;
- Rats moved faster during 50-kHz ultrasonic presentations;
- Rats slowed down right after 22-kHz ultrasonic presentations;
- Rats approached the speaker during and following 50-kHz and 22-kHz playbacks;
- The approach was more pronounced during and following 50-kHz playback;
- HR levels declined during the whole experimental session;
- Rats' HR increase after exposure to 50-kHz playback;
- Rats' HR decrease when exposed to 22-kHz playback;
- The difference in HR following 50-kHz vs. 22-kHz playback lasts for at least 3 min;
- Both 50- and 22-kHz sounds evoked an ultrasonic response, mainly in the 50-kHz range;
- 50-kHz USV emitted in response to 50-kHz playback were longer and of higher frequency;
- Rats vocalized more often during and following 50-kHz playback than 22-kHz playback;
- In general, rats reacted in a similar way to both natural and artificial ultrasonic playback.

Rats emitted a high amount of 50-kHz USV in response to our ultrasonic presentation in contrast to other playback studies [29–33], possibly due to the housing-like conditions during recording. We also observed behavioral changes in animals when faced with different playbacks, which included approaching the speaker. It was shown before that playback of 50-kHz calls causes approach behavior in rats, however 22-kHz presentation was repeatedly reported to elicit behavioral inhibition and no-social-approach [33–40] and hiding [41], while we observed speaker-side preference following 22-kHz playback (Figure 1, Table S2). The preference could again be a result of the low-stress, home-cage-like experimental conditions. Please note that there was still a decrease in locomotor activity following 22-kHz playback [39]. Another reason for the discrepancy could be the difference in playback duration, which was only 10 s in our case, while others have investigated up to 10–15 min (e.g., [41,42]).

Along with visible behavioral alternations, we observed changes in rats' HR with a striking difference in HR response to 50-kHz and 22-kHz playback. These changes could be explained by the emotional arousal evoked by the two different call types, which activate specific limbic and cortical areas of the brain. These are predominantly the frontal and motor cortices and nucleus accumbens for 50-kHz calls as well as the perirhinal cortex, basolateral amygdala, and periaqueductal gray—for 22-kHz USV [25,29]. The nucleus accumbens is responsible for modulating appetitive behaviors and is regulated by dopaminergic afferent fibers. A sudden increase in HR correlated with an emergent approach behavior following playback of 50-kHz USV [43], while activation of the periaqueductal gray, which is regarded as a defense–response center, following playback of 22-kHz USV was accompanied by reduced locomotor activity and freezing [44,45]. According to the polyvagal theory, physiological changes such as the regulation of HR, respiratory rhythm along with several behaviors, for example, vocal emissions, are intrinsically linked via a common signaling pathway—the vagus nerve [46–48]. Therefore sensory stimulation by 50-kHz or 22-kHz USV most likely lead to system-wide physiological changes including cardiovascular, locomotor and vocal reactions.

The electric shock protocol is used to study physiological associative aversive memory and/or to imitate traumatic events leading to pathological conditions. Notably, exposure to even a single foot-shock session was shown to induce long-lasting inhibition of activity in the shock-context and in unknown environments that markedly differ from the shock context. This effect is known as fear generalization [49]. Delivery of a higher number of shocks with higher amperage, frequency and longer shock duration represents a more severe traumatic stressor than a lower number of shorter, lower amperage shocks; see Figure 2 in [3]. In our study, a single 1 mA, 1 s shock served as the weakest stimulus while repeating it six and ten times increased the severity of the treatment. Rats that received six and ten shocks had higher freezing levels than those receiving only one.

To the best of our knowledge, USV emission in reaction to USV playback has never been studied in the context of previously experienced shock. We propose that the observed reactions can broadly be interpreted as a sign of hypervigilance. Our conditioned rats showed higher locomotor activity during the 50-kHz playback and a more significant decrease in activity following the 22-kHz playback. Increased locomotor activity during appetitive playback and decreased activity immediately following the aversive playback were previously observed [26]; however, these reactions are intensified in FC rats (Figure 2A,C,E,G).

Similarly, induced hypervigilance with exaggerated reactions has been previously observed in rat models of PTSD. For example, previously shocked rats showed increased avoidance reactions and unnecessary crossings after cessation of foot-shock [50]. Previously shocked rats buried unfamiliar objects, while control animals did not [51]. Correspondingly, PTSD patients showed physiological and behavioral hyperreactivity to environmental stressors even if they are not related to the traumatic situation, for example, exaggerated acoustic startle responses [52]. Furthermore, increased startle has been reported during experimental induction of fear in healthy individuals, especially in high-fear subjects [53].

In this study, a sign of hypervigilance was more frequent vocalization by previously-shocked rats compared to control animals. Their 50-kHz USV were longer and of higher frequency as well (Figure 2B,D,F,H and Figure 3B; Table 1). Interestingly, a similar finding was observed in single-reared rats that vocalize more often than paired ones to ultrasonic playback [26]. It is worth noting that the rearing of rats in isolation causes both anxiety-like and depression-like symptoms [54,55]. It was hypothesized before that the peak frequency, along with the number of calls per time unit, is involved in coding the quantitative aspect of 50-kHz calls [56]. A recent study reported on a higher frequency of 50-kHz USV appearing in a foot-shock paradigm where one animal in a pair was witnessing the other receiving the aversive stimuli. A fraction of high-frequency (>75 kHz) USV were observed in pairs where the observer animal was naïve to the testing conditions and the foot-shocks were preceded by an audible cue [57]. Finally, the aforementioned dopamine system also modulates the mechanisms underlying fear and anxiety and supports the acquisition of conditioned fear with a potential key role of dopamine receptors in supporting amygdaloid synaptic plasticity underlying the consolidation of the CS–US association [58]. Dopamine receptors' antagonism was demonstrated to result in reduced call rate, increased latency to call, decreased duration, intensity, bandwidth and peak frequency [59].

The electric shock affected HR levels as well. However, the effects were not linear, that is, they did not correlate with shock intensity. In particular, 1-Trial rats showed lower HR levels and a smaller response to 50-kHz playback in HR increase than control and 6- and 10-Trial rats (Figure S2, Table S8). The stress-induced changes in HR have been shown before not to be directly proportional to the intensity of the stressor and have been postulated to be a function of stress severity and duration, which translate into a differential stimulation and dynamic balance between sympathetic and parasympathetic (vagal) tones in the heart [12,60,61].

For example, a reduction of HR, which lasted five days, was observed following sub-chronic fear conditioning. This effect was shown to be predominantly vagally-mediated, that is, the post-stress vagal tone was higher compared with the prestress level [13]. Similarly, FC rats showed less pronounced tachycardia when compared with control animals, which was attributed to simultaneous activation of the sympathetic nervous system and parasympathetic nervous system. In contrast, in non-shocked controls, a predominant sympathetic nervous system activation results in a more significant increase in HR [12].

In humans, vagal tone dominates in healthy resting conditions. In this study, the increase in the vagal tone may cause an HR decrease observed in mildly shocked 1-Trial rats. On the other hand, chronically stressed humans are often characterized by anomalies in the autonomic regulation of HR, such as elevated sympathetic and reduced vagal tone, which can induce tachycardia [13]. The increased HR observed in 6- and 10-Trial rats might model emotional trauma and anxiety observed in humans to correlate with lowering the vagal tone in the heart and increasing the sympathetic tone [60]. PTSD patients were shown to have more significant cardiac responses to startling sounds and idiosyncratic trauma reminders [62,63].

Our results are of limited generalizability, since only male rats were used in the experiments. Estimates from community studies suggest that women are two to three times more likely to develop PTSD than men, while USA prevalence estimates of lifetime PTSD from the National Comorbidity Survey Replication are 9.7% for women and 3.6% for men (https://www.ptsd.va.gov/understand/common/common_adults.asp, assessed on 30 June 2021). Therefore, relevant future research should include female rats.

5. Conclusions

Our ultrasonic playback–answer behavioral paradigm in rats combined with the previously applied electric shock can serve as a model of hypervigilance associated with past trauma and PTSD syndrome according to DSM-5 [64]. Therefore, the detection of increased vocalization can serve as a valuable new measure of hypervigilance for the behavioral animal modeling of PTSD. Future pharmacological evaluation of this measurement should be considered.

Supplementary Materials: The following are available online at https://www.mdpi.com/article/10.3390/brainsci11080970/s1, Figure S1: Emitted USV, heart rate and freezing levels of rats before conditioning and test freezing levels to context and context with cue, Figure S2: Assessment of locomotor activity, heart rate and the number of USV emitted of all animals during the first 10 min of silence in the experimental cage, Figure S3: Effect of 50-kHz ultrasonic playback sounds (vocalizations vs. tones) on distance traveled, time spent in the speaker's half of the cage, heart rate (HR) and USV emission, Figure S4: Effect of 22-kHz ultrasonic playback sounds (vocalizations vs. tones) on distance traveled, time spent in the speaker's half of the cage, heart rate (HR) and USV emission, Table S1: Evaluation of changes in distance traveled around playback of ultrasounds, Table S2: Evaluation time spent in the speaker's half values in comparison with 50% chance level, Table S3: Evaluation of changes in time spent in the speaker's half around playback of ultrasounds as well as during the control intervals, Table S4: Evaluation of changes in time spent in the speaker's half before vs. during playback vs. after the playback, Table S5: Evaluation of changes in heart rate around playback of ultrasounds as well as during the control intervals, Table S6: Evaluation of changes in heart rate around the time of ultrasonic playback i.e. between before or immediately before vs. during playback vs. after the playback, Table S7: Evaluation of comparisons between the effects of 50-kHz vs. 22-kHz playbacks on heart rate, Table S8: Comparisons of heart rate changes in 1-Trial group vs. other groups, Table S9: Evaluation of changes in the number of USV emitted around playback of ultrasounds as well as during the control intervals, Table S10: Evaluation of changes in the number of USV around the time of ultrasonic playback i.e. between before vs. during playback vs. after the playback, Table S11: Evaluation of comparisons between the effects of 50-kHz vs. 22-kHz playbacks on the number of emitted vocalizations, Table S12: Comparison of number of USV of different types and selected characteristics of 50-kHz USV emitted in response to 50- vs. 22-kHz playback, Table S13: Comparison of number of USV of different types and selected characteristics of 50-kHz USV emitted in response to USV- vs. tone-playback, Table S14: Differences in time spent in the speaker's half in response to USV- vs. tone playbacks.

Author Contributions: R.K.F. and K.H.O. conceptualized the study; K.H.O. designed the experiments; K.H.O., R.P., A.D.W., A.W.G. performed the experiments. K.H.O. and R.P. analyzed the data with the contribution of A.D.W.; K.H.O. and R.P. drafted the manuscript; R.K.F. wrote the final version of the manuscript. All authors have read and agreed to the published version of the manuscript.

Funding: This research was funded by the National Science Centre, Poland, grant OPUS no. 2015/19/B/NZ4/03393.

Institutional Review Board Statement: The study was conducted according to the guidelines of the Declaration of Helsinki, and approved by the Second Warsaw Local Ethics Committee for Animal Experimentation (WAW2/093/2019) on 28 June 2019.

Data Availability Statement: Raw data, analyzed herein, have been deposited to Mendeley Data at https://data.mendeley.com/datasets/3pbnnxjzv7/1.

Acknowledgments: We are grateful to Jakub Zieliński for his help in statistical analysis.

Conflicts of Interest: The authors declare no conflict of interest.

References

1. Bali, A.; Jaggi, A.S. Electric foot shock stress: A useful tool in neuropsychiatric studies. *Rev. Neurosci.* **2015**, *26*, 655–677. [CrossRef]
2. Careaga, M.B.L.; Girardi, C.E.N.; Suchecki, D. Understanding posttraumatic stress disorder through fear conditioning, extinction and reconsolidation. *Neurosci. Biobehav. Rev.* **2016**, *71*, 48–57. [CrossRef]
3. Aliczki, M.; Haller, J. Electric shock as model of post-traumatic stress disorder in rodents. In *Comprehensive Guide to Post-Traumatic Stress Disorder*; Martin, C.R., Preedy, V.R., Patel, V.B., Eds.; Springer International Publishing: Basel, Switzerland, 2014.
4. Verbitsky, A.; Dopfel, D.; Zhang, N. Rodent models of post-traumatic stress disorder: Behavioral assessment. *Transl. Psychiatry* **2020**, *10*, 132. [CrossRef]
5. Asok, A.; Kandel, E.R.; Rayman, J.B. The Neurobiology of Fear Generalization. *Front. Behav. Neurosci.* **2018**, *12*, 329. [CrossRef]
6. Fanselow, M.S. Conditioned and unconditional components of post-shock freezing. *Pavlov. J. Biol. Sci.* **1980**, *15*, 177–182. [CrossRef]

7. Frankland, P.W.; Cestari, V.; Filipkowski, R.K.; McDonald, R.J.; Silva, A.J. The dorsal hippocampus is essential for context discrimination but not for contextual conditioning. *Behav. Neurosci.* **1998**, *112*, 863–874. [CrossRef]
8. Blanchard, R.J.; Blanchard, D.C. Passive and active reactions to fear-eliciting stimuli. *J. Comp. Physiol. Psychol.* **1969**, *68*, 129–135. [CrossRef]
9. Katz, R.J.; Roth, K.A.; Carroll, B.J. Acute and chronic stress effects on open field activity in the rat: Implications for a model of depression. *Neurosci. Biobehav. Rev.* **1981**, *5*, 247–251. [CrossRef]
10. Goncalves, R.; Rodrigues, H.; Novaes, F.; Arbol, J.; Volchan, E.; Coutinho, E.S.; Figueira, I.; Ventura, P. Listening to the heart: A meta-analysis of cognitive behavior therapy impact on the heart rate of patients with anxiety disorders. *J. Affect. Disord.* **2015**, *172*, 231–240. [CrossRef]
11. Pittig, A.; Arch, J.J.; Lam, C.W.; Craske, M.G. Heart rate and heart rate variability in panic, social anxiety, obsessive-compulsive, and generalized anxiety disorders at baseline and in response to relaxation and hyperventilation. *Int. J. Psychophysiol.* **2013**, *87*, 19–27. [CrossRef]
12. Nijsen, M.J.; Croiset, G.; Diamant, M.; Stam, R.; Delsing, D.; de Wied, D.; Wiegant, V.M. Conditioned fear-induced tachycardia in the rat: Vagal involvement. *Eur. J. Pharmacol.* **1998**, *350*, 211–222. [CrossRef]
13. Carnevali, L.; Bondarenko, E.; Sgoifo, A.; Walker, F.R.; Head, G.A.; Lukoshkova, E.V.; Day, T.A.; Nalivaiko, E. Metyrapone and fluoxetine suppress enduring behavioral but not cardiac effects of subchronic stress in rats. *Am. J. Physiol. Regul. Integr. Comp Physiol.* **2011**, *301*, R1123–R1131. [CrossRef]
14. Resstel, L.B.; Alves, F.H.; Reis, D.G.; Crestani, C.C.; Correa, F.M.; Guimaraes, F.S. Anxiolytic-like effects induced by acute reversible inactivation of the bed nucleus of stria terminalis. *Neuroscience* **2008**, *154*, 869–876. [CrossRef]
15. Resstel, L.B.; Correa, F.M.; Guimaraes, F.S. The expression of contextual fear conditioning involves activation of an NMDA receptor-nitric oxide pathway in the medial prefrontal cortex. *Cereb. Cortex* **2008**, *18*, 2027–2035. [CrossRef]
16. Antoniadis, E.A.; McDonald, R.J. Discriminative fear conditioning to context expressed by multiple measures of fear in the rat. *Behav. Brain Res.* **1999**, *101*, 1–13. [CrossRef]
17. Carrive, P. Conditioned fear to environmental context: Cardiovascular and behavioral components in the rat. *Brain Res.* **2000**, *858*, 440–445. [CrossRef]
18. Zhang, W.N.; Murphy, C.A.; Feldon, J. Behavioural and cardiovascular responses during latent inhibition of conditioned fear: Measurement by telemetry and conditioned freezing. *Behav. Brain Res.* **2004**, *154*, 199–209. [CrossRef]
19. Meerlo, P.; Sgoifo, A.; De Boer, S.F.; Koolhaas, J.M. Long-lasting consequences of a social conflict in rats: Behavior during the interaction predicts subsequent changes in daily rhythms of heart rate, temperature, and activity. *Behav. Neurosci.* **1999**, *113*, 1283–1290. [CrossRef]
20. Burgdorf, J.S.; Brudzynski, S.M.; Moskal, J.R. Using rat ultrasonic vocalization to study the neurobiology of emotion: From basic science to the development of novel therapeutics for affective disorders. *Curr. Opin. Neurobiol.* **2020**, *60*, 192–200. [CrossRef]
21. Simola, N.; Granon, S. Ultrasonic vocalizations as a tool in studying emotional states in rodent models of social behavior and brain disease. *Neuropharmacology* **2019**, *159*, 107420. [CrossRef]
22. Wohr, M.; Schwarting, R.K. Affective communication in rodents: Ultrasonic vocalizations as a tool for research on emotion and motivation. *Cell Tissue Res.* **2013**, *354*, 81–97. [CrossRef] [PubMed]
23. Jelen, P.; Soltysik, S.; Zagrodzka, J. 22-kHz ultrasonic vocalization in rats as an index of anxiety but not fear: Behavioral and pharmacological modulation of affective state. *Behav. Brain Res.* **2003**, *141*, 63–72. [CrossRef]
24. Wohr, M.; Borta, A.; Schwarting, R.K. Overt behavior and ultrasonic vocalization in a fear conditioning paradigm: A dose-response study in the rat. *Neurobiol. Learn Mem.* **2005**, *84*, 228–240. [CrossRef]
25. Brudzynski, S.M. Ethotransmission: Communication of emotional states through ultrasonic vocalization in rats. *Curr. Opin. Neurobiol.* **2013**, *23*, 310–317. [CrossRef]
26. Olszynski, K.H.; Polowy, R.; Malz, M.; Boguszewski, P.M.; Filipkowski, R.K. Playback of Alarm and Appetitive Calls Differentially Impacts Vocal, Heart-Rate, and Motor Response in Rats. *iScience* **2020**, *23*, 101577. [CrossRef]
27. Pestana-Oliveira, N.; Nahey, D.B.; Johnson, T.; Collister, J.P. Development of the Deoxycorticosterone Acetate (DOCA)-salt Hypertensive Rat Model. *Bio Protoc.* **2020**, *10*, e3708. [CrossRef]
28. Lindquist, D.H.; Jarrard, L.E.; Brown, T.H. Perirhinal cortex supports delay fear conditioning to rat ultrasonic social signals. *J. Neurosci.* **2004**, *24*, 3610–3617. [CrossRef]
29. Sadananda, M.; Wohr, M.; Schwarting, R.K. Playback of 22-kHz and 50-kHz ultrasonic vocalizations induces differential c-fos expression in rat brain. *Neurosci. Lett.* **2008**, *435*, 17–23. [CrossRef] [PubMed]
30. Berg, E.L.; Copping, N.A.; Rivera, J.K.; Pride, M.C.; Careaga, M.; Bauman, M.D.; Berman, R.F.; Lein, P.J.; Harony-Nicolas, H.; Buxbaum, J.D.; et al. Developmental social communication deficits in the Shank3 rat model of phelan-mcdermid syndrome and autism spectrum disorder. *Autism Res.* **2018**, *11*, 587–601. [CrossRef]
31. Seffer, D.; Rippberger, H.; Schwarting, R.K.; Wohr, M. Pro-social 50-kHz ultrasonic communication in rats: Post-weaning but not post-adolescent social isolation leads to social impairments-phenotypic rescue by re-socialization. *Front. Behav. Neurosci.* **2015**, *9*, 102. [CrossRef]
32. Wohr, M.; Schwarting, R.K. Ultrasonic communication in rats: Effects of morphine and naloxone on vocal and behavioral responses to playback of 50-kHz vocalizations. *Pharmacol. Biochem. Behav.* **2009**, *94*, 285–295. [CrossRef]

33. Wohr, M.; Schwarting, R.K. Ultrasonic communication in rats: Can playback of 50-kHz calls induce approach behavior? *PLoS ONE* **2007**, *2*, e1365. [CrossRef] [PubMed]
34. Burman, O.; Owen, D.; Abouismail, U.; Mendl, M. Removing individual rats affects indicators of welfare in the remaining group members. *Physiol. Behav.* **2008**, *93*, 89–96. [CrossRef] [PubMed]
35. Seffer, D.; Schwarting, R.K.; Wohr, M. Pro-social ultrasonic communication in rats: Insights from playback studies. *J. Neurosci. Methods* **2014**, *234*, 73–81. [CrossRef]
36. Snoeren, E.M.; Agmo, A. The incentive value of males' 50-kHz ultrasonic vocalizations for female rats (Rattus norvegicus). *J. Comp. Psychol.* **2014**, *128*, 40–55. [CrossRef] [PubMed]
37. Davidson, N.B.; Hurst, J.L. Testing the potential of 50 kHz rat calls as a species-specific rat attractant. *PLoS ONE* **2019**, *14*, e0211601. [CrossRef] [PubMed]
38. Schonfeld, L.M.; Zech, M.P.; Schable, S.; Wohr, M.; Kalenscher, T. Lesions of the rat basolateral amygdala reduce the behavioral response to ultrasonic vocalizations. *Behav. Brain Res.* **2020**, *378*, 112274. [CrossRef]
39. Brudzynski, S.M.; Chiu, E.M. Behavioural responses of laboratory rats to playback of 22 kHz ultrasonic calls. *Physiol. Behav.* **1995**, *57*, 1039–1044. [CrossRef]
40. Wohr, M.; Seffer, D.; Schwarting, R.K. Studying Socio-Affective Communication in Rats through Playback of Ultrasonic Vocalizations. *Curr. Protoc. Neurosci.* **2016**, *75*, 8–35. [CrossRef]
41. Inagaki, H.; Ushida, T. The effect of playback of 22-kHz and 50-kHz ultrasonic vocalizations on rat behaviors assessed with a modified open-field test. *Physiol. Behav.* **2021**, *229*, 113251. [CrossRef] [PubMed]
42. Demaestri, C.; Brenhouse, H.C.; Honeycutt, J.A. 22 kHz and 55 kHz ultrasonic vocalizations differentially influence neural and behavioral outcomes: Implications for modeling anxiety via auditory stimuli in the rat. *Behav. Brain Res.* **2019**, *360*, 134–145. [CrossRef] [PubMed]
43. Willuhn, I.; Tose, A.; Wanat, M.J.; Hart, A.S.; Hollon, N.G.; Phillips, P.E.; Schwarting, R.K.; Wohr, M. Phasic dopamine release in the nucleus accumbens in response to pro-social 50 kHz ultrasonic vocalizations in rats. *J. Neurosci.* **2014**, *34*, 10616–10623. [CrossRef]
44. Vianna, D.M.; Brandao, M.L. Anatomical connections of the periaqueductal gray: Specific neural substrates for different kinds of fear. *Braz. J. Med. Biol. Res.* **2003**, *36*, 557–566. [CrossRef] [PubMed]
45. Koba, S.; Inoue, R.; Watanabe, T. Role played by periaqueductal gray neurons in parasympathetically mediated fear bradycardia in conscious rats. *Physiol. Rep.* **2016**, *4*. [CrossRef]
46. Porges, S.W. Orienting in a defensive world: Mammalian modifications of our evolutionary heritage. A Polyvagal Theory. *Psychophysiology* **1995**, *32*, 301–318. [CrossRef]
47. Porges, S.W. The polyvagal perspective. *Biol. Psychol.* **2007**, *74*, 116–143. [CrossRef] [PubMed]
48. Porges, S.W. The polyvagal theory: New insights into adaptive reactions of the autonomic nervous system. *Cleve Clin. J. Med.* **2009**, *76*, S86–S90. [CrossRef]
49. Daviu, N.; Fuentes, S.; Nadal, R.; Armario, A. A single footshock causes long-lasting hypoactivity in unknown environments that is dependent on the development of contextual fear conditioning. *Neurobiol. Learn Mem.* **2010**, *94*, 183–190. [CrossRef]
50. Sawamura, T.; Shimizu, K.; Nibuya, M.; Wakizono, T.; Suzuki, G.; Tsunoda, T.; Takahashi, Y.; Nomura, S. Effect of paroxetine on a model of posttraumatic stress disorder in rats. *Neurosci. Lett.* **2004**, *357*, 37–40. [CrossRef]
51. Mikics, E.; Baranyi, J.; Haller, J. Rats exposed to traumatic stress bury unfamiliar objects—A novel measure of hyper-vigilance in PTSD models? *Physiol. Behav.* **2008**, *94*, 341–348. [CrossRef] [PubMed]
52. Armario, A.; Escorihuela, R.M.; Nadal, R. Long-term neuroendocrine and behavioural effects of a single exposure to stress in adult animals. *Neurosci. Biobehav. Rev.* **2008**, *32*, 1121–1135. [CrossRef]
53. Grillon, C.; Ameli, R.; Foot, M.; Davis, M. Fear-potentiated startle: Relationship to the level of state/trait anxiety in healthy subjects. *Biol. Psychiatry* **1993**, *33*, 566–574. [CrossRef]
54. Djordjevic, J.; Djordjevic, A.; Adzic, M.; Radojcic, M.B. Effects of chronic social isolation on Wistar rat behavior and brain plasticity markers. *Neuropsychobiology* **2012**, *66*, 112–119. [CrossRef]
55. Robbins, T.W. Neurobehavioural sequelae of social deprivation in rodents revisited: Modelling social adversity for developmental neuropsychiatric disorders. *J. Psychopharmacol.* **2016**, *30*, 1082–1089. [CrossRef]
56. Brudzynski, S.M. Principles of rat communication: Quantitative parameters of ultrasonic calls in rats. *Behav. Genet.* **2005**, *35*, 85–92. [CrossRef] [PubMed]
57. Karwicka, W.; Wiatrowska, M.; Kondrakiewicz, K.; Knapska, E.; Kursa, M.B.; Hamed, A. Relaying Aversive Ultrasonic Alarm Calls Depends on Previous Experience. Empathy, Social Buffering, or Panic? *Brain Sci.* **2021**, *11*, 759. [CrossRef]
58. Pezze, M.A.; Feldon, J. Mesolimbic dopaminergic pathways in fear conditioning. *Prog. Neurobiol.* **2004**, *74*, 301–320. [CrossRef]
59. Ringel, L.E.; Basken, J.N.; Grant, L.M.; Ciucci, M.R. Dopamine D1 and D2 receptor antagonism effects on rat ultrasonic vocalizations. *Behav. Brain Res.* **2013**, *252*, 252–259. [CrossRef] [PubMed]
60. Carnevali, L.; Sgoifo, A. Vagal modulation of resting heart rate in rats: The role of stress, psychosocial factors, and physical exercise. *Front. Physiol.* **2014**, *5*, 118. [CrossRef]
61. Armario, A.; Labad, J.; Nadal, R. Focusing attention on biological markers of acute stressor intensity: Empirical evidence and limitations. *Neurosci. Biobehav. Rev.* **2020**, *111*, 95–103. [CrossRef]

62. Buckley, T.C.; Kaloupek, D.G. A meta-analytic examination of basal cardiovascular activity in posttraumatic stress disorder. *Psychosom. Med.* **2001**, *63*, 585–594. [CrossRef] [PubMed]
63. Pole, N. The psychophysiology of posttraumatic stress disorder: A meta-analysis. *Psychol. Bull.* **2007**, *133*, 725–746. [CrossRef]
64. American Psychiatric Association. *Diagnostic and Statistical Manual of Mental Disorders*, 5th ed.; American Psychiatric Association: Arlington, VA, USA, 2013.

brain sciences

Brief Report

Extracellular Dopamine Levels in Nucleus Accumbens after Chronic Stress in Rats with Persistently High vs. Low 50-kHz Ultrasonic Vocalization Response

Kadri Kõiv [1], Kai Tiitsaar [1], Karita Laugus [2] and Jaanus Harro [2,*]

1 Institute of Psychology, University of Tartu, Ravila 14A, 50411 Tartu, Estonia; Kadri.Koiv@ut.ee (K.K.); Kai.Tiitsaar@ut.ee (K.T.)
2 Chair of Neuropsychopharmacology, Institute of Chemistry, University of Tartu, Ravila 14A, 50411 Tartu, Estonia; Karita.Laugus@ut.ee
* Correspondence: Jaanus.Harro@ut.ee; Tel.: +372-737-6657

Abstract: Fifty-kHz ultrasonic vocalizations (USVs) in response to an imitation of rough-and-tumble play ('tickling') have been associated with positive affective states and rewarding experience in the rat. This USV response can be used as a measure of inter-individual differences in positive affect. We have previously shown that rats with persistently low positive affectivity are more vulnerable to the effects of chronic variable stress (CVS). To examine whether these differential responses are associated with dopaminergic neurotransmission in the nucleus accumbens (NAc), juvenile male Wistar rats were categorized as of high or low positive affectivity (HC and LC, respectively), and after reaching adulthood, extracellular dopamine (DA) levels in the NAc shell were measured using in vivo microdialysis after three weeks of CVS. Baseline levels of DA were compared as well as the response to K^+-induced depolarization and the effect of glial glutamate transporter EAAT2 inhibition by 4 mM l-trans-pyrrolidine-2,4-dicarboxylate (PDC). DA baseline levels were higher in control LC-rats, and stress significantly lowered the DA content in LC-rats. An interaction of stress and affectivity appeared in response to depolarization where stress increased the DA output in HC-rats whereas it decreased it in LC-rats. These results show that NAc-shell DA is differentially regulated in response to stress in animals with high and low positive affect.

Keywords: 50-khz ultrasonic vocalizations (USVs); individual differences; chronic variable stress (CVS); dopamine; nucleus accumbens; microdialysis

check for updates

Citation: Kõiv, K.; Tiitsaar, K.; Laugus, K.; Harro, J. Extracellular Dopamine Levels in Nucleus Accumbens after Chronic Stress in Rats with Persistently High vs. Low 50-kHz Ultrasonic Vocalization Response. *Brain Sci.* **2021**, *11*, 470. https://doi.org/10.3390/brainsci11040470

Academic Editor: Stefan M. Brudzynski

Received: 28 February 2021
Accepted: 3 April 2021
Published: 8 April 2021

1. Introduction

Rats emit and perceive ultrasonic vocalizations (USVs) to communicate emotional states and manage social contacts with conspecifics [1–3]. The 50-kHz type of USVs occur during various appetitive situations such as mating, rough-and-tumble play, and experimenter-administered playful tactile stimulation that mimics the natural rough-and-tumble play, also called 'tickling' [4,5]. The production of 50-kHz USVs is strongly related to the activity in the ascending mesolimbic dopaminergic projections from the ventral tegmental area (VTA) to the nucleus accumbens (NAc) [1,6], and it is modulated by several neurochemical systems, e.g., noradrenergic [7], serotonergic [8], opioidergic [9], adenosinergic [10], and glutamatergic [11] receptors.

Animal models for positive emotionality are scarce but understanding the neurobiology of positive affect could make a unique contribution to drug development for affective disorders [12]. We have previously demonstrated that when juvenile rats are individually housed after weaning, daily tickling for two weeks reveals stable inter-individual differences in the level of 50-kHz USVs that persist into adulthood [13]. Male rats with low 50-kHz USV response (LC-rats) are more susceptible to chronic variable stress (CVS) than rats with persistently high 50-kHz USV response (HC-rats) as reflected by changes in behavior, oxidative metabolism in brain regions relevant for emotion processing, higher corticosterone response, and reduced sensitivity to amphetamine treatment [14–16].

There is ample evidence that stress modifies the functioning of the mesolimbic dopamine system. Acute stressors such as restraint stress, forced swimming, or intermittent foot-shock increase dopamine (DA) release in the NAc [17–20] showing that the mesolimbic DA system is activated by stress. Chronic stress regimens have been shown to reduce the basal and to blunt the stimulated overflow of DA in the shell part of NAc [21] and to blunt DA release in response to feeding of a palatable food [22,23]. Also, resilience to stress is associated with dopamine function in the NAc (e.g., [24]). How the role of accumbal DA in stress resilience relates to the positive affect is not known. Interestingly, pharmacological inhibition of corticosterone synthesis reduced 50-kHz USVs elicited either by social contact or amphetamine treatment, even though corticosterone itself did not lead to 50-kHz USV emission [25], but reversed the stress-induced decrease in 50-kHz USVs in another study [26].

Several neural pathways relate to the dopamine reward signals in the NAc, with glutamatergic excitatory innervation possibly exerting the broadest influence. The NAc receives dense glutamatergic innervation from the prefrontal cortex, amygdala, and hippocampus, among many other regions, and distinct glutamatergic inputs may serve different roles in regulating reward and aversion [23,27]. Interaction of dopamine and glutamate in the NAc regulates the shifting between pleasurable activities and responses to stressful environments [28]. Previous studies on the 50-kHz USV production have suggested that glutamate-elicited USVs are dependent on DA [29] and social-encounter-elicited vocalizations cause simultaneous alterations in VTA glutamate and dopamine [30]. It is thus of interest whether glutamate levels contribute to any eventual role of accumbal dopamine in the interaction of stress response with positive affectivity.

In the present study we compared the effect of chronic stress on basal and depolarization-dependent dopamine levels in the NAc shell in rats with higher vs. lower traits of positive affect. Further, as this area also receives glutamatergic innervations, glutamate-mediated modulation of DA levels using a blockade of glutamate transporters by retrodialysis of l-trans-pyrrolidine-2,4-dicarboxylate (PDC) was used.

2. Materials and Methods

2.1. General Procedure

The animals were bred at the local animal house, using parent Wistar rats (Harlan Laboratories, the Netherlands). Rat pups were weaned at 21 days of age (experiment timeline in Figure 1) and single-housed in standard transparent polypropylene cages with wood-chip bedding in a temperature-controlled colony room (20–22 °C) under a 12:12 light/dark cycle, with lights on at 08:00 h. Rats had free access to tap water and food pellets (diet R70, Lactamin AB, Sweden), except during testing. Tickling started the day after single-housing and only male pups were used for the experiment. The rats were group-housed (four per group) three weeks after the end of the tickling sessions, which had lasted for 14 days, and they remained so until the day of surgery. The CVS regimen started when the rats were 11 weeks old. The rats were weighed before stress, 11 days into stress, and after the end of the CVS regimen. One day after the end of the stress period, the microdialysis probe was implanted and the next day, the microdialysis experiment in awake and freely moving animals was carried out. The experimental protocol was approved by the Animal Experimentation Committee at the Estonian Ministry of Agriculture (7.2–11/10).

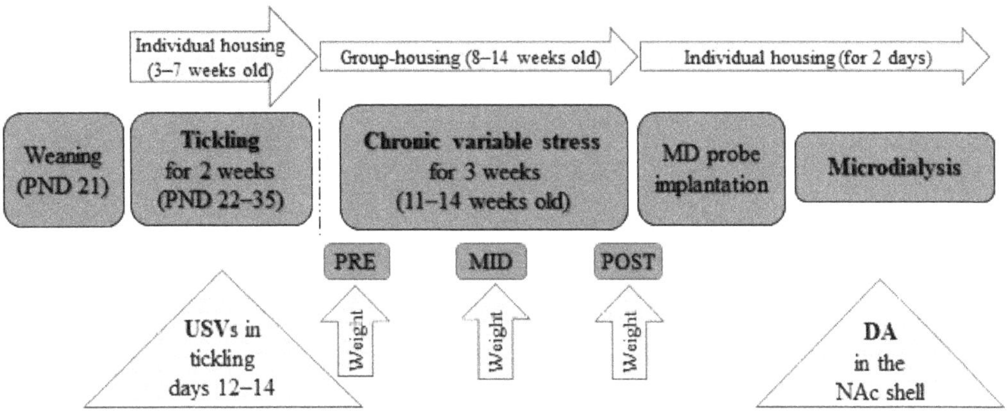

Figure 1. The general timeline of experimental procedures. DA, dopamine; MD, microdialysis; NAc, nucleus accumbens; PND, postnatal day; USVs, ultrasonic vocalizations.

2.2. Tickling

Tickling procedure was carried out on postnatal days 22–35, as previously [13,16]. Animals were individually transported to the adjacent experiment room, transferred from their home cage into a smaller (30 × 15 × 13 cm) plastic cage without bedding and given 15 s for habituation. This was followed by 15 s of tickling, which is a playful manual stimulation resembling natural rough-and-tumble play in juvenile rats, by the experimenter. Separated by 15 s pauses, altogether four 15 s sessions of tactile stimulation were given over 2 min, after which animals were returned to their home cage and the test cage was cleaned with a damp tissue. The tickling session consisted of stimulating the rat with one hand and included vigorous alternating finger movements on the back and scruff, rapidly turning the animal over on the back whilst stimulation was administered on the ventral surface, followed by release after a few seconds [1,5].

During the tickling sessions, an ultrasound microphone (Avisoft Ultra Sound Gate 116–200, Avisoft Bioacoustics, Berlin, Germany) was located about 20 cm from the floor of the tickling cage, recording USVs with a sampling rate of 300 kHz in 16-bit format on a computer hard drive. The files were later analyzed with Avisoft SASLab Pro (Avisoft Bioacoustics, Berlin, Germany) software, creating spectrograms using the Fast Fourier Transform algorithm (1024 FFT length, 75% frame, Hamming window, and 75% time window overlap). Spectral data of the sound recordings were manually cleaned from noise after which automatic scoring was applied to count the number of 50-kHz USVs with frequencies over 38 kHz during the tickling stimulation time (altogether 1 min). The rats were divided into groups with high and low levels of 50-kHz USVs by the median split of the average response on days 12–14 of tickling, providing the HC and LC groups. The HC-rats emitted on average 1.7 times more 50-kHz vocalizations compared to the LC-rats (HC, 268 ± 5; LC, 159 ± 9; $F(1,27) = 114.1$, $p < 0.0001$)

2.3. Chronic Variable Stress Regimen (CVS)

The chronic variable stress (CVS) regimen was applied to half of the animals for 3 weeks, as carried out previously [16]. In brief, the protocol comprised of six short and five long moderately unpleasant environmental and social stressors that were used intermittently, one per day, each one maximally once per week. The control rats lived undisturbed in the colony room. The CVS-regimen was applied by one group-housing cage at a time (four rats) and timed so that the 21-day stress regimen could be followed by microdialysis. The stressors were (in order of occurrence), movement restriction in a small plastic compartment (25 × 9 cm, 2 h), cage tilt at 45° (24 h), placement on a round

10-cm diameter platform elevated 75 cm from ground with strong illumination (900 lx, 30 min), wet bedding (24 h), cold environment (4 °C, 1 h), overcrowding (eight rats in a novel home-cage, 24 h), short immobilization with a thick glove (1 min), stroboscopic light during the predicted dark period (10–50 Hz; 12 h), tail-pinch with a clothespin placed 1 cm distal from the base of tail (5 min), strong illumination (900 lx) during predicted dark phase (12 h), and loud white noise (1 h). These stressors were administered in a separate room during the light phase of the cycle (except for the stressors that lasted overnight), while the control rats stayed undisturbed in the colony room.

2.4. Microdialysis Procedure

Microdialysis was carried out essentially as previously described [31]. The animals were anaesthetized with combined solution of ketamine and medetomidine (45 mg/kg and 0.3 mg/kg, IP, respectively) and mounted in a Kopf stereotactic frame while being kept on a heating-pad. A self-made concentric Y-shaped microdialysis probe with 8-mm shaft length and 1.5-mm active tip was implanted at 20° angle aimed at the left NAc shell according to the following coordinates: anterior-posterior (AP) +1.7; medial-lateral (ML) −2.0; dorsal-ventral (DV) −8.0, according to the brain atlas by Paxinos and Watson [32]. The dialysis membrane used was polyacrylonitrile/sodium methallyl sulphonate copolymer (Filtral 12; innerdiameter(ID) 0.22 mm; outerdiameter (OD) 0.31 mm; AN 69, Hospal, Bologna, Italy). Two stainless steel screws and dental cement were used to fix the probe to the scull, with 1% lidocaine being used for wound infiltration anesthesia, and the rats were placed in 21 cm × 36 cm × 18 cm individual cages in which they remained throughout the microdialysis experiment. The rats were left to recover and on the next day, the microdialysis procedure was conducted in awake, freely moving animals. The microdialysis probe was connected to a syringe pump (SP101, World Precision Instruments, Inc., Sarasota, USA) and cooling microfraction collector (4 °C, CMA/142, CMA Microdialysis AB, Solna, Sweden) and perfused with Ringer solution (140 mM NaCl, 4 mM KCl, 1.2 mM $CaCl_2$, 1 mM $MgCl_2$, 1 mM Na_2HPO_4, 0.2 mM NaH_2PO_4, pH = 7.21) at a constant rate of 1.5 µL/min. Connections to the infusion pump and sample collection loop or microfraction collector were made with flexible FEP tubing (i.d. 0.12 mm, AgnTho's AB, Lidingö, Sweden). After connecting the animal to the microdialysis system, the perfusate was discarded during the first 60 min to allow stabilization. Twenty microdialysis samples were collected with 20 min intervals.

During microdialysis, two manipulations were used to stimulate the dopamine output in the NAc shell. KCl stimulation imitates neural excitation and can be applied to study the potential to release a neurotransmitter from the neurons in response to stimulation. The concentration of KCl appropriate for depolarization was based on previous work [33,34]. In order to investigate the role of glutamate in dopamine release after stress, we used glutamate re-uptake inhibition by perfusion with 4 mM l-trans-pyrrolidine-2,4-dicarboxylate (PDC), an effective EAAT2 inhibitor [35–37].

Four baseline samples were collected, followed by 1-h perfusion with 50 mM KCl solution (samples 5–7). Next, Ringer solution was used for perfusion in samples 8–12, after which 4 mM PDC was infused for an hour (samples 13–15). For the final five samples, the system was switched back to Ringer solution. Then, 22.5 µL of the dialysate buffered in 7.5 µL of 0.02 M acetic acid was used and the samples were kept at −80 °C until measurement. At the end of the microdialysate collection, the rats were decapitated and the brains were removed and kept at −80 °C. The brains were sectioned on a cryostatic microtome (Microm GmbH, Walldorf, Germany) and probe placements were determined according to the atlas of Paxinos and Watson [32]. Only animals with correct probe placements were included in the HPLC analysis (n = 33).

2.5. Quantification of Dopamine in Microdialysates

The quantity of dopamine in the microdialysis samples was determined by high performance liquid chromatography (HPLC) with electrochemical detection. The chromatography system consisted of a Shimadzu LC-10AD series solvent delivery pump (Shimadzu Corporation, Kyoto, Japan), a Luna C18(2) 5 µm column (150 × 2 mm) (Phenomenex Inc., Torrane, CA, USA) kept at 30 °C and a Decade II digital electrochemical amperometric detector (Antec Scientific, Zoeterwoude, The Netherlands) with electrochemical flow cell VT-03 (2 mm GC WE, ISAAC reference electrode, Antec Scientific, Zoeterwoude, The Netherlands). The mobile phase consisted of 0.05 M sodium citrate buffered to pH 5.3, 2 mM KCl, 0.02 mM EDTA, 3.5 mM sodium octyl sulfonate, and 14% acetonitrile. The mobile phase was filtered through a 0.22-µm pore size filter (MilliporeSigma, Burlington, MA, USA) and was pumped through the column at a rate of 0.2 mL/min. DA eluted from the column was measured with a glassy carbon working electrode maintained at a potential of +0.4 V versus Ag/AgCl reference electrode. Dopamine content (fmol/22.5 µL) was calculated using external standard solutions (Sigma, Buchs, Switzerland).

2.6. *Data Analysis*

Microdialysis data were analyzed with repeated measures (time) ANOVA with affectivity (HC/LC) and stress (stress vs. control) as independent variables. Baseline dopamine levels in the NAc shell were calculated for every individual animal as the average (mean ± standard error of the mean, SEM) of the samples 2–4. (The first baseline sample was excluded owing to the large variability, hence making the effective stabilization period 80 min.) In order to estimate the magnitude of stimulation with KCl and PDC, the area under the curve (AUC) for samples 5–11 and 13–20 was calculated for every individual animal, respectively, using OriginPro 9 (OriginLab Corporation, Northampton, MA, USA). Body weight data, data about the baseline levels of accumbal dopamine, and AUC data were analyzed with two-factor ANOVA with affectivity (HC vs. LC) and stress (stress vs. control) as independent variables. Where appropriate, Fisher's Protected Least Significant Difference (PLSD) test was used as a post hoc test. Statistical analysis was performed using StatView 5.0. (SAS Institute Inc., Cary, NC, USA) and Statistica 8.0. (StatSoft Inc., Tulsa, OK, USA). The data from four animals were left out of the analysis as their DA levels differed from their group averages by more than three standard deviations.

3. Results

3.1. *Stress Effect on Body Weight Gain*

On average, the three-week-long stress regimen decelerated weight gain of comparable magnitude in HC- and LC-rats, by 23% in HC- and by 31% in LC-rats (main effect of stress $F(1,25) = 18.23$, $p < 0.001$, Figure 2).

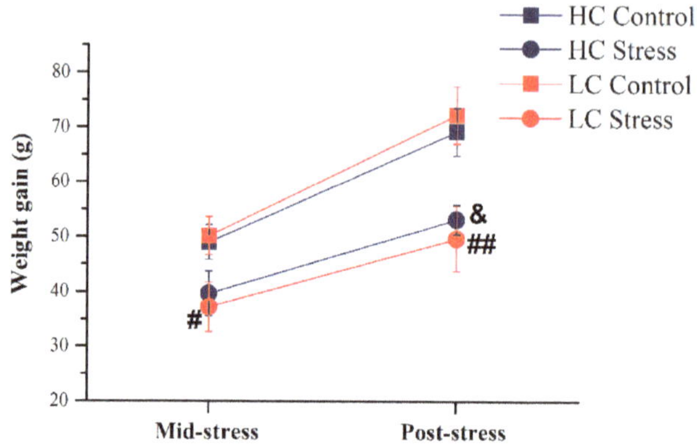

Figure 2. Cumulative weight gain in the middle (day 11 of stress) and after the three-week stress regimen (average ± SEM). #, ## $p < 0.05$, 0.01 vs. LC-control; and $p < 0.05$ vs. HC-control. HC-controls: $n = 8$; HC-stress, $n = 8$; LC-controls, $n = 7$; and LC-stress, $n = 6$. HC, high 50-kHz USV rats, LC, low 50-kHz USV rats.

3.2. *Dopamine Content in Microdialysates*

Dopamine levels in microdialysate samples differed group-wise across the experiment and the effect of stress on DA levels was dependent on the affectivity phenotype (repeated measures ANOVA main effect of time ($F(18,450) = 8.0$, $p < 0.001$), interaction of affectivity × stress ($F(1,25) = 5.8$, $p < 0.05$), and interaction of affectivity × stress × time ($F(18,450) = 1.8$, $p < 0.05$), Figure 3).

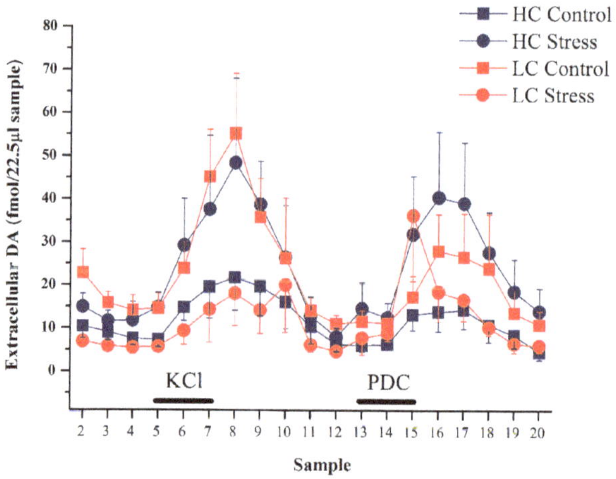

Figure 3. Extracellular dopamine levels in the NAc shell as measured during microdialysis (average + or-SEM). Baseline (samples 2–4), perfusion with 50 mM KCl solution (samples 5–7), and perfusion with 4 mM l-trans-pyrrolidine-2,4-dicarboxylate (PDC) (samples 13–15). HC-controls, $n = 8$; HC-stress, $n = 8$; LC-controls, $n = 7$; and LC-stress, $n = 6$. HC, high 50-kHz USV rats and LC, low 50-kHz USV rats.

In detail, the baseline extracellular levels of dopamine (average in samples 2–4) in the NAc shell were as follows: HC-controls, 9.0 ± 2.0 fmol/22.5 μL sample ($n = 8$) and HC-stress, 12.8 ±3.0 fmol/22.5 μL sample ($n = 8$); LC-controls, 17.6 ± 3.0 fmol/22.5 μL sample ($n = 7$) and LC-stress, 6.1 ± 0.9 fmol/22.5 μL sample ($n = 6$). LC-controls had higher baseline levels compared to HC-rats, and stress reduced baseline DA levels only in the LC-rats while rather tending to increase dopamine levels in the HC-rats (interaction of affectivity × stress $F(1,25) = 9.3$; $p < 0.01$, Figure 4A).

There was an increase in DA levels in response both to KCl and PDC (Figure 4B,C). To compare the effect of KCl on dopamine release, for every rat the area under the curve was calculated for samples 5–11. Again, affectivity and stress interaction on dopamine output was revealed, as stress heightened the DA levels for HC whereas it lowered it in LC-animals ($F(1,25) = 4.2$; $p < 0.05$, Figure 4B). However, post hoc tests failed to show significant differences between individual groups, because while the difference between groups was almost perfectly reversed by stress, the within-group variability was substantial. After PDC retrodialysis (samples 13–20), a tendency for higher output of DA in stressed HC-rats appeared, but this effect failed to reach statistical significance (NS; Figure 4C).

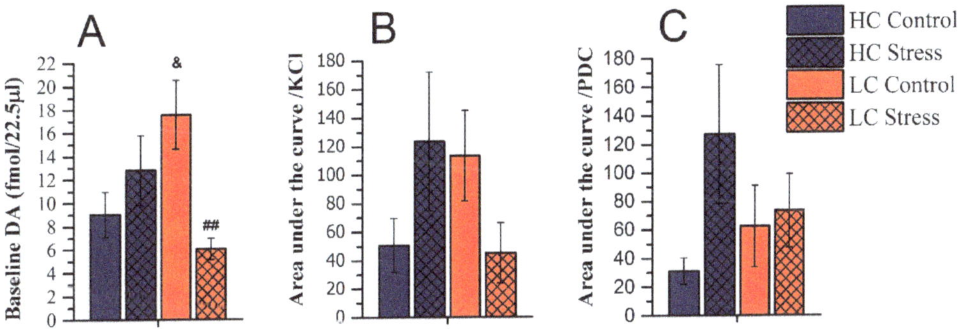

Figure 4. Extracellular dopamine levels in the NAc shell as measured during microdialysis (average ± SEM). (**A**) Average in baseline dopamine content (samples 2–4), (**B**) area under the curve (AUC) during and after perfusion with 50 mM KCl solution (samples 5–11), (**C**) area under the curve (AUC) during and after perfusion with 4 mM PDC (samples 13–20). ## $p < 0.01$ vs. LC-control; and $p < 0.05$ vs. HC-control. HC-controls, $n = 8$; HC-stress, $n = 8$; LC-controls, $n = 7$; and LC-stress: $n = 6$. HC, high 50-kHz USV rats and LC, low 50-kHz USV rats.

4. Discussion

The results show that the positive affectivity trait is associated with dopamine levels in the NAc, and that chronic stress can affect the dopamine output in the NAc shell differently depending on the positive affectivity phenotype as measured by tickling-induced 50-kHz vocalizations.

Stress prevented weight gain significantly in both HC- and LC-rats, showing the effectiveness of the chronic stress regimen used in this study. In our previous experiments stressed LC-rats gained weight significantly less as compared to the HC-rats [14,15]. In the present experiment the reduction of weight gain by stress in the LC-rats and HC-rats was not significantly different, while a tendency to lower weight gain in comparison with the corresponding control group was observed at mid-stress. In the present experiment weight gain was somewhat lower in all groups, possibly leading to reduced capacity to discriminate between LC- and HC-rats at this relatively mild stress regimen.

Overall, without stress, extracellular levels of DA were higher in LC-rats than in HC-rats. Individual differences in DA release in the NAc shell have been reported by many studies with variable design [38,39]. The mechanisms that have been found to underlie such differences include the size of the presynaptic pools of DA, the tyrosine hydroxylase activity, and the concentration of vesicular monoamine transporters, as well as the level of function of the noradrenergic neurotransmission [40]. It will need to be established in further experiments which mechanism(s) could underlie the higher extracellular DA levels in LC-rats. Microinjection of DA into the NAc shell elicited 50-kHz USVs [41]. This study, however, also reported that direct application of DA to the accumbens was less efficacious in evoking 50-kHz USVs than amphetamine if DA uptake was not additionally inhibited. Speculatively, LC-rats may have constitutively lower DA reuptake, causing functional desensitization of postsynaptic response to DA by constitutively higher extracellular levels of dopamine. At least three subtypes of DA receptors, D_1, D_2, and D_3, are involved in the emission of 50-kHz USVs [42]. Interestingly, administration of amphetamine into the NAc shell did elicit 50-kHz USVs in the Long Evans rat line with high vocalization rate in response to social stimulation, but not in the corresponding line with low response [43]. This may suggest that in LC-rats, the 50-kHz USV response can be uncoupled of DA-ergic neurotransmission in the NAc.

Chronic stress reduced extracellular DA levels in LC-rats. In our previous studies, chronic variable stress produced, in male LC-rats 22-kHz USVs, lower sucrose consumption, higher increase in immobility in the forced swimming test, and increase of oxidative metabolism in several brain areas (but not the accumbens) [13], higher levels of corticosterone in an acute stressful condition, robustly higher extracellular levels of 5-HT in the hippocampus if 5-HT reuptake was blocked [15], and larger reduction of amphetamine stimulation of 50-kHz USVs and locomotion [16]. Altogether these findings support the notion that low positive affectivity presents a vulnerability to chronic stress. Other investigators using different chronic stress paradigms have also observed reduction of extracellular DA in the NAc shell [21–23], so this mechanism appears to serve as a common ground for vulnerability to stress. However, the fact that in HC-rats the stress effect on extracellular DA strongly tended to be exactly the opposite requires an explanation. It may be relevant to consider that in the present experiment all rats had been tickled in early age. Several behavioral differences can be found in both LC- and HC-rats in comparison with not-tickled control rats [13]. In general, tickled rats express less fear and anxiety in adulthood. The HC-rats have particularly reduced anxiety as revealed in the elevated plus-maze test. Thus, a potential for developing higher extracellular DA levels in the NAc in stressful conditions must be acknowledged in the context of combination of low anxiety and high positive affectivity. In one study rats were classified on the basis of behavioral activity in the Porsolt's forced swimming test and the passive animals were found to produce more 50-kHz USVs in a social encounter [44]; in these rats chronic stress reduced in the NAc shell the levels of the cocaine- and amphetamine-regulated transcript peptide that is known to reduce DA release in this region [45]. While this experiment has substantial differences it appears to associate higher 50-kHz production with the potential of increased DA release after chronic stress.

Amongst the possible mediators of the inverse relationship of extracellular dopamine with stress in LC- vs. HC-rats, the role of the serotonergic system should be considered. Serotonin potentiates the cocaine-enhanced extracellular DA levels in the NAc shell [46] and chronic stress can interfere with the action of serotonin on DA release in the mesoaccumbens pathway [47]. In mice, serotonin controls DA release through 5-HT_{2C} receptors in the NAc core [48], and in the shell, alcohol exposure increased basal extracellular levels of both DA and serotonin, and 5-HT_{2C} receptor blockade reduced alcohol intake [49]. In our previous study, ex vivo measurements in the NAc revealed that while chronic variable stress tended to increase 5-HT turnover in HC-rats, it was the opposite in the case of LC-rats [16]. Serotonin turnover was reduced in the NAc shell by fluoxetine if rats had previously been treated with corticotropin-releasing factor, a major stress signal [50]. Thus, the high behavioral sensitivity to stress of the LC-rats may in part relate to differences in serotonergic function.

Perfusion of PDC increased DA levels in the NAc shell as has been shown before [51,52]. Projections from the VTA to the NAc shell release both DA and glutamate, and glutamatergic inputs are received by the medium spiny neurons of the NAc shell from several other brain regions [53]. Owing to large variability in the individual levels no group-wise difference was statistically significant. It seems, however, worth noting that LC-rats had twice the level of DA increase by inhibition of glutamate uptake as compared to unstressed HC-rats, and no stress effect was observed in LC-rats while the average increase in stressed HC-rats was several-fold as compared to the respective control. A hypothesis that animals with high positive affect will develop a reduction of glutamatergic input to accumbens as an adaptation to stress should be tested.

Limitations of this study include the small number of animals per group, owing to the complex study design. For this reason liberal post-hoc comparisons were applied after statistically significant interaction effects had been revealed. The period of individual housing applied in the study design may raise questions, as single housing has been shown to be stressful to young animals, especially females, depending on the duration of the isolation period [54,55]. In the present experiment, single-housing was employed as it has been shown to substantially increase the ultrasonic vocalization levels in juveniles in response to tickling [4]. During two of the single-housing weeks, a daily session of imitated rough-and-tumble play was applied and this should lessen the effects of stress. Resocialization was applied as soon as the HC/LC groups were assigned based on their 50-kHz vocalization response. A further limitation is the use of only male rats. We have previously studied both female and male rat response to tickling and shown that the separation to HC- and LC-rats applies to both [13]; however, female HC-rats are more prone to elicit 22-kHz ultrasonic vocalizations and their response to chronic stress significantly differs from males [14]. Future studies should address the USV response after stress and dopamine levels in the NAc in females.

A further consideration is the contrast between the stress responses observed in two different behavioral traits relevant to reward. Thus, we have consistently observed that sensitivity to chronic stress is higher in rats with low positive affect as measured by the expression level of 50-kHz USVs [14–16]. In the same general laboratory setting we have also consistently found rats that consume less sucrose to be less sensitive to chronic stress [56,57]. This trait-wise dissociation of stress resilience remains to be replicated in animals characterized for both phenotypes, and the underlying mechanisms identified.

Author Contributions: Conceptualization, J.H. and K.T.; methodology, K.T., K.K., and K.L.; validation, K.T., K.K., and K.L.; formal analysis, K.K.; investigation, K.T., K.K., and K.L.; resources, J.H.; original draft preparation, K.K. and J.H.; review and editing, J.H.; visualization, K.K.; supervision, J.H.; project administration, K.T.; funding acquisition, J.H. All authors have read and agreed to the published version of the manuscript.

Funding: This research was funded by the Hope for Depression Research Foundation and the Institute for the Study of Affective Neuroscience, and the Estonian Research Council (PRG1213).

Institutional Review Board Statement: The experimental protocol was approved by the Animal Experimentation Committee at the Estonian Ministry of Agriculture (7.2–11/10).

Conflicts of Interest: The authors declare no conflict of interest.

References

1. Brudzynski, S.M. Ethotransmission: Communication of emotional states through ultrasonic vocalization in rats. *Curr. Opin. Neurobiol.* **2013**, *23*, 310–317. [CrossRef]
2. Wöhr, M.; Schwarting, R.K.W. Affective communication in rodents: Ultrasonic vocalizations as a tool for research on emotion and motivation. *Cell Tissue Res.* **2013**, *354*, 81–97. [CrossRef]
3. Wöhr, M. Ultrasonic communication in rats: Appetitive 50-kHz ultrasonic vocalizations as social contact calls. *Behav. Ecol. Sociobiol.* **2018**, *72*, 1–12. [CrossRef]
4. Panksepp, J.; Burgdorf, J. 50-kHz chirping (laughter?) in response to conditioned and unconditioned tickle-induced reward in rats: Effects of social housing and genetic variables. *Behav. Brain Res.* **2000**, *115*, 25–38. [CrossRef]
5. Burgdorf, J.; Panksepp, J. Tickling induces reward in adolescent rats. *Physiol. Behav.* **2001**, *72*, 167–173. [CrossRef]
6. Burgdorf, J.; Wood, P.L.; Kroes, R.A.; Moskal, J.R.; Panksepp, J. Neurobiology of 50-kHz ultrasonic vocalizations in rats: Electrode mapping, lesion, and pharmacology studies. *Behav. Brain Res.* **2007**, *182*, 274–283. [CrossRef] [PubMed]
7. Grant, L.M.; Barth, K.J.; Muslu, C.; Kelm-Nelson, C.A.; Bakshi, V.P.; Ciucci, M.R. Noradrenergic receptor modulation influences the acoustic parameters of pro-social rat ultrasonic vocalizations. *Behav. Neurosci.* **2018**, *132*, 269–283. [CrossRef]
8. Wöhr, M.; van Gaalen, M.M.; Schwarting, R.K.W. Affective communication in rodents. *Behav. Pharmacol.* **2015**, *26*, 506–521. [CrossRef] [PubMed]

9. Wöhr, M.; Schwarting, R.K.W. Ultrasonic communication in rats: Effects of morphine and naloxone on vocal and behavioral responses to playback of 50-kHz vocalizations. *Pharmacol. Biochem. Behav.* **2009**, *94*, 285–295. [CrossRef]
10. Simola, N.; Costa, G.; Morelli, M. Activation of adenosine A2A receptors suppresses the emission of pro-social and drug-stimulated 50-kHz ultrasonic vocalizations in rats: Possible relevance to reward and motivation. *Psychopharmacology* **2016**, *233*, 507–519. [CrossRef]
11. Fu, X.W.; Brudzynski, S.M. High-frequency ultrasonic vocalization induced by intracerebral glutamate in rats. *Pharmacol. Biochem. Behav.* **1994**, *49*, 835–841. [CrossRef]
12. Harro, J. Animal models of depression vulnerability. *Curr. Top. Behav. Neurosci.* **2013**, *14*, 29–54. [CrossRef]
13. Mällo, T.; Matrov, D.; Herm, L.; Kõiv, K.; Eller, M.; Rinken, A.; Harro, J. Tickling-induced 50-kHz ultrasonic vocalization is individually stable and predicts behaviour in tests of anxiety and depression in rats. *Behav. Brain Res.* **2007**, *184*, 57–71. [CrossRef] [PubMed]
14. Mällo, T.; Matrov, D.; Kõiv, K.; Harro, J. Effect of chronic stress on behavior and cerebral oxidative metabolism in rats with high or low positive affect. *Neuroscience* **2009**, *164*, 963–974. [CrossRef]
15. Raudkivi, K.; Mällo, T.; Harro, J. Effect of chronic variable stress on corticosterone levels and hippocampal extracellular 5-HT in rats with persistent differences in positive affectivity. *Acta Neuropsychiatr.* **2012**, *24*, 208–214. [CrossRef]
16. Kõiv, K.; Metelitsa, M.; Vares, M.; Tiitsaar, K.; Raudkivi, K.; Jaako, K.; Vulla, K.; Shimmo, R.; Harro, J. Chronic variable stress prevents amphetamine-elicited 50-kHz calls in rats with low positive affectivity. *Eur. Neuropsychopharmacol.* **2016**, *26*. [CrossRef]
17. Puglisi-Allegra, S.; Imperato, A.; Angelucci, L.; Cabib, S. Acute stress induces time-dependent responses in dopamine mesolimbic system. *Brain Res.* **1991**, *554*, 217–222. [CrossRef]
18. Yadid, G.; Overstreet, D.H.; Zangen, A. Limbic dopaminergic adaptation to a stressful stimulus in a rat model of depression. *Brain Res.* **2001**, *896*, 43–47. [CrossRef]
19. Imperato, A.; Puglisi-Allegra, S.; Casolini, P.; Angelucci, L. Changes in brain dopamine and acetylcholine release during and following stress are independent of the pituitary-adrenocortical axis. *Brain Res.* **1991**, *538*, 111–117. [CrossRef]
20. Abercrombie, E.D.; Keefe, K.A.; DiFrischia, D.S.; Zigmond, M.J. Differential effect of stress on in vivo dopamine release in striatum, nucleus accumbens, and medial frontal cortex. *J. Neurochem.* **1989**, *52*, 1655–1658. [CrossRef] [PubMed]
21. Gambarana, C.; Masi, F.; Tagliamonte, A.; Scheggi, S.; Ghiglieri, O.; De Montis, M.G. A chronic stress that impairs reactivity in rats also decreases dopaminergic transmission in the nucleus accumbens: A microdialysis study. *J. Neurochem.* **1999**, *72*, 2039–2046. [CrossRef]
22. Di Chiara, G.; Tanda, G. Blunting of reactivity of dopamine transmission to palatable food: A biochemical marker of anhedonia in the cms model? *Psychopharmacology* **1997**, *134*, 351–353. [CrossRef]
23. Di Chiara, G.; Loddo, P.; Tanda, G. Reciprocal changes in prefrontal and limbic dopamine responsiveness to aversive and rewarding stimuli after chronic mild stress: Implications for the psychobiology of depression. *Biol. Psychiatry* **1999**, *46*, 1624–1633. [CrossRef]
24. Żurawek, D.; Faron-Górecka, A.; Kuśmider, M.; Kolasa, M.; Gruca, P.; Papp, M.; Dziedzicka-Wasylewska, M. Mesolimbic dopamine D2 receptor plasticity contributes to stress resilience in rats subjected to chronic mild stress. *Psychopharmacology* **2013**, *227*, 583–593. [CrossRef]
25. Simola, N.; Paci, E.; Serra, M.; Costa, G.; Morelli, M. Modulation of Rat 50-kHz Ultrasonic Vocalizations by Glucocorticoid Signaling: Possible Relevance to Reward and Motivation. *Int. J. Neuropsychopharmacol.* **2018**, *21*, 73–83. [CrossRef]
26. Popik, P.; Kos, T.; Pluta, H.; Nikiforuk, A.; Rojek, K.; Ryguła, R. Inhibition of the glucocorticoid synthesis reverses stress-induced decrease in rat's 50-kHz ultrasonic vocalizations. *Behav. Brain Res.* **2014**, *260*, 53–57. [CrossRef]
27. Russo, S.J.; Nestler, E.J. The brain reward circuitry in mood disorders. *Nat. Rev. Neurosci.* **2013**, *14*, 609–625. [CrossRef]
28. Richard, J.M.; Berridge, K.C. Nucleus accumbens dopamine/glutamate interaction switches modes to generate desire versus dread: D1 alone for appetitive eating but D1 and D2 together for fear. *J. Neurosci.* **2011**, *31*, 12866–12879. [CrossRef]
29. Wintink, A.J.; Brudzynski, S.M. The related roles of dopamine and glutamate in the initiation of 50-kHz ultrasonic calls in adult rats. *Pharmacol. Biochem. Behav.* **2001**, *70*, 317–323. [CrossRef]
30. Hamed, A.; Kursa, M.B. Social deprivation substantially changes multi-structural neurotransmitter signature of social interaction: Glutamate concentration in amygdala and VTA as a key factor in social encounter-induced 50-kHz ultrasonic vocalization. *Eur. Neuropsychopharmacol.* **2020**, *37*, 82–99. [CrossRef]
31. O'Leary, A.; Kõiv, K.; Raudkivi, K.; Harro, J. Antidepressants differentially affect striatal amphetamine-stimulated dopamine and serotonin release in rats with high and low novelty-oriented behaviour. *Pharmacol. Res.* **2016**, *113*, 739–746. [CrossRef] [PubMed]
32. Paxinos, G.; Watson, C. *The Rat Brain in Stereotaxic Coordinates*, 6th ed.; Academic Press: London, UK, 2007.
33. Ripley, T.L.; Jaworski, J.; Randall, P.K.; Gonzales, R.A. Repeated perfusion with elevated potassium in in vivo microdialysis—A method for detecting small changes in extracellular dopamine. *J. Neurosci. Methods* **1997**, *78*, 7–14. [CrossRef]
34. Häidkind, R.; Kivastik, T.; Eller, M.; Kolts, I.; Oreland, L.; Harro, J. Denervation of the locus coeruleus projections by treatment with the selective neurotoxin DSP-4 [N (2-chloroethyl)-N-ethyl-2-bromobenzylamine] reduces dopamine release potential in the nucleus accumbens shell in conscious rats. *Neurosci. Lett.* **2002**, *332*, 79–82. [CrossRef]
35. Del Arco, A.; Segovia, G.; Prieto, L.; Mora, F. Endogenous glutamate-taurine interaction in striatum and nucleus accumbens of the freely moving rat: Studies during the normal process of aging. *Mech. Ageing Dev.* **2001**, *122*, 401–414. [CrossRef]

36. Behrens, P.F.; Franz, P.; Woodman, B.; Lindenberg, K.S.; Landwehrmeyer, G.B. Impaired glutamate transport and glutamate–glutamine cycling: Downstream effects of the Huntington mutation. *Brain* **2002**, *125*, 1908–1922. [CrossRef] [PubMed]
37. Bridges, R.J.; Stanley, M.S.; Chamberlin, A.R.; Anderson, M.W.; Cotman, C.W. Conformationally defined neurotransmitter analogues. Selective inhibition of glutamate uptake by one pyrrolidine-2,4-dicarboxylate diastereomer. *J. Med. Chem.* **1991**, *34*, 717–725. [CrossRef] [PubMed]
38. Giorgi, O.; Piras, G.; Corda, M.G. The psychogenetically selected Roman high- and low-avoidance rat lines: A model to study the individual vulnerability to drug addiction. *Neurosci. Biobehav. Rev.* **2007**, *31*, 148–163. [CrossRef]
39. Shimamoto, A.; Holly, E.N.; Boyson, C.O.; Debold, J.F.; Miczek, K.A. Individual differences in anhedonic and accumbal dopamine responses to chronic social stress and their link to cocaine self-administration in female rats. *Psychopharmacology* **2015**, *232*, 825–834. [CrossRef] [PubMed]
40. Verheij, M.M.M.; Cools, A.R. Twenty years of dopamine research: Individual differences in the response of accumbal dopamine to environmental and pharmacological challenges. *Eur. J. Pharmacol.* **2008**, *585*, 228–244. [CrossRef]
41. Mulvihill, K.G.; Brudzynski, S.M. Effect of microinjections of dopamine into the nucleus accumbens shell on emission of 50 kHz USV: Comparison with effects of D-amphetamine. *Pharmacol. Biochem. Behav.* **2019**, *176*, 23–32. [CrossRef]
42. Simola, N.; Brudzynski, S.M. Rat 50-kHz ultrasonic vocalizations as a tool in studying neurochemical mechanisms that regulate positive emotional states. *J. Neurosci. Methods* **2018**, *310*, 33–44. [CrossRef] [PubMed]
43. Brudzynski, S.M.; Silkstone, M.; Komadoski, M.; Scullion, K.; Duffus, S.; Burgdorf, J.; Kroes, R.A.; Moskal, J.R.; Panksepp, J. Effects of intraaccumbens amphetamine on production of 50kHz vocalizations in three lines of selectively bred Long-Evans rats. *Behav. Brain Res.* **2011**, *217*, 32–40. [CrossRef]
44. Wisłowska-Stanek, A.; Płaźnik, A.; Kołosowska, K.; Skórzewska, A.; Turzyńska, D.; Liguz-Lęcznar, M.; Krząścik, P.; Gryz, M.; Szyndler, J.; Sobolewska, A.; et al. Differences in the dopaminergic reward system in rats that passively and actively behave in the Porsolt test. *Behav. Brain Res.* **2019**, *359*, 181–189. [CrossRef]
45. Rakovska, A.; Baranyi, M.; Windisch, K.; Petkova-Kirova, P.; Gagov, H.; Kalfin, R. Neurochemical evidence that cocaine- and amphetamine-regulated transcript (CART) 55–102 peptide modulates the dopaminergic reward system by decreasing the dopamine release in the mouse nucleus accumbens. *Brain Res. Bull.* **2017**, *134*, 246–252. [CrossRef] [PubMed]
46. Bubar, M.J.; McMahon, L.R.; De Deurwaerdère, P.; Spampinato, U.; Cunningham, K.A. Selective serotonin reuptake inhibitors enhance cocaine-induced locomotor activity and dopamine release in the nucleus accumbens. *Neuropharmacology* **2003**, *44*, 342–353. [CrossRef]
47. Amato, J.L.; Bankson, M.G.; Yamamoto, B.K. Prior exposure to chronic stress and MDMA potentiates mesoaccumbens dopamine release mediated by the 5-HT1B receptor. *Neuropsychopharmacology* **2007**, *32*, 946–954. [CrossRef] [PubMed]
48. Cathala, A.; Devroye, C.; Maitre, M.; Piazza, P.V.; Abrous, D.N.; Revest, J.M.; Spampinato, U. Serotonin2C receptors modulate dopamine transmission in the nucleus accumbens independently of dopamine release: Behavioral, neurochemical and molecular studies with cocaine. *Addict. Biol.* **2015**, *20*, 445–457. [CrossRef]
49. Yoshimoto, K.; Watanabe, Y.; Tanaka, M.; Kimura, M. Serotonin 2C receptors in the nucleus accumbens are involved in enhanced alcohol-drinking behavior. *Eur. J. Neurosci.* **2012**, *35*, 1368–1380. [CrossRef] [PubMed]
50. Lowry, C.A.; Hale, M.W.; Plant, A.; Windle, R.J.; Shanks, N.; Wood, S.A.; Ingram, C.D.; Renner, K.J.; Lightman, S.L.; Summers, C.H. Fluoxetine inhibits corticotropin-releasing factor (CRF)-induced behavioural responses in rats. *Stress* **2009**, *12*, 225–239. [CrossRef]
51. Segovia, G.; Del Arco, A.; Mora, F. Effects of aging on the interaction between glutamate, dopamine, and GABA in striatum and nucleus accumbens of the awake rat. *J. Neurochem.* **1999**, *73*, 2063–2072. [CrossRef]
52. Segovia, G.; Mora, F. Involvement of NMDA and AMPA/kainate receptors in the effects of endogenous glutamate on extracellular concentrations of dopamine and GABA in the nucleus accumbens of the awake rat. *Brain Res. Bull.* **2001**, *54*, 153–157. [CrossRef]
53. Yu, J.; Ishikawa, M.; Wang, J.; Schlüter, O.M.; Sesack, S.R.; Dong, Y. Ventral tegmental area projection regulates glutamatergic transmission in nucleus accumbens. *Sci. Rep.* **2019**, *9*. [CrossRef]
54. Krupina, N.A.; Shirenova, S.D.; Khlebnikova, N.N. Prolonged Social Isolation, Started Early in Life, Impairs Cognitive Abilities in Rats Depending on Sex. *Brain Sci.* **2020**, *10*, 799. [CrossRef] [PubMed]
55. Noschang, C.; Lampert, C.; Krolow, R.; de Almeida, R.M.M. Social isolation at adolescence: A systematic review on behaviour related to cocaine, amphetamine and nicotine use in rats and mice. *Psychopharmacology* **2021**, *238*, 927–947. [CrossRef] [PubMed]
56. Kõiv, K.; Vares, M.; Kroon, C.; Metelitsa, M.; Tiitsaar, K.; Laugus, K.; Jaako, K.; Harro, J. Effect of chronic variable stress on sensitization to amphetamine in high and low sucrose-consuming rats. *J. Psychopharmacol.* **2019**, *33*. [CrossRef]
57. Kanarik, M.; Alttoa, A.; Matrov, D.; Kõiv, K.; Sharp, T.; Panksepp, J.; Harro, J. Brain responses to chronic social defeat stress: Effects on regional oxidative metabolism as a function of a hedonic trait, and gene expression in susceptible and resilient rats. *Eur. Neuropsychopharmacol.* **2011**, *21*, 92–107. [CrossRef]

brain sciences

MDPI

Article

Association between Novel Object Recognition/Spontaneous Alternation Behavior and Emission of Ultrasonic Vocalizations in Rats: Possible Relevance to the Study of Memory

Giulia Costa, Marcello Serra and Nicola Simola *

Department of Biomedical Sciences, University of Cagliari, 09042 Monserrato, Italy; gcosta@unica.it (G.C.); marcelloserra@unica.it (M.S.)

* Correspondence: nicola.simola@unica.it; Tel.: +39-070-6758687

Abstract: Rats emit ultrasonic vocalizations (USVs) in situations with emotional valence, and USVs have also been proposed as a marker for memories conditioned to those situations. This study investigated whether USV emissions can predict and/or be associated with the behavior of rats in tests that evaluate unconditioned memory. To this end, rats were subjected to "tickling", a procedure of heterospecific play that has emotional valence and elicits the emission of USVs, and afterwards evaluated in the novel object recognition test (NOR) and in the single trial continuous spontaneous alternation behavior (SAB) test in a Y-maze. The number of 22-kHz USVs (aversive) and 50-kHz USVs (appetitive) emitted in response to tickling and during NOR and SAB tests were scored, and the correlations among them and with rats' behavior evaluated. Rats emitted 50-kHz USVs, but not 22-kHz USVs, during the NOR and SAB tests, and such calling behavior was not linked with the behavioral readouts indicative of memory function in either test. However, rats that prevalently emitted 22-kHz USVs in response to tickling displayed an impaired NOR performance. These findings suggest that measuring the emission of USVs could be of interest in studies of unconditioned memory, at least with regard to 22-kHz USVs.

Keywords: 22-kHz calls; 50-kHz calls; affect; aversion; NOR; reward; working memory; Y maze

Citation: Costa, G.; Serra, M.; Simola, N. Association between Novel Object Recognition/Spontaneous Alternation Behavior and Emission of Ultrasonic Vocalizations in Rats: Possible Relevance to the Study of Memory. *Brain Sci.* **2021**, *11*, 1053. https://doi.org/10.3390/brainsci11081053

Academic Editors: Stefan M. Brudzynski, Jeffrey Burgdorf and Anne-Marie Mouly

Received: 7 July 2021
Accepted: 6 August 2021
Published: 9 August 2021

1. Introduction

Rats emit ultrasonic vocalizations (USVs) in response to and/or anticipation of various stimuli that possess emotional valence [1–3]. Two families of USVs have been characterized in rats, which possess different acoustic features and carry different information about the emotional valence of stimuli and situations [4]. Situations that possess negative emotional valence for rats (i.e., encounters with or detection of aggressive conspecifics, predators or unfamiliar humans) elicit the emission of the so-called 22-kHz USVs, which have a long duration (longer than 300 ms, although 22-kHz of short duration have also been described), sound frequency between 20–35 kHz, and scarce frequency modulation [4]. Conversely, situations that possess positive emotional valence for rats (i.e., non-aggressive interactions with conspecifics or familiar humans and administration of drugs with rewarding properties) elicit the emission of the so-called 50-kHz USVs, which have a short duration (generally 30–40 ms), sound frequency between 35–80 kHz (which can be even higher), and may possess marked frequency modulation [4]. On these bases, measuring the emission of USVs is increasingly emerging as a useful marker in rat studies that evaluate emotional behavior and the factors that may modify the emotional state, both in physiological conditions and experimental models of brain disease [5,6].

Of particular interest in USV research is the evidence that rats may call in anticipation of stimuli that possess emotional valence. Thus, rats may emit either anticipatory 22-kHz USVs when re-exposed to an environment where they received aversive electric footshocks [7,8], or anticipatory 50-kHz USVs in response to environmental cues previously associated with pharmacological or non-pharmacological rewards [9–17]. Taken together, these findings have suggested that calling behavior may be a marker of conditioned memories for stimuli/situations that possess emotional valence. Notably, a very recent study has confirmed and extended this hypothesis, by demonstrating that rats tested in a fear conditioning paradigm with electric shocks of low intensity displayed minimal freezing behavior, but showed a decreased emission of frequency-modulated (FM) 50-kHz USVs that persisted even after the extinction of freezing behavior [18]. Elucidating the relationship between the emission of USVs and memory in rats is of interest, since experimental evidence exists to suggest the presence of an interplay between modifications in the emotional state and changes in memory function [19,20]. Moreover, measuring USVs could potentially complement the classical behavioral markers that are evaluated in rat studies of memory (i.e., preference for items, patterns of arm exploration in mazes), thus increasing the quantity of information that can be collected in those studies.

The present study was performed to elucidate whether the emission of USVs can be predictive of and/or associated with the behavior of rats in tests that are used to evaluate memory and do not employ conditioned stimuli and primary reinforcers, but rely on the animals' preference for novelty [21,22]. To this end, rats were first subjected to "tickling", a procedure that consists of the interaction with the hands of a familiar human that resembles "rough and tumble" play of juvenile rats, and which may be associated with changes in the emotional state and emission of USVs [23]. Afterwards, rats were evaluated in the novel object recognition test (NOR) as well as in the single trial continuous spontaneous alternation behavior (SAB) test in a Y-maze. The number of 22-kHz USVs (aversive) and 50-kHz USVs (appetitive) emitted in response to tickling as well as during NOR and SAB tests were scored, and the existence of a relationship between the magnitude/type of calling behavior and the performance of rats in each of the memory tests carried out (i.e., preference for novel objects, SAB) was investigated.

2. Materials and Methods

2.1. Subjects

A total of 40 male Sprague–Dawley rats (Envigo, Italy) were involved in the study. Rats weighed 100–125 g (aged 28–35 days) at the beginning of the experiments and were housed 5 animals per cage in standard polycarbonate cages (L 48 cm × H 21 cm × W 38 cm) under a 12-h light/dark cycle (lights on at 08:00 h). Rats had free access to standard laboratory chow and water, except during the experiments that were performed between 10:00 and 16:00 h. The study was carried out in accordance with the guidelines for animal experimentation of the EU directives (2010/63/EU, L.276; 22/09/2010), and with the guidelines issued by the Committee for Animal Welfare (OPBA) of the University of Cagliari. All the appropriate procedures were followed to minimize animal discomfort and number of animals used.

2.2. Experimental Plan

All experiments were performed in a quiet room under an illumination of 40 lx and the experimental plan is described in Figure 1. Rats were gently handled daily (5 min) for 7 consecutive days; thereafter, they were evaluated a single time for their basal emission of USVs (day 8). Afterwards, starting immediately after the evaluation of basal calling behavior, all rats were subjected to tickling for 5 consecutive days. Finally, starting 3 days after the completion of tickling, all rats were evaluated in the NOR test and in the single trial continuous SAB test in a Y-maze. Evaluations were performed in a counterbalanced manner by dividing the rats in two groups of 20 subjects: one group was evaluated first in the NOR test and then in the SAB test, while the opposite was done in the other group. Evaluations were separated by a 7-day interval and the emission of USVs was recorded throughout the experiments.

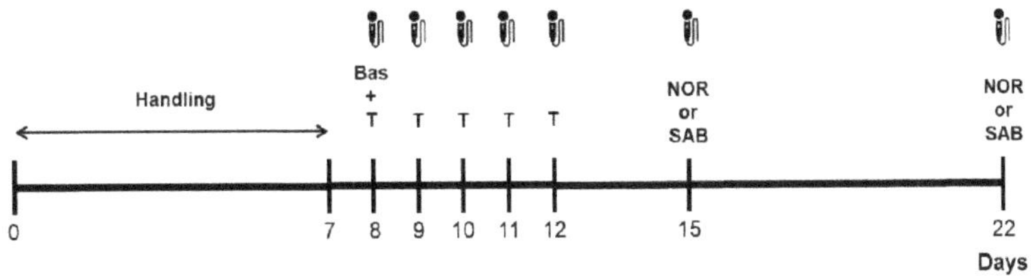

Figure 1. Experimental plan. A total of 40 rats were gently handled daily for 7 consecutive days. The day after, rats were evaluated a single time for their basal calling behavior and, starting the same day, subjected to tickling for 5 consecutive days. On days 15 and 22, rats were arranged in two groups of 20 subjects and tested once either in the novel object recognition test or in the single trial continuous spontaneous alternation behavior test in a Y maze. Testing was done in a counterbalanced manner and the emission of ultrasonic vocalizations was recorded throughout the experiments. Bas = recording of basal calling behavior before the beginning of tickling; NOR = novel object recognition test; SAB = spontaneous alternation behavior; T = tickling.

2.3. Evaluation of Basal Calling Behavior

Rats were individually placed in medium-sized polycarbonate cages (L 42 cm × H 14 cm × W 14 cm) without bedding and left to explore the environment freely for 5 min, during which the emission of USVs was recorded.

2.4. Tickling

Beginning immediately after the evaluation of basal calling behavior, tickling was performed in medium-sized polycarbonate cages (L 42 cm × H 14 cm × W 14 cm) without bedding, according to the procedure previously described [23,24]. Briefly, tickling was done with one hand and consisted of scaled-down and rapid movements of fingers and hand (i.e., "belly tickle", "flip over", "hand chase", "neck tickle", and "push and drill"). Each rat was subjected to 1 cycle of tickling × day × 5 consecutive days, and each cycle of tickling consisted of blocks of 15 s of baseline with no tactile stimulation, followed by blocks of 15 s of tactile stimulation, repeated for a total of 2 min [23]. The experimenter's hand was kept still inside the cage during the baseline blocks.

2.5. Novel Object Recognition Test

The NOR test is a behavioral paradigm that allows to evaluate non-spatial working memory in rodents in the absence of emotional and learning components [21]. NOR tests were performed in polycarbonate cages (L 42 cm × W 14 cm × H 30 cm) that were enclosed by cardboard walls (H 50 cm) and had their bottom covered with an elevated grid (3 cm) having a handful of sawdust below it [25]. The objects to be discriminated in the NOR test were plastic-made, had different shape and color, and were devoid of genuine significance and emotional valence for rats. The experimental procedure consisted of three sessions: habituation (S0), acquisition (S1), and testing (S2), and rats were individually evaluated in each of these sessions. For habituation, each rat was placed in the test cage in the absence of objects and left to explore the environment freely for a single trial (5 min). Twenty-four hours after S0, acquisition (S1) was performed by placing each rat in the test cage together with two copies of an object (familiar objects); the rat was allowed to freely explore the objects for 3 min. The testing session (S2) was performed 60 min after S1 and consisted of the exposure of each rat for 3 min to a pair of objects made of one copy of the objects already encountered in S1 and of an object that the rat had never encountered before (novel object). Objects were always placed in the vicinity of the two adjacent corners along the long side of the cage, leaving a distance between the objects and between the objects and the walls of the cage that allowed the rats to turn around the objects. The exploration times and the inter-session times were selected based on previous studies of our group, demonstrating that such times are adequate to reveal the presence of memory impairment in rodents that are evaluated in the NOR test [25–27]. Object exploration was scored when a rat sniffed, bit, or touched a specific object, whereas object exploration was not scored when rats circled around the objects and/or sat on them. The objects were cleaned at the end of each session to take away olfactory traces and counterbalanced for location (right or left side of the cage) and status (old or novel). The behavior of rats was videotaped during S1 and S2, and later evaluated to determine the following parameters: (a) seconds spent in object exploration during S1 and S2; (b) percentage of time spent exploring the novel and old objects during S2.

2.6. Single-Trial Continuous Spontaneous Alternation Behavior Test in a Y Maze

Evaluation of continuous SAB in a Y maze is an experimental paradigm that allows to assess the sensory/attentional functions and spatial working memory in rodents, and does not rely on conditioned stimuli and primary reinforcers [22]. The Y maze used was made of black PVC and had three equally sized symmetrical arms (L 50 cm × W 20 cm × H 35 cm) that converged onto a central triangular area; moreover, the maze had its bottom covered with sawdust. For testing, each rat was individually placed in the central triangular area and left to explore freely the entire maze for a single 8 min trial. In order to remove olfactory cues, the sawdust was changed and the maze was cleaned in between each rat. The rats' performance was videotaped to score for: (i) number of arm entries; (ii) sequence of arm entries, to calculate spontaneous alternation. A rat was considered inside an arm when it had all its four paws inside a specific arm, and spontaneous alternation was defined as successive entries into all the three arms of the maze in overlapping triplet sets, and expressed as the percentage of actual to possible alternations (defined as the total number of entries in arms − 2) × 100 [26].

2.7. Recording of Ultrasonic Vocalizations

The emission of USVs was recorded in each step of the experimental protocol by means of ultrasonic microphones (CM16/CMPA, Avisoft, Berlin, Germany) that were connected to an ultrasound recording device (Ultrasound Gate 116 Hb, Avisoft, Berlin, Germany). Intensity gain was kept at a constant level throughout recordings. For recording in cages (i.e., basal emission of USVs, tickling, NOR test), a single microphone was hung on a support and centered above the cage at a distance of 40 cm from the bottom. For recording in the Y maze, three microphones were used; each microphone was hung sideways from the center of the right wall of each arm of the maze at a distance of 30 cm from the bottom. The duration of USV recordings matched that of behavioral evaluation in each step of the experimental protocol.

2.8. Data Collection and Statistical Analyses

USV recordings were converted into spectrograms by means of the software SASLab Pro 4.52 (Avisoft, Berlin, Germany), which was also used to count the number of 22-kHz and 50-kHz USVs isolated in each spectrogram and their acoustic parameters (duration, maximum frequency, minimum frequency) [28]. In addition, 50-kHz calls were categorized into flat and FM according to the criteria proposed by Wright and coworkers [29]. Examples of USVs recorded in the present study are provided in Figure 2.

Means ± S.E.M. were calculated for the following parameters: (i) number and acoustic parameters of 22-kHz and 50-kHz USVs (total and categorized, when appropriate) emitted at each step of the experimental protocol; (ii) percentages of time (seconds) spent exploring the objects in S1 and S2 of the NOR test; (iii) percentages of SAB and number of entries in the arms of the Y maze. USV data were analyzed by an experimenter blind to the conditions of recording. All data obtained in the present study were tested for normality with the Kolmogorov–Smirnov test and analyzed accordingly with one of the following tests: (i) one-way or two-way analysis of variance (ANOVA) followed by Bonferroni's multiple comparisons test, when appropriate; (ii) Kruskal–Wallis test followed by Dunn's multiple comparisons test, when appropriate; (iii) Student's t-test; (iv) Mann–Whitney

U test or Wilcoxon test. Moreover, and when appropriate, Spearman's test was used to correlate the behavioral readouts obtained in the NOR and SAB tests (i.e., preference for novel objects, SAB, number of arm entries) with the number of 22-kHz and 50-kHz USVs emitted either in response to tickling or during memory testing. Bonferroni correction for multiple comparisons was applied to correlation analysis, when appropriate. Finally, in order to further clarify whether the emission of USVs in response to tickling could predict the behavior of rats in the NOR and SAB tests, an additional analysis was performed by dividing the rats in three groups according to the prevalent type of calls emitted in response to tickling, calculated as the total number of calls emitted over the five sessions of tickling. Subdivision of rats was performed as follows: (i) rats that emitted only 50-kHz calls, (ii) rats that emitted more 50-kHz calls than 22-kHz calls, (iii) rats that emitted more 22-kHz calls than 50-kHz calls. Statistical analysis was performed with Prism 8 (GraphPad, San Diego, CA, USA) for Windows. Significance was set at $p < 0.05$ for each analysis. Two rats were excluded from the analysis of USVs emitted during Y maze exploration due to a loss of spectrograms. The acoustic parameters of the USVs recorded in the present study were in the range of those previously reported for 22-kHz and 50-kHz calls [4] (data not shown).

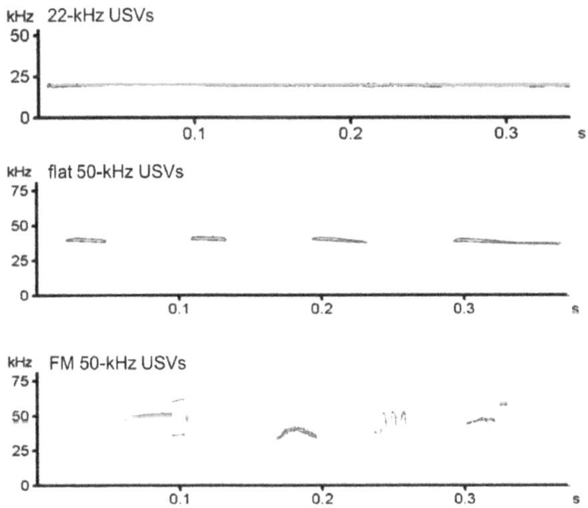

Figure 2. Example of spectrograms of 22-kHz and 50-kHz ultrasonic vocalizations recorded in the present study. The vocalizations reported in the figure are independent calls emitted by different rats. FM = frequency-modulated; USVs = ultrasonic vocalizations.

3. Results

3.1. Basal Levels of Calling Behavior

Rats emitted a very low number of 50-kHz USVs when exposed to a novel cage in the absence of bedding (average number of calls = 0.23 ± 0.05 in 5 min of recording), and no emission of 22-kHz USVs was observed in the same situation.

3.2. Emission of Ultrasonic Vocalizations in Response to Tickling

Tickling significantly stimulated the emission of USVs in rats. The Kruskal–Wallis test showed the presence of significant modifications in the cumulative number of calls emitted over the 5 days of tickling (K = 53.22, $p < 0.01$), and Dunn's test for multiple comparisons revealed that a significant increase in the emission of 50-kHz USVs ($p < 0.01$), but not 22-kHz USVs, occurred during tickling sessions, compared with baseline sessions when the hand of the experimenter was passively left inside the cage (Figure 3A). Moreover, the Kruskal–Wallis test showed the presence of significant changes in the number of calls emitted in each day of tickling (day 8, K = 46.47, $p < 0.01$; day 9, K = 27.63, $p < 0.01$; day 10, K = 46.90, $p < 0.01$; day 11, K = 82.90, $p < 0.01$; day 12, K = 57.86, $p < 0.01$), and

Dunn's test for multiple comparisons revealed that a significant increase in the emission of 50-kHz USVs (days 8–12, $p < 0.01$), but not 22-kHz USVs, occurred during tickling sessions, compared with baseline sessions (Figure 3B).

The Kruskal–Wallis test showed the presence of significant modifications in the cumulative number of categorized 50-kHz USVs emitted over the 5 days of tickling (K = 71.43, $p < 0.01$), and Dunn's test for multiple comparisons revealed that a significant increase in the emission of FM ($p < 0.01$) and flat ($p < 0.01$) calls occurred during tickling sessions, compared with baseline sessions when the hand of the experimenter was passively left inside the cage (Figure 3C). Moreover, the Kruskal–Wallis test showed the presence of significant changes in the number of categorized 50-kHz USVs emitted on each day of tickling (day 8, K = 61.24, $p < 0.01$; day 9, K = 45.80, $p < 0.01$; day 10, K = 57.45, $p < 0.01$; day 11, K = 43.80, $p < 0.01$; day 12, K = 40.24, $p < 0.01$). Dunn's test for multiple comparisons revealed that a significant increase occurred during tickling sessions for the emission of FM calls in days 8–12 of tickling ($p < 0.01$ for all days), and for the emission of flat calls in days 8–11 of tickling (days 8, 10, and 11, when the hand of the experimenter was passively left inside the cage, $p < 0.01$; day 9, $p < 0.05$), all compared with baseline sessions (Figure 3D).

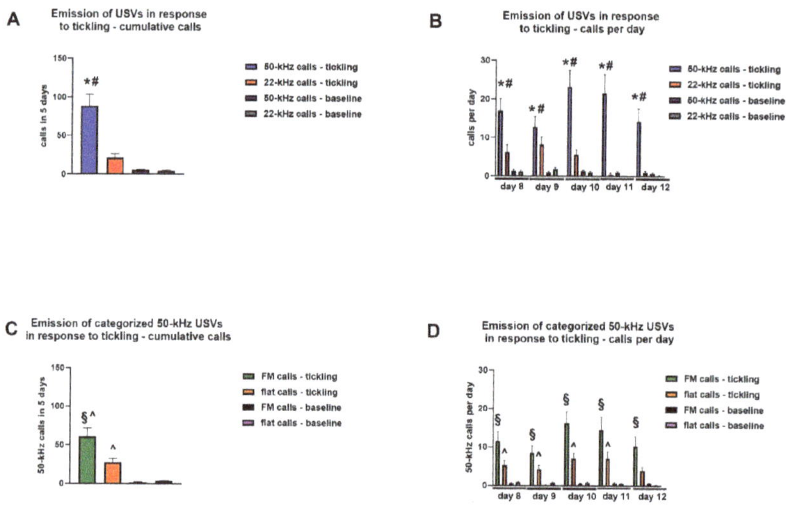

Figure 3. Emission of ultrasonic vocalizations in response to tickling. Rats underwent 1 cycle of tickling × day × 5 consecutive days, and each cycle of tickling consisted in blocks where no tactile stimulation was performed (baseline) alternated with blocks where tactile stimulation was performed (tickling). The emission of ultrasonic vocalizations was recorded throughout each cycle of tickling performed. Panel (**A**) demonstrates the cumulative numbers of 22-kHz and 50-kHz calls emitted over the 5 sessions of tickling. Panel (**B**) demonstrates the numbers of 22-kHz and 50-kHz calls emitted in each session of tickling. Panel (**C**) demonstrates the cumulative numbers of categorized 50-kHz calls emitted over the 5 sessions of tickling. Panel (**D**) demonstrates the numbers of categorized 50-kHz calls emitted in each session of tickling. * indicates a significant difference vs. 50-kHz calls—baseline. # indicates a significant difference vs. 22-kHz calls—tickling. § indicates a significant difference vs. FM calls—baseline. ^ indicates a significant difference vs. flat calls—baseline. FM = frequency modulated; USVs = ultrasonic vocalizations. $n = 40$.

3.3. Novel Object Recognition Test and Emission of Ultrasonic Vocalizations

Within-group analysis performed in all rats revealed the presence of a preference for novel objects (Figure 4A), since during S2, rats spent a significantly higher percentage of time exploring the novel objects than the old objects (paired *t*-test, t = 4.67, df = 39, $p < 0.01$). Counterbalancing of NOR and Y maze testing did not influence the preference for novel objects in the NOR test. Indeed, the percentage of time spent in novel object exploration was comparable between rats that were first evaluated in the NOR test and then in the Y maze (65.29 ± 4.33) and rats that were first evaluated in the Y maze and then in the NOR test (62.81 ± 4.01).

During the different phases of the NOR test, rats displayed a very scarce emission of 50-kHz USVs (S0, average number of calls = 2.68 ± 0.8 in 5 min of recording; S1, average number of calls = 2.03 ± 0.35 in 3 min of recording; S2, average number of calls = 1.58 ± 0.36 in 3 min of recording) and no emission of 22-kHz USVs was observed. Nevertheless, the Kruskal–Wallis test revealed that the number of 50-kHz USVs emitted per minute during the NOR test was significantly higher compared with that recorded during the evaluation of basal calling behavior performed before the beginning of tickling (K = 12.03; $p < 0.05$), and Dunn's test for multiple comparisons revealed that this effect occurred during S1 ($p < 0.05$) (Figure 4B).

Differences in object discrimination were observed when rats were divided in three groups based on the prevalent type of USVs (i.e., 22-kHz or 50-kHz calls) that were emitted in response to tickling. Within-group analysis revealed a significant increase in the percentage of time spent exploring the novel objects during S2 in the group of rats that emitted only 50-kHz USVs in response to tickling (paired *t*-test, t = 6.17, df = 20, $p < 0.01$), as well as in the group of rats that emitted more 50-kHz calls than 22-kHz calls in response to tickling (Wilcoxon test, W = 12, $p < 0.05$). Conversely, within-group analysis revealed that rats that emitted more 22-kHz calls than 50-kHz calls in response to tickling spent comparable percentages of time exploring the novel and old objects during S2 (paired t-test, t = 0.12, df = 6, $p = 0.91$) (Figure 4C).

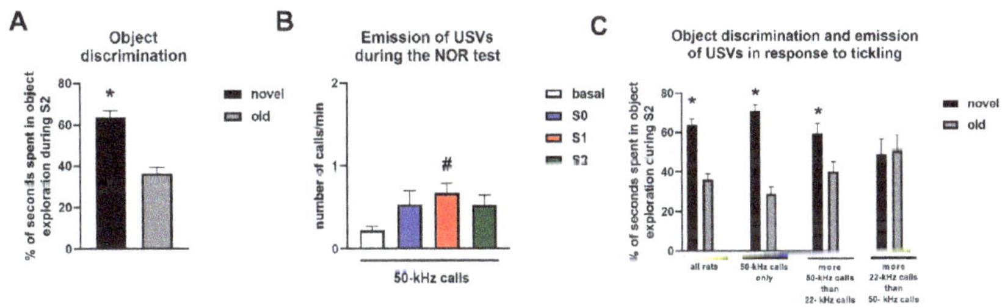

Figure 4. Object discrimination and emission of ultrasonic vocalizations during the novel object recognition test. Panel (**A**) reports the percentage of time spent exploring the novel and old objects in all the rats tested. Panel (**B**) demonstrates the emission of 50-kHz ultrasonic vocalizations in the different sessions of the novel object recognition test. Panel (**C**) reports the percentages of time spent exploring the novel and old objects when rats were grouped according to the prevalent type of ultrasonic vocalizations (i.e., 22-kHz or 50-kHz calls) emitted in response to tickling. * Indicates a significant difference vs. old objects. # Indicates a significant difference vs. basal. NOR = novel object recognition; S0, S1, and S2 = session 0, 1, and 2 of the novel object recognition test. $n = 40$ for panels A and B and for the group "all rats" in panel (**C**); $n = 21$ for the group "50-kHz calls only" in panel (**C**); $n = 12$ for the group "more 50-kHz calls than 22-kHz calls" in panel (**C**); $n = 7$ for the group "more 22-kHz calls than 50-kHz calls" in panel (**C**).

The results of the within-group analysis were confirmed by the between-group analysis with two-way ANOVA, which revealed a significant effect of object ($F_{1,152} = 41.32$, $p < 0.001$) and a significant interaction object × group ($F_{3,152} = 5.49$, $p = <0.01$), but no significant effect of group ($F_{3,152} = 0.01$, $p = 0.99$). Bonferroni's multiple comparisons test revealed that a significant discrimination between novel and old objects occurred in all rats ($p < 0.01$), in the group of rats that emitted only 50-kHz USVs in response to tickling ($p < 0.01$), as well as in the group of rats that emitted more 50-kHz calls than 22-kHz calls in response to tickling ($p = 0.036$), but not in the group of rats that emitted more 22-kHz calls than 50-kHz calls in response to tickling ($p = 0.99$) (Figure 4C). Bonferroni's multiple comparisons test revealed no significant group differences in the percentages of time spent in novel object exploration (Figure 4C).

No group differences in the cumulative times of object exploration during S1 and S2 were observed when rats were grouped according to the prevalent type of USVs emitted in response to tickling (Table 1).

Table 1. Cumulative times of object exploration during sessions 1 and 2 of the novel object recognition test in rats grouped according to the prevalent type of ultrasonic vocalizations emitted in response to tickling. Exploration times are reported as the average number of seconds ± S.E.M. S1 = session 1; S2 = session 2.

	Cumulative Time of Object Exploration during S1	Cumulative Time of Object Exploration during S2
all rats	13.59 ± 1.29	15.33 ± 1.21
50-kHz calls only	13.21 ± 1.43	14.99 ± 1.38
more 50-kHz calls than 22-kHz calls	12.27 ± 1.96	14.75 ± 2.26
more 22-kHz calls than 50-kHz calls	16.99 ± 5.15	17.34 ± 4.19

Finally, Spearman's test revealed that the percentages of time spent in novel object exploration during S2 were not significantly correlated with: (i) the overall number of 50-kHz USVs (total calls) (r = 0.08, $p = 0.63$); (ii) the number of flat 50-kHz calls; (r = 0.03, $p = 0.85$); (iii) the number of FM 50-kHz calls (r = 0.10, $p = 0.53$), emitted in the five sessions of tickling (Figure 5A–C). Nevertheless, Spearman's test revealed that a significant negative correlation existed between the number of 22-kHz USVs emitted in the five sessions of tickling and the percentages of time spent in novel object exploration during S2 (r = −0.36, $p = 0.02$), although the statistical significance of this correlation no longer persisted after Bonferroni's correction for multiple comparisons (Figure 5D).

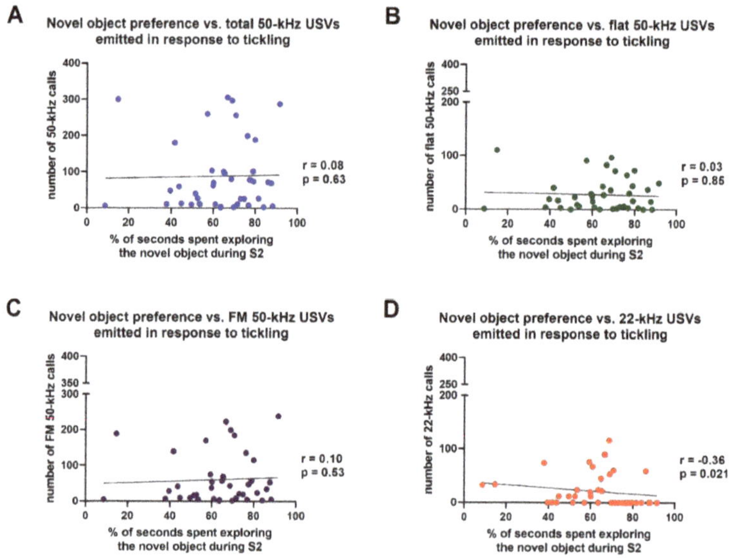

Figure 5. Results of Spearman's correlation test of the percentages of time spent exploring the novel objects during S2 of the novel object recognition test and the number of total 50-kHz ultrasonic vocalizations (**A**), flat 50-kHz ultrasonic vocalizations (**B**), frequency-modulated 50-kHz ultrasonic vocalizations (**C**), and 22-kHz ultrasonic vocalizations (**D**) emitted in response to tickling. FM = frequency-modulated; S2 = session 2 of the novel object recognition test; USVs = ultrasonic vocalizations. $n = 40$.

3.4. Single-Trial Continuous Spontaneous Alternation Behavior Test in a Y Maze and Emission of Ultrasonic Vocalizations

Figure 6A demonstrates the average percentage of SAB and the average number of entries in the arms of the Y maze calculated in all rats. During Y maze exploration, rats emitted a number of 50-kHz USVs per minute that were significantly higher compared with those recorded during the evaluation of basal calling behavior performed before the beginning of tickling (average number of calls = 28.03 ± 6.62 in 8 min of recording, Mann–Whitney U test, U = 276, $p < 0.01$) (Figure 6B), whereas no emission of 22-kHz calls occurred in the same situation. Counterbalancing of NOR and Y maze testing did not affect SAB during Y maze exploration. Indeed, the percentage of SAB was comparable between rats that were first evaluated in the NOR test and then in the Y maze (62.28 ± 2.56) and rats that were first evaluated in the Y maze and then in the NOR test (61.12 ± 3.65).

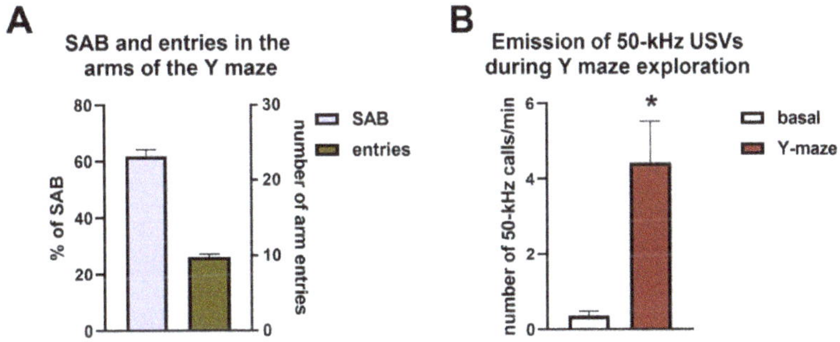

Figure 6. Panel A demonstrates the percentage of spontaneous alternation behavior (left Y axis) and the number of entries in the arms of the Y maze (right Y axis) for all the rats tested. Panel (**B**) demonstrates the emission of 50-kHz ultrasonic vocalizations recorded during Y maze exploration in all the rats tested. * Indicates a significant difference vs. basal. SAB = spontaneous alternation behavior; USVs = ultrasonic vocalizations. $n = 40$ for panel (**A**); $n = 38$ for panel B.

Spearman's test revealed the existence of positive but not significant correlations between the percentages of SAB and: (i) the overall number of 50-kHz USVs (total calls); (ii) the number of flat 50-kHz calls; (iii) the number of FM 50-kHz calls, emitted during Y maze exploration (total calls: r = 0.23, $p = 0.16$; flat calls: r = 0.12, $p = 0.44$; FM calls: r = 0.25, $p = 0.13$) (Figure 7A–C). Conversely, Spearman's test revealed that significant positive correlations existed between the number of entries in the arms of the Y maze and: (i) the overall number of 50-kHz USVs (total calls); (ii) the number of flat 50-kHz calls; (iii) the number of FM 50-kHz calls, emitted during Y maze exploration (total calls: r = 0.55, $p < 0.01$; flat calls: r = 0.49, $p < 0.01$; FM calls: r = 0.54, $p < 0.01$) (Figure 7D–F).

One-way ANOVA revealed no differences in both the percentages of SAB ($F3,76 = 0.68$, $p > 0.05$) (Figure 8A) and the number of entries in the arms of the Y maze ($F3,76 = 1.49$, $p > 0.05$) (Figure 8B) when rats were divided in three groups based on the prevalent type of USVs (i.e., 22-kHz or 50-kHz calls) emitted in response to tickling. Nevertheless, rats that emitted more 22-kHz calls than 50-kHz calls in response to tickling displayed a trend towards a reduction in the number of arm entries.

Spearman's test revealed that the percentages of SAB were not significantly correlated with: (i) the overall number of 50-kHz USVs (total calls) (r = −0.005, $p > 0.05$); (ii) the number of flat 50-kHz calls (r = −0.004, $p > 0.05$); (iii) the number of FM 50-kHz calls (r = −0.009, $p > 0.05$); or iv) the number of 22-kHz USVs (r = −0.02, $p > 0.05$) emitted in the five sessions of tickling (Figure 9).

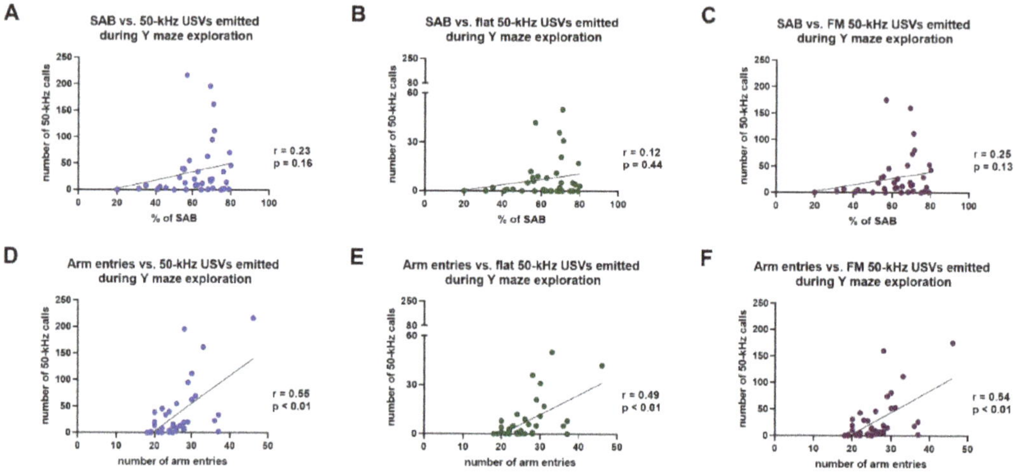

Figure 7. Results of Spearman's correlation test of the percentages of spontaneous alternation behavior (**A–C**) or the number of entries in the arms of the Y maze (**D–F**) and the number of total, flat, or frequency-modulated 50-kHz ultrasonic vocalizations emitted during the exploration of the Y maze. FM = frequency-modulated; SAB = spontaneous alternation behavior; USVs = ultrasonic vocalizations. $n = 38$.

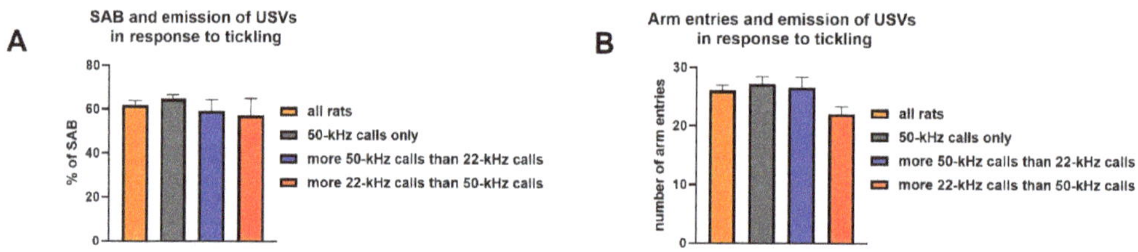

Figure 8. Percentages of spontaneous alternation behavior (**A**) and the number of entries in the arms of the Y maze (**B**) in rats grouped according to the prevalent type of ultrasonic vocalizations (i.e., 22-kHz or 50-kHz calls) emitted in response to tickling. SAB = spontaneous alternation behavior; USVs = ultrasonic vocalizations. $n = 40$ for the groups "all rats"; $n = 21$ for the groups "50-kHz calls only"; $n = 12$ for the groups "more 50-kHz calls than 22-kHz calls"; $n = 7$ for the groups "more 22-kHz calls than 50-kHz calls".

Conversely, Spearman's test revealed that the number of entries in the arms of the Y maze were significantly correlated with the number of FM 50-kHz USVs (r = 0.32, $p < 0.05$) emitted in the five sessions of tickling (Figure 10C), although the statistical significance of this correlation no longer persisted after Bonferroni's correction for multiple comparisons. Finally, Spearman's test revealed that the number of entries in the arms of the Y maze were not significantly correlated with the number of: total 50-kHz USVs (r = 0.29, $p > 0.05$), flat 50-kHz calls (r = 0.29, $p > 0.05$), or 22-kHz calls (r = −0.16, $p > 0.05$) emitted in the five sessions of tickling (Figure 10A,B,D).

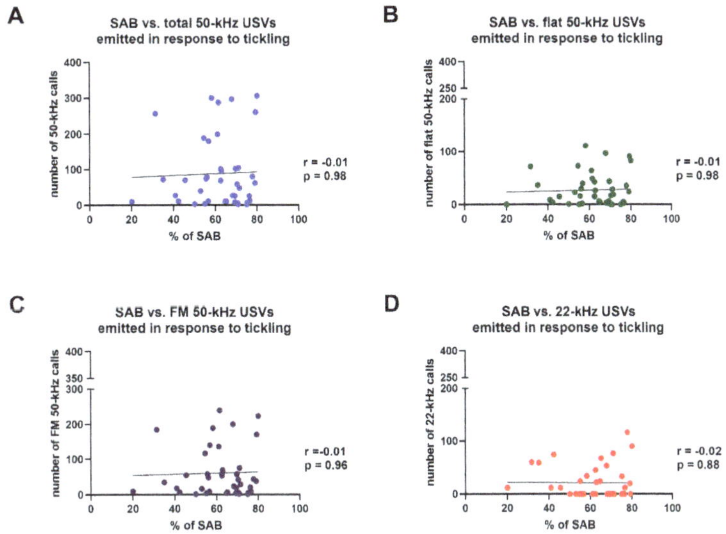

Figure 9. Results of Spearman's correlation test between the percentages of spontaneous alternation behavior and the numbers of total 50-kHz ultrasonic vocalizations (**A**), flat 50-kHz ultrasonic vocalizations (**B**), frequency modulated 50-kHz ultrasonic vocalizations (**C**) or 22-kHz ultrasonic vocalizations (**D**) emitted in response to tickling. FM = frequency modulated; SAB = spontaneous alternation behavior; USVs = ultrasonic vocalizations. $n = 40$.

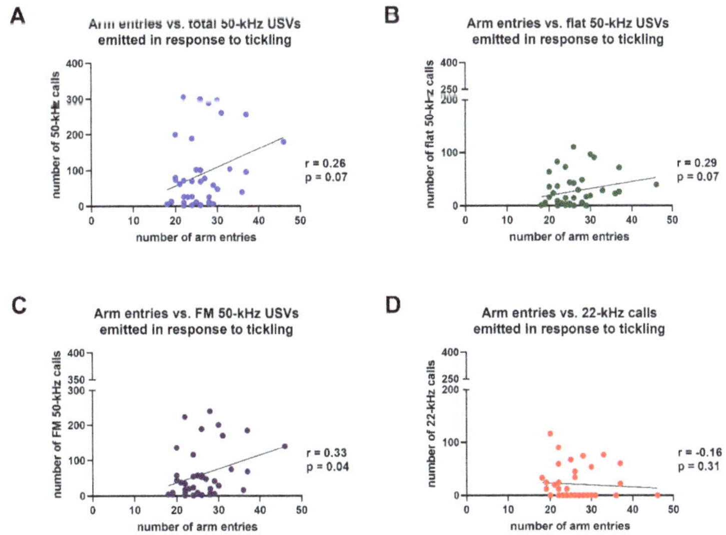

Figure 10. Results of Spearman's correlation test between the numbers of entries in the arms of the Y maze and the numbers of total 50-kHz ultrasonic vocalizations (**A**), flat 50-kHz ultrasonic vocalizations (**B**), frequency modulated 50-kHz ultrasonic vocalizations (**C**) or 22-kHz ultrasonic vocalizations (**D**) emitted in response to tickling. FM = frequency modulated; USVs = ultrasonic vocalizations. $n = 40$.

4. Discussion

In the present study, we measured the emission of 22-kHz and 50-kHz USVs in rats that were subjected to the NOR test and the single-trial continuous SAB test in a Y maze, two experimental paradigms that are used to evaluate memory and do not rely on conditioned stimuli and primary reinforcers. An increased emission of 50-kHz USVs, but not 22-kHz USVs, was observed during the NOR and SAB tests. However, calling behavior during the NOR test was scarce and occurred only in a specific phase of testing. Moreover, the emission of 50-kHz USVs during Y maze exploration was not correlated with the SAB of rats. Nevertheless, differences in object discrimination in the NOR test were observed when rats were grouped according to the prevalent type of USVs emitted in response to tickling, which was performed before the beginning of memory testing.

The NOR test is a behavioral paradigm used to evaluate non-spatial memory in rodents, and we here found that, when considered globally, rats effectively discriminated between novel and old objects, consistent with previous results [21]. However, rats emitted a very low number of 50-kHz USVs during the NOR test and a significant increase in the magnitude of calling behavior was observed in the acquisition phase (S1, when two identical objects were present), but not in the testing phase (S2, when two different objects to be discriminated were present), which indicates that no association existed between object discrimination and modifications in calling behavior. These results indicate that measuring the emission of 50-kHz USVs during testing may be not a useful behavioral marker that reflects the presence of an altered memory function evaluated in experiments of NOR. Nevertheless, additional results obtained in the present study indicate that an interplay may exist between calling behavior and rats' performance in the NOR test. In fact, differences in object discrimination were observed when rats were grouped according to the prevalent type of USVs (i.e., 22-kHz or 50-kHz calls) emitted in response to tickling that was performed before the beginning of NOR and SAB testing. Tickling is a procedure that may alter the emotional state and that may robustly stimulate calling behavior in rats [12]. Two subpopulations of rats have been characterized that respond differently to tickling in terms of 22-kHz and 50-kHz USV emissions, reflecting the presence of differential dispositional tendency for positive and negative affectivity [30]. In the present study, we replicated these previous findings by showing that the majority of the rats responded to tickling by emitting only or mainly 50-kHz USVs, which may reflect the presence of positive affectivity, and that the remaining minority of rats tested emitted mainly 22-kHz USVs in response to tickling, which may indicate the presence of negative affectivity [30]. Interestingly, the rats that emitted only 50-kHz USVs or that emitted more 50-kHz USVs than 22-kHz USVs in response to tickling effectively discriminated between novel and old objects in the NOR test. Conversely, the rats that emitted more 22-kHz USVs than 50-kHz USVs in response to tickling did not show significant object discrimination in the NOR test.

We may speculate that the abovementioned differences in object discrimination reflect the presence of alterations in memory function rather than in object exploration, since rats spent comparable amounts of time exploring the objects during S1 and S2 of the NOR test, irrespective of the prevalent type of USVs emitted in response to tickling. Moreover, we found that the preference for novel objects in the NOR test displayed a trend towards negative correlation with the emission of 22-kHz USVs in response to tickling, but did not correlate with the overall emission of 50-kHz USVs as well as the emission of FM and flat 50-kHz calls in response to tickling. Taken together, these findings suggest that the emission of 22-kHz USVs in response to tickling may be a behavioral marker potentially predictive of the performance of rats that are subsequently evaluated in the NOR test, and we may propose two explanatory hypotheses in this regard. Based on the previous findings by Burgdorf and coworkers [30], we may speculate that rats that emitted mostly 22-kHz USVs in response to tickling were characterized by a disposition towards negative affectivity, which may have influenced the memory for objects, resulting, in turn, in an impaired performance in the NOR test. This hypothesis would be consistent with previous preclinical studies showing that rats bred for low levels of positive affectivity in response to tickling displayed abnormalities in tests of social behavior and associative learning [30,31], and it would also agree with the results of clinical investigations demonstrating that several abnormalities exist in the cognitive domain of patients suffering from mood disorders [19,20].

Alternatively, we may speculate that rats that emitted mostly 22-kHz USVs in response to tickling were unable to discriminate between objects in the NOR test not because they had a disposition towards negative affectivity, but because they had a dysfunction in specific, yet undefined, brain regions that regulate the emission of 22-kHz USVs as well as item recognition. In this regard, it is worth considering that the emission of 22-kHz USVs can be initiated by the activation of cholinergic transmission at the level of the lateral septum [32,33], and that an impaired object discrimination has been reported in rats that were evaluated in the NOR test after the infusion of pregnenolone in the lateral septum [34]. Hence, we may speculate that an altered function of the lateral septum could be a common mechanism that may explain why rats that emitted mostly 22-kHz USVs in response to tickling also displayed an impaired performance in the NOR test, although the neurochemical events underlying this possible mechanism appear ill defined. In this regard, it is also important to consider that while several cortical and subcortical regions are known to regulate object discrimination in the NOR test [35,36], limited information is available on the brain regions and neurochemical mechanisms that initiate and modulate the emission of 22-kHz USVs. Therefore, clarifying these aspects of the neurobiology of 22-kHz USVs may help to elucidate if an interplay exists between the emission of 22-kHz USVs and unconditioned memories, and how the changes in the emission of these calls relate to the presence of altered memory function evaluated in the NOR test. Furthermore, additional studies will be necessary to clarify whether the emission of 22-kHz USVs may be predictive of and/or associated with the behavioral readouts that are evaluated in other behavioral paradigms that are used to assess non-spatial memory in rats.

The single-trial continuous SAB test in a Y maze is a behavioral paradigm that is used to evaluate non-spatial working memory in rodents, and we here found that rats displayed percentages of SAB in the range of those previously reported [37,38]. Moreover, rats exhibited a significant increase in the emission of 50-kHz USVs during Y maze exploration. Nevertheless, SAB was not significantly correlated with the emission of 50-kHz USVs (total and categorized) recorded during Y maze exploration, although a positive correlation was observed, nor was SAB correlated with the emission of 50-kHz USVs (total and categorized) in response to tickling. These results indicate that the emission of 50-kHz USVs is not a behavioral marker that may be associated with or predict the SAB of rats tested in a Y maze.

On the other hand, positive and significant correlations were found between the entries in the arms of the Y maze and the emission of 50-kHz USVs (total and categorized) recorded either during Y maze exploration or in response to tickling performed before memory testing. The number of arm entries in tests of continuous SAB in a Y maze can provide a measure of locomotor activity [22], which could suggest that an interplay exists between the emission of 50-kHz USVs and locomotion in rats during Y maze testing. However, several lines of evidence indicate that the emission of 50-kHz USVs cannot be simply considered a byproduct of locomotion, but it rather reflects the presence of arousal/positive affect [39,40]. On these bases, one hypothesis that could explain the increased emission of 50-kHz USVs during Y maze exploration is that this situation elicited arousal/positive affect in rats and, accordingly, calling behavior. A possible mechanism that could underlie the emission of 50-kHz USVs during Y maze exploration is the curiosity towards a novel environment. Indeed, curiosity may be associated with increased arousal [41], and curiosity is thought to be a factor that drives the exploratory behavior of rats exposed to a Y maze [22]. Moreover, the existence of an interplay between curiosity and emission of USVs has been proposed by studies that evaluated calling behavior during social contacts in mice [6,42]. Nevertheless, it is noteworthy that rats may also emit 50-kHz USVs in situations that are not necessarily pleasurable or appetitive for them [43]. Hence, an alternative hypothesis may be that the emission of 50-kHz USVs during Y maze exploration did not stem from changes in the affective state of rats but from other, yet undefined, mechanisms. Furthermore, it is noteworthy that we found a trend to negative correlation between the entries in the arms of the Y maze and the emission of 22-kHz USVs recorded in response to tickling. Considering all the findings, further investigations are warranted to clarify whether the interplay observed here between calling behavior and entries in the arm of a Y maze has behavioral significance, and whether it may have any relevance as a behavioral marker of spatial memory. Indeed, it has to be remarked that in the continuous SAB test in a Y maze, it is the sequence of arm entries, rather than the number of arm entries, that is used as a behavioral readout to assess spatial working memory [22].

Earlier investigations by us and others have demonstrated an increased emission of 50-kHz USVs in rats upon the presentation of environmental stimuli that were previously paired with either a social reward or the administration of drugs that possess rewarding properties [9–17,44,45]. Moreover, a very recent study has demonstrated a persistently decreased emission of FM 50-kHz USVs in rats subjected to fear conditioning [18]. Based on these findings, the changes in the emission of 50-kHz USVs may be regarded as a behavioral marker of conditioned memories, and it has been suggested that such calling behavior may capture the affective component of conditioned memories [18]. In the present study, we obtained evidence to suggest that measuring the emission of USVs may deserve further consideration as a potential behavioral marker also in studies of memory based on the use of paradigms that do not rely on conditioned stimuli/primary reinforcers. More specifically, according to the presented data, we propose that the emission of 50-kHz USVs may be not a useful behavioral marker associated with, or predictive of, object discrimination in rats tested in the NOR and of SAB in rats tested in a Y maze. However, the results of this study suggest that an interplay may exist between the emission of 22-kHz USVs and object discrimination in the NOR test, although a more detailed investigation of the relationship between the emission of 22-kHz USVs and unconditioned non-spatial memory is needed.

The present study may have a potential limitation in that it did not include a pure control group of rats that were not subjected to tickling. Our results indicate that prior tickling experience did not affect memory performance of rats in the NOR test, at least in those animals that emitted only or mostly 50-kHz USVs, which were found to effectively discriminate between novel and old objects. Moreover, prior tickling experience did not affect the pattern of Y maze exploration, since rats displayed percentages of SAB that were in the range of those previously described. Nevertheless, we cannot rule out the possibility that prior tickling experience may have elicited an enduring influence on rats' vocal behavior, which eventually resulted in a dissimilar emission of USVs between the NOR and SAB tests. In this regard, it is also noteworthy that the lack of USV emissions observed here during the NOR test may appear unexpected, and in contrast to the increased calling behavior recorded during Y maze exploration. In fact, previous studies have demonstrated that the exposure to novel environments and situations of novelty, as may be the case for the presentation of objects during the NOR test, may stimulate the emission of USVs in rats [43,46,47]. In this connection, we cannot exclude the possibility that methodological issues related to test implementation contributed, at least in part, to the differences in calling behavior observed here between the NOR and SAB tests. Indeed, during the NOR test, rats were not in direct contact with sawdust, since the bottom of the test cage was covered with an elevated grid that had a handful of sawdust below it. Importantly, it has been demonstrated that contact with sawdust is a factor that facilitates calling behavior in rats [47]. Conversely, the Y maze used in this study had the bottom covered with sawdust and also had the walls painted black: these factors may have resulted in a more favorable environment for rats, and in turn could have facilitated the emission of USVs. Based on these considerations, we suggest that methodological issues are carefully examined in future studies, since doing so will help to elucidate the behavioral significance of the USVs that are recorded in rats subjected to tests of memory.

5. Conclusions

Elucidating the networking between the emission of USVs and memory function in rats appears of interest in the consideration of the evidence, suggesting that a relationship may exist between changes in the affective state and alterations in memory function [19,20], and in light of the evidence that the emission of USVs is a behavioral marker of affect in rats [39,40]. Accordingly, further clarification of how the emission of USVs varies in magnitude and type (i.e., 22-kHz vs. 50-kHz calls) in experimental paradigms that evaluate different forms of memory (i.e., non-spatial vs. spatial, long-term vs. short term) is of interest, as it may potentially contribute to increase the amount of information that can be collected in studies of memory in rats.

Author Contributions: Conceptualization, N.S.; formal analysis, G.C., M.S. and N.S., investigation, G.C., M.S. and N.S., resources, N.S.; writing—original draft preparation, G.C. and M.S.; writing—review and editing, N.S.; visualization; supervision, N.S.; funding acquisition, N.S. All authors have read and agreed to the published version of the manuscript.

Funding: This research was funded by Intramural Funds (FIR) from the University of Cagliari and by the Fondazione di Sardegna (Progetti Biennali di Ateneo—2017).

Institutional Review Board Statement: All applicable international, national, and/or institutional guidelines for the care and use of animals were followed. All experimental procedures using animal subjects were performed in accordance with the guidelines issued by the Committee for Animal Welfare (OPBA) of the University of Cagliari in compliance with Italian Law (4 March 2014) and European Law (2010/63/EU). Study protocol 1236/2015.

Data Availability Statement: Data will be made available by the corresponding author upon reasonable request.

Conflicts of Interest: The authors declare no conflict of interest. The funders had no role in the design of the study; in the collection, analyses, or interpretation of data; in the writing of the manuscript, or in the decision to publish the results.

References

1. Brudzynski, S.M. Ethotransmission: Communication of emotional states through ultrasonic vocalization in rats. *Curr. Opin. Neurobiol.* **2013**, *23*, 310–317. [CrossRef] [PubMed]
2. Burgdorf, J.; Panksepp, J.; Moskal, J.R. Frequency-modulated 50kHz ultrasonic vocalizations: A tool for uncovering the molecular substrates of positive affect. *Neurosci. Biobehav. Rev.* **2011**, *35*, 1831–1836. [CrossRef]
3. Wöhr, M.; Schwarting, R.K.W. Affective communication in rodents: Ultrasonic vocalizations as a tool for research on emotion and motivation. *Cell Tissue Res.* **2013**, *354*, 81–97. [CrossRef]
4. Simola, N.; Brudzynski, S.M. Repertoire and Biological Function of Ultrasonic Vocalizations in Adolescent and Adult Rats. In *Handbook of Ultrasonic Vocalization: A Window into the Emotional Brain*; Elsevier Academic Press: Cambridge, MA, USA, 2018; Volume 25, pp. 177–186.
5. Simola, N.; Brudzynski, S.M. Rat 50-kHz ultrasonic vocalizations as a tool in studying neurochemical mechanisms that regulate positive emotional states. *J. Neurosci. Methods* **2018**, *310*, 33–44. [CrossRef]
6. Simola, N.; Granon, S. Ultrasonic vocalizations as a tool in studying emotional states in rodent models of social behavior and brain disease. *Neuropharmacology* **2019**, *159*, 107420. [CrossRef]
7. Kassai, F.; Gyertyán, I. Effects of Selective Serotonin Reuptake Inhibitors on the Shock-Induced Ultrasonic Vocalization of Rats in Different Experimental Designs. In *Handbook of Ultrasonic Vocalization: A Window into the Emotional Brain*; Elsevier Academic Press: Cambridge, MA, USA, 2018; Volume 25, pp. 309–316.
8. Sánchez, C. Stress-induced vocalisation in adult animals. A valid model of anxiety? *Eur. J. Pharmacol.* **2003**, *463*, 133–143. [CrossRef]
9. Costa, G.; Serra, M.; Marongiu, J.; Morelli, M.; Simola, N. Influence of dopamine transmission in the medial prefrontal cortex and dorsal striatum on the emission of 50-kHz ultrasonic vocalizations in rats treated with amphetamine: Effects on drug-stimulated and conditioned calls. *Prog. Neuro-Psychopharmacol. Biol. Psychiatry* **2020**, *97*, 109797. [CrossRef]
10. Ma, S.T.; Maier, E.Y.; Ahrens, A.M.; Schallert, T.; Duvauchelle, C.L. Repeated intravenous cocaine experience: Development and escalation of pre-drug anticipatory 50-kHz ultrasonic vocalizations in rats. *Behav. Brain Res.* **2010**, *212*, 109–114. [CrossRef]
11. Hamed, A.; Kursa, M.B. Inter-individual differences in serotonin and glutamate co-transmission reflect differentiation in context-induced conditioned 50-kHz USVs response after morphine withdrawal. *Brain Struct. Funct.* **2018**, *223*, 3149–3167. [CrossRef]
12. Panksepp, J.; Burgdorf, J. 50-kHz chirping (laughter?) in response to conditioned and unconditioned tickle-induced reward in rats: Effects of social housing and genetic variables. *Behav. Brain Res.* **2000**, *115*, 25–38. [CrossRef]

13. Simola, N.; Frau, L.; Plumitallo, A.; Morelli, M. Direct and long-lasting effects elicited by repeated drug administration on 50-kHz ultrasonic vocalizations are regulated differently: Implications for the study of the affective properties of drugs of abuse. *Int. J. Neuropsychopharmacol.* **2014**, *17*, 429–441. [CrossRef]
14. Simola, N.; Costa, G.; Morelli, M. Activation of adenosine A2A receptors suppresses the emission of pro-social and drug-stimulated 50-kHz ultrasonic vocalizations in rats: Possible relevance to reward and motivation. *Psychopharmacology* **2016**, *233*, 507–519. [CrossRef]
15. Simola, N.; Serra, M.; Marongiu, J.; Costa, G.; Morelli, M. Increased emissions of 50-kHz ultrasonic vocalizations in hemiparkinsonian rats repeatedly treated with dopaminomimetic drugs: A potential preclinical model for studying the affective properties of dopamine replacement therapy in Parkinson's disease. *Prog. Neuro-Psychopharmacol. Biol. Psychiatry* **2021**, *108*, 110184. [CrossRef]
16. Simola, N.; Morelli, M. Repeated amphetamine administration and long-term effects on 50-kHz ultrasonic vocalizations: Possible relevance to the motivational and dopamine-stimulating properties of the drug. *Eur. Neuropsychopharmacol.* **2015**, *25*, 343–355. [CrossRef]
17. Tripi, J.A.; Dent, M.L.; Meyer, P.J. Individual differences in food cue responsivity are associated with acute and repeated cocaine-induced vocalizations, but not cue-induced vocalizations. *Psychopharmacology* **2017**, *234*, 437–446. [CrossRef]
18. Reyes, K.A.E.; Kudva, P.S.; Heckler, B.; Gonzalez, A.E.; Sorg, B.A. Rat ultrasonic vocalizations as an index of memory. *Neurosci. Lett.* **2021**, *741*, 135458. [CrossRef]
19. Anacker, C.; Hen, R. Adult hippocampal neurogenesis and cognitive flexibility—linking memory and mood. *Nat. Rev. Neurosci.* **2017**, *18*, 335–346. [CrossRef] [PubMed]
20. Panchal, P.; Kaltenboeck, A.; Harmer, C.J. Cognitive emotional processing across mood disorders. *CNS Spectr.* **2019**, *24*, 54–63. [CrossRef]
21. Ennaceur, A.; Delacour, J. A new one-trial test for neurobiological studies of memory in rats. 1: Behavioral data. *Behav. Brain Res.* **1988**, *31*, 47–59. [CrossRef]
22. Hughes, R.N. The value of spontaneous alternation behavior (SAB) as a test of retention in pharmacological investigations of memory. *Neurosci. Biobehav. Rev.* **2004**, *28*, 497–505. [CrossRef]
23. Burgdorf, J.; Panksepp, J. Tickling induces reward in adolescent rats. *Physiol. Behav.* **2001**, *72*, 167–173. [CrossRef]
24. Schwarting, R.K.W. Ultrasonic vocalization in juvenile and adult male rats: A comparison among stocks. *Physiol. Behav.* **2018**, *191*, 1–11. [CrossRef]
25. Cadoni, C.; Pisanu, A.; Simola, N.; Frau, L.; Porceddu, P.F.; Corongiu, S.; Dessì, C.; Sil, A.; Plumitallo, A.; Wardas, J.; et al. Widespread reduction of dopamine cell bodies and terminals in adult rats exposed to a low dose regimen of MDMA during adolescence. *Neuropharmacology* **2017**, *123*, 385–394. [CrossRef] [PubMed]
26. Costa, G.; Morelli, M.; Simola, N. Involvement of glutamate NMDA receptors in the acute, long-term, and conditioned effects of amphetamine on rat 50 kHz ultrasonic vocalizations. *Int. J. Neuropsychopharmacol.* **2015**, *18*, pyv057. [CrossRef] [PubMed]
27. Simola, N.; Bustamante, D.; Pinna, A.; Pontis, S.; Morales, P.; Morelli, M.; Herrera-Marschitz, M. Acute perinatal asphyxia impairs non-spatial memory and alters motor coordination in adult male rats. *Exp. Brain Res.* **2008**, *185*, 595–601. [CrossRef] [PubMed]
28. Simola, N.; Fenu, S.; Costa, G.; Pinna, A.; Plumitallo, A.; Morelli, M. Pharmacological characterization of 50-kHz ultrasonic vocalizations in rats: Comparison of the effects of different psychoactive drugs and relevance in drug-induced reward. *Neuropharmacology* **2012**, *63*, 224–234. [CrossRef]
29. Wright, J.M.; Gourdon, J.C.; Clarke, P.B.S. Identification of multiple call categories within the rich repertoire of adult rat 50-kHz ultrasonic vocalizations: Effects of amphetamine and social context. *Psychopharmacology* **2010**, *211*, 1–13. [CrossRef]
30. Burgdorf, J.; Panksepp, J.; Brudzynski, S.M.; Beinfeld, M.C.; Cromwell, H.C.; Kroes, R.A.; Moskal, J.R. The effects of selective breeding for differential rates of 50-kHz ultrasonic vocalizations on emotional behavior in rats. *Dev. Psychobiol.* **2009**, *51*, 34–46. [CrossRef]
31. Webber, E.S.; Harmon, K.M.; Beckwith, T.J.; Peña, S.; Burgdorf, J.; Panksepp, J.; Cromwell, H.C. Selective breeding for 50kHz ultrasonic vocalization emission produces alterations in the ontogeny and regulation of rough-and-tumble play. *Behav. Brain Res.* **2012**, *229*, 138–144. [CrossRef]
32. Bihari, A.; Hrycyshyn, A.W.; Brudzynski, S.M. Role of the mesolimbic cholinergic projection to the septum in the production of 22 kHz alarm calls in rats. *Brain Res. Bull.* **2003**, *60*, 263–274. [CrossRef]
33. Silkstone, M.; Brudzynski, S.M. Dissimilar interaction between dopaminergic and cholinergic systems in the initiation of emission of 50-kHz and 22-kHz vocalizations. *Pharmacol. Biochem. Behav.* **2020**, *188*, 172815. [CrossRef]
34. Nanfaro, F.; Cabrera, R.; Bazzocchini, V.; Laconi, M.; Yunes, R. Pregnenolone sulfate infused in lateral septum of male rats impairs novel object recognition memory. *Pharmacol. Rep.* **2010**, *62*, 265–272. [CrossRef]
35. Chao, O.Y.; Huston, J.P.; Li, J.S.; Wang, A.L.; de Souza Silva, M.A. The medial prefrontal cortex-lateral entorhinal cortex circuit is essential for episodic-like memory and associative object-recognition. *Hippocampus* **2016**, *26*, 633–645. [CrossRef]
36. Nelson, A.J.D.; Thur, K.E.; Marsden, C.A.; Cassaday, H.J. Dissociable Roles of Dopamine Within the Core and Medial Shell of the Nucleus Accumbens in Memory for Objects and Place. *Behav. Neurosci.* **2010**, *124*, 789–799. [CrossRef] [PubMed]
37. Maurice, T.; Su, T.P.; Privat, A. Sigma1 (σ1) receptor agonists and neurosteroids attenuate b25-35-amyloid peptide-induced amnesia in mice through a common mechanism. *Neuroscience* **1998**, *83*, 413–428. [CrossRef]
38. Meunier, J.; Gué, M.; Récasens, M.; Maurice, T. Attenuation by a sigma 1 (σ 1) receptor agonist of the learning and memory deficits induced by a prenatal restraint stress in juvenile rats. *Br. J. Pharmacol.* **2004**, *142*, 689–700. [CrossRef]

39. Knutson, B.; Burgdorf, J.; Panksepp, J. Ultrasonic vocalizations as indices of affective states in rats. *Psychol. Bull.* **2002**, *128*, 961–977. [CrossRef]
40. Burgdorf, J.; Panksepp, J. The neurobiology of positive emotions. *Neurosci. Biobehav. Rev.* **2006**, *30*, 173–187. [CrossRef]
41. Berlyne, D.E. The arousal and satiation of perceptual curiosity in the rat. *J. Comp. Physiol. Psychol.* **1955**, *48*, 238–246. [CrossRef] [PubMed]
42. Coura, R.S.; Cressant, A.; Xia, J.; De Chaumont, F.; Olivo-Marin, J.C.; Pelloux, Y.; Dalley, J.W.; Granon, S. Nonaggressive and adapted social cognition is controlled by the interplay between noradrenergic and nicotinic receptor mechanisms in the prefrontal cortex. *FASEB J.* **2013**, *27*, 4343–4354. [CrossRef]
43. Wöhr, M.; Houx, B.; Schwarting, R.K.W.; Spruijt, B. Effects of experience and context on 50-kHz vocalizations in rats. *Physiol. Behav.* **2008**, *93*, 766–776. [CrossRef] [PubMed]
44. Simola, N.; Costa, G. Emission of categorized 50-kHz ultrasonic vocalizations in rats repeatedly treated with amphetamine or apomorphine: Possible relevance to drug-induced modifications in the emotional state. *Behav. Brain Res.* **2018**, *347*, 88–98. [CrossRef]
45. Garcia, E.J.; McCowan, T.J.; Cain, M.E. Harmonic and frequency modulated ultrasonic vocalizations reveal differences in conditioned and unconditioned reward processing. *Behav. Brain Res.* **2015**, *287*, 207–214. [CrossRef]
46. Schwarting, R.K.W.; Jegan, N.; Wöhr, M. Situational factors, conditions and individual variables which can determine ultrasonic vocalizations in male adult Wistar rats. *Behav. Brain Res.* **2007**, *182*, 208–222. [CrossRef] [PubMed]
47. Natusch, C.; Schwarting, R.K.W. Using bedding in a test environment critically affects 50-kHz ultrasonic vocalizations in laboratory rats. *Pharmacol. Biochem. Behav.* **2010**, *96*, 251–259. [CrossRef] [PubMed]

brain sciences

MDPI

Article

Rat 50 kHz Trill Calls Are Tied to the Expectation of Social Interaction

Candace J. Burke [1], Mariya Markovina [2], Sergio M. Pellis [1] and David R. Euston [1,*]

[1] Department of Neuroscience, University of Lethbridge, Lethbridge, AB T1K 3M4, Canada; cj.burke@uleth.ca (C.J.B.); pellis@uleth.ca (S.M.P.)
[2] Department of Psychology, University of Calgary, 2500 University Drive NW, Calgary, AB T2N 1N4, Canada; mariyamarkovina@gmail.com
* Correspondence: euston@uleth.ca

Abstract: Rats emit a variety of calls in the 40–80 kHz range (50 kHz calls). While these calls are generally associated with positive affect, it is unclear whether certain calls might be used selectively in certain contexts. To examine this, we looked at ultrasonic calls in 30–40 day old male rats during the expectation of either play or food, both of which are reinforcing. Behavior and vocalizations were recorded while rats were in a test chamber awaiting the arrival of a play partner or food over seven days of testing. Control groups were included for the non-specific effects of food deprivation and social isolation. Play reward led to an increase in 50 kHz vocalizations, generally, with specific increases in trill and "trill with jump" calls not seen in other groups. Expectation of food reward did not lead to a significant increase in vocalizations of any type, perhaps due to the young age of our study group. Further, rats that were food deprived for the food expectation study showed markedly lower calls overall and had a different profile of call types compared to rats that were socially isolated. Taken together, the results suggest that trill-associated calls may be used selectively when rats are socially isolated and/or expecting a social encounter.

Keywords: ultrasonic vocalizations; communication; play; rats; reward; food

check for updates

Citation: Burke, C.J.; Markovina, M.; Pellis, S.M.; Euston, D.R. Rat 50 kHz Trill Calls Are Tied to the Expectation of Social Interaction. *Brain Sci.* **2021**, *11*, 1142. https://doi.org/10.3390/brainsci11091142

Academic Editors: Stefan M. Brudzynski and Jeffrey Burgdorf

Received: 14 July 2021
Accepted: 27 August 2021
Published: 28 August 2021

1. Introduction

A predominate theory about the purpose of ultrasonic vocalizations (USVs) in rats is that these calls signal the affective state of the animal [1,2]. Two main categories of calls have been described: 50 kHz calls associated with appetitive situations and positive affect and 22 kHz calls associated with threatening situations and negative affect [2]. While 22 kHz calls are mainly long and flat, 50 kHz calls come in a variety of shapes, including trills, ramps and jumps [3]. Whether the different types of 50 kHz calls have different functional roles is a topic of active research [3,4].

Vocalizations of the 50 kHz type are strongly associated with non-social rewarding stimuli. There is a significant increase in 50 kHz vocalizations emitted when rats are placed in a chamber in which they have received amphetamine (AMPH) [5,6]. Interestingly, the amount of AMPH administered has a direct relationship with the amount of 50 kHz USVs produced [7]. Anticipation of self-administration of electrical stimulation to brain reward centers, such as the ventral tegmental area (VTA) and lateral hypothalamic area, also elicits high rates of 50 kHz calls [8]. The animal in that study showed a marked increase in 50 kHz USVs to cues associated with the electrical stimulation as well as to the stimulation itself. Finally, 50 kHz calls have also been associated with cues indicating food reward [8–10] or during anticipation of daily feeding [11]. These findings suggest that 50 kHz USVs signal positive affective states associated with rewarding contexts, independent of social context.

Fifty kHz USVs are also emitted during, and in anticipation of, a variety of rewarding social interactions. Significant increases in 50 kHz calls have been found in males during the anticipatory period before introduction of a female [12]. During copulation, both male and female rats produce 50 kHz vocalizations [13,14]. Interestingly, the number of 50 kHz vocalizations appears to relate to the level of sexual motivation in the respective vocalizing party [12,15]. Juvenile male rats will also emit 50 kHz vocalizations when anticipating the presence of a conspecific, and these vocalizations will increase over days of testing in rats that are socially isolated before testing [16,17]. At least one study has failed to find anticipatory calling in juveniles, but that study used only limited social isolation [18]. Rats will also emit these vocalizations when entering an area frequently visited by other rats [19].

One social context that is known to be particularly rewarding and associated with high numbers of 50 kHz vocalizations is rough and tumble play in juvenile rats. The calls are most common before contact is made [20]. Further, these calls have also been elicited by rats tickled by human handlers, and are more common in isolated than socially housed animals, possibly reflecting the greater value of this hetero-species contact when other social interactions are lacking [14,21]. Rats will also produce 50 kHz vocalizations when introduced to an immobilized and, therefore, easily approachable conspecific and when being introduced to a conspecific after separation [22].

In summary, the 50 kHz USVs are emitted during acquisition and anticipation of non-social and social rewards and also elicit a response from conspecifics. To complicate matters, these calls are also elicited during negative social contexts such as during aggression and when a resident initially meets an intruder [23,24]. Rats also emit 50 kHz calls when a companion is taken away [25]. One explanation for the variety of usage is the 14 potential categories of calls existing in the 50 kHz range [3]. Indeed, the specific calls have been linked to anticipation of play behaviors [16], to mitigate aggression [26], signal play [4], feeding [27] and social contact signaling [25]. Thus, rather than signaling a general positive state, different 50 kHz calls may serve different functional roles.

In this study, we sought to contrast anticipatory calling in juvenile rats to both social and non-social stimuli using play and food, respectively. Two recent studies have attempted similar comparisons. Willey et al. [28] compared vocalizations in male rats to the presence of either food or a female rat on the other side of a wire mesh barrier. The social stimulus elicited far more vocalizations than the food reward. Similarly, Mulvihill and Brudzynski [29] compared vocalizations in males to food reward and to exploration of space recently vacated by an estrous female. The estrous female elicited an increase in 50 kHz calls, especially trill calls, whereas the food reward did not cause an overall change in vocalization rate, but rats did produce more flat calls in the 50 kHz range. This latter finding is consistent with previous reports that feeding is associated with flat calls in the 40 kHz range [27]. These studies show clearly that social stimuli elicit more calls than food reward, but a detailed comparison of calls during *anticipation* of both food and social reward has not yet been reported.

To investigate if anticipation of different types of reward elicited different patterns of calling, we compared the vocalizations of food restricted animals anticipating food to socially isolated animals anticipating play. To ensure that the vocalizations were not due to the restrictions or to the chamber, we had control animals, who were either socially isolated or food restricted, run in the same paradigm as the test subjects but without food or play reward. If a particular 50 kHz call communicates positive affect, we would expect to see elevated rates of this particular call type during anticipation of both food and play. Trill calls are a likely candidate, given their frequency and strong association with drug reward [3]. Any differences in call types or usage, on the other hand, would indicate that 50 kHz vocalizations are more nuanced, signaling specific features of the anticipated reward.

2. Materials and Methods

2.1. Subjects

Thirty juvenile, male Long Evans, aged 30–40 days, obtained from Charles River (Kingston, NY, USA) at 22 days old were used. These animals were pair housed and given five days to acclimatize to the facility. Eighteen animals were used in the anticipation of play paradigm, 6 in the Play Reward group, who received a play partner after a two minute waiting period, 6 in the Play Control group, who similarly waited for a partner that never came, and 6 as play partners for the Play Reward group. The remaining 12 animals were used for the anticipation of food paradigm, 6 in the Food Reward group, which received food in the test chamber after a two minute waiting period, and 6 in the Food Control group who did not receive food. All animals were maintained on the Lab Diet Enriched Rat Chow (Lab Diet, St. Louis, MO, USA). Housing rooms were lit during the day and dark at night and all testing occurred during the day.

2.2. Behavioral Procedure

The testing enclosure was a Plexiglas box (50 × 50 × 50 cm), which was situated inside a soundproof chamber (61 × 61 × 83 cm) lined with acoustic foam. The floor of the chamber was covered with 2 cm of paper-based bedding (Care Fresh, Ferndale, WA, USA) which we found to facilitate play while causing very low levels of ultrasonic interference. Ultrasonic vocalizations were collected using a specialized microphone (Model 4939, Brüel & Kjaer, Denmark) with a frequency response of 4 Hz to 100 kHz. The microphone was located in the ceiling of the chamber and was approximately 15 cm above the center of the play enclosure. The microphone was connected to a Soundconnect™ amplifier (Listen, Inc., Boston, MA, USA) and sound waves were recorded at 195,313 Hz using 16-bit resolution via a multifunction processor (model RX6, Tucker-Davis Technologies, Alachua, FL, USA). Video was recorded using a USB webcam (Microsoft Lifecam Studio, Redmond, WA, USA) with its infrared filter removed, positioned directly above the animal

2.3. Anticipation of Play Test

Data presented were taken from a 2 min anticipation period during which a target animal either waited in the testing enclosure for the arrival of a familiar play partner (i.e., his former cage mate) or received no partner. For the Play Reward group, once the play partner was introduced, animals were allowed to play for 10 min, following previously established methods [30]. After testing, rats were returned to their original home cages for an additional hour of play and then separated. The Play Control animals, who received no partner, waited in the chamber for 10 min, and then were placed back in their home cage. One hour later, these animals were introduced to their former cage mate for 1 h and 12 min of play before separation. Prior to testing, animals were individually habituated to the enclosure for 10 min each day for 3 consecutive days. On the 3rd day all subjects were socially isolated from their cage mates for 24 h prior to play testing and isolation continued until after all 7 days of testing were complete, in order to increase overall playfulness [31–33]. Both habituation and testing sessions were conducted in complete darkness, as this has been shown to facilitate USV production [17]. Audio and video recordings began after the target rat was placed in the test enclosure. Because both audio and video data were recorded on separate devices, a custom-made beeper with an LED light was used to emit a simultaneous light/sound cue at the beginning and end of each recording session and these times were used to synchronize audio and video recordings during subsequent analysis. Following each session, the apparatus was thoroughly cleaned with Virkon, a broad-spectrum disinfectant (Virkon Disinfectant Technologies, Sudbury, United Kingdom), and bedding was replaced to avoid any odors from other subjects. The data analyzed comes from day 1 and day 7 in all animals with the exception of one animal in the Play group who was not separated from his cage mate after testing on day 6. For this one animal, we use data from day 6 instead.

2.4. Food Restriction

In order to food restrict animals at such a young age, we used the play animals as weight controls. Each food-restricted animal was matched based on weight to a play animal when handling started. The target weight was calculated based on that of the play animal. The food restriction animals were restricted to maintain 85% of the weight of play controls.

The animals were separated for three hours to eat the appropriate amount of food and then were placed back in with their cage mate, with any food remaining in their isolated chambers being placed in with both the animals.

2.5. Anticipation of Food Test

The Food Reward group consisted of 6 animals who anticipated food reward in the chamber for 2 min and subsequently received half a semi-sweet chocolate chip each 30 s for 10 min. The chocolate chips were dropped by the experimenter from the top of the sound chamber. The animals were then brought back to their individual feeding cages, which had their allocated food, and then were given 3 h to eat before being return to their shared cages. The remaining 6 animals, the Food Control group, were placed in the chamber for 12 min while the experimenter remained in the room; however, no food was given. These animals were similarly isolated and given 3 h to eat before being return to their shared cages. One hour into this period, the Food Control rats were given 10 chocolate chips so as to equate both the quantity and type of food eaten each day between the Food Reward and Food Control groups.

2.6. Ethics

All procedures were in accordance with the University of Lethbridge institutional animal care and use committee and Canadian Council on Animal Care recommendations and guidelines.

2.7. Behavioral Analyses

The 2 min anticipatory period was analyzed in each group. The behaviors were coded using recorded video sequences and were evaluated at normal speed, slow motion and frame-by-frame to manually code these behaviors [23,34]. To capture the range of possible actions, behavior patterns associated with anticipation were scored (Table 1).

Table 1. Description of the anticipatory behaviors that were scored.

Behavior	Description
Step	Removal of at least two paws from the ground in an alternating manner
Walk	Removal of all four limbs off the ground in an alternating manner (left paw and right hind limb move simultaneously followed by right paw and left hind limb) OR significant shift from one location to another (if all limbs are not visible)
Run	Only two limbs touch the ground at any given time; the rat may alternate two limbs at a time (as is seen during walking behavior) OR the rat may move two paws followed by two hind limbs at any given time; such movement is accompanied by the extension of the torso as the front limbs reach forward followed by flexion of the torso as the hind limbs are removed from the ground and placed under the body
Jump	Up jump: the two front limbs leave the ground followed by the hind limbs while body is lifted into the air, then all limbs touch the ground simultaneously or closely one after the otherForward jump: the two front limbs are extended forward and removed from the ground followed by the removal of the hind limbs from the ground; this behavior is accompanied by the extension of the torso as the front limbs reach forward followed by flexion of the torso as the hind limbs are removed from the ground
Turn	Turn with one or both front limbs at a 45-, 90- or 180-degree angle OR turn with three or more limbs at a 360-degree angle. Turning may also be preceded by a stepping or walking pattern or followed by a rear (see below for the operational definition of rearing behavior)
Explore	Immobile; may extend one front limb; turning of head so as to examine the surrounding area

Table 1. *Cont.*

Behavior	Description
Dig	Vigorous forward and backward motion of front limbs while significantly displacing bedding
Rear	Standing on rear limbs with both front paws off ground (either free standing or against wall)
Shake	Vigorous side-to-side shudder of head, neck and trunk
Groom	Licking of paws; wipes/rubs face and nose; wipes behind ears, neck and/or downward to either side of the body; may grab fur and nibble with teeth. Grooming may consist of a variation of these behaviors many consecutive times. However, grooming is typically initiated by wiping of the nose or face and followed by grooming of the neck and body
Scratch	Rapid movement of hind limb with the claws rubbing against head, neck or side
Rest	Immobile; may turn head, but significantly less than is seen during exploration

Both the type of behavior and duration of that particular behavior were scored manually. Importantly, we assigned a behavioral category at every video frame, so that no time was left unaccounted for. This meant that the video frame of the termination of each behavior was the beginning of the next behavior. Frame-by-frame analysis of video was performed using Avidmux software, and the behaviors scored are shown in Table 1.

2.8. Vocalization Analyses

Acoustic data were analyzed using Raven Pro 1.4 software (Bioacoustics Research Program, Cornell Lab of Ornithology, Ithaca, NY, USA). The Raven Pro software generated spectrograms with a 256-sample Hann window from which the experimenter manually selected 50 kHz vocalizations. The 14 different 50 kHz vocalizations characterized by Wright et al. [3] were scored as distinct calls as we have done previously [16,26]. The occurrence of these calls was used to compare rates of calling, types of calling and whether different types of calls were associated with particular types of actions.

To compute the proportion of each call category emitted by each group, we summed the total number of each call for that group (e.g., total number of trills across all six Food Reward rats) and divided by the total number of all calls emitted by that group and expressed the result as a percentage. Analysis was based on the entire 2 min anticipation period. We also analyzed the change in vocalization rate from day 1 to day 7 for each call category. For this analysis, we first computed the rate of calling on day 1 and day 7 for each rat for each call, and then expressed this as a difference score (i.e., day 7 rate—day 1 rate). Difference scores were then averaged to compare the change in vocalization rate for each call category for each group. A similar method was used to analyze the vocalization rates for each call when the two play and two food groups were combined, except that all data was from day 1 and results are shown as raw vocalization rates.

2.9. Statistical Analyses

To evaluate the associations between all the behaviors and call types a Monte Carlo Shuffling method was used [16]. We first counted the number of co-occurrences of each vocalization type with each of the coded behavioral categories. A vocalization was counted as occurring during a particular behavior if the mid-point of the call occurred between the start and stop time of the behavior. To allow for small errors in coding of the start and stop times of behaviors, the window for counting a call as associated with a behavior was extended to 200 milliseconds before the start of the behavior and 200 milliseconds after the end of the behavior. Shuffling was achieved by assigning each vocalization a random time within the duration of the 2 min observation period. Hence, the relative frequency of vocalizations was kept the same for each shuffle. This shuffling was done 10,000 times and the total number of co-occurrences of each vocalization type with each type of behavior was tabulated. Based on the distribution of these counts, a z-score was calculated for each of the actual co-occurrences of each call–behavior pairing. The higher the z-score, the more likely that specific combination of call and behavior could have occurred by chance (i.e., for $p \leq 0.05$ the z score is +1.96 and for $p \leq 0.01$ the z score is +2.58). Large negative z-scores, on the other hand, indicate that the call and behavior are associated much less than expected by chance. Shuffling was performed separately for each animal and the z-scores averaged across animals in the same group to generate the final, average z- score values.

3. Results

3.1. Vocalizations

3.1.1. Vocalization Counts

To gage anticipation, we calculated the average vocalizations produced during the two minute period of anticipation for each group. As is evident in Figure 1, by day 7 of testing the Play Reward group had a significantly greater average vocal production than the other conditions. A repeated measures ANOVA was conducted on the influence of group (Play Reward, Play Control, Food Reward and Food Control) on the vocalizations produced on day 1 and 7 of testing. The effect for testing day was not significant $F(1, 4) = 0.233$, $p = 0.654$, partial $\eta^2 = 0.055$, but the group $F(3, 12) = 9.10$, $p = 0.002$, partial $\eta^2 = 0.695$ and testing day X group interaction were significant $F(3, 12) = 5.49$, $p = 0.013$, partial $\eta^2 = 0.578$. The Play Reward group increased vocalizations production in the anticipatory period, explaining the significant effect of group, however, overall the other groups did not show an increase; in fact, the two control groups actually decreased over days.

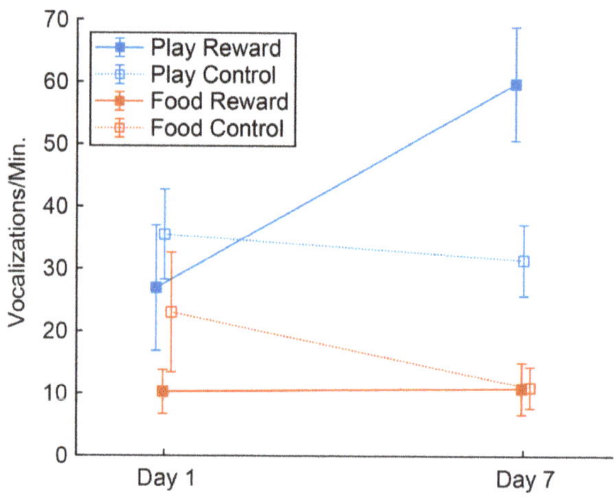

Figure 1. The average rate of 50 kHz vocalizations produced on day 1 and 7 of anticipatory testing. All error bars are standard error of the mean.

3.1.2. Vocalization Analyzed by Category

To assess if anticipation of different rewards impacted the type of vocalizations produced, we calculated the average number of each call subtype emitted over the entire 2 min anticipation period in each condition and expressed this as a proportion of all calls (Play Reward, Play Control, Food Reward, Food Control). To assess if the calls emitted changed over days of testing we performed this analysis on both day 1 of testing, when the animals had been habituated to the chamber but had not experienced reward, and day 7 when the reward groups had received 7 days of experience with rewards and the chamber and the control groups had 7 days experience with the test chamber. The analysis, shown in Figure 2, reveals several interesting patterns. First, both food deprivation and social deprivation appear to influence the types of calls produced on both days. Secondly, the pattern of calling on day 1 is similar for the Play Reward and Play Control groups, as is the pattern of calling in the Food Reward and Food Control groups, but the play and food conditions differ markedly. In particular, the rats in the food conditions exhibited a much higher proportion of flat and upramp calls and proportionally fewer trills than the rats in the play conditions. Thirdly, while the rats in the play groups mostly had minor changes in call distribution from day 1 to day 7, the rats in both food groups had a large increase in trill and trill with jumps and a reduction in flat calls. In fact, by day 7, the majority of calls from these rats were trills, trills with jumps and upramps. The rats in the play conditions, in contrast, had a wider variety of call types on day 7. This reduction in the variety of calls in the two food conditions on day 7 was also validated by a comparison of Gini coefficients [35]. On day 1 in the Play Reward and Play Control conditions, the Gini coefficients were 0.60 and 0.58, respectively, while on day 7, they were barely different at 0.58 and 0.54. In contrast, in the Food Reward and Food Control conditions on day 1, the Gini coefficients were 0.54 and 0.58, respectively, but then increased to 0.70 and 0.79.

To quantify these effects, we also compared the change in the average number of vocalizations of each type from day 1 to day 7, computing a change score for each vocalization. In Figure 3A, it is apparent that the Play Reward group showed increases in trills, trills with jumps and upramps. A two-tailed t-test was used to compare these change scores for control and reward groups for each vocal category. Compared to the Play Control group, the increase in calls was significant for both the trill ($t(10) = 2.81$, $p = 0.019$) and trill with jump ($t(10) = 2.70$, $p = 0.022$) calls. In contrast, Figure 3B shows that the Food Reward group showed an increase in trills and a decrease in upramps and flats, but none of these were statistically different compared to the changes in the Food Control group. Hence, the anticipation of social reward seems to lead to an increase in calls with trills (trills and trills with jumps), whereas the anticipation of food does not cause unique changes in the number of any types of calls.

As previously mentioned, our qualitative analysis (Figure 2) revealed dramatic differences between play and food groups in the types of calls used on day 1. To examine this effect in more detail, we combined the day 1 data from the Food Reward and Food Control groups and separately combined the data from the Play Reward and Play Control groups. As none of these groups had yet to experience the associated reward, there is no reason to suspect differences within the Play Reward/Play Control or Food Reward/Food Control supergroups. Hence, the only difference between the play and food supergroups is that one was socially isolated (play groups) and the other food deprived (food groups). Figure 4 shows a comparison of the average number of calls emitted by each group during the anticipation period, broken down by category. A two-tailed t-test was used to compute the probability of a difference between the play groups and food groups for each call category. Both groups emitted more trills than any other calls, but the play groups emitted far more trills than the food groups ($t(22) = 3.10$, $p = 0.005$). Significant differences were also seen between play and food groups in trills with jumps ($t(22) = 2.39$, $p = 0.026$) and short calls ($t(22) = 3.08$, $p = 0.005$) although the strength of this latter effect is due to the fact that there were zero short calls emitted by the food deprived animals on day 1. In sum, when placed in a new environment, rats that have been socially isolated emitted more trill, trill with jumps and short calls compared to rats that had been food deprived but not socially isolated.

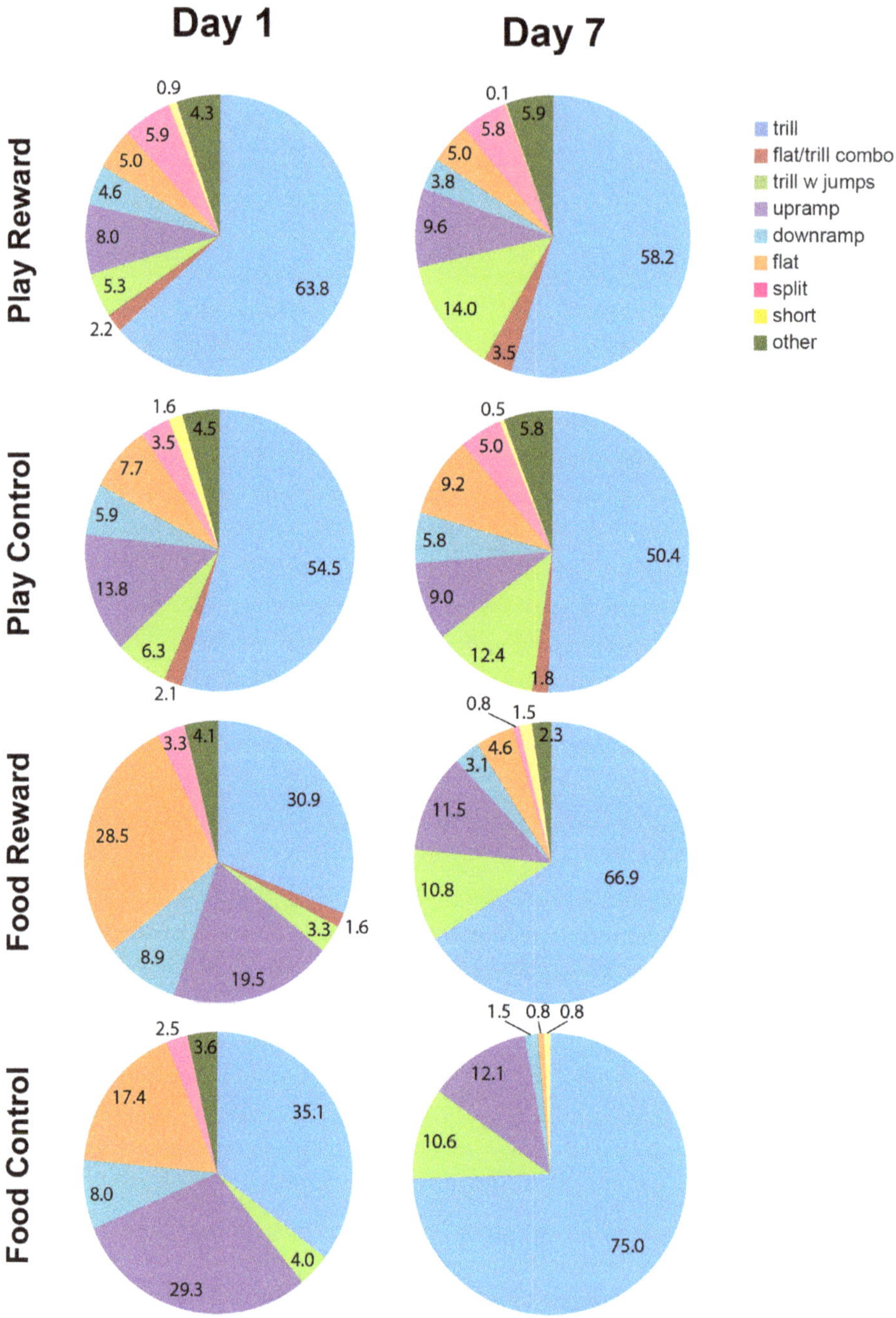

Figure 2. The average proportion of all commonly used call categories for each experimental group (Play Reward, Play Control, Food Reward, Food Control) on day 1 and 7 of testing are shown. Numbers show percentage of all calls for that group and testing day.

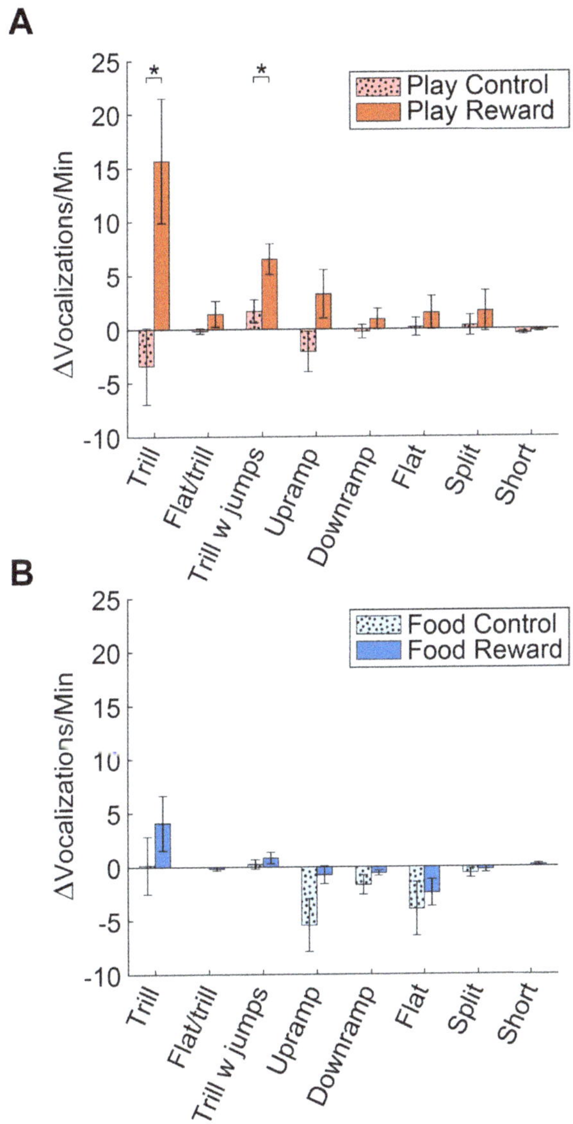

Figure 3. Comparison of the increase/decrease in call rates from day 1 to day 7 for each category of commonly produced calls. (**A**) Comparison of the change in call rates for Play Reward and Play Control groups. Asterisks denote comparisons that were statistically significant ($p < 0.05$). (**B**) Comparison of the change in call rates for Food Reward and Food Control groups.

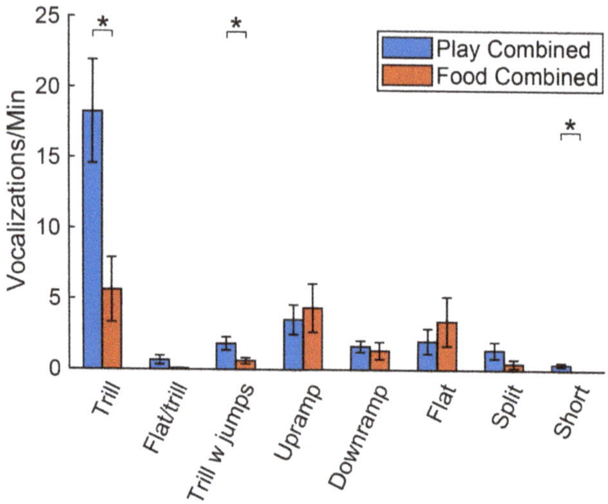

Figure 4. Comparison of the rates of calls on day 1 for all food and play groups. Play includes both Play Reward and Play Control while food includes both Food Reward and Food Control. Asterisks denote comparisons that were statistically significant ($p < 0.05$).

3.2. *Behavior*

We compared the mean time spent in each of the coded behaviors on day 1 and day 7 for each of the four treatment groups. We then grouped these measurements into slow locomotion (step, turn or walk) and fast locomotion (run or jump). The latter is of particular relevance because it could indicate the level of arousal. As shown in Figure 5, the two food groups showed no change in the average time spent in slow locomotion from day 1 to day 7, while both play groups showed a slight decrease. A two-way ANOVA with between-subjects factor group (Play Reward, Play Control, Food Reward, Food Control) and repeated-subjects factor day (1 or 7) showed only an effect of group ($F(3, 20) = 3.64$, $p = 0.03$, partial $\eta^2 = 0.35$). Tukey's HSD tests for multiple comparisons showed that the primary reason for this group effect was a significant difference between the Food Control and Play Control groups ($p = 0.046$). With fast locomotion, all groups showed an upward trend from day 1 to day 7 and this was borne out in a two-way ANOVA (group x day), which showed a significant effect of day ($F(1,20) = 24.20$, $p < 0.001$, partial $\eta^2 = 0.548$). There were also differences between groups ($F(3,20) = 4.458$, $p = 0.015$, partial $\eta^2 = 0.401$), which Tukey's HSD tests revealed were primarily due to a significant difference between the Food Control and Play Control groups ($p = 0.026$). Hence, we see effects of group on both slow and fast locomotion with play groups showing less slow locomotion and more fast locomotion but no significant differences in the rate of change of either variable from day 1 to day 7.

Two other behaviors, rearing and exploring, stood out because they apparently showed different patterns for the Play Reward group compared to the other groups. For exploration, the Play Reward group showed a reduction in duration from day 1 to day 7, but a two-way ANOVA (group × day) showed no significant effects of group, day or their interaction. Similarly, rearing duration increased from day 1 to day 7 only in the Play Reward group, but a two-way ANOVA (group × day) failed to show any significant differences, although the day X group interaction was close, with a p-value of 0.055.

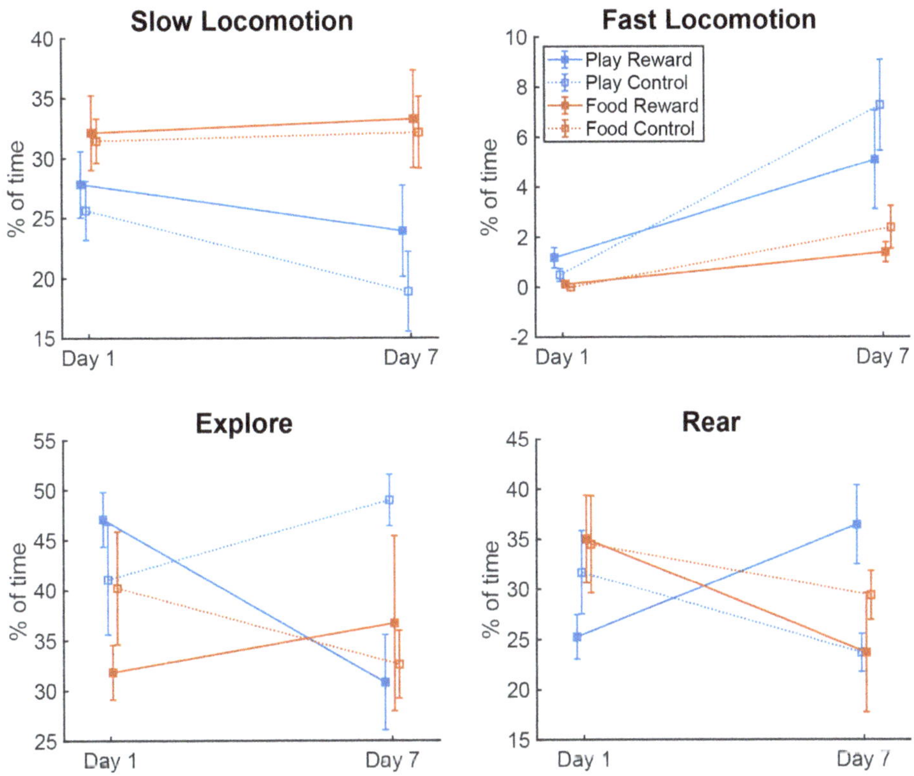

Figure 5. The average proportion of time spent in each of several key behaviors on day 1 and day 7 for each experimental group. Upper left shows the proportion of the 2 min test period spend in slow locomotion (single step, turn or walk). Upper right shows proportion of time spent in fast locomotion (running or jumping). Lower left shows time spent in exploratory behaviors. Lower right shows time spent rearing on hind legs. All error bars are standard error of the mean.

3.3. Vocal-Behavioral Associations

The vocal-behavior correlations shown in Figure 6 demonstrate several differences both between groups and days tested. First, on day 1, the Play Reward and Play Control groups had very similar profiles, with the strongest associations being between trill calls and walking, and between downramp and running. When comparing between groups on day 7, the Play Reward and Play Control groups were less similar. For the Play Reward group, the majority of the strong behavior-call associations involving running and jumping, whereas for the Play Control group, the majority of strong associations were with walking and running. More specifically, in the Play Reward group there were strong associations of downramp, flat and split with jumps. Trills, flat/trill combinations and trills with jumps were associated with running. In the Play Control group, on the other hand, the strongest associations were between trills and walking, and between downramps and running. On day 1, the Food Reward and Food Control also had similar profiles, with the trill-walk, upramp-walk and flat-walk being the predominant associations. Arguably the most interesting finding is that by day 7, both the Food Reward and Food Control groups did not have any significant vocal-behavior associations.

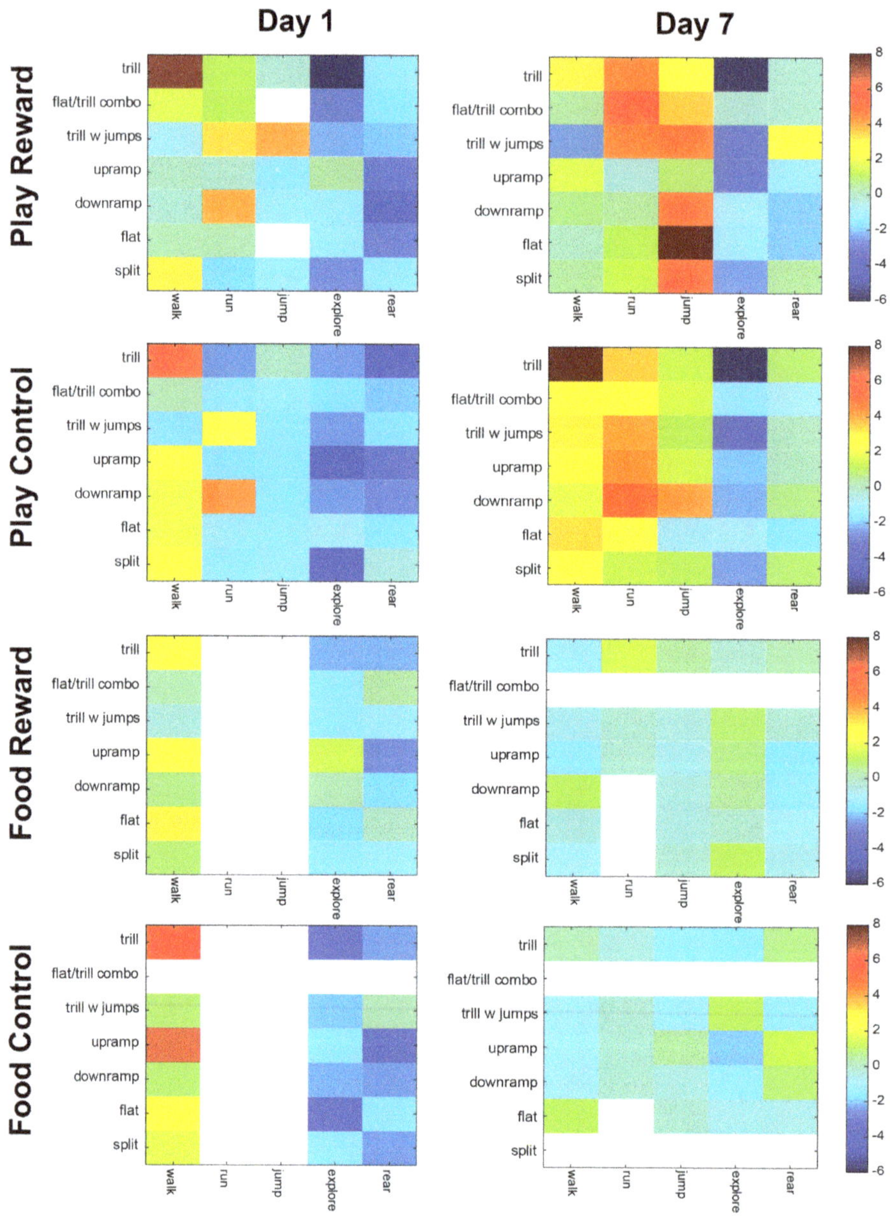

Figure 6. Association between types of calls and types of behavior are shown for the four groups on day 1 and 7 of the anticipation trials. Each matrix shows the strength of association, as a z-score, for each combination of behavior (x-axis) and vocalization category (y-axis). Deep red indicates the strongest positive association and deep blue the strongest negative association. The white sections indicate that either the behavior or the vocalization in that category did not have sufficient instances to run the analysis.

4. Discussion

The primary goal of this study was to compare the vocalizations emitted by male rats in anticipation of two types of reward: food and social play. Care was taken to equate the age of the animals being tested, as social behavior and vocalizations change dramatically with age [36]. We also included controls for the effects of social isolation and food deprivation, as both might be expected to affect the production of vocalizations irrespective of the presence of rewards. Over seven days, repeated pairing of the recording chamber with the reward of a play partner led to an increase in 50 kHz vocalizations, a change not present in the social isolation control group. There were also trends towards greater high-intensity movement, less exploration and increased rearing in the group rewarded with play, although none of these behavioral changes were statistically significant. Examining each call category individually showed that both the trill and trill with jump calls increased with increased training, suggesting that these two calls in particular may have a social role. In contrast, repeated pairing of the reward chamber with food did not lead to any discernable changes in either vocalizations or behavior.

A secondary finding was the robust, but different effect of social isolation and food deprivation on vocalizations. Hungry rats produced fewer and different calls than socially isolated ones. The lower number of calls in the food-deprived animals is apparent in Figure 1 and this finding is consistent with previous reports that food-deprived animals call less [37,38]. Differences in the distribution of calls is apparent in Figure 2 on both days 1 and 7, where distributions look similar within the two play conditions and within the two food conditions, but very different between the food and play super-groups. A quantitative comparison of call rates on the first day of testing showed that the primary difference between play and food groups was in trills and trills with jumps, the same two calls whose prevalence correlates with the expectation of social reward. This adds to the evidence that these calls have a social role. The increased drive for play induced by social isolation increases their prevalence and the expectation of the arrival of a play partner increases their prevalence further.

Our finding that trills and trills with jumps are tied to the expectation of social reward is broadly consistent with the findings of others. Earlier studies have shown that the expectation of social play in juveniles increases the emission of 50 kHz calls, generally [17]. While this early study did not categorize calls, a later study showed that frequency-modulated 50 kHz calls predominate during play itself, at least among juvenile males [14]. Our results are also similar to the pattern reported by Wright et al. [3] who found that two male adult rats placed in a chamber together after saline injection emitted, as the most frequent call categories, 20% trills, 17% flat/trills and 20% flats. The increased frequency of flat-containing calls in that study may be either because of recent injection or because adult males tend to have more aggressive encounters, which have been linked to 50 kHz flat calls [14]. More recently, Mulvihill and Brudzynski [29] showed that trill calls, in particular, are more common as male rats explore a space recently vacated by an estrous female. Mating is also commonly associated with 50 kHz calls in the period before ejaculation and 22 kHz calls afterwards [39]. It would be interesting to compare the types of 50 kHz calls emitted during sex with those during play and other affiliative behaviors, but most studies of sexual vocalizations were conducted well before the common use spectroscopic analysis. One relatively recent report shows that the calls during sex are frequency modulated, but does not categorize calls any further [14]. In studies very similar to the present one, we have previously found high rates of both trills and trills with jumps in male rats anticipating social reward (but there we did not demonstrate that rates were modulated by social expectation) [16,40]. Further, when male juvenile rats play, the most commonly emitted calls are trills and trills with jumps [4]. Taken together, the data suggest that trills and possibly trills with jumps play a role in calling to other rats, possibly to broadcast a general state of positive affect and/or to attract them [1,2,41,42]. Trills, in particular, are the most common call detected in many studies, suggesting that, although trill rate is modulated by social expectation, rats may be set, by default, for constant social signaling [3,16,43].

The lack of anticipatory vocalizations or behavior in our food condition is puzzling, especially in light of several other studies showing an increase in 50 kHz calls during the expectation of food. Willey and Spear [28] showed elevated 50 kHz calling when male rats were placed in a chamber with food on the other side of a barrier. Rats also show an increase in 50 kHz calls in the 15 min before their daily feeding [11]. Several studies have shown that 50 kHz vocalizations increase after presentation of a tone or light cue that predicts food delivery [8–10,38]. The one exception to this pattern was a study by Tripi et al. [44], which showed that Pavlovian conditioning with lever, light and food did not lead to elevated cue-related calling, though the anticipatory period was relatively short (8 s). The weight of the evidence, however, suggests that rats will elevate their calling when context or cues predict the arrival of food.

Why did others find increases in 50 kHz vocalizations associated with food expectation while we did not? The contrast is most striking with Burgdorf et al. [8], upon which our study was modelled. Both studies used the same strain of rat, the same 2 min expectation window, and similar methods of food deprivation. Burgdorf et al. did use a mix of male and female rats while we used solely males, but previous studies have shown that sex differences in vocalizations are either subtle or non-existent [45–47] (see below). Hence, the most notable difference was that we used juvenile rats while they, and all the other studies cited above, used adults. The idea that anticipatory vocalizations for food are age-specific is consistent with a previous report that male adolescent rats show lower levels of food-associated vocalizations than adults [28].

Another explanation for our lack of food expectancy calls is that our rats were not as motivated by food. The use of juveniles was necessary because we wanted to compare the effects of food and play in similar groups of animals and juveniles exhibit a pronounced peak in play activity between 30–40 days of age [33,48]. Unfortunately, this imposed constrains on our ability to food-deprive our animals, because prolonged caloric restriction at this age leads to stunted growth. In our study, a control group given free access to food (in this case, the two play groups) were used as a control to set weight targets for rats as they grew. However, as the food-deprived rats probably did slow their growth, the freely fed rats may have served as an overly generous target for our 15% weight reduction. As a consequence, the motivation to seek out food reward may have been reduced in our food groups, resulting in the lack of anticipatory behavior for food reward. On the other hand, we used a highly palatable food reward (chocolate chips) that at least one other study showed was sufficient to induce 50 kHz vocalizations even in rats that were not food deprived [28].

A third possible explanation for our lack of food expectancy calls is that we simply lacked the statistical power to detect increased vocalizations for food reward. There are intriguing non-significant differences in the change in vocalizations from day 1 to day 7 between the Food Reward and Food Control groups (Figure 3B) that suggest that a larger number of subjects might have allowed detection of some differences. On the other hand, even with the low number of animals, there were very clear differences in the Play Reward and Play Control conditions (Figure 3A), suggesting that low power was not a critical limitation. To us, the most likely explanation is simply that rats at this age, due to some biological programming, simply care far more about social activities than they do about food.

While our study did not allow us to determine which calls are tied to food reward, there is considerable evidence from other studies. Many studies use only broad categories of calls. For example, Opiol et al. [11], showed an increase in frequency modulated calls tied to a tone that predicted the delivery of food. Other studies are more specific. In one, male rats were trained for 24 days to expect food after a light cue [9]. In the 2 min anticipatory window, the calls related to the expectation of food were "other frequency modulated" (which excludes calls with trills), "step frequency modulated" (which look like Wright et al.'s category, split) and 50 kHz flat calls. Similarly, Brenes and Schwarting [10] found an increase in step calls over flat and trill calls during cued anticipation of food reward. In a recent study that directly examined the different calls elicited by food and social stimuli, male rats were allowed to explore a space with either a highly palatable food reward or an empty space recently occupied by an estrous female [29]. Flat calls were more common in the food group while calls with trills were more common in response to the female. The elevated flat calls are consistent with previous reports of 40 kHz flat calls related to food consumption [11,27]. Taken together, the evidence suggests that the expectation of food elicits flat, step, split and other frequency-modulated calls but notably, not calls with trills. This is a striking contrast with the social-related calls we observed, both of which include trills.

Further insights into the function of different 50 kHz calls can be gleaned from studies with amphetamine, a highly reinforcing drug. Amphetamine induces a robust increase in 50 kHz vocalizations both from acute administration or to contextual cues associated with the drug [5,49–51]. Some studies have found that amphetamine increases all types of 50 kHz calls [52]. Another study by the same first author found increases in flat, trill, complex, inverted U, short, step up, multistep, upward ramp calls [43]. Other studies have found more selective effects on specific vocalizations. Two separate studies showed that injection of amphetamine causes a selective increase in trills and a decrease in 50 kHz flat calls [3,53]. In sum, amphetamine induces trills and possibly other frequency-modulated calls. From this perspective, amphetamine elicits vocalizations very similar to those associated with social reward while food seems to elicit a non-overlapping set consisting of flat, step and other calls without trills. This suggests that vocalizations may be specific to certain types of reward, but more study is clearly needed.

We have previously provided evidence that certain categories of rat vocalizations are selectively emitted when rats are performing specific actions [4,16]. In the current study, we provide further evidence of the selective emission of calls with respect to behavior (Figure 6). The data are roughly consistent with findings from our previous study with the same strain of rats on the anticipatory period before play [16]. Admittedly, the plots are not identical because certain calls were omitted from each analysis due to low numbers, and the omitted calls were different in the two studies. Looking at the data from Day 7 in the Play Reward condition, we can see that walking is associated with trills, running with any trill call (trill, flat/trill combo and trill with jumps) and jumping is associated with a wide range of calls, most notably split, flat and downramp. In contrast, the data from Burke et al., [16], Figure 3C, largely agree for walk and partially agree for runs. In that study, running was associated with composite and trill with jumps, both of which have trills. The current associations between split calls and jumps agrees with Burke et al. [16] but that study included composite and multi-step, neither of which was common enough in the present dataset to analyze. Both studies agree that exploration and rearing are negatively correlated with vocalizations. Finally, it is interesting to note that the food-deprived groups show a marked lack of correlation between vocalizations and specific behaviors, especially on Day 7 (Figure 6, bottom two rows). Our data suggest that food deprivation not only reduces the number of vocalizations, but also may desynchronize their association with behaviors. We cannot rule out the possibility that this de-synchronization is due to low numbers of calls, but our shuffling method is robust and, if anything, tends to overestimate associations when vocal counts are low (which is why many rows and columns in that data are left blank). Taken together, this data strengthens the case that specific calls are tied to specific behaviors, but studies with a much higher number of vocalizations may be needed to iron out the specifics.

One limitation of our study is that it was restricted to male rats, as were the majority of studies cited above. Studies of the effects of sex differences suggest that, at least in juvenile play, sex differences in vocalizations are, if anything, subtle. Certainly, juvenile male rats play more than female rats [54,55]. It is hence not surprising that one recent study in Sprague-Dawley rats found lower overall vocalizations in juvenile females [47]. The authors also found decreases in specific calls (flat and step) but not others (trill). On the other hand, Gzielo et al. [46], also working with adolescent Sprague-Dawley rats, found no sex differences in either the number of vocalizations or their structure (duration, frequency and bandwidth). Similarly, we have recently compared vocalization sub-types in juvenile Long Evans rats during play and found no sex differences [45]. Adding a female group would certainly be valuable; however, as our study also used Long Evans rats, we have reasonable grounds to assume that their results would be substantively the same as those reported here.

5. Conclusions

In conclusion, our data add to the evidence that calls with trills are associated with social reward. It may not be possible to do a strict apples-to-apples comparison between food and social reward because the motivation for social interaction peaks in adolescence while food deprivation studies work better with adults. However, future studies that compare vocalizations to both forms of reward in rats of different ages but keeping all other parameters equivalent would be helpful to elicit exactly which calls are tied to social reward and which to food reward.

Author Contributions: Conceptualization, C.J.B., D.R.E. and S.M.P.; investigation, C.J.B. and M.M., methodology, C.J.B., D.R.E. and S.M.P.; analysis, C.J.B. and D.R.E.; resources, D.R.E.; software, C.J.B. and D.R.E.; data curation, C.J.B.; writing—original draft preparation, C.J.B., D.R.E. writing—review and editing, D.R.E., S.M.P.; supervision, D.R.E. and S.M.P.; project administration, C.J.B. and D.R.E.; funding acquisition, D.R.E. and S.M.P. All authors have read and agreed to the published version of the manuscript.

Funding: The work was supported by grants from the Natural Science and Engineering Research Council of Canada to DRE and SMP, grant numbers 2018-05777 and 2018-03706, respectively and an Alberta Innovates Health Solutions grant to DRE.

Institutional Review Board Statement: The study was conducted according to the guidelines of the Declaration of Helsinki, and approved by the Institutional Animal Welfare Committee of the University of Lethbridge (protocol #1201 approved 2 February 2012).

Informed Consent Statement: Not applicable.

Data Availability Statement: Data available upon request.

Acknowledgments: We kindly thank the animal care staff and veterinarian for their help with the care of the rats.

Conflicts of Interest: The authors declare no conflict of interest.

References

1. Knutson, B.; Burgdorf, J.; Panksepp, J. Ultrasonic vocalizations as indices of affective states in rats. *Psychol. Bull.* **2002**, *128*, 961–977. [CrossRef]
2. Brudzynski, S.M. Ethotransmission: Communication of emotional states through ultrasonic vocalization in rats. *Curr. Opin. Neurobiol.* **2013**, *23*, 310–317. [CrossRef] [PubMed]
3. Wright, J.M.; Gourdon, J.C.; Clarke, P.B. Identification of multiple call categories within the rich repertoire of adult rat 50-kHz ultrasonic vocalizations: Effects of amphetamine and social context. *Psychopharmacology* **2010**, *211*, 1–13. [CrossRef] [PubMed]
4. Burke, C.J.; Kisko, T.M.; Euston, D.R.; Pellis, S.M. Do juvenile rats use specific ultrasonic calls to coordinate their social play? *Anim. Behav.* **2018**, *140*, 81–92. [CrossRef]
5. Knutson, B.; Burgdorf, J.; Panksepp, J. High-frequency ultrasonic vocalizations index conditioned pharmacological reward in rats. *Physiol. Behav.* **1999**, *66*, 639–643. [CrossRef]
6. Burgdorf, J.; Knutson, B.; Panksepp, J.; Ikemoto, S. Nucleus accumbens amphetamine microinjections unconditionally elicit 50-kHz ultrasonic vocalizations in rats. *Behav. Neurosci.* **2001**, *115*, 940–944. [CrossRef]
7. Wintink, A.J.; Brudzynski, S.M. The related roles of dopamine and glutamate in the initiation of 50-kHz ultrasonic calls in adult rats. *Pharmacol. Biochem. Behav.* **2001**, *70*, 317–323. [CrossRef]
8. Burgdorf, J.; Knutson, B.; Panksepp, J. Anticipation of rewarding electrical brain stimulation evokes ultrasonic vocalization in rats. *Behav. Neurosci.* **2000**, *114*, 320–327. [CrossRef]
9. Buck, C.L.; Vendruscolo, L.F.; Koob, G.F.; George, O. Dopamine D1 and μ-opioid receptor antagonism blocks anticipatory 50 kHz ultrasonic vocalizations induced by palatable food cues in Wistar rats. *Psychopharmacology* **2014**, *231*, 929–937. [CrossRef]

10. Brenes, J.C.; Schwarting, R.K. Individual differences in anticipatory activity to food rewards predict cue-induced appetitive 50-kHz calls in rats. *Physiol. Behav.* **2015**, *149*, 107–118. [CrossRef]
11. Opiol, H.; Pavlovski, I.; Michalik, M.; Mistlberger, R.E. Ultrasonic vocalizations in rats anticipating circadian feeding schedules. *Behav. Brain. Res.* **2015**, *284*, 42–50. [CrossRef]
12. Bialy, M.; Rydz, M.; Kaczmarek, L. Precontact 50-kHz vocalizations in male rats during acquisition of sexual experience. *Behav. Neurosci.* **2000**, *114*, 983–990. [CrossRef]
13. Thomas, D.A.; Barfield, R.J. Ultrasonic vocalization of the female rat (*Rattus norvegicus*) during mating. *Anim. Behav.* **1985**, *33*, 720–725. [CrossRef]
14. Burgdorf, J.; Kroes, R.A.; Moskal, J.R.; Pfaus, J.G.; Brudzynski, S.M.; Panksepp, J. Ultrasonic vocalizations of rats (*Rattus norvegicus*) during mating, play, and aggression: Behavioral concomitants, relationship to reward, and self-administration of playback. *J. Comp. Psychol.* **2008**, *122*, 357–367. [CrossRef] [PubMed]
15. Börner, A.; Hjemdahl, R.; Götz, T.; Brown, G.R. Ultrasonic vocalizations of female Norway rats (*Rattus norvegicus*) in response to social partners. *J. Comp. Psychol.* **2016**, *130*, 76. [CrossRef] [PubMed]
16. Burke, C.J.; Kisko, T.M.; Swiftwolfe, H.; Pellis, S.M.; Euston, D.R. Specific 50-kHz vocalizations are tightly linked to particular types of behavior in juvenile rats anticipating play. *PLoS ONE* **2017**, *12*, e0175841. [CrossRef] [PubMed]
17. Knutson, B.; Burgdorf, J.; Panksepp, J. Anticipation of play elicits high-frequency ultrasonic vocalizations in young rats. *J. Comp. Psychol.* **1998**, *112*, 65–73. [CrossRef]
18. Willey, A.R.; Spear, L.P. Development of anticipatory 50 kHz USV production to a social stimuli in adolescent and adult male Sprague-Dawley rats. *Behav. Brain Res.* **2012**, *226*, 613–618. [CrossRef] [PubMed]
19. Brudzynski, S.M.; Pniak, A. Social contacts and production of 50-kHz short ultrasonic calls in adult rats. *J. Comp. Psychol.* **2002**, *116*, 73–82. [CrossRef]
20. Himmler, B.T.; Kisko, T.M.; Euston, D.R.; Kolb, B.; Pellis, S.M. Are 50-kHz calls used as play signals in the playful interactions of rats? I. Evidence from the timing and context of their use. *Behav. Process.* **2014**, *106*, 60–66. [CrossRef]
21. Burgdorf, J.; Panksepp, J. Tickling induces reward in adolescent rats. *Physiol. Behav.* **2001**, *72*, 167–173. [CrossRef]
22. Blanchard, R.J.; Yudko, E.B.; Blanchard, D.C.; Taukulis, H.K. High-frequency (35–70 kHz) ultrasonic vocalizations in rats confronted with anesthetized conspecifics: Effects of gepirone, ethanol, and diazepam. *Pharmacol. Biochem. Behav.* **1993**, *44*, 313–319. [CrossRef]
23. Kisko, T.M.; Euston, D.R.; Pellis, S.M. Are 50-khz calls used as play signals in the playful interactions of rats? III. The effects of devocalization on play with unfamiliar partners as juveniles and as adults. *Behav. Process.* **2015**, *113*, 113–121. [CrossRef]
24. Brudzynski, S.M. Communication of adult rats by ultrasonic vocalization: Biological, sociobiological, and neuroscience approaches. *ILAR J.* **2009**, *50*, 43–50. [CrossRef]
25. Wohr, M.; Houx, B.; Schwarting, R.K.; Spruijt, B. Effects of experience and context on 50-kHz vocalizations in rats. *Physiol. Behav.* **2008**, *93*, 766–776. [CrossRef]
26. Burke, C.J.; Kisko, T.M.; Pellis, S.M.; Euston, D.R. Avoiding escalation from play to aggression in adult male rats: The role of ultrasonic calls. *Behav. Process.* **2017**, *144*, 72–81. [CrossRef] [PubMed]
27. Takahashi, N.; Kashino, M.; Hironaka, N. Structure of rat ultrasonic vocalizations and its relevance to behavior. *PLoS ONE* **2010**, *5*, e14115. [CrossRef]
28. Willey, A.R.; Spear, L.P. The effects of pre-test social deprivation on a natural reward incentive test and concomitant 50 kHz ultrasonic vocalization production in adolescent and adult male Sprague-Dawley rats. *Behav. Brain Res.* **2013**, *245*, 107–112. [CrossRef] [PubMed]
29. Mulvihill, K.G.; Brudzynski, S.M. Non-pharmacological induction of rat 50kHz ultrasonic vocalization: Social and non-social contexts differentially induce 50kHz call subtypes. *Physiol. Behav.* **2018**, *196*, 200–207. [CrossRef]
30. Himmler, B.T.; Pellis, S.M.; Kolb, B. Juvenile play experience primes neurons in the medial prefrontal cortex to be more responsive to later experiences. *Neurosci. Lett.* **2013**, *556*, 42–45. [CrossRef]
31. Niesink, R.J.; Van Ree, J.M. Involvement of opioid and dopaminergic systems in isolation-induced pinning and social grooming of young rats. *Neuropharmacology* **1989**, *28*, 411–418. [CrossRef]
32. Panksepp, J.; Beatty, W.W. Social deprivation and play in rats. *Behav. Neural. Biol.* **1980**, *30*, 197–206. [CrossRef]
33. Pellis, S.M.; Pellis, V.C. Differential rates of attack, defense, and counterattack during the developmental decrease in play fighting by male and female rats. *Dev. Psychobiol.* **1990**, *23*, 215–231. [CrossRef]
34. Kisko, T.M.; Himmler, B.T.; Himmler, S.M.; Euston, D.R.; Pellis, S.M. Are 50-kHz calls used as play signals in the playful interactions of rats? II. Evidence from the effects of devocalization. *Behav. Process.* **2015**, *111*, 25–33. [CrossRef] [PubMed]
35. Lengwiler, Y. Gini Coefficient and the Lorentz Curve. MATLAB Central File Exchange: 2021. Available online: https://www.mathworks.com/matlabcentral/fileexchange/28080-gini-coefficient-and-the-lorentz-curve (accessed on 23 August 2021).
36. Pellis, S.M.; Burke, C.J.; Kisko, T.M.; Euston, D.R. 50-kHz Vocalizations, Play and the Development of Social Competence. In *Handbook of Behavioral Neuroscience*; Brudzynski, S.M., Ed.; Elsevier: Amsterdam, The Netherlands, 2018; Volume 25, pp. 117–126.
37. Brenes, J.C.; Lackinger, M.; Hoglinger, G.U.; Schratt, G.; Schwarting, R.K.; Wohr, M. Differential effects of social and physical environmental enrichment on brain plasticity, cognition, and ultrasonic communication in rats. *J. Comp. Neurol.* **2015**. [CrossRef]
38. Brenes, J.C.; Schwarting, R.K. Attribution and expression of incentive salience are differentially signaled by ultrasonic vocalizations in rats. *PLoS ONE* **2014**, *9*, e102414. [CrossRef]

39. Barfield, R.J.; Thomas, D.A. The role of ultrasonic vocalizations in the regulation of reproduction in rats. *Ann. N. Y. Acad. Sci.* **1986**, *474*, 33–43. [CrossRef] [PubMed]
40. Burke, C.J.; Modlinska, K.; Mauro, M.H.; Aleksandrova, L.R.; Pellis, S.M.; Phillips, A.G.; Euston, D.R. A naturalistic method to test depression: Anticipation of play. *Behav. Brain Res.* **2021**, *398*, 112975. [CrossRef] [PubMed]
41. Burgdorf, J.; Panksepp, J.; Moskal, J.R. Frequency-modulated 50 kHz ultrasonic vocalizations: A tool for uncovering the molecular substrates of positive affect. *Neurosci. Biobehav. Rev.* **2011**, *35*, 1831–1836. [CrossRef] [PubMed]
42. Wöhr, M. Ultrasonic communication in rats: Appetitive 50-kHz ultrasonic vocalizations as social contact calls. *Behav. Ecol. Sociobiol* **2018**, *72*, 14. [CrossRef]
43. Simola, N.; Costa, G. Emission of categorized 50-kHz ultrasonic vocalizations in rats repeatedly treated with amphetamine or apomorphine: Possible relevance to drug-induced modifications in the emotional state. *Behav. Brain Res.* **2018**, *347*, 88–98. [CrossRef] [PubMed]
44. Tripi, J.A.; Dent, M.L.; Meyer, P.J. Individual differences in food cue responsivity are associated with acute and repeated cocaine-induced vocalizations, but not cue-induced vocalizations. *Psychopharmacology* **2017**, *234*, 437–446. [CrossRef] [PubMed]
45. Stark, R.A.; Ramkumar, R.; Pellis, S.M. Deficient play-derived experiences in juvenile Long Evans rats reared with a Fischer 344 partner: A deficiency shared by both sexes. *Int. J. Comp. Psychol.* **2021**, in press.
46. Gzielo, K.; Potasiewicz, A.; Holuj, M.; Litwa, E.; Popik, P.; Nikiforuk, A. Valproic acid exposure impairs ultrasonic communication in infant, adolescent and adult rats. *Eur. Neuropsychopharmacol.* **2020**, *41*, 52–62. [CrossRef]
47. Kisko, T.M.; Schwarting, R.K.W.; Wohr, M. Sex differences in the acoustic features of social play-induced 50-kHz ultrasonic vocalizations: A detailed spectrographic analysis in wild-type Sprague-Dawley and Cacna1c haploinsufficient rats. *Dev. Psychobiol.* **2021**, *63*, 262–276. [CrossRef]
48. Pellis, S.M.; Pellis, V.C. The prejuvenile onset of play fighting in laboratory rats (*Rattus norvegicus*). *Dev. Psychobiol.* **1997**, *31*, 193–205. [CrossRef]
49. Ma, S.T.; Maier, E.Y.; Ahrens, A.M.; Schallert, T.; Duvauchelle, C.L. Repeated intravenous cocaine experience: Development and escalation of pre-drug anticipatory 50-kHz ultrasonic vocalizations in rats. *Behav. Brain. Res.* **2010**, *212*, 109–114. [CrossRef]
50. Mu, P.; Fuchs, T.; Saal, D.B.; Sorg, B.A.; Dong, Y.; Panksepp, J. Repeated cocaine exposure induces sensitization of ultrasonic vocalization in rats. *Neurosci. Lett.* **2009**, *453*, 31–35. [CrossRef]
51. Simola, N.; Frau, L.; Plumitallo, A.; Morelli, M. Direct and long-lasting effects elicited by repeated drug administration on 50-kHz ultrasonic vocalizations are regulated differently: Implications for the study of the affective properties of drugs of abuse. *Int. J. Neuropsychopharmacol.* **2014**, *17*, 429–441. [CrossRef]
52. Simola, N.; Fenu, S.; Costa, G.; Pinna, A.; Plumitallo, A.; Morelli, M. Pharmacological characterization of 50-kHz ultrasonic vocalizations in rats: Comparison of the effects of different psychoactive drugs and relevance in drug-induced reward. *Neuropharmacology* **2012**, *63*, 224–234. [CrossRef] [PubMed]
53. Willadsen, M.; Best, L.M.; Wohr, M.; Clarke, P.B.S. Effects of anxiogenic drugs on the emission of 22- and 50-kHz ultrasonic vocalizations in adult rats. *Psychopharmacology* **2018**, *235*, 2435–2445. [CrossRef] [PubMed]
54. Pellis, S.M.; Field, E.F.; Smith, L.K.; Pellis, V.C. Multiple differences in the play fighting of male and female rats. Implications for the causes and functions of play. *Neurosci. Biobehav. Rev.* **1997**, *21*, 105–120. [CrossRef]
55. Auger, A.P.; Olesen, K.M. Brain Sex Differences and the Organisation of Juvenile Social Play Behaviour. *J. Neuroendocrinol.* **2009**, *21*, 519–525. [CrossRef] [PubMed]

brain sciences

MDPI

Article

Nucleus Accumbens Chemogenetic Inhibition Suppresses Amphetamine-Induced Ultrasonic Vocalizations in Male and Female Rats

Kate A. Lawson †, Abigail Y. Flores †, Rachael E. Hokenson †, Christina M. Ruiz and Stephen V. Mahler *

Department of Neurobiology & Behavior, University of California, Irvine. 1203 McGaugh Hall, Irvine, CA 92697, USA; lawsonk1@uci.edu (K.A.L.); abigaiyf@uci.edu (A.Y.F.); rhokenso@uci.edu (R.E.H.); cruiz1@uci.edu (C.M.R.)
* Correspondence: mahlers@uci.edu
† Denotes equal contributions.

Abstract: Adult rats emit ultrasonic vocalizations (USVs) related to their affective states, potentially providing information about their subjective experiences during behavioral neuroscience experiments. If so, USVs might provide an important link between invasive animal preclinical studies and human studies in which subjective states can be readily queried. Here, we induced USVs in male and female Long Evans rats using acute amphetamine (2 mg/kg), and asked how reversibly inhibiting nucleus accumbens neurons using designer receptors exclusively activated by designer drugs (DREADDs) impacts USV production. We analyzed USV characteristics using "Deepsqueak" software, and manually categorized detected calls into four previously defined subtypes. We found that systemic administration of the DREADD agonist clozapine-n-oxide, relative to vehicle in the same rats, suppressed the number of frequency-modulated and trill-containing USVs without impacting high frequency, unmodulated (flat) USVs, nor the small number of low-frequency USVs observed. Using chemogenetics, these results thus confirm that nucleus accumbens neurons are essential for production of amphetamine-induced frequency-modulated USVs. They also support the premise of further investigating the characteristics and subcategories of these calls as a window into the subjective effects of neural manipulations, with potential future clinical applications.

Keywords: 50 kHz vocalizations; 22 kHz vocalizations; amphetamine; nucleus accumbens; chemogenetics; clozapine-n-oxide; UMAP; males; females

check for updates

Citation: Lawson, K.A.; Flores, A.Y.; Hokenson, R.E.; Ruiz, C.M.; Mahler, S.V. Nucleus Accumbens Chemogenetic Inhibition Suppresses Amphetamine-Induced Ultrasonic Vocalizations in Male and Female Rats. *Brain Sci.* **2021**, *11*, 1255. https://doi.org/10.3390/brainsci11101255

Academic Editors: Stefan M. Brudzynski and Jeffrey Burgdorf

Received: 28 July 2021
Accepted: 17 September 2021
Published: 22 September 2021

1. Introduction

Rats emit vocalizations in the ultrasonic range, but the fundamental function of these emissions is still controversial. These rat ultrasonic vocalizations (USVs) seem to be related to affective states experienced by the animal, and may also have socially communicative functions [1–11]. We are most interested in the potential usefulness of USVs for querying the subjective states of rats, such as those produced by addictive drugs, in behavioral neuroscience experiments. By carefully measuring and analyzing effects of neural manipulations on the quantity and quality of USVs produced, we hope to gain new information about how defined neural populations help generate rats' affective states. Along with the intrinsic scientific importance of this question, we hope this approach may also help inform future development of brain intervention approaches for treating psychiatric disease in humans, including addiction.

Adult rats emit USVs in two main ranges that are thought to correspond roughly to negative and positive emotional or anticipatory states [2,4,12,13]: Lower-frequency USVs (LF; 18–30 kHz; also called "22 kHz" vocalizations) are linked to stress and fear states [14–19]; higher-frequency USVs (30–100 kHz) are linked instead to reward anticipation or experience [6–8,20–23]. LF vocalizations are generally not frequency-modulated, and can occur with short or long durations, potentially reflecting the intensity of negative effects [24–26]. In contrast, high-frequency USVs (also called "50 kHz" vocalizations), are more complex, with at least 14 distinct patterns [27]. They are emitted at a range of principal frequencies (30–100 kHz), durations, and combinations of acoustic elements. Of particular note, some vocalizations have a rapid oscillating component called "trills." Other high-frequency USVs have a consistent principal frequency (Flat), and others have a variety of non-trill, but still frequency-modulated patterns (FM). Although many reports do not distinguish between these patterns of high-frequency USVs, there is some evidence that they are produced under distinct circumstances, and subserved by different neural substrates [28–31].

In order to determine whether potentially useful information is contained in rat USVs, and to determine the neural substrates responsible, USVs must first be accurately recorded and analyzed. Advances in microphone technology and tools for visualizing USVs allow for accurate analysis of their presence and characteristics, greatly facilitating this objective. However, this approach has been limited by the necessity of manually categorizing and quantifying calls, which is very time consuming. Therefore, several groups and companies have developed USV detection and analysis software in recent years [32–37]. Here, we employed one of these, a machine learning-driven detection and categorization tool called Deepsqueak [33], which detects rat USVs and extracts parameters of them with reasonable accuracy. We contend that optimizing this, or similar, approaches may hold the key to extracting rich information about rat subjective states from USV data, thus facilitating future studies of neural substrates in rats.

In a step toward this aim, we here asked how reversibly inhibiting nucleus accumbens (NAc) neurons affects emission of rat USVs. NAc is a key anatomical substrate for high frequency USV production [22,38–41], and here, for the first time in the context of affective USV measurement, we inhibit neurons there using designer receptors exclusively activated by designer drugs (DREADDs). DREADDs are engineered G-protein-coupled receptors which do not respond to endogenous neurotransmitters and are thus inert—unless recruited by the experimenter via systemic application of the DREADD agonist clozapine n-oxide (CNO). In this way, DREADDs allow experimental "remote control" of neural populations during behavior [42–44]. We used a viral vector to express inhibitory DREADDs in NAc neurons under a synapsin promoter. This allows recruitment of inhibitory Gi/o-coupled signaling cascades in neurons, selectively inhibiting their activity without affecting glia, as could occur with conventional reversible inactivation approaches such as GABA receptor agonist-induced inhibition. We here asked whether NAc neuron DREADD inhibition impacts production of amphetamine-induced USVs.

We found that chemogenetic suppression of NAc neural activity with DREADDs suppresses amphetamine-induced trill and non-trill frequency modulated, high-frequency USVs. NAc inhibition did not affect high-frequency flat calls that were also increased by amphetamine, and did not induce low-frequency calls, which were rarely observed. These findings demonstrate that NAc is essential for some but not all amphetamine-induced USVs, showing that NAc neuron activity mediates amphetamine-induced subjective states in rats.

2. Materials and Methods

All procedures were approved by the Institutional Animal Care and Use Committee at UC Irvine, and are in accordance with the NIH Guide for the Care and Use of Animals [45].

Subjects: Long Evans rats (n = 8 males, n = 8 females) were bred in-house, and housed as adults in pairs in ventilated tub cages with corncob bedding and ad libitum chow and water. Rats were at least 75 days old at the start of experiments. Rats were housed in reverse 12:12 h lighting, and behavior experiments took place during the dark cycle.

Drugs: CNO was graciously provided by the NIDA Drug Supply Program, stored in desiccated, opaque powder aliquots at 4 °C, and prepared daily, mixed in 5% dimethyl sulfoxide (DMSO) in saline solution. D-amphetamine hemisulfate salt was attained from Sigma and mixed in saline at 2 mg/mL.

Viral Vector and Surgery: A previously validated [46] AAV2 hSyn-hM4Di-eGFP vector (titer $\geq 7 \times 10^{12}$ vg/mL) was attained from AddGene. Rats were anesthetized with ketamine (56.5 mg/kg) and xylazine (8.7 mg/kg), with saline and meloxicam (1.0 mg/kg) analgesic, then stereotaxically injected via glass pipette and Picospritzer with 0.4 μL of the vector bilaterally, aimed at the nucleus accumbens core (relative to bregma (mm): +1.45 AP, ±1.3 ML, −7.6 DV). Pipettes were left in place for 5 min to reduce spread prior to removal. Rats were allowed at least 25 days to recover following surgery before handling and habituation. Vector expression is shown in Figure 1.

Testing Protocol: Rats were handled daily for a week prior to habituation to the testing chamber, which occurred 1 h/day over 2 consecutive days. On each of the two subsequent test days, they received counterbalanced i.p. injections of CNO (5 mg/kg) or its vehicle (veh; 5% DMSO in saline), then were placed in the testing chamber for recording of USVs. 30 min later, they were injected with amphetamine (2 mg/kg) and returned to the recording chamber for another hour. At least 48 h later, the second test was conducted under the same protocol with the alternative CNO/veh treatment.

USV Recording Apparatus: Testing was done in one of four clear acrylic tub cages (25 cm × 46 cm × 20 cm), with paper fiber bedding and a wire cage top. These testing chambers were enclosed within a wood sound-attenuating box (120 cm × 60 cm × 60 cm), which contained four bays in which testing cages could be placed, separated from each other by wood walls. These sound attenuating boxes helped exclude unwanted electrical and other noise in the 20–100 kHz range from recordings. The top of the sound-attenuating box was comprised of clear acrylic with a 3 cm hole drilled in the center of each bay, allowing placement of a USV microphone 33 cm above tub cages above the testing chambers, aimed at the center of each of the 4 testing cages. Avisoft Bioacoustics ultrasonic sensitive microphones (model CM16/CMPA; frequency range: 10–200 kHz), receivers (model 816H; sampling rate 250 kHz; 16-bit resolution), and PC software (Version 3.4.4) were employed, and recordings were done on a laptop running Windows.

Algorithm-Detected USV Analyses: DeepSqueak software ([33]; DeepSqueak 2.0 with MATLAB) was employed to detect USVs and identify the detailed characteristics of human-verified USVs from this dataset. Audio files were run though DeepSqueak's All Short Call neural network, then a post-hoc denoiser trained on noise inherent to the experimental setup automatically excluded non-USVs from the dataset. Call statistics for accepted USVs (20–100 kHz) were calculated using the spectrotemporal contours output by the detection network. These statistics included each call's (1) principal frequency (average frequency over the call), (2) change in frequency over the course of the call, (3) sinuosity (length of the path between the first and last points on the contour, divided by the Euclidean distance between the first and last points), and (4) duration.

Hand-Labeled USV Analysis: USVs identified by Deepsqueak were verified by a trained observer, blind to experimental conditions, using a previously reported scheme [24]. The observer verified around 95% of DeepSqueak-identified calls as valid, with 4160 calls out of 77,454 rejected as noise. USVs were visualized in spectrograms, which allowed inspection of call frequency over time. Putative calls were greater than 2 ms and were confirmed by the observer listening to them on headphones (frequency was transduced into the audible range by playing calls at 0.05 speed for this confirmatory evaluation). Identified calls were manually categorized into 1 of 4 categories: (1) Low-Frequency (LF) calls, which were between 20–30 kHz, (2) Flat high-frequency calls, which were between 30–100 kHz in principal frequency, and which lacked visible frequency modulation, (3) Frequency-Modulated (FM) calls, which had visible frequency modulation without a trill component, and (4) Trills, which had the same definitional criteria as FMs, but were at least 8 ms and contained 2 cycles of rapid oscillating frequency.

Analyses and Statistics: For analyses of NAc inhibition effects on the number of each USV type emitted, USVs were quantified in 10 min bins, both in the 30 min baseline period (after veh/CNO but before amphetamine), and in the 1 h post-amphetamine testing period, on both test days (veh or CNO; Figure 2). ANOVAs with within-subjects Drug (veh and CNO days) and Time (three 10 min baseline bins, or six 10 min post-amphetamine bins) factors, and a between-subjects Sex (female/male) factor, with Tukey post hoc tests and Greenhouse-Geisser correction for normality assumption violations when needed, were employed.

Call parameters were analyzed in detail for USVs emitted in the 1 h post-amphetamine session, on the vehicle pre-treatment day. USVs were analyzed for each of 4 measured call parameters (principal frequency, change in frequency, sinuosity, and duration). Averages for each rat on each parameter, categorized based on manually assigned USV types (LF, Flat, FM, Trill), were calculated, and group averages are shown in Figure 3A. For each measured parameter, call types were compared to one another using one-way ANOVA, with USV type treated as a between-subjects variable, since not all call types were emitted by all rats. This analysis served to confirm that Deepsqueak-derived USV parameters were logically related to human-assigned call categories, i.e., LF calls had low principal frequency, FMs and Trill had greater change in frequency than Flats, etc. To explore the underlying structure of amphetamine-induced USVs (on vehicle treatment day), we used Uniform Manifold Approximation and Projection (UMAP; [47]) to faithfully represent all four acoustic parameters derived by Deepsqueak in two dimensions, with results shown in Figure 3C. To determine NAc inhibition effects on characteristics of FM and Trill calls (those suppressed by CNO), for each parameter, t-tests comparing per-rat averages for USV parameters on vehicle or CNO amphetamine tests were used (Figure 2).

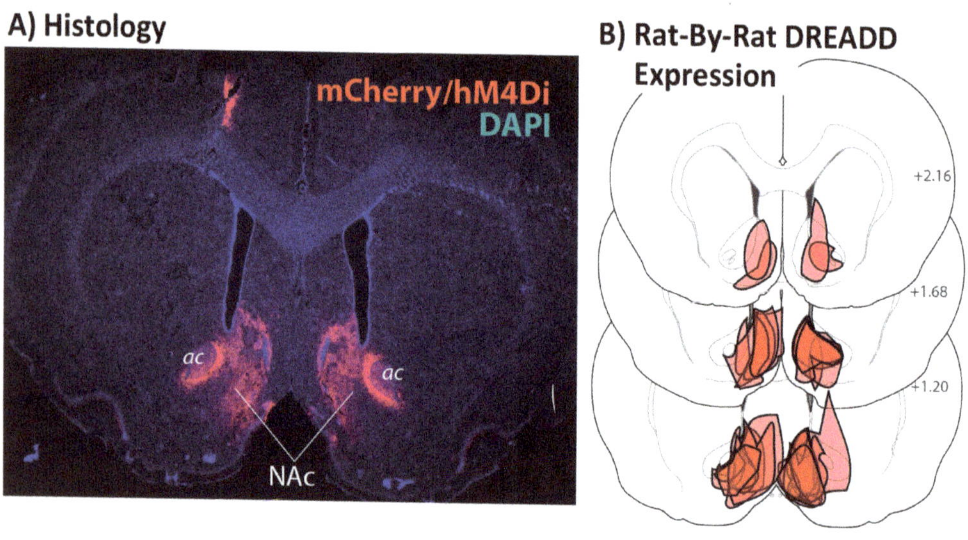

Figure 1. Nucleus Accumbens DREADD Expression: (**A**) Expression of the AAV2-hSyn-hM4Di-mCherry vector is depicted in a typical rat, shown in a coronal plane. mCherry and hM4Di are co-expressed in NAc neurons (red stain), and DAPI (4′,6-diamidino-2-phenylindole, in blue) defines anatomical landmarks, including anterior commissure (ac). (**B**) Viral expression localization is shown in each of the 16 tested animals in 3 coronal views of rat brain [48], with numbers indicating the approximate bregma-relative anterior coordinate.

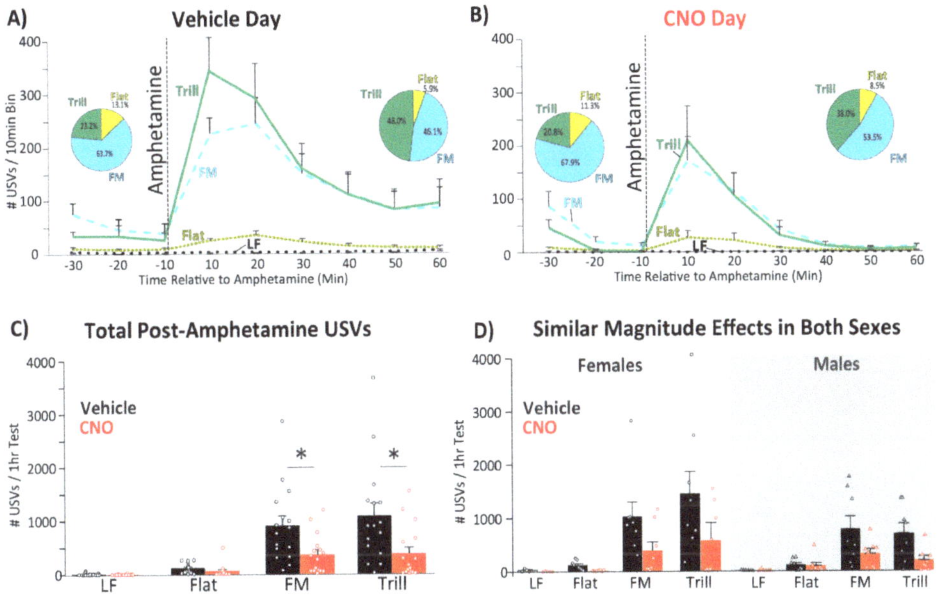

Figure 2. Nucleus Accumbens Inhibition Suppresses the Number of Frequency Modulated and Trill USVs: (**A**) USVs of each call type are depicted in 10 min bins in the post-vehicle, but pre-amphetamine baseline (left), and for 1 h after amphetamine (right; amphetamine injection occurred at the dotted vertical line). Pie charts show relative prevalence of each call subtype shown for pre-amphetamine baseline (left) and for 1 h after amphetamine (right). (**B**) CNO day data is shown in the same manner as in panel A. (**C**) Summary data are shown for the entire 1 h post-amphetamine session for the vehicle day (black) and CNO day (red), depicting main effects of this treatment. Dots depict data from individual animals. * $p < 0.05$. (**D**) Post-amphetamine summary data is shown in the same manner as in panel C, but separately in females (left) and males (right). Statistically similar CNO-suppression of the number of FM and Trill, but not Flat, USVs is seen in both sexes.

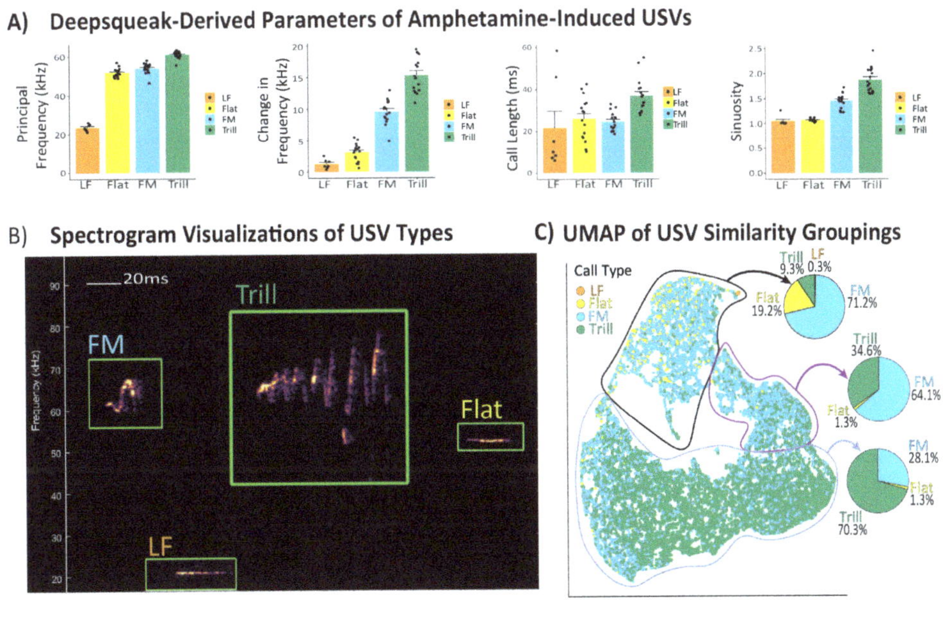

Figure 3. *Cont.*

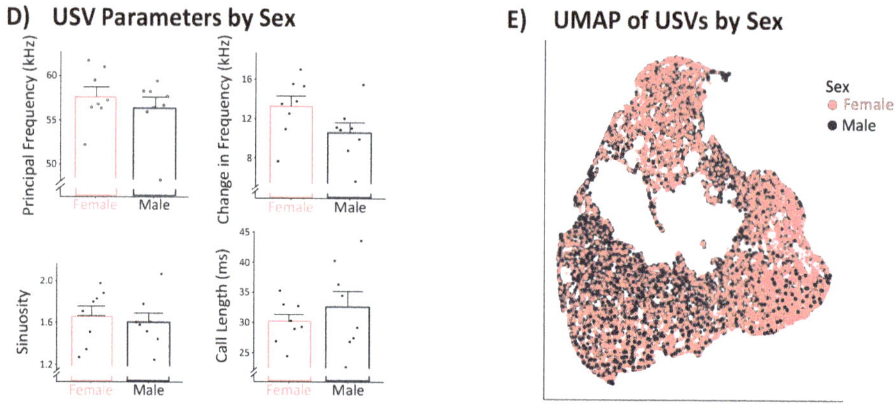

Figure 3. Characteristics of Quantified USVs: All USVs emitted on the vehicle + amphetamine test day were categorized by an observer into one of four categories: Low-frequency (LF), Flat, non-trill-containing frequency-modulated (FM), and trill-containing (Trill). Parameters (principal frequency, frequency modulation, duration, and sinuosity) for each emitted vocalization were quantified by Deepsqueak, and per-animal averages computed for each call type. (**A**) The mean and SEM of these per-rat average values for each call type are shown for each parameter, with dots indicating the mean value for each tested rat. (**B**) Examples of each of the 4 call types are shown. Scale bar indicates 20 ms. (**C**) UMAP representation of how calls relate to one another on the 4 measured parameters is depicted. Each emitted call is represented by a dot, which was color-coded based on manually assigned call type (LF = orange, Flat = yellow, FM = blue, Trill = green). Calls with similar characteristics in the 4 dimensions are represented within a cluster. To illustrate the proportion of call types present within visually determined clusters (circled), pie charts show the percentage of calls in each cluster that were assigned to each call type. Though each cluster differs in the types of calls contained within them, clusters do not neatly parse calls into our 4 pre-defined categories, which were based on the literature. This suggests that the 4-category system we used to define calls into subcategories may not fully capture underlying call subtypes present in the data, which could have functional significance. (**D**) Per-rat averages on each USV parameter, collapsed across call types, is shown broken down by the emitting animal's sex. No significant sex differences were found, suggesting that rats do not differ based on sex in the fundamental features of emitted USVs. (**E**) Further demonstrating this point, the same UMAP plot is shown, with calls color-coded by sex rather than call type, demonstrating that both sexes emit calls throughout the entirety of 4-dimensional space, and in each of the data-derived clusters depicted.

3. Results

3.1. Localization of DREADD Expression

Each brain was examined for expression of mCherry, the DREADD tag in the employed vector (Figure 1A). In most brains, expression was largely localized to the nucleus accumbens core and shell, with some expression also seen in nearby regions such as the ventral caudate putamen, and lateral and medial septum, as shown in Figure 1B.

3.2. Amphetamine Increases the Number of High Frequency USVs Emitted in Both Sexes

On vehicle day, amphetamine induced increases in all types of USVs except LFs, as determined by t-test comparing the 10 min prior to amphetamine (20–30 min after veh) to the first 10 min after amphetamine on that day (LF: $t_{15} = 0.1$; Flat: $t_{15} = 3.5$, $p = 0.003$; FM: $t_{15} = 7.86$, $p < 0.001$; Trill: $t_{15} = 4.92$, $p < 0.001$). The only major sex difference seen after amphetamine was that females produced more Trill calls than males (main effect of Sex: $F_{1,14} = 5.5$, $p = 0.034$). No effects of Sex on other call types, or interactions of Sex with pre versus post-amphetamine epochs were observed. Characteristic examples of each call type are shown in Figure 3B.

3.3. CNO Effects on the Number of USVs Emitted Prior to Amphetamine

To examine whether inhibition of NAc suppressed spontaneously emitted USVs even prior to amphetamine, we examined USVs that occurred in the 30 min after veh/CNO injection, during which period CNO effects were expected to emerge. Indeed, CNO suppressed certain USVs, but only late in the baseline period after it had time to reach the brain, as shown with interactions between Drug (veh/CNO) and Time (three 10 min bins) seen on FM ($F_{2,28} = 7.0$, $p = 0.003$) and Trill ($F_{2,28} = 6.52$, $p = 0.005$), but not LF ($F_{2,18\cdot 2} = 1.2$, $p = 0.31$) or Flat USVs ($F_{2,28} = 1.45$, $p = 0.25$; Figure 2A,B). Main effects of Time also indicated that most calls declined over the baseline period independent of veh/CNO treatment (LF: $F_{2,28} = 8.61$, $p = 0.001$; Flat: $F_{2,28} = 3.03$, $p = 0.06$; FM: $F_{2,28} = 5.89$, $p = 0.007$; Trill: $F_{2,28} = 3.82$, $p = 0.034$). No significant main effects of Sex (ps > 0.075), nor interactions of Sex with other variables (ps > 0.2) were seen for any call type during baseline. This suggests that NAc inhibition modestly suppressed even the low levels of FM and Trill USVs seen prior to amphetamine administration.

3.4. CNO Effects on the Number of Amphetamine-Induced USVs

NAc inhibition with DREADDs strongly and selectively suppressed amphetamine-induced FM and Trill calls (main effect of Drug; FM: $F_{1,14} = 5.38$, $p = 0.036$; Trill: $F_{1,14} = 7.11$, $p = 0.018$), but not Flat ($F_{1,14} = 2.47$, $p = 0.14$) or LF calls ($F_{1,14} = 2.78$, $p = 0.12$; Figure 2C). Since FM and Trill calls were similarly suppressed by CNO, we also analyzed data with these call types combined, and again found a robust main effect of Drug ($F_{1,14} = 7.9$, $p = 0.014$) and Time ($F_{1\cdot 5,20\cdot 97} = 13.46$, $p < 0.001$), as well as a main effect of Sex, with females emitting more total FM + Trill vocalizations ($F_{1,14} = 4.95$, $p = 0.043$). However, no interactions of Sex with Drug or Time were seen (ps > 0.52), indicating that inhibiting NAc had similar effects in both sexes (Figure 2D). Flat, FM and Trill calls were all maximal in the first 20 min after amphetamine, and decreased subsequently across the next 40 min in a similar manner after both veh and CNO (Main effect of Time; Flat: $F_{1\cdot 43,20\cdot 1} = 5.36$, $p = 0.021$; FM: $F_{1\cdot 8,25\cdot 5} = 14.0$, $p < 0.001$; Trill: $F_{1\cdot 5,20\cdot 4} = 10.68$, $p < 0.001$; Figure 2A,B). This suggests that CNO's effect was to suppress the number of FM/Trill USVs seen, without impacting the timecourse of the USV-promoting effects of amphetamine. LF calls were rare after amphetamine, were almost exclusively "short," non-canonical "22 kHz alarm calls," and did not show a similar time course of amphetamine-induced increase to the other call types ($F_{1\cdot 35,18\cdot 91} = 1.35$, $p = 0.25$). A Wilcoxon rank sum test revealed that administration of amphetamine increased the proportion of Trills produced (W = 40, $p < 0.001$; Figure 2A), while reducing the proportion of both FM and Flat calls (FM: W = 212, $p < 0.002$; Flat: W = 187, $p < 0.03$). CNO administration did not change call subtype prevalence (ps > 0.05; Figure 2B).

3.5. Amphetamine-Induced USV Characteristics, and Sex Differences

Analysis of the acoustic characteristics of each of the four call types (examples shown in Figure 3B) revealed significant differences between amphetamine-induced USV types in each of the measured parameters (vehicle day; 1 h post-amphetamine period data were analyzed). USV types differed predictably (Figure 3A) in their principal frequency ($F_{3,51} = 450.8$, $p < 0.0001$) with LFs being lower-frequency than other call types (ps < 0.0001), and Trills having a higher principal frequency than the other calls ($p < 0.0001$). Also as expected, calls also differed in the degree to which frequency changes over the course of the call ($F_{3,51} = 132.3$ $p < 0.0001$), with Trills and FMs having a higher frequency variation than other calls (Trills: ps < 0.0001; FMs: ps < 0.0001), and Trills also being more modulated than FMs ($p < 0.0001$). Call types also differed in length ($F_{3,51} = 5.42$, $p < 0.01$), with Trills being longer than Flats, FMs, and LFs (ps < 0.03). For sinuosity, USV types also differed ($F_{3,51} = 80.58$, $p < 0.0001$), in that Trills had higher values than all other call types (ps < 0.0001), and FMs had higher values than Flats and LFs (ps < 0.0001). Collapsing across USV types, we saw no overt sex differences in the 4 measured call parameters, implying that amphetamine-induced USV characteristics are likely to be fundamentally similar regardless of sex (Figure 3D). In sum, though we are not convinced that Deepsqueak-derived USV parameters are precise in all cases, we nonetheless conclude based on these analyses that the software performs acceptably for roughly characterizing USV features under the recording conditions employed here.

Therefore, we next conducted an exploratory, data-driven analysis examining how USV features relate to manually categorized USV subtypes and to sex (Figure 3C,E). Using UMAP [47,49–52], an advanced method of analyzing underlying features of large datasets such as this one, we plotted a projection of all 33,370 USVs quantified during the vehicle/amphetamine session in 4-dimensional space into 2 dimensions for easy visualization. In Figure 3C, each dot represents an emitted USV, color-coded based on manually assigned call type. Calls that are similar in all 4 dimensions cluster together, and groups of similar calls are represented in neighboring clusters. We found approximately 3 clusters of USV calls, each of which contained distinct proportions of each USV call type (Figure 3C). These data imply that Deepsqueak-quantified call parameters do not line up 1:1 with human-categorized USV categories, suggesting that these USV data contain an underlying architecture that is not captured by a simple 4 category quantification scheme, as employed here and in many other reports [24–26,53]. Finally, we took the same UMAP representation and color-coded USVs based on the sex of the rat that produced them (Figure 3E); we interpret results to show that both male and female rats show similar distribution of calls across all 3 clusters.

3.6. Effects of CNO on USV Characteristics

Inhibiting NAc did not affect the primary quantitative characteristics of FM and Trill USVs, though as described above, the number of these calls were suppressed (Figure 2A,B). Relative to vehicle, CNO failed to impact any of the parameters of FM (ts < 2.05; ps < 0.05), or Trill calls (ts < 1.49; ps < 0.15; Figure 4).

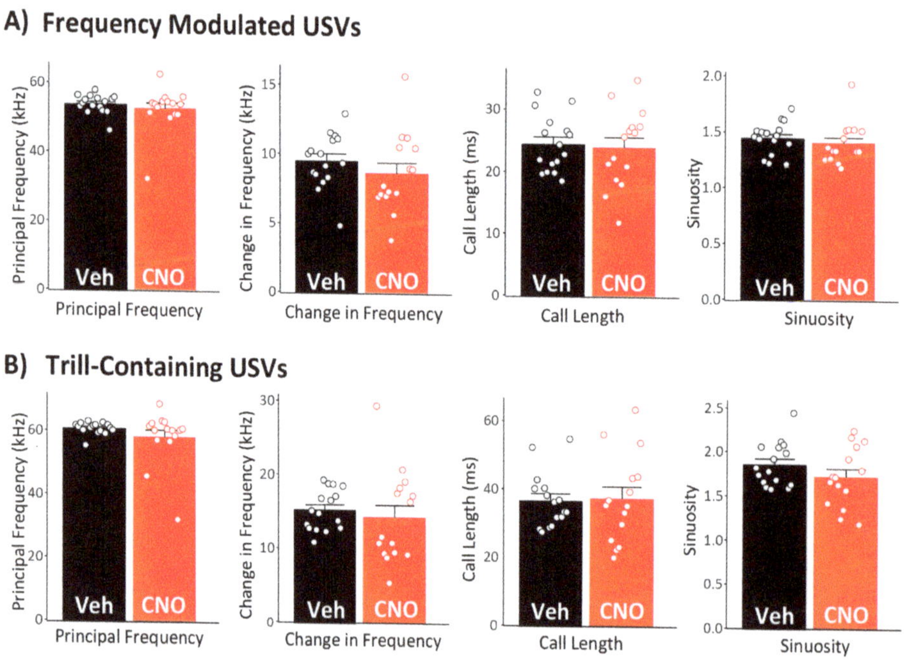

Figure 4. Nucleus Accumbens Inhibition Does Not Affect the Characteristics of Amphetamine-Induced USVs: Relative to vehicle day (black), USVs on CNO day (red) did not differ in principal frequency, frequency modulation, duration, or sinuosity for either (**A**) frequency-modulated non-trill, or (**B**) trill-containing USVs. Average values for each rat are shown with dots within bars.

4. Discussion

To our knowledge, this is the first investigation on neural substrates of rodent ultrasonic vocalizations (USVs) employing a reversable chemogenetic inhibition approach. Results confirm that the nucleus accumbens (NAc) is a key locus in brain circuits underlying USVs emitted after systemic amphetamine injection. We show a preferential suppression of frequency-modulated vocalizations, both with and without trill components, after chemogenetic NAc neuronal inhibition with DREADDs. Although females emitted more FM/Trill USVs than males overall, DREADD inhibition suppressed these frequency-modulated, amphetamine-induced USVs to a similar degree in both sexes. Interestingly, non-frequency modulated high-frequency Flat USVs were induced mildly but significantly by amphetamine, but NAc inhibition did not alter their production, nor did it induce low-frequency, aversion-related USVs. Aversive low-frequency vocalizations are often hundreds of milliseconds long, and have been recorded after exposure to predator odor, foot shocks, and social isolation [14–16,18,19]. When we did see LF vocalizations here, they were much shorter, lasting less than 150 ms. These short LF calls, which are also seen during rat self-administration of cocaine and methamphetamine, may reflect less severe aversion than prototypical "22 kHz" alarm calls [24–26,54,55]. Regardless, amphetamine very rarely induced LFs of any duration, nor did NAc inhibition induce their production. Though NAc inhibition suppressed the quantity of emitted FM and Trill USVs, the quality of these calls as reflected in their quantitative parameters was not affected. Together, these data speak to the complexities of rat USVs both in their characteristics and their neural substrates, and show that detailed quantitative analyses of rat USVs could inform our understanding of the subjective effects of neural manipulations in preclinical behavioral neuroscience experiments.

We show here that inhibition of NAc using an established chemogenetic inhibition approach [46,56–58] markedly suppresses amphetamine-induced USVs. Prior reports have shown that administration of amphetamine to the NAc shell increases local dopamine signaling and 50 kHz call rate [2,39,59,60], with similar increases in calling resulting from systemic or intracerebral injection of other dopamine agonists such as quinpirole [5,54,61–63]. Playback of 50 kHz vocalizations has also been shown to increase dopamine levels in the NAc and elicit approach behavior [41]. Disruption of dopamine signaling in the NAc through administration of dopamine antagonists or lesions to the ventral tegmental area suppresses 50 kHz calling [21,31,64–66]. Here, we used a viral vector causing expression of inhibitory DREADDs only in NAc neurons, without impacting activity of other NAc cell types which could be influenced by lesion or pharmacological inactivation approaches. When the DREADD agonist CNO is applied systemically, neurons are inhibited in their firing and neurotransmitter release [56,67]. We also found that NAc inhibition alone suppressed FM and Trill calls in a baseline period, prior to amphetamine administration. When amphetamine was injected, FM, Trill and Flat USVs increased markedly on both vehicle and CNO treatment days. Yet on the CNO day, FM and Trill calls (but not Flats) were suppressed relative to vehicle day, indicating that normal NAc neural activity is required for emission of these calls when either spontaneously emitted, or induced by a catecholamine-enhancing addictive drug. Further studies using more targeted chemo- or optogenetic manipulation approaches should explore further the mechanisms of this effect, such as the NAc neuron types, and the wider circuits interactions responsible for these effects.

In addition to showing that NAc inhibition suppressed the number of USVs produced, we also sought to determine how quantitative analysis of USV parameters might inform our understanding of how these vocalizations could be used to interrogate the subjective states of rats, and the neural mechanisms thereof. Towards this aim, we employed a newly developed, machine learning-based software package for analyzing rat USVs, Deepsqueak [33]. We found Deepsqueak very useful for detecting USVs in our sound-attenuated setup. We did notice that some instances of noise were also registered as potential calls in our experiment, so we verified each auto-identified call using visual and audio inspection by a trained, blinded observer. This observer also manually categorized each call into one of 4 pre-defined categories, high frequency Flat, FM, and Trill calls, as well as aversion-related low frequency calls, based on prior work [24,26,68]. These pre-defined categories seemed to be supported by Deepsqueak-generated parameters, with logical characteristics generally captured (e.g. LF calls have lower frequency than high frequency calls, FM/Trill calls had more frequency modulation than Flat/LF calls, etc.). When we used these parameters to explore potential underlying relationships between emitted calls based on data structure rather than experimenter-determined categories, we found that our pre-set categories were imperfectly reflected in the data architecture, suggesting that hidden structures may exist in these data which is not completely captured by our literature-based USV categories. Moreover, we found little evidence for either sex or NAc inhibition affecting the quality of USVs produced. We strongly feel that further optimization of software solutions to accurately capture USV parameters, and examining them under a range of pharmacological, neural, and behavioral circumstances could hold the key to understanding the deeper meaning of rat USVs.

Our study has a number of limitations: Sample size is relatively low ($n = 8$/sex), so although NAc inhibition effects were very clear, more subtle effects of sex or hormonal contributions may have been missed. Our viral injections were targeted at the NAc, although as is common in such experiments, not all DREADD expression was entirely restricted to the NAc, let alone NAc core or shell subregions. We also did not include a non-DREADD expressing group to control for potential off-target effects of CNO, which have been reported previously [69–72]. We therefore cannot exclude the possibility that CNO alone, even in the absence of DREADDs, could have impacted the results we observed. Additionally, we did not examine the impacts of NAc inhibition on saline-, rather than amphetamine-induced calling, which would have allowed better examination of chemogenetic impacts on non-pharmacologically-induced USV production. We thus focused our quantitative analyses of USVs on data generated after vehicle injection followed by amphetamine, since we were concerned that the relatively low number of USVs seen in the baseline period, and potential subtle effects of NAc inhibition on baseline or post-amphetamine responses could impact these analyses. Further work should involve recording of USVs under a variety of behavioral circumstances, neural states, and after dose-dependent pharmacological manipulations.

In sum, these results confirm using a neuron-specific chemogenetic inactivation approach, that NAc is a key node in the neural circuits underlying rat USVs. Intriguingly, NAc seems to mediate frequency-modulated USVs in particular, but not high-frequency unmodulated calls that were also induced by amphetamine, nor the rarely observed, potentially aversion-related low-frequency USVs. Our results also point to the remaining mysteries surrounding these effect-relevant, but still poorly understood vocalizations, since the four pre-selected USV types did not clearly map onto clusters revealed by running UMAP on four quantitative USV parameters. We hope that with further refinement of USV recording and analysis methods, we may finally extract information about rat subjective states from their calls, which would be of tremendous benefit to preclinical behavioral neuroscience experiments.

Author Contributions: Conceptualization: S.V.M., C.M.R. and R.E.H.; Data collection: R.E.H., A.Y.F. and C.M.R.; Data analysis: K.A.L. and S.V.M.; Writing: K.A.L., A.Y.F., R.E.H. and S.V.M. All authors have read and agreed to the published version of the manuscript.

Funding: Funding was provided by NIH grant P50 DA044118, the UCI School of Biological Sciences, and the Hellman Fellows Foundation.

Institutional Review Board Statement: The study was conducted according to the guidelines of the Declaration of Helsinki, and approved by the Institutional Animal Use and Care Committee of the University of California Irvine (protocol 2015-3153, date of approval 12 February 2017).

Data Availability Statement: Processed data and code are available on the Mahler Lab GitHub (https://github.com/Mahler-Lab/NAc_USV accessed on 17 September 2021). Data will be made available upon request.

Acknowledgments: We thank Jordan Andrada for assistance with USV quantification, and the NIDA Drug Supply Program for CNO. We also thank Ronald E. See for intellectual and material support.

Conflicts of Interest: The authors declare no conflict of interest.

References

1. Best, L.M.; Zhao, L.L.; Scardochio, T.; Clarke, P.B.S. Effects of repeated morphine on ultrasonic vocalizations in adult rats: Increased 50-kHz call rate and altered subtype profile. *Psychopharmacology* **2017**, *234*, 155–165. [CrossRef]
2. Brudzynski, S.M. Ethotransmission: Communication of emotional states through ultrasonic vocalization in rats. *Curr. Opin. Neurobiol.* **2013**, *23*, 310–317. [CrossRef]
3. Burgdorf, J.; Brudzynski, S.; Moskal, J. Using rat ultrasonic vocalization to study the neurobiology of emotion: From basic science to the development of novel therapeutics for affective disorders. *Curr. Opin. Neurobiol.* **2020**, *60*, 192–200. [CrossRef]
4. Burgdorf, J.; Kroes, R.A.; Moskal, J.R.; Pfaus, J.G.; Brudzynski, S.M.; Panksepp, J. Ultrasonic vocalizations of rats (*Rattus norvegicus*) during mating, play, and aggression: Behavioral concomitants, relationship to reward, and self-administration of playback. *J. Comp. Psychol.* **2008**, *122*, 357–367. [CrossRef]
5. Engelhardt, K.-A.; Fuchs, E.; Schwarting, R.K.W.; Wöhr, M. Effects of amphetamine on pro-social ultrasonic communication in juvenile rats: Implications for mania models. *Eur. Neuropsychopharmacol.* **2017**, *27*, 261–273. [CrossRef] [PubMed]
6. Himmler, B.T.; Kisko, T.M.; Euston, D.R.; Kolb, B.; Pellis, S.M. Are 50-kHz calls used as play signals in the playful interactions of rats? I. Evidence from the timing and context of their use. *Behav. Process.* **2014**, *106*, 60–66. [CrossRef] [PubMed]
7. Kisko, T.M.; Himmler, B.T.; Himmler, S.M.; Euston, D.R.; Pellis, S.M. Are 50-kHz calls used as play signals in the playful interactions of rats? II. Evidence from the effects of devocalization. *Behav. Process.* **2015**, *111*, 25–33. [CrossRef] [PubMed]
8. Kisko, T.M.; Euston, D.R.; Pellis, S.M. Are 50-khz calls used as play signals in the playful interactions of rats? III. The effects of devocalization on play with unfamiliar partners as juveniles and as adults. *Behav. Process.* **2015**, *113*, 113–121. [CrossRef] [PubMed]
9. Wöhr, M.; Schwarting, R.K.W. Ultrasonic Communication in Rats: Can Playback of 50-kHz Calls Induce Approach Behavior? *PLoS ONE* **2007**, *2*, e1365. [CrossRef] [PubMed]
10. Seffer, D.; Schwarting, R.K.W.; Wöhr, M. Pro-social ultrasonic communication in rats: Insights from playback studies. *J. Neurosci. Methods* **2014**, *234*, 73–81. [CrossRef] [PubMed]
11. Tasaka, G.-I.; Feigin, L.; Maor, I.; Groysman, M.; DeNardo, L.A.; Schiavo, J.K.; Froemke, R.C.; Luo, L.; Mizrahi, A. The Temporal Association Cortex Plays a Key Role in Auditory-Driven Maternal Plasticity. *Neuron* **2020**, *107*, 566–579.e7. [CrossRef] [PubMed]
12. Brudzynski, S.M. Communication of Adult Rats by Ultrasonic Vocalization: Biological, Sociobiological, and Neuroscience Approaches. *ILAR J.* **2009**, *50*, 43–50. [CrossRef]
13. Wöhr, M.; Schwarting, R.K.W. Affective communication in rodents: Ultrasonic vocalizations as a tool for research on emotion and motivation. *Cell Tissue Res.* **2013**, *354*, 81–97. [CrossRef] [PubMed]
14. Blanchard, R.J.; Blanchard, D.C.; Agullana, R.; Weiss, S.M. Twenty-two kHz alarm cries to presentation of a predator, by laboratory rats living in visible burrow systems. *Physiol. Behav.* **1991**, *50*, 967–972. [CrossRef]
15. Brudzynski, S.M.; Ociepa, D. Ultrasonic vocalization of laboratory rats in response to handling and touch. *Physiol. Behav.* **1992**, *52*, 655–660. [CrossRef]
16. Borta, A.; Wöhr, M.; Schwarting, R.K.W. Rat ultrasonic vocalization in aversively motivated situations and the role of individual differences in anxiety-related behavior. *Behav. Brain Res.* **2006**, *166*, 271–280. [CrossRef]
17. Chang, S.; Fan, Y.; Shin, J.H.; Ryu, Y.; Kim, M.S.; Steffensen, S.C.; Kim, H.K.; Kim, J.M.; Lee, B.H.; Jang, E.Y.; et al. Unpleasant Sound Elicits Negative Emotion and Reinstates Drug Seeking. *Mol. Neurobiol.* **2019**, *56*, 7594–7607. [CrossRef]
18. Inagaki, H. Chapter 32—Sex Differences in Ultrasonic Vocal Expression of Negative Emotional States in Rats. In *Handbook of Behavioral Neuroscience*; Brudzynski, S.M., Ed.; Handbook of Ultrasonic Vocalization; Elsevier: Amsterdam, The Netherlands, 2018; Volume 25, pp. 337–344.
19. Kroes, R.A.; Burgdorf, J.; Otto, N.J.; Panksepp, J.; Moskal, J.R. Social defeat, a paradigm of depression in rats that elicits 22-kHz vocalizations, preferentially activates the cholinergic signaling pathway in the periaqueductal gray. *Behav. Brain Res.* **2007**, *182*, 290–300. [CrossRef] [PubMed]
20. Burgdorf, J.; Panksepp, J.; Moskal, J.R. Frequency-modulated 50 kHz ultrasonic vocalizations: A tool for uncovering the molecular substrates of positive affect. *Neurosci. Biobehav. Rev.* **2011**, *35*, 1831–1836. [CrossRef]
21. Burgdorf, J.; Wood, P.L.; Kroes, R.A.; Moskal, J.R.; Panksepp, J. Neurobiology of 50-kHz ultrasonic vocalizations in rats: Electrode mapping, lesion, and pharmacology studies. *Behav. Brain Res.* **2007**, *182*, 274–283. [CrossRef]
22. Hori, M.; Shimoju, R.; Tokunaga, R.; Ohkubo, M.; Miyabe, S.; Ohnishi, J.; Murakami, K.; Kurosawa, M. Tickling increases dopamine release in the nucleus accumbens and 50 kHz ultrasonic vocalizations in adolescent rats. *NeuroReport* **2013**, *24*, 241–245. [CrossRef] [PubMed]

23. Redecker, T.M.; Kisko, T.M.; Schwarting, R.K.W.; Wöhr, M. Effects of Cacna1c haploinsufficiency on social interaction behavior and 50-kHz ultrasonic vocalizations in adult female rats. *Behav. Brain Res.* **2019**, *367*, 35–52. [CrossRef] [PubMed]
24. Mahler, S.V.; Moorman, D.E.; Feltenstein, M.W.; Cox, B.M.; Ogburn, K.B.; Bachar, M.; McGonigal, J.T.; Ghee, S.M.; See, R.E. A rodent "self-report" measure of methamphetamine craving? Rat ultrasonic vocalizations during methamphetamine self-administration, extinction, and reinstatement. *Behav. Brain Res.* **2013**, *236*, 78–89. [CrossRef] [PubMed]
25. Pereira, M.; Andreatini, R.; Schwarting, R.K.W.; Brenes, J.C. Amphetamine-induced appetitive 50-kHz calls in rats: A marker of affect in mania? *Psychopharmacology* **2014**, *231*, 2567–2577. [CrossRef] [PubMed]
26. Schwarting, R.K.W.; Jegan, N.; Wöhr, M. Situational factors, conditions and individual variables which can determine ultrasonic vocalizations in male adult Wistar rats. *Behav. Brain Res.* **2007**, *182*, 208–222. [CrossRef]
27. Wright, J.M.; Gourdon, J.C.; Clarke, P.B.S. Identification of multiple call categories within the rich repertoire of adult rat 50-kHz ultrasonic vocalizations: Effects of amphetamine and social context. *Psychopharmacology* **2010**, *211*, 1–13. [CrossRef]
28. Ahrens, A.M.; Ma, S.T.; Maier, E.Y.; Duvauchelle, C.L.; Schallert, T. Repeated intravenous amphetamine exposure: Rapid and persistent sensitization of 50-kHz ultrasonic trill calls in rats. *Behav. Brain Res.* **2009**, *197*, 205–209. [CrossRef]
29. Wright, J.M.; Dobosiewicz, M.R.S.; Clarke, P.B.S. α- and β-Adrenergic Receptors Differentially Modulate the Emission of Spontaneous and Amphetamine-Induced 50-kHz Ultrasonic Vocalizations in Adult Rats. *Neuropsychopharmacology* **2012**, *37*, 808–821. [CrossRef]
30. Buck, C.L.; Vendruscolo, L.F.; Koob, G.F.; George, O. Dopamine D1 and μ-opioid receptor antagonism blocks anticipatory 50 kHz ultrasonic vocalizations induced by palatable food cues in Wistar rats. *Psychopharmacology* **2014**, *231*, 929–937. [CrossRef]
31. Ciucci, M.R.; Ahrens, A.M.; Ma, S.T.; Kane, J.R.; Windham, E.B.; Woodlee, M.T.; Schallert, T. Reduction of dopamine synaptic activity: Degradation of 50-kHz ultrasonic vocalization in rats. *Behav. Neurosci.* **2009**, *123*, 328–336. [CrossRef]
32. Barker, D.J.; Herrera, C.; West, M.O. Automated Detection of 50-kHz Ultrasonic Vocalizations Using Template Matching in XBAT. *J. Neurosci. Methods* **2014**, *236*, 68–75. [CrossRef]
33. Coffey, K.R.; Marx, R.G.; Neumaier, J.F. DeepSqueak: A deep learning-based system for detection and analysis of ultrasonic vocalizations. *Neuropsychopharmacology* **2019**, *44*, 859–868. [CrossRef]
34. Premoli, M.; Baggi, D.; Bianchetti, M.; Gnutti, A.; Bondaschi, M.; Mastinu, A.; Migliorati, P.; Signoroni, A.; Leonardi, R.; Memo, M.; et al. Automatic classification of mice vocalizations using Machine Learning techniques and Convolutional Neural Networks. *PLoS ONE* **2021**, *16*, e0244636. [CrossRef]
35. Reno, J.M.; Marker, B.; Cormack, L.K.; Schallert, T.; Duvauchelle, C.L. Automating Ultrasonic Vocalization Analyses: The WAAVES Program. *J. Neurosci. Methods* **2013**, *219*, 155–161. [CrossRef] [PubMed]
36. Van Segbroeck, M.; Knoll, A.T.; Levitt, P.; Narayanan, S. MUPET—Mouse Ultrasonic Profile ExTraction: A Signal Processing Tool for Rapid and Unsupervised Analysis of Ultrasonic Vocalizations. *Neuron* **2017**, *94*, 465–485.e5. [CrossRef] [PubMed]
37. Zala, S.M.; Reitschmidt, D.; Noll, A.; Balazs, P.; Penn, D.J. Automatic mouse ultrasound detector (A-MUD): A new tool for processing rodent vocalizations. *PLoS ONE* **2017**, *12*, e0181200. [CrossRef] [PubMed]
38. Hamed, A.; Daszczuk, P.; Kursa, M.B.; Turzyńska, D.; Sobolewska, A.; Lehner, M.; Boguszewski, P.M.; Szyndler, J. Non-parametric analysis of neurochemical effects and Arc expression in amphetamine-induced 50-kHz ultrasonic vocalization. *Behav. Brain Res.* **2016**, *312*, 174–185. [CrossRef] [PubMed]
39. Mulvihill, K.G.; Brudzynski, S.M. Effect of microinjections of dopamine into the nucleus accumbens shell on emission of 50 kHz USV: Comparison with effects of d-amphetamine. *Pharmacol. Biochem. Behav.* **2019**, *176*, 23–32. [CrossRef]
40. Simola, N.; Brudzynski, S.M. Rat 50-kHz ultrasonic vocalizations as a tool in studying neurochemical mechanisms that regulate positive emotional states. *J. Neurosci. Methods* **2018**, *310*, 33–44. [CrossRef] [PubMed]
41. Willuhn, I.; Tose, A.; Wanat, M.J.; Hart, A.S.; Hollon, N.G.; Phillips, P.E.M.; Schwarting, R.K.W.; Wöhr, M. Phasic dopamine release in the nucleus accumbens in response to pro-social 50 kHz ultrasonic vocalizations in rats. *J. Neurosci. Off. J. Soc. Neurosci.* **2014**, *34*, 10616–10623. [CrossRef]
42. Armbruster, B.N.; Li, X.; Pausch, M.H.; Herlitze, S.; Roth, B.L. Evolving the lock to fit the key to create a family of G protein-coupled receptors potently activated by an inert ligand. *Proc. Natl. Acad. Sci. USA* **2007**, *104*, 5163–5168. [CrossRef]
43. Rogan, S.C.; Roth, B.L. Remote Control of Neuronal Signaling. *Pharmacol. Rev.* **2011**, *63*, 291–315. [CrossRef]
44. Smith, K.S.; Bucci, D.J.; Luikart, B.W.; Mahler, S.V. DREADDs: Use and Application in Behavioral Neuroscience. *Behav. Neurosci.* **2016**, *130*, 137–155. [CrossRef] [PubMed]
45. National Research Council (US) Committee for the Update of the Guide for the Care and Use of Laboratory Animals. *Guide for the Care and Use of Laboratory Animals*, 8th ed.; National Academies Press (US): Washington, DC, USA, 2011.
46. Farrell, M.R.; Ruiz, C.M.; Castillo, E.; Faget, L.; Khanbijian, C.; Liu, S.; Schoch, H.; Rojas, G.; Huerta, M.Y.; Hnasko, T.S.; et al. Ventral pallidum is essential for cocaine relapse after voluntary abstinence in rats. *Neuropsychopharmacology* **2019**, *44*, 2174–2185. [CrossRef] [PubMed]
47. McInnes, L.; Healy, J.; Melville, J. UMAP: Uniform Manifold Approximation and Projection for Dimension Reduction. *arXiv* **2020**, arXiv:1802.03426.
48. Paxinos, G.; Watson, C. *The Rat Brain in Stereotaxic Coordinates: Hard Cover Edition*; Elsevier: Amsterdam, The Netherlands, 2006; p. 451.
49. Becht, E.; McInnes, L.; Healy, J.; Dutertre, C.-A.; Kwok, I.W.H.; Ng, L.G.; Ginhoux, F.; Newell, E.W. Dimensionality reduction for visualizing single-cell data using UMAP. *Nat. Biotechnol.* **2019**, *37*, 38–44. [CrossRef]

50. Bender, N.K.; Heilig, C.E.; Droll, B.; Wohlgemuth, J.; Armbruster, F.P.; Heilig, B. Immunogenicity, efficacy and adverse events of adalimumab in RA patients. *Rheumatol. Int.* **2007**, *27*, 269–274. [CrossRef]
51. Dorrity, M.W.; Saunders, L.M.; Queitsch, C.; Fields, S.; Trapnell, C. Dimensionality reduction by UMAP to visualize physical and genetic interactions. *Nat. Commun.* **2020**, *11*, 1537. [CrossRef]
52. Sánchez-Rico, M.; Alvarado, J.M. A Machine Learning Approach for Studying the Comorbidities of Complex Diagnoses. *Behav. Sci.* **2019**, *9*, 122. [CrossRef]
53. Kisko, T.M.; Schwarting, R.K.W.; Wöhr, M. Sex differences in the acoustic features of social play-induced 50-kHz ultrasonic vocalizations: A detailed spectrographic analysis in wild-type Sprague–Dawley and Cacna1c haploinsufficient rats. *Dev. Psychobiol.* **2021**, *63*, 262–276. [CrossRef]
54. Barker, D.J.; Root, D.H.; Ma, S.; Jha, S.; Megehee, L.; Pawlak, A.P.; West, M.O. Dose-dependent differences in short ultrasonic vocalizations emitted by rats during cocaine self-administration. *Psychopharmacology* **2010**, *211*, 435–442. [CrossRef] [PubMed]
55. Brudzynski, S.M.; Bihari, F.; Ociepa, D.; Fu, X.-W. Analysis of 22 kHz ultrasonic vocalization in laboratory rats: Long and short calls. *Physiol. Behav.* **1993**, *54*, 215–221. [CrossRef]
56. Mahler, S.V.; Vazey, E.M.; Beckley, J.T.; Keistler, C.R.; McGlinchey, E.M.; Kaufling, J.; Wilson, S.P.; Deisseroth, K.; Woodward, J.J.; Aston-Jones, G. Designer receptors show role for ventral pallidum input to ventral tegmental area in cocaine seeking. *Nat. Neurosci.* **2014**, *17*, 577–585. [CrossRef] [PubMed]
57. Farrell, M.R.; Esteban, J.S.D.; Faget, L.; Floresco, S.B.; Hnasko, T.S.; Mahler, S.V. Ventral Pallidum GABA Neurons Mediate Motivation Underlying Risky Choice. *J. Neurosci.* **2021**, *41*, 4500–4513. [CrossRef] [PubMed]
58. McGlinchey, E.M.; James, M.H.; Mahler, S.V.; Pantazis, C.; Aston-Jones, G. Prelimbic to Accumbens Core Pathway Is Recruited in a Dopamine-Dependent Manner to Drive Cued Reinstatement of Cocaine Seeking. *J. Neurosci.* **2016**, *36*, 8700–8711. [CrossRef]
59. Thompson, B.; Leonard, K.C.; Brudzynski, S.M. Amphetamine-induced 50kHz calls from rat nucleus accumbens: A quantitative mapping study and acoustic analysis. *Behav. Brain Res.* **2006**, *168*, 64–73. [CrossRef]
60. Wöhr, M. Measuring mania-like elevated mood through amphetamine-induced 50-kHz ultrasonic vocalizations in rats. *Br. J. Pharmacol.* **2021**, 1–19. [CrossRef]
61. Wintink, A.J.; Brudzynski, S.M. The related roles of dopamine and glutamate in the initiation of 50-kHz ultrasonic calls in adult rats. *Pharmacol. Biochem. Behav.* **2001**, *70*, 317–323. [CrossRef]
62. Simola, N.; Fenu, S.; Costa, G.; Pinna, A.; Plumitallo, A.; Morelli, M. Pharmacological characterization of 50-kHz ultrasonic vocalizations in rats: Comparison of the effects of different psychoactive drugs and relevance in drug-induced reward. *Neuropharmacology* **2012**, *63*, 224–234. [CrossRef]
63. Burgdorf, J.; Knutson, B.; Panksepp, J.; Ikemoto, S. Nucleus accumbens amphetamine microinjections unconditionally elicit 50-kHz ultrasonic vocalizations in rats. *Behav. Neurosci.* **2001**, *115*, 940–944. [CrossRef]
64. Wright, J.M.; Dobosiewicz, M.R.S.; Clarke, P.B.S. The role of dopaminergic transmission through D1-like and D2-like receptors in amphetamine-induced rat ultrasonic vocalizations. *Psychopharmacology* **2013**, *225*, 853–868. [CrossRef]
65. Brudzynski, S.M. Pharmacology of Ultrasonic Vocalizations in adult Rats: Significance, Call Classification and Neural Substrate. *Curr. Neuropharmacol.* **2015**, *13*, 180–192. [CrossRef]
66. Ciucci, M.R.; Ma, S.T.; Kane, J.R.; Ahrens, A.M.; Schallert, T. Limb use and complex ultrasonic vocalization in a rat model of Parkinson's disease: Deficit-targeted training. *Parkinsonism Relat. Disord.* **2008**, *14*, S172–S175. [CrossRef] [PubMed]
67. Mahler, S.V.; Brodnik, Z.D.; Cox, B.M.; Buchta, W.C.; Bentzley, B.S.; Quintanilla, J.; Cope, Z.A.; Lin, E.C.; Riedy, M.D.; Scofield, M.D.; et al. Chemogenetic Manipulations of Ventral Tegmental Area Dopamine Neurons Reveal Multifaceted Roles in Cocaine Abuse. *J. Neurosci.* **2019**, *39*, 503–518. [CrossRef] [PubMed]
68. Barker, D.J.; Simmons, S.J.; Servilio, L.C.; Bercovicz, D.; Ma, S.; Root, D.H.; Pawlak, A.P.; West, M.O. Ultrasonic Vocalizations: Evidence for an affective opponent process during cocaine self-administration. *Psychopharmacology* **2014**, *231*, 909–918. [CrossRef]
69. Gomez, J.L.; Bonaventura, J.; Lesniak, W.; Mathews, W.B.; Sysa-Shah, P.; Rodriguez, L.A.; Ellis, R.J.; Richie, C.T.; Harvey, B.K.; Dannals, R.F.; et al. Chemogenetics revealed: DREADD occupancy and activation via converted clozapine. *Science* **2017**, *357*, 503–507. [CrossRef]
70. MacLaren, D.A.A.; Browne, R.W.; Shaw, J.K.; Radhakrishnan, S.K.; Khare, P.; España, R.A.; Clark, S.D. Clozapine N-Oxide Administration Produces Behavioral Effects in Long–Evans Rats: Implications for Designing DREADD Experiments. *eNeuro* **2016**, *3*. [CrossRef] [PubMed]
71. Mahler, S.V.; Aston-Jones, G. CNO Evil? Considerations for the Use of DREADDs in Behavioral Neuroscience. *Neuropsychopharmacology* **2018**, *43*, 934–936. [CrossRef]
72. Manvich, D.F.; Webster, K.A.; Foster, S.L.; Farrell, M.S.; Ritchie, J.C.; Porter, J.H.; Weinshenker, D. The DREADD agonist clozapine N-oxide (CNO) is reverse-metabolized to clozapine and produces clozapine-like interoceptive stimulus effects in rats and mice. *Sci. Rep.* **2018**, *8*, 3840. [CrossRef]

brain sciences

MDPI

Review

Rat Models of Vocal Deficits in Parkinson's Disease

Maryann N. Krasko [1,2], Jesse D. Hoffmeister [1,2], Nicole E. Schaen-Heacock [1,2], Jacob M. Welsch [1], Cynthia A. Kelm-Nelson [1] and Michelle R. Ciucci [1,2,3,*]

1 Department of Surgery, University of Wisconsin-Madison, Madison, WI 53792, USA; krasko@surgery.wisc.edu (M.N.K.); hoffmeister@surgery.wisc.edu (J.D.H.); nschaen@wisc.edu (N.E.S.-H.); jmwelsch@wisc.edu (J.M.W.); cakelm@wisc.edu (C.A.K.-N.)
2 Department of Communication Sciences and Disorders, University of Wisconsin-Madison, Madison, WI 53706, USA
3 Neuroscience Training Program, University of Wisconsin-Madison, Madison, WI 53705, USA
* Correspondence: ciucci@surgery.wisc.edu

Abstract: Parkinson's disease (PD) is a progressive, degenerative disorder that affects 10 million people worldwide. More than 90% of individuals with PD develop hypokinetic dysarthria, a motor speech disorder that impairs vocal communication and quality of life. Despite the prevalence of vocal deficits in this population, very little is known about the pathological mechanisms underlying this aspect of disease. As such, effective treatment options are limited. Rat models have provided unique insights into the disease-specific mechanisms of vocal deficits in PD. This review summarizes recent studies investigating vocal deficits in 6-hydroxydopamine (6-OHDA), alpha-synuclein overexpression, *DJ1-/-*, and *Pink1-/-* rat models of PD. Model-specific changes to rat ultrasonic vocalization (USV), and the effects of exercise and pharmacologic interventions on USV production in these models are discussed.

Keywords: Parkinson's disease; rat; ultrasonic vocalization; USV; alpha-synuclein; 6-OHDA; *Pink1*; *DJ1*; exercise; pharmacology; pathology

Citation: Krasko, M.N.; Hoffmeister, J.D.; Schaen-Heacock, N.E.; Welsch, J.M.; Kelm-Nelson, C.A.; Ciucci, M.R. Rat Models of Vocal Deficits in Parkinson's Disease. *Brain Sci.* **2021**, *11*, 925. https://doi.org/10.3390/brainsci11070925

Academic Editors: Stefan M. Brudzynski and Jeffrey Burgdorf

Received: 4 June 2021
Accepted: 9 July 2021
Published: 13 July 2021

1. Introduction

Parkinson's disease (PD) is a progressive, degenerative disorder that affects 10 million people worldwide [1,2]. While the disease is known for hallmark motor signs including a resting tremor, bradykinesia, and rigidity that arise as a result of nigrostriatal dopamine depletion, other signs of disease appear years prior to diagnosis, including changes to voice [3–6]. More than 90% of individuals with PD develop hypokinetic dysarthria, a motor speech disorder that greatly impairs vocal communication [7,8]. Vocal deficits include decreased loudness, monotone pitch, imprecise articulation, and overall decreased intelligibility [9–12]. This negatively impacts vocal quality and overall quality of life [12]. Pharmacological treatments for PD typically target dopamine pathways by increasing neurotransmitter levels or as dopamine receptor agonists [13,14]. These treatments, however, are not effective in alleviating voice dysfunction, suggesting pathology for voice differs in important ways from classical limb motor alterations [13,14]. Similarly, surgical treatments, like deep brain stimulation, improve limb motor signs, yet do not improve vocal communication and may in fact worsen deficits [15–21]. Despite the prevalence of hypokinetic dysarthria in PD, pharmacological and surgical treatment options remain limited. Behavioral therapies continue to be the gold standard in treating voice disorders in this population [22]. While research investigating the efficacy of speech-language interventions for PD-related voice dysfunction has grown, a robust understanding of the underlying biological mechanisms responsible for the onset, progression, and treatment-related improvement in vocal dysfunction is limited. Furthermore, while about 10% of PD cases are familial in nature, a vast majority are deemed idiopathic [23]. There are differences among patients with regard to phenotypic expression of PD, including but not limited to akinetic (freezing), tremor-predominant, young onset, etc. Variability is also noted regarding the presence and severity of signs and symptoms, age of onset, and rate of progression of the disease [24]. This extends to vocal deficits, which often present variably. As such, optimizing treatment remains a universal challenge.

With no yet known etiology and such heterogeneity in the presentation of disease, animal models, including rodent, non-human primate, and non-mammalian, have been used to study different aspects of PD on both behavioral and pathophysiological levels. Rat model systems have allowed for a greater level of experimental control, the ability to study deficits in the prodromal (preclinical) stage of disease, and the means of correlating behavior to neurochemical findings. Specifically, when considering PD-related vocal deficits, neurotoxin, alpha-synuclein, and genetic rodent models have been studied most-extensively and continue to show promise regarding this aspect of disease. The use of animal models has contributed to the PD-voice literature over the last few decades, including characterization of vocal communication in the prodromal stage of disease, assessment of associated neurobiology, especially in extra-dopaminergic pathways, and the development of a training paradigm to study exercise effects on vocal rescue.

The study of ultrasonic vocalizations (USVs) in rat models of PD has increased understanding of vocal communication changes that occur with PD. Similar to humans, rats are highly social animals, generate sound within the larynx, and produce vocalizations that are semiotic in nature [25–29]. USVs are typically categorized by two call types—22-kilohertz (kHz) alarm calls and 50-kHz calls [30–35]. 22-kHz calls occur in response to aversive conditions or in negative affective states and are initiated via activation of the ascending cholinergic system [30,34,36–38]. 50-kHz calls occur in response to activity in the mesolimbic dopaminergic system originating in the ventral tegmental area, and are produced in social, nonaggressive, positive affective states [30–35]. They represent purposeful affiliative vocalizations, are highly relevant to human communication, and as such, are commonly studied and will be the focus of this review. 50-kHz calls are also more complex, varying by acoustic parameters, such as duration (ms), intensity (dB), bandwidth (Hz), and peak frequency (Hz), as well as non-acoustic parameters, such as complexity (%), call rate (calls/s), latency to call (s), and call type (categorical). There are many different approaches to categorizing call type and categories should correspond to the research question [39–43]. Generally, 50-kHz calls are defined as simple or complex and, depending on the research group, can have sub-categories. Simple calls have constant, non-modulating frequency, and complex calls contain two or more directional changes in frequency of at least 3 kHz each [39,40]. Commonly described complex calls include frequency modulated (FM) calls (frequency changes within a call) and harmonic calls (calls with a fundamental frequency near 30 kHz with a visible harmonic one octave above) [40]. In contrast to human voice, 50-kHz USV production does not involve the vibration of vocal folds [44,45]. USV production shares characteristics with human vocalization including the generation of airflow via buildup of lung pressure, the activation of intrinsic laryngeal muscles, and the modulation of the vocal tract during egressive airflow [28,29]. As such, USVs are used to study vocal sensorimotor control in models of PD.

The purpose of this focused review is to highlight the changes that occur to 50-kHz vocal communication in various models of PD. This review also expands on model-specific findings regarding targeted exercise and pharmacological interventions for the treatment of vocal deficits in PD, as well as the strengths and weaknesses of these models in studying PD. See Figure 1 for organization of the manuscript and Figure 2 for a summary of pathologic changes across models discussed.

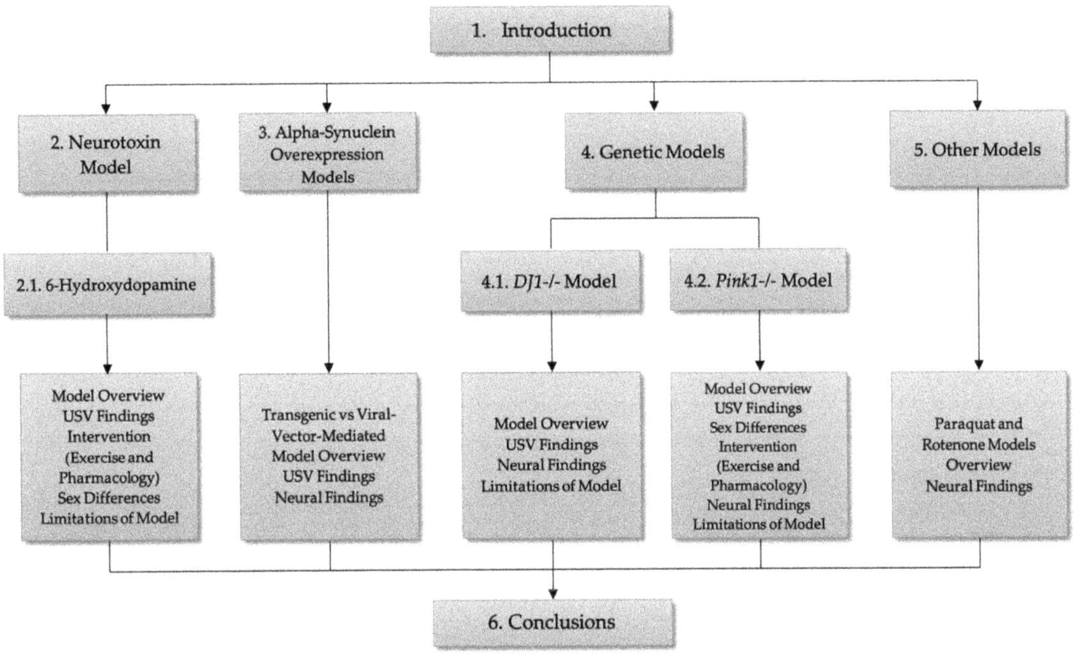

Figure 1. Manuscript overview.

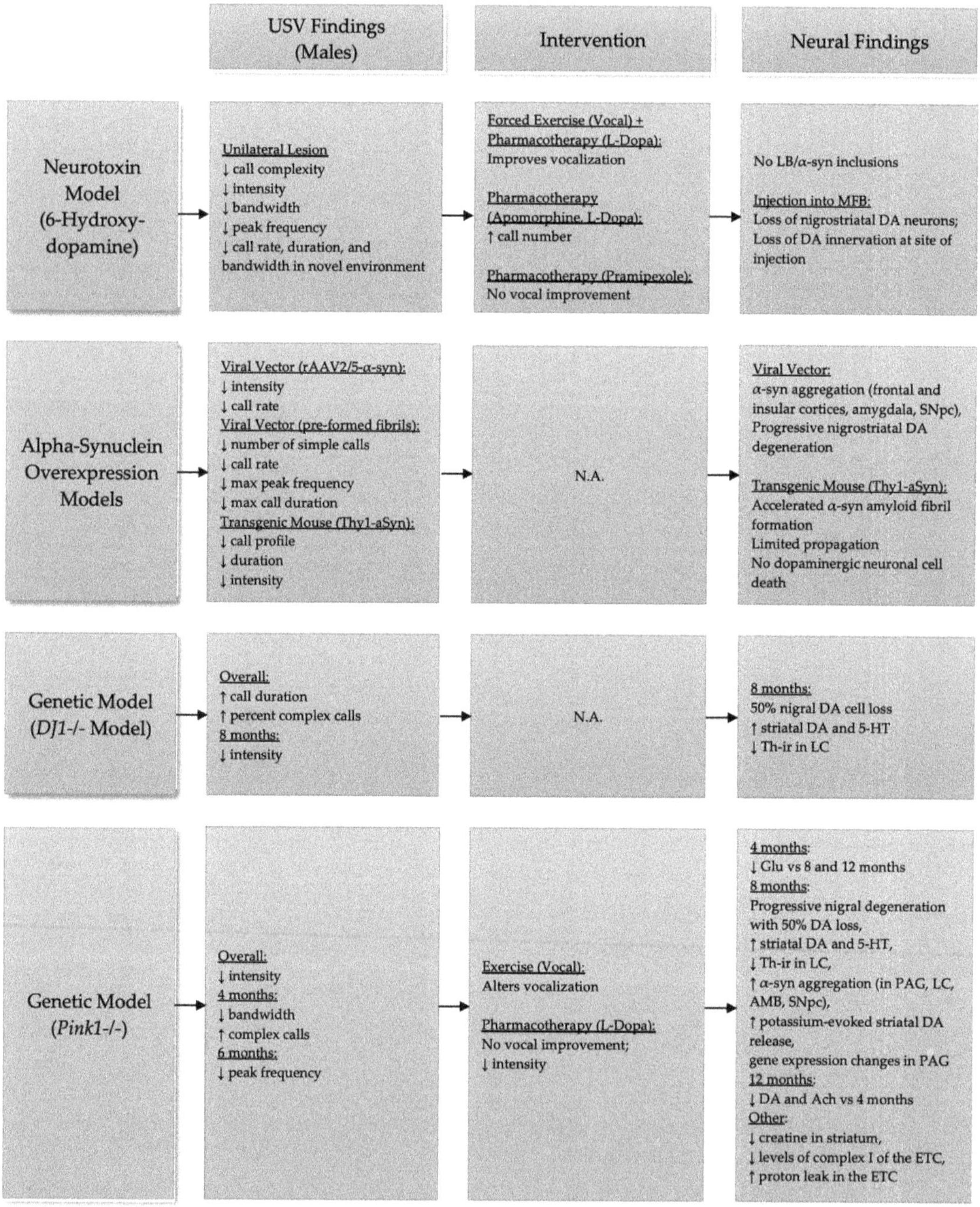

Figure 2. Summary of model-specific USV, intervention, and neural findings in male rats. Ach = acetylcholine, AMB = nucleus ambiguus, α-syn = alpha-synuclein, DA = dopamine, ETC = electron transport chain, Glu = glutamate, LB = Lewy bodies, LC = locus coeruleus, L-Dopa = levodopa, MFB = medial forebrain bundle, PAG = periaqueductal gray, SNpc = substantia nigra pars compacta, Th-ir = tyrosine hydroxylase immunoreactivity, 5-HT = serotonin, ↓ = decrease, ↑ = increase. Underlined text notes category.

2. Neurotoxin Models of PD

6-Hydroxydopamine

Oxidopamine, or 6-hydroxydopamine (6-OHDA), is a catecholaminergic neurotoxin classically used to model PD by inducing significant neurodegeneration of the nigrostriatal dopamine system by unilateral or bilateral infusion to the medical forebrain bundle or the striatum [46–53]. The well-established 6-OHDA rat model has been used to study behavioral changes, mechanisms of cell death, and therapies that could potentially improve PD signs [48,54–58]. Deficits in this model are widespread. In addition to affecting limb movements [59–63], unilateral lesions to the medial forebrain bundle or the striatum have been shown to reduce tongue force, lick force, and lick frequency [64–66], as well as chewing behaviors [67], suggesting that nigrostriatal dopaminergic systems may contribute, at least in part, to oral sensorimotor dysfunction.

Nigrostriatal dopamine depletion via unilateral 6-OHDA infusion into the medial forebrain bundle leads to significant changes in USV production. Rat 50-kHz USVs show decreased call intensity, amplitude, and bandwidth [25,68,69]. Additionally, call complexity degrades as a result of the unilateral 6-OHDA lesion. Specifically, there are fewer FM calls and more simple/flat calls without change to the total number of USVs produced (Table 1) [68]. Of all call types (simple, FM, and harmonic), harmonic calls were produced the least frequently; however, this was observed regardless of dopamine depletion [68]. Subsequent work has largely supported these findings, and further showed decreases in call rate, call duration, and bandwidth when tested in a novel cage environment, suggesting that environment can have a significant impact on behavioral outcomes [69]. Observed decreases in complexity and intensity of calls are analogous to hypophonia noted in individuals with PD, thereby demonstrating utility of USVs in assessing phonatory deficits [70]. The effect of time post-lesion on USV production was also studied at acute (72 h) and chronic (4 weeks) timepoints. Results show that after 72 h, call complexity, bandwidth, and intensity of FM calls correlate with striatal dopamine loss. After 4 weeks, bandwidth, intensity of simple calls, and duration of FM calls were correlated with measures of dopamine depletion. Call complexity was less affected at 4 weeks and was only significantly correlated with percent of tyrosine hydroxylase loss [71]. The 6-OHDA model itself does not fully embody the progressive nature of PD. While dopamine loss may play a role in vocal dysfunction, particularly around the time of diagnosis when dopamine has significantly depleted in the substantia nigra pars compacta (SNpc), other systems may be implicated earlier in disease progression that cannot be fully captured with a 6-OHDA model.

Table 1. Summary of main effects of dopamine condition and call type on USV production.

Publication	Sex	Dopamine Conditions (Independent Variable)	Call Type (Independent Variable)	USV (Dependent Variables)	Effect	Finding
Ciucci et al., 2007 [72]	M	6-OHDA, Haloperidol, Control	n.a.	bandwidth	main effect of DA condition	6-OHDA bandwidth ↓
				number of calls	n.s.	
Ciucci et al., 2008 [25]	M	6-OHDA, Haloperidol, Control	n.a.	peak amplitude	significant	6-OHDA peak amplitude ↓

Table 1. *Cont.*

Publication	Sex	Dopamine Conditions (Independent Variable)	Call Type (Independent Variable)	USV (Dependent Variables)	Effect	Finding
Ciucci et al., 2009 [68]	M	6-OHDA, Haloperidol, Control	Simple, FM	percent simple calls	significant	Percent of simple calls > harmonic
				percent FM calls	significant	Percent FM calls > harmonic
				percent call type	significant	Simple was most frequent in 6-OHDA
						FM was most frequent in haloperidol and controls
				total number of calls	n.s.	
				duration	n.s.	
				bandwidth	main effect of DA condition	6-OHDA bandwidth $\downarrow$
				maximum frequency	main effect of call type	Maximum frequency in FM > simple
				maximum intensity	main effect of DA condition	6-OHDA intensity $\downarrow$
					main effect of call type	Maximum intensity in FM > simple

DA = dopamine, FM = frequency modulated, M = male, n.a. = not applicable, n.s. = not significant, USV = ultrasonic vocalization, $\downarrow$ = decrease.

Studies have also investigated the potential for exercise and/or pharmacotherapies to rescue deficits induced by the unilateral 6-OHDA lesion. The capacity for forced motor exercise to be neuroprotective against behavioral (motor asymmetry) and neurochemical deficits has yielded mixed results. Forced exercise (casting) implemented prior to induction of toxin-related PD has been shown to be neuroprotective [73], and exercise pre- and post-induced 6-OHDA PD resulted in the prevention of motor deficits and reduced striatal dopamine depletion in adult rats [74–77]. An intensive 4-week vocal exercise paradigm, in which rats were trained to produce a greater amount of calls with increased complexity and intensity, also rescued decline in call complexity, intensity, and bandwidth compared to no-exercise in the unilateral 6-OHDA model [78]. Lack of physical activity following the 6-OHDA infusion showed exacerbation of behavioral and neurochemical deficits, thereby suggesting dose-dependent relationships between exercise and severity of disease and associated behavioral deficits [79].

The effects of vocal exercise, pharmacotherapy, and a combination of vocal exercise and pharmacotherapy (dopamine replacement, e.g., L-Dopa) on USVs produced in familiar and novel environments have also been studied. Vocal exercise post-infusion of 6-OHDA increased maximum peak frequency of USVs; however, this was likely influenced by the testing environment, as the increase was seen in the home cage (compared to novel cage) [69]. Additionally, a combination of vocal exercise and L-Dopa was found to be the most effective treatment approach to increasing peak frequency (as opposed to exercise or L-Dopa alone) and this increase was noted in the novel cage. Findings of this study highlight the benefits of combined interventions, and also suggest the influence that testing environment can have on vocal communication [69].

In addition to L-Dopa, other pharmacotherapies have been trialed for their effects on vocalizations. Drugs that modulate dopamine bioavailability are commonly prescribed to patients with PD in an effort to combat motor deficits throughout disease progression; however, their prolonged use can impact affective state and potentially lead to iatrogenic psychiatric disturbances [80–82]. Furthermore, while drugs that target dopamine systems improve motor deficits in humans and animal models of PD [69,83], minimal effect has been observed in rescuing PD-related vocal dysfunction [84]. Simola and colleagues recently observed the effects of three different drugs—apomorphine, L-Dopa, and pramipexole—on USVs in the 6-OHDA rat model. Each drug resulted in different patterns of USV emissions, which is not surprising given their different mechanism of action, with L-Dopa converting into dopamine, pramipexole serving as a dopamine agonist, and apomorphine serving as dopamine receptor agonist. Apomorphine and L-Dopa significantly increased the total number of 50-kHz calls after repeated administration, while pramipexole did not. These findings contrast with previous reports that L-Dopa does not alter the number of calls produced [69]; however, this may be due to differences in the timing of the studies employed between drug administration and USV recording. Overall, results of this study demonstrate that vocal behavior is complex and suggests that the underlying mechanisms driving vocal production are not solely motor in nature [82].

Glial cell-line derived neurotrophic factor (GDNF) has also been trialed in an attempt to establish a neuroprotective therapy and rescue deficits post-lesion. Administration of equine infectious anaemia virus (EIAV) vector coupled with GDNF (EIAV-GDNF) to achieve GDNF expression in the substantia nigra and striatum has been shown to protect dopaminergic neurons in the presence of neurotoxin administration [85]. Furthermore, EIAV-GDNF administration rescued 6-OHDA-induced motor deficits, including rotational asymmetry and spontaneous contralateral motor functions [85,86]. Parkin, the gene product of the PARK2 gene (mutations in which are a major cause of early-onset familial PD), has demonstrated to be neuroprotective in the 6-OHDA model, as well [87]. Overexpressing parkin correlated with improved motor functioning in both cylinder and amphetamine-induced rotation tests [87]. Pharmacotherapies administered prior to or after 6-OHDA infusion have demonstrated utility in rescuing motor deficits; however, further research is necessary to determine the role pharmacotherapies play in rescuing vocalization-related deficits.

Although PD disproportionately affects males, important sex-differences have been identified using animal models of PD. Unilateral 6-OHDA lesioned males and females have shown differences in maintenance of posture, coordination, and initiation of movements [88]. Furthermore, female rats are less susceptible to 6-OHDA than males, specifically, by experiencing significantly less dopaminergic cell loss compared to males after injury [89]. Estrogen has been found to have anti-inflammatory and anti-apoptotic properties on nigrostriatal dopaminergic neurons and, as such, has been suggested to be neuroprotective in females for developing PD [90]. Furthermore, estrogen's activation of adaptive mechanisms in the nigrostriatal system has been shown to decrease neuronal loss [91] and decrease microglial activation and density, further reducing progression of degeneration [90]. Despite these findings, however, no studies using 6-OHDA to date have focused on the role of estrogen or sex in vocal behavior.

While neurotoxin models can be helpful in investigating behavioral deficits and related neuropathology related to damage in the nigrostriatal pathway, they are not without limitations. Although the 6-OHDA model demonstrates acute neurodegenerative properties, it lacks the ability to induce age-dependent, progressive deficits of PD [92]. Additionally, this model lacks the presence of Lewy bodies, a pathological hallmark in human patients with PD. While unilateral lesions to the medial forebrain bundle are the most common in this model, a disadvantage of this approach is that both A9 (nigrostriatal) and A10 (mesolimbic) cell groups comprise the medial forebrain bundle, and therefore lesioning this site implicates the latter axons [93]. Further, different lesion site selections introduce ambiguity into the model with respect to assessing which site most appropriately approximates clinical PD [93]. Lesion site and unilateral vs. bilateral selections for the model create variation amongst the studied animals; however, 6-OHDA-induced lesions can still provide valuable insight into pathological and related behavioral investigations of PD [93,94].

The control of vocalization is complex, involving multiple sensorimotor, cognitive, and limbic brain regions [95,96]. The basal ganglia are certainly implicated in the initiation and modulation of vocalizations. Disrupting nigrostriatal pathways disrupts the quality of vocalization because of altered input to the striatum and consequently the complex circuitry of the basal ganglia and related brain areas. The 6-OHDA lesion to nigrostriatal pathways models one aspect of this complex disease.

3. Alpha-Synuclein Overexpression Models of PD

PD pathology is characterized by the loss of dopaminergic neuronal cells and the formation of misfolded proteins, of which fibrillar alpha-synuclein are the most common, that form Lewy neurites and Lewy bodies in surviving neurons [97]. Mutations include duplications, triplications, or point mutation of the SNCA gene, where alpha-synuclein is the product, which causes autosomal dominant forms of PD [98–102]. Several studies have shown that alpha-synuclein can aggregate and spread, suggesting that it plays a central role in PD progression [103–105]. Two common model systems of achieving alpha-synuclein overexpression are genetic modification and viral transduction. Two point mutations of the alpha-synuclein expressing gene (A53T, A30P) have been linked to autosomal-dominant, early onset PD [98,106] and have been shown to accelerate alpha-synuclein amyloid fibril formation [107,108]. Interestingly, transgenic models demonstrate non-motor signs of disease such as olfactory and digestive deficits [109,110]. Additionally, these models show progressive sensorimotor deficits in the absence of dopamine depletion in the striatum, further suggesting alternative mechanisms responsible for these behavioral changes [111]. On the other hand, overexpressing alpha-synuclein using viral vectors models nigrostriatal pathology by injecting within or near the SNpc. In contrast to transgenic models, overexpression via viral vector allows for induction at different timepoints, allows for the targeting of a defined region of the brain, and results in rapid degeneration of nigrostriatal neurons [112]. Furthermore, viral-vector mediated models also show the presence of limb motor deficits [113–116]. Until recently, vocal deficits were not studied in alpha-synuclein overexpressing models. This is still a largely understudied area, with only two articles discussing vocal deficits in viral-vector-mediated rat models.

Gombash and colleagues examined USVs of male rats 8 weeks post-administration of 5.9×10^{13} rAAV2/5-α-syn injections and compared findings to controls. Duration, bandwidth, intensity, and peak frequency of both simple and FM calls were assessed, as well as call rate and latency to call. Of all acoustic and non-acoustic parameters, only intensity and call rate were found to differ between groups. Specifically, rAAV2/5-α-syn rats produced simple and FM calls with a lower intensity and were found to call at a lower rate compared to controls [117]. Results of this study demonstrated that targeted unilateral nigrostriatal alpha-syn overexpression led to some deficits in USV production.

The effects of striatal injection of fibrillized mouse alpha-synuclein on USV production have also been studied. Paumier and colleagues demonstrated that rats treated with preformed fibrils (PFF) injections produced a lower number of simple calls, had an overall lower call rate, and a significantly reduced maximum peak frequency compared to controls. Furthermore, PFF-treated and recombinant alpha-synuclein rats had shorter maximum call durations compared to their naïve counterparts (Table 2) [118]. However, this study did not directly correlate USV findings to brain neurochemistry.

Table 2. Summary of effects of alpha-synuclein treatment on USV production.

Publication	Sex	Treatment Group	USV (Dependent Variables)	Effect	Finding
Gombash et al., 2013 [117]	M	rAAV2/5-α-syn, controls	call type	n.s.	
			duration	n.s.	
			bandwidth	n.s.	
			intensity	significant	rAAV2/5-α-syn intensity ↓
			peak frequency	n.s.	
			call rate	significant	rAAV2/5-α-syn call rate ↓
			latency to call	n.s.	

Table 2. *Cont.*

Publication	Sex	Treatment Group	USV (Dependent Variables)	Effect	Finding
Paumier et al., 2015 [118]	M	recombinant α-syn, α-syn PFF, controls	number of calls	treatment x call type interaction effect	α-syn PFF number of simple calls ↓
			call rate	main effect of treatment	α-syn PFF call rate ↓
			duration	main effect of treatment	recombinant α-syn and α-syn PFF duration ↓
			peak frequency	main effect of treatment	α-syn PFF peak frequency ↓
			intensity	n.s.	
			latency to call	n.s.	

α-syn = alpha-synuclein, M = male, n.s. = not significant, USV = ultrasonic vocalization, ↓ = decrease.

Mouse models overexpressing alpha-synuclein [111] have also shown relationships between nigrostriatal alpha-synuclein overexpression and early and progressive decline in behavior. Although not widely studied in the context of vocalization, one additional study characterized vocal deficits in mice overexpressing human wild-type alpha-synuclein under a broad neuronal promoter (Thy1-aSyn) [119]. Grant (2014) found call profile of Thy1-aSyn mice to be significantly different compared to wildtype (WT; healthy) controls. The percent of two-cycle calls and jump down calls was significantly reduced in the Thy1-aSyn model at 2–3 months and 6–7 months, respectively. Furthermore, at 2–3 months, the average duration of calls was significantly decreased (for harmonic, jump down, half cycle, and cycle calls) and at 6–7 months, intensity was significantly reduced in the Thy1-aSyn group. Immunohistochemical findings also revealed alpha-synuclein aggregates in the periaqueductal gray at 5 months in the Thy1-aSyn mice [119]. These deficits coincided with previously reported early sensorimotor deficits, deficits in olfaction, circadian rhythm, and gastrointestinal functioning, and high extracellular striatal dopamine levels [119,120]. Similar to alpha-synuclein overexpressing rat models, mice show early and progressive vocal deficits compared to WTs, suggesting similar underlying mechanisms between both species. Results from these studies indicate that vocal deficits can be induced by alpha-synuclein overexpression, in the absence of dopamine depletion.

4. Genetic Models of PD

4.1. DJ1-/- Model

PD is also frequently studied using genetic models. In this model, 2–8 months of age is considered analogous to prodromal to early stage PD in humans. A deletion and missense mutation in the DJ1 (PARK7, Chromosome 1p36) leads to an inherited, autosomal recessive, early onset form of PD [121], that presents with dyskinesia, rigidity, tremors, and later decline in cognitive function. The DJ1 mutation is the second most common identifiable genetic PD etiology after Parkin mutations [122]. Though its role in the pathogenesis of PD is not yet fully understood, DJ1 has been shown to neutralize reactive oxygen species, regulate transcription as well as chaperone, protease, and mitochondrial homeostasis, inhibit alpha-synuclein aggregation, and prevent excessive oxidative stress in the cell [123–125]. Thus, deletion of this gene results in a number of behavioral dysfunctions similar to those manifested in human familial PD.

This review concentrates on the only published paper with regard to vocalization in this rat model. The *DJ1* knockout (*DJ1-/-*) model demonstrates early onset and progressive limb motor, oromotor, and cranial sensorimotor deficits, including decreased limb, tongue/chewing, and vocalization functions. Yang and colleagues (2018) assessed vocalization abilities in the *DJ1-/-* rat model in prodromal to early timepoints of disease (2–8 months of age) and correlated findings to noradrenergic cell loss within the locus coeruleus. Compared to WT controls, *DJ1-/-* rats were found to develop early and progressive ultrasonic vocalization deficits. Specifically, *DJ1-/-* rats produced longer average and maximum calls, and a greater overall percentage of complex calls (Table 3). At 8 months of age, *DJ1-/-* rats showed a lower average intensity of calls, a deficit analogous to the decreased vocal loudness (i.e., hypophonia) PD patients typically experience. Findings also revealed that at 8 months of age, *DJ1-/-* rats demonstrated loss of tyrosine hydroxylase-immunoprotective cells in the locus coeruleus, a brainstem region responsible for the synthesis and regulation of noradrenaline. With widespread connections to the central nervous system, including projections into the prefrontal cortex, striatum, hippocampus, and thalamus, the locus coeruleus has a large impact on PD pathology. Disruptions in the central noradrenergic system are associated with motor and non-motor signs of PD, including vocalization [126]. Tyrosine hydroxylase-positive cells in the locus coeruleus were also found to be negatively correlated with tongue force, suggesting that the greater the loss of neurons within the locus coeruleus, the greater the disruption to oromotor functioning [127]. Whether the loss of these neurons is progressive, however, is still unknown. Overall, noradrenaline has been shown to have widespread implications for PD pathology, including vocalization deficits.

Table 3. Summary of interaction effects and main effects of genotype and age on USV production between *DJ1-/-* rats and WT controls.

Publication	Sex	Genotypes (Independent Variable)	Ages (mo) (Independent Variable)	USV (Dependent Variables)	Effect	Finding
Yang et al., 2018 [127]	M	*DJ1-/-*, WT	2, 4, 6, 8	percent complex calls	main effect of genotype	*DJ1-/-* percent complex calls ↑
					main effect of age	Percent complex calls at 6 and 8 mo > 2 mo;
						Percent complex calls at 6 and 8 mo > 4 mo
				maximum duration	main effect of genotype	*DJ1-/-* maximum duration ↑ (longer)
					main effect of age	Maximum duration at 4 mo > 2 mo
				maximum bandwidth	main effect of age	Maximum bandwidth at 4 mo > 2 mo
				maximum intensity	main effect of age	Maximum intensity at 4, 6, and 8 mo > 2 mo
				maximum peak frequency	n.s.	
				average duration	main effect of genotype	*DJ1-/-* average duration ↑
				average bandwidth	main effect of age	Average bandwidth at 4 mo > 2 mo
				average intensity	genotype x age interaction	At 8, *DJ1-/-* average intensity ↓
				average peak frequency	main effect of age	Average peak frequency at 8 mo < 2 mo

M = male, mo = months, n.s. = not significant, USV = ultrasonic vocalization, WT = wildtype, ↓ = decrease, ↑ = increase.

Other neurotransmitters are also implicated in the *DJ1-/-* rat model [128–130]. At 8 months of age, this model shows decreased glutamate release in the striatum and at 4 and 8 months, increased acetylcholine release compared to WTs [130]. Authors of this study, however, speculate that perhaps the difference in neurotransmitter concentrations may have more to do with a change in reuptake rather than release [130]. Studies have also shown that *DJ1-/-* rats are impacted by progressive dopaminergic degeneration in the SNpc as early as 6 months of age, with 50% cell death by the age of 8 months. Interestingly, limb motor deficits including hindlimb grip strength and motor coordination are present before a significant loss of dopaminergic cells. This suggests that the loss of DJ1 may lead to a period of dopamine cell dysfunction that contributes to cognitive impairments in early PD that proceeds cell death [128]. However, the role these systems play in vocal behavior has not yet been studied in this model.

In contrast to the neurotoxin model, the *DJ1-/-* rat is an early and progressive model, demonstrating unique advantages for the study of vocal deficits in PD. This model also highlights degeneration that occurs outside the classical nigrostriatal dopaminergic system [127], lending to a more robust understanding of the neurobiology underlying voice dysfunction. Despite work by Yang and colleagues, that has characterized vocal behavior in this genetic model, no other studies have looked at *DJ1-/-* vocalizations and little research has focused on the underlying neuropathology. Furthermore, rats used in the majority of studies ranged from 4 to 8 months of age, a period associated with early/prodromal PD. As such, rats in this model may be manifesting their earliest signs of motor and non-motor disruption [121,131], warranting further work that focuses on later time points in PD progression. Similarly, the study of vocalization in this model was completed with males only, underrepresenting sex-specific differences in PD. General exercise has been shown to improve motor function in the *DJ1-/-* model [132], but no exercise paradigm has been evaluated specifically for the improvement of vocalization.

4.2. *Pink1-/- Model*

Another gene implicated in early-onset recessive PD cases is *Pink1* [PTEN (phosphatase and tensin homologue)-induced kinase 1; *PARK6*], mutations of which manifest in signs of disease clinically identical to sporadic PD [133,134]. The gene encodes PINK1, a serine/threonine kinase protein that plays a role in autophagic clearance of dysfunctional mitochondria. Deletion or mutation of *Pink1* results in decreased mitochondrial protection against oxidative stress, nigrostriatal dopamine cell death with consequent motor deficits, and non-motor dysfunction, potentially due to extra-dopaminergic breakdown [135–138]. It is currently the second-most common known cause of autosomal recessive familial PD [139,140].

The *Pink1-/-* rat is a well-established model for the study of PD-related behavioral deficits. Rats ages 2–8 months of age represent prodromal and early PD. Similar to *DJ1-/-* rats, *Pink1-/-* rats show early and progressive cranial sensorimotor signs, including vocalization deficits, and associated pathology in the central nervous system and periphery [128]. Of all the models discussed in this review, the *Pink1-/-* rat has been most extensively studied with regard to vocal deficits. As early as 2 months of age (representing prodromal stage), male *Pink1-/-* rats show reductions in vocal intensity [141,142]. This is one of the most commonly disrupted acoustic outcomes in this model and is highly analogous to the reduced vocal loudness almost always seen in patients with PD [8,9,143]. Compared to WTs, *Pink1-/-* rats also demonstrate significantly decreased bandwidth of calls at 2 months of age that progressively worsen to 10 months of age. Furthermore, average peak frequency, which is important for conspecific communication [144], decreases from 2 and 4 months of age to 6 and 8 months of age [141,142,145]. At 6 months of age, peak frequency, maximum intensity, and bandwidth of USVs are significantly decreased in *Pink1-/-* males compared to WT controls. In a playback study of male vocalizations, female rats preferred WT calls and even background noise to *Pink1-/-* calls [146]. By 10 months of age (representing mid-stage), *Pink1-/-* rats continue to demonstrate significant vocal differences compared to WTs. Interestingly, vocal intensity is significantly increased compared to WTs, due to increased motor variability. Motor variability is a hallmark feature of mid-stage PD. Vocalization was also shown to be related to respiratory function. Specifically, elevated minute ventilation, elevated tidal volumes, and lower breathing frequencies were associated with reduced peak frequencies of USVs in *Pink1-/-* rats [145].

Until recently, the study of vocal impairments in progressive genetic animal models of PD has been limited to males. Marquis et al. (2019) were the first to highlight sex-specific differences in the *Pink1-/-* rat by assessing limb sensorimotor, affective state, and vocalization changes in females while also accounting for estrous cycle [142]. *Pink1-/-* females did not demonstrate a decline in limb motor functioning with disease progression, indicating that perhaps *Pink1-/-* females follow a different time-course for motor dysfunction, compared to males. In terms of affective state, anxiety was heightened in *Pink1-/-* females by 8 months of age. Similarly, by 8 months, vocal loudness for complex (FM) calls decreased in *Pink1-/-* females; however, similar studies in *Pink1-/-* males demonstrated significantly worse vocal deficits across multiple acoustic variables at the 8-month and earlier timepoints (Table 4), suggesting that there may be a sex-specific difference in vocal degradation, with female vocalization breakdown occurring at a slower rate.

A number of studies have attempted to restore vocalization in *Pink1-/-* rats through behavioral and pharmacologic interventions. Similar to the lack of improvement in vocal communication in humans who are prescribed medications, levodopa has a minimal impact on rat vocalization, and can even further reduce vocal intensity [147]. This provides support to the evolving hypothesis that neuropathology associated with vocalization deficits in PD is at least partially extra-dopaminergic. In contrast, behavioral intervention (vocal exercise) rescues some acoustic aspects of USVs, however, these changes are most effective in early stages of disease progression [148], and do not necessarily "normalize" to WT-like calls [149].

Among the most-important advantages of the *Pink1-/-* rat model of PD is the ability to study both neural tissue and peripheral disease pathology (cranial muscles and nerves) and to correlate findings to vocalization behavior. While findings of nigrostriatal dopamine depletion are inconsistent in this model [150], deficits in other brain regions and neurotransmitter systems are common and reproducible [138]. The number of tyrosine hydroxylase immunoreactive cell bodies in the locus coeruleus is reduced in *Pink1-/-* rats, and these cell counts are correlated with vocalization intensity deficits [141]. In *Pink1-/-* rats, protein concentration of norepinephrine in the locus coeruleus is reduced and catechol-o-methyltransferase gene expression in the locus coeruleus is increased. Furthermore, catechol-o-methyltransferase mRNA expression is associated with a percentage of complex calls [147]. Increased levels of striatal serotonin (5-HT) have been reported in 8-month-old *Pink1-/-* rats compared to age-matched controls [128,151], similar to findings of prior studies in other models of PD [152,153]. However, other studies have not found significant differences in striatal [130] or dorsal raphe 5-HT levels between *Pink1-/-* rats and WTs [145]. Further exploration is warranted.

Pink1-/- rats show significant differences in vocal brain regions and in laryngeal and tongue muscles. For example, 8-month old *Pink1-/-* rats have increased alpha-synuclein in the periaqueductal gray yet show no significant mRNA expression of alpha-synuclein compared to WT controls. Instead, *Pink1-/-* rats demonstrate decreased expression of *Atp13a2*, a lysosomal P-type transport ATPase, in the periaqueductal gray [154], suggesting a possible mechanism for alpha-synuclein aggregation. Furthermore, Glass and colleagues recently demonstrated that ex-vivo thyroarytenoid muscles of *Pink1-/-* rats produce decreased force levels in response to 1-Hz and 20-Hz stimulations and show significantly different proportions of myosin heavy chain isoforms relative to WTs, namely an increase in 2L and a reduction in 2X isoforms [155]; work relating this to progression of vocal deficits is ongoing.

Finally, Kelm-Nelson and Gammie recently used high-throughput RNA sequencing to identify differences in gene expression in the periaqueductal gray (a vocal modulatory region) in male and female *Pink1-/-* rats compared to WT controls [156]. Subsequent weighted gene co-expression network analysis identified correlations between relevant gene modules and vocalization in female *Pink1-/-* and WT rats. Differentially expressed genes for both male and female rats mapped to human PD datasets, suggesting that the rat model closely aligns to human PD. This work highlights the potential for the *Pink1-/-* rat to be used to identify targeted pharmacologic interventions for vocal deficits in PD.

Table 4. Summary of interaction effects and main effects of genotype and age on USV production between *Pink1-/-* rats and WT controls.

Publication	Sex	Genotypes (Independent Variable)	Ages (mo) (Independent Variable)	USV (Dependent Variables)	Effect	Finding
Grant et al., 2015 [141]	M	*Pink1-/-*, *Pink1+/-*, WT	2, 4, 6, 8	average intensity of FM calls	main effect of genotype	*Pink1-/-* average intensity ↓ vs. WT and *Pink1+/-*
					main effect of age	Average intensity at 4 mo > 2 mo
				average bandwidth of FM calls	genotype x age interaction	At 4 and 6 mo, *Pink1-/-* average bandwidth ↓ vs. WT;
						Pink1-/- average bandwidth at 4, 6, and 8 mo < 2 mo
				average peak frequency of FM calls	genotype x age interaction	At 6 and 8 mo, *Pink1-/-* average peak frequency ↓ vs. *Pink1+/-*;
						Pink1-/- average peak frequency at 6 and 8 mo < 2 and 4 mo;
						At 6 and 8 mo, *Pink1+/-* average peak frequency ↑ vs. WT
				average duration of FM calls	main effect of age	Average duration at 4, 6, and 8 mo > 2 mo
				percent complex calls	genotype x age interaction	All rats produced more complex calls over time;
						Pink1-/- complex calls at 4, 6, and 8 mo > 2 mo;
						WT complex calls at 8 mo > 4 mo;
						Pink1+/- and WT complex calls at 6 and 8 mo > 2 and 4 mo
Johnson et al., 2020 [145]	M	*Pink1-/-*, WT	10	intensity	main effect of genotype	*Pink1-/-* intensity ↑
				peak frequency	main effect of genotype	*Pink1-/-* peak frequency ↓
				average duration of FM calls	n.s.	
				average bandwidth	n.s.	
Marquis et al., 2020 [142]	F	*Pink1-/-*, WT	2, 4, 6, 8	number of calls	genotype x age interaction	At 2 mo, *Pink1-/-* number of calls ↑;
						Pink1-/- and WT number of calls at 2 mo > 4, 6, and 8 mo
				percent complex calls	main effect of age	All rats produced ↓ percent complex calls over time;
						Significant differences between 2-4, 2-8, 4-6, 4-8, and 6-8 mo
				duration of simple calls	main effect of age	Duration of simple calls significantly different at 4 mo vs. 2, 6, 8 mo
				duration of complex calls	main effect of age	Duration of FM calls at 8 mo > 2 and 4 mo
				bandwidth of simple calls	main effect of age	Bandwidth of simple calls at 4 mo < 2, 6, 8 mo
				average intensity of FM calls	main effect of genotype	*Pink1-/-* average intensity of FM calls ↓

F = female, FM = frequency modulated, M = male, mo = months, n.s. = not significant, USV = ultrasonic vocalization, WT = wildtype, ↓ = decrease, ↑ = increase.

5. Other Models

PD etiology remains largely unknown. With such a high number of cases being deemed idiopathic, other models have been developed to study how environment contributes to pathogenesis. Pesticides and herbicides play a role in the development of PD. Paraquat, an herbicide commonly used in agriculture, has been linked to PD via experimental work in rodents. Exposure leads to alpha-synuclein upregulation, increased alpha-synuclein aggregation, microglial activation, oxidative stress, dose-dependent loss of TH-positive striatal fibers and SNpc neurons, and reduced motor activity [157–159]. As reviewed by Nandipati and Litvan (2016), several studies have also explored the relationship between paraquat and PD in humans; while research is not entirely consistent, most findings show that exposure to paraquat is linked to PD [160].

Another environmental toxin used to study PD is rotenone—a lipophilic insecticide and herbicide that crosses the blood–brain barrier and serves as a complex I inhibitor in cellular respiration. Experimental animals exposed to rotenone show robust signs of PD, including alpha-synuclein aggregation in the brain and periphery, inflammation and activation of microglia, mitochondrial dysfunction, oxidative stress, and behavioral deficits, including motor, postural, and gastrointestinal dysfunction [161–163]. While rotenone has a rather selective toxicity toward dopaminergic cells [164,165], rotenone also causes neurodegeneration of striatal serotonergic and cholinergic cells, as well as noradrenergic cells in the locus coeruleus. Additionally, administration of rotenone results in impairment of locomotor activity, providing evidence that other neurotransmitters in addition to dopamine may be implicated in behavioral changes [165].

Although vocalizations have not been assayed in models of environmental toxins, given the number of parallels observed in these models to idiopathic PD, studying vocalization could provide critical insights into the mechanisms responsible for vocal decline in idiopathic PD. Findings could also contribute to earlier disease identification, and guide the development of new behavioral and pharmacological interventions for PD-related voice disorders.

6. Conclusions

Rat models have contributed to our understanding of PD. While hallmark motor deficits are relatively well-understood, certain signs of PD, including vocal deficits, remain poorly understood due to their prodromal onset and complex pathology. As such, multiple complementary models are necessary to provide insights into the progression and pathophysiological underpinnings of communication deficits. In this paper, we discussed neurotoxin, alpha-synuclein mutation, and genetic rat models that have recently been used to interrogate mechanisms of vocal communication impairment in PD. We reviewed model-specific changes to USV production and associated neurochemistry, and reviewed the role of exercise and pharmacological interventions in vocal rescue. Each of the different models of PD have unique advantages and limitations. Neurotoxin models such as 6-OHDA are useful for the study of mid- to late-stage PD associated with nigrostriatal dopamine depletion, and demonstrate widespread deficits; however, this model shows minimal alpha-synuclein aggregation and does not account for the progressive nature of the disease. In contrast, genetic models like *DJ1-/-* and *Pink1-/-* allow for the study of disease progression, as well as the study of intervention at early, prodromal, and later timepoints. However, genetic mutations make up only a small subset of PD cases and may not capture the subtle differences associated with the pathogenesis of other forms of PD. Paraquat and rotenone models display many signs of PD; however, the effect of these environmental toxins on vocalization is still unknown. Although no one model fully captures the complexity of PD, these models serve as a valuable tool for expanding our understanding of the disease and translating findings to human populations to advance identification and treatment.

Author Contributions: Conceptualization, M.N.K., J.D.H., and M.R.C.; methodology, M.N.K.; resources, M.N.K., J.D.H., N.E.S.-H., and J.M.W.; data curation, M.N.K.; writing—original draft preparation, M.N.K., J.D.H., N.E.S.-H., and J.M.W.; writing—review and editing, M.N.K., C.A.K.-N., and M.R.C.; visualization, M.N.K.; supervision, M.N.K. and M.R.C.; project administration, M.N.K.; funding acquisition, M.N.K., J.D.H., C.A.K.-N., and M.R.C. All authors have read and agreed to the published version of the manuscript.

Funding: This review was funded by the National Institutes of Health, grant number T32DC009401 (Krasko), R21 DC016135 (Kelm-Nelson); R01 NS117469 (Kelm-Nelson), R01 DC018584 (Ciucci), R01 DC014358 (Ciucci). The APC was funded by R01 DC018584 (Ciucci).

Institutional Review Board Statement: Not applicable.

Informed Consent Statement: Not applicable.

Acknowledgments: The authors of this review would like to acknowledge and thank Ciucci Lab personnel for their feedback.

Conflicts of Interest: The authors declare no conflict of interest.

References

1. Brogley, J.E. DaTQUANT: The Future of Diagnosing Parkinson Disease. *J. Nucl. Med. Technol.* **2019**, *47*, 21–26. [CrossRef] [PubMed]
2. Kowal, S.L.; Dall, T.M.; Chakrabarti, R.; Storm, M.V.; Jain, A. The current and projected economic burden of Parkinson's disease in the United States. *Mov. Disord.* **2013**, *28*, 311–318. [CrossRef]
3. Pfeiffer, R.F. Non-motor symptoms in Parkinson's disease. *Park. Relat. Disord.* **2016**, *22*, S119–S122. [CrossRef] [PubMed]
4. Hlavnika, J.; Cmejla, R.; Tykalová, T.; Šonka, K.; Ruzicka, E.; Rusz, J. Automated analysis of connected speech reveals early biomarkers of Parkinson's disease in patients with rapid eye movement sleep behaviour disorder. *Sci. Rep.* **2017**, *7*, 1–13. [CrossRef] [PubMed]
5. Postuma, R.B.; Lang, A.E.; Gagnon, J.F.; Pelletier, A.; Montplaisir, J.Y. How does parkinsonism start? Prodromal parkinsonism motor changes in idiopathic REM sleep behaviour disorder. *Brain* **2012**, *135*, 1860–1870. [CrossRef] [PubMed]
6. Harel, B.T.; Cannizzaro, M.S.; Cohen, H.; Reilly, N.; Snyder, P.J. Acoustic characteristics of Parkinsonian speech: A potential biomarker of early disease progression and treatment. *J. Neurolinguistics* **2004**, *17*, 439–453. [CrossRef]
7. Ramig, L.O.; Countryman, S.; Thompson, L.L.; Horii, Y. Comparison of two forms of intensive speech treatment for Parkinson disease. *J. Speech Hear. Res.* **1995**, *38*, 1232–1251. [CrossRef]
8. Ho, A.K.; Iansek, R.; Marigliani, C.; Bradshaw, J.L.; Gates, S. Speech impairment in a large sample of patients with Parkinson's disease. *Behav. Neurol.* **1998**, *11*, 131–137. [CrossRef] [PubMed]
9. Plowman-Prine, E.K.; Okun, M.S.; Sapienza, C.M.; Shrivastav, R.; Fernandez, H.H.; Foote, K.D.; Ellis, C.; Rodriguez, A.D.; Burkhead, L.M.; Rosenbek, J.C. Perceptual characteristics of parkinsonian speech: A comparison of the pharmacological effects of levodopa across speech and non-speech motor systems. *NeuroRehabilitation* **2009**, *24*, 131–144. [CrossRef]
10. Darley, F.L.; Aronson, A.E.; Brown, J.R. Differential diagnostic patterns of dysarthria. *J. Speech Hear. Res.* **1969**, *12*, 246–269. [CrossRef]
11. Ramig, L.O.; Fox, C.; Sapir, S. Speech treatment for Parkinson's disease. *Expert Rev. Neurother.* **2008**, *8*, 297–309. [CrossRef]
12. Sapir, S.; Ramig, L.; Fox, C. Speech and swallowing disorders in Parkinson disease. *Curr. Opin. Otolaryngol. Head Neck Surg.* **2008**, *16*, 205–210. [CrossRef]
13. Wang, E.; Kompoliti, K.; Jiang, J.J.; Goetz, C.G. An instrumental analysis of laryngeal responses to apomorphine stimulation in Parkinson disease. *J. Med. Speech Lang. Pathol.* **2000**, *8*, 175–186.
14. Kompoliti, K.; Wang, Q.E.; Goetz, C.G.; Leurgans, S.; Raman, R. Effects of central dopaminergic stimulation by apomorphine on speech in Parkinson's disease. *Neurology* **2000**, *54*, 458–462. [CrossRef] [PubMed]
15. Taha, J.M.; Janszen, M.A.; Favre, J. Thalamic deep brain stimulation for the treatment of head, voice, and bilateral limb tremor. *J. Neurosurg.* **1999**, *91*, 68–72. [CrossRef] [PubMed]
16. Ghika, J.; Villemure, J.G.; Fankhauser, H.; Favre, J.; Assal, G.; Ghika-Schmid, F. Efficiency and safety of bilateral contemporaneous pallidal stimulation (deep brain stimulation) in levodopa-responsive patients with parkinson's disease with severe motor fluctuations: A 2-year follow-up review. *J. Neurosurg.* **1998**, *89*, 713–718. [CrossRef] [PubMed]
17. Putzke, J.D.; Wharen, R.E.; Wszolek, Z.K.; Turk, M.F.; Strongosky, A.J.; Uitti, R.J. Thalamic deep brain stimulation for tremor-predominant Parkinson's disease. *Park. Relat. Disord.* **2003**, *10*, 81–88. [CrossRef] [PubMed]
18. Krack, P.; Batir, A.; Van Blercom, N.; Chabardes, S.; Fraix, V.; Ardouin, C.; Koudsie, A.; Limousin, P.D.; Benazzouz, A.; LeBas, J.F.; et al. Five-Year Follow-up of Bilateral Stimulation of the Subthalamic Nucleus in Advanced Parkinson's Disease. *N. Engl. J. Med.* **2003**, *349*, 1925–1934. [CrossRef]
19. Romito, L.M.; Scerrati, M.; Contarino, M.F.; Iacoangeli, M.; Bentivoglio, A.R.; Albanese, A. Bilateral high frequency subthalamic stimulation in Parkinson's disease: Long-term neurological follow-up. *J. Neurosurg. Sci.* **2003**, *47*, 119–128.

20. Hariz, M.I.; Johansson, F.; Shamsgovara, P.; Johansson, E.; Hariz, G.M.; Fagerlund, M. Bilateral subthalamic nucleus stimulation in a parkinsonian patient with preoperative deficits in speech and cognition: Persistent improvement in mobility but increased dependency: A case study. *Mov. Disord.* **2000**, *15*, 136–139. [CrossRef]
21. Kleiner-Fisman, G.; Herzog, J.; Fisman, D.N.; Tamma, F.; Lyons, K.E.; Pahwa, R.; Lang, A.E.; Deuschl, G. Subthalamic nucleus deep brain stimulation: Summary and meta-analysis of outcomes. *Mov. Disord.* **2006**, *21*, S290–S304. [CrossRef]
22. Schulz, G.M.; Grant, M.K. Effects of speech therapy and pharmacologic and surgical treatments on voice and speech in Parkinson's disease: A review of the literature. *J. Commun. Disord.* **2000**, *33*, 59–88. [CrossRef]
23. Klein, C.; Westenberger, A. Genetics of Parkinson's disease. *Cold Spring Harb. Perspect. Med.* **2012**, *2*, a008888. [CrossRef] [PubMed]
24. Foltynie, T.; Brayne, C.; Barker, R.A. The heterogeneity of idiopathic Parkinson's disease. *J. Neurol.* **2002**, *249*, 138–145. [CrossRef] [PubMed]
25. Ciucci, M.R.; Ma, S.T.; Kane, J.R.; Ahrens, A.M.; Schallert, T. Limb use and complex ultrasonic vocalization in a rat model of Parkinson's disease: Deficit-targeted training. *Park. Relat. Disord.* **2008**, *14*, S172–S175. [CrossRef] [PubMed]
26. Brudzynski, S.M.; Pniak, A. Social contacts and production of 50-kHz short ultrasonic calls in adult rats. *J. Comp. Psychol.* **2002**, *116*, 73–82. [CrossRef]
27. Brudzynski, S.M. Principles of rat communication: Quantitative parameters of ultrasonic calls in rats. *Behav. Genet.* **2005**, *35*, 85–92. [CrossRef]
28. Riede, T. Subglottal pressure, tracheal airflow, and intrinsic laryngeal muscle activity during rat ultrasound vocalization. *J. Neurophysiol.* **2011**, *106*, 2580–2592. [CrossRef]
29. Johnson, A.M.; Ciucci, M.R.; Russell, J.A.; Hammer, M.J.; Connor, N.P. Ultrasonic output from the excised rat larynx. *J. Acoust. Soc. Am.* **2010**, *128*, EL75–EL79. [CrossRef]
30. Brudzynski, S.M. Communication of adult rats by ultrasonic vocalization: Biological, sociobiological, and neuroscience approaches. *ILAR J.* **2009**, *50*, 43–50. [CrossRef]
31. Portfors, C.V. Types and functions of ultrasonic vocalizations in laboratory rats and mice. *J. Am. Assoc. Lab. Anim. Sci.* **2007**, *46*, 28–34. [PubMed]
32. Schwarting, R.K.W.; Wöhr, M. On the relationships between ultrasonic calling and anxiety-related behavior in rats. *Braz. J. Med. Biol. Res.* **2012**, *45*, 337–348. [CrossRef]
33. Wöhr, M.; Schwarting, R.K.W. Affective communication in rodents: Ultrasonic vocalizations as a tool for research on emotion and motivation. *Cell Tissue Res.* **2013**, *354*, 81–97. [CrossRef]
34. Brudzynski, S.M. Ethotransmission: Communication of emotional states through ultrasonic vocalization in rats. *Curr. Opin. Neurobiol.* **2013**, *23*, 310–317. [CrossRef] [PubMed]
35. Brudzynski, S. Pharmacology of Ultrasonic Vocalizations in adult Rats: Significance, Call Classification and Neural Substrate. *Curr. Neuropharmacol.* **2015**, *13*, 180–192. [CrossRef] [PubMed]
36. Knutson, B.; Burgdorf, J.; Panksepp, J. Ultrasonic vocalizations as indices of affective states in rats. *Psychol. Bull.* **2002**, *128*, 961–977. [CrossRef]
37. Blanchard, R.J.; Agullana, R.; McGee, L.; Weiss, S.; Blanchard, D.C. Sex differences in the incidence and sonographic characteristics of antipredator ultrasonic cries in the laboratory rat (Rattus norvegicus). *J. Comp. Psychol.* **1992**, *106*, 270–277. [CrossRef]
38. Inagaki, H. Sex Differences in Ultrasonic Vocal Expression of Negative Emotional States in Rats. In *Handbook of Behavioral Neuroscience*; Elsevier B.V.: Amsterdam, The Netherlands, 2018; Volume 25, pp. 337–344. ISBN 9780128096000.
39. Wright, J.M.; Gourdon, J.C.; Clarke, P.B.S. Identification of multiple call categories within the rich repertoire of adult rat 50-kHz ultrasonic vocalizations: Effects of amphetamine and social context. *Psychopharmacology* **2010**, *211*, 1–13. [CrossRef]
40. Johnson, A.M.; Ciucci, M.R.; Connor, N.P. Vocal training mitigates age-related changes within the vocal mechanism in old rats. *J. Gerontol. Ser. A Biol. Sci. Med. Sci.* **2013**, *68*, 1458–1468. [CrossRef]
41. Simola, N.; Costa, G. Emission of categorized 50-kHz ultrasonic vocalizations in rats repeatedly treated with amphetamine or apomorphine: Possible relevance to drug-induced modifications in the emotional state. *Behav. Brain Res.* **2018**, *347*, 88–98. [CrossRef]
42. Barker, D.J.; Root, D.H.; Ma, S.; Jha, S.; Megehee, L.; Pawlak, A.P.; West, M.O. Dose-dependent differences in short ultrasonic vocalizations emitted by rats during cocaine self-administration. *Psychopharmacology* **2010**, *211*, 435–442. [CrossRef]
43. Ahrens, A.M.; Ma, S.T.; Maier, E.Y.; Duvauchelle, C.L.; Schallert, T. Repeated intravenous amphetamine exposure: Rapid and persistent sensitization of 50-kHz ultrasonic trill calls in rats. *Behav. Brain Res.* **2009**, *197*, 205–209. [CrossRef] [PubMed]
44. Roberts, L.H. The rodent ultrasound production mechanism. *Ultrasonics* **1975**, *13*, 83–88. [CrossRef]
45. Roberts, L.H. Evidence for the laryngeal source of ultrasonic and audible cries of rodents. *J. Zool.* **1975**, *175*, 243–257. [CrossRef]
46. Berger, K.; Przedborski, S.; Cadet, J.L. Retrograde degeneration of nigrostriatal neurons induced by intrastriatal 6-hydroxydopamine injection in rats. *Brain Res. Bull.* **1991**, *26*, 301–307. [CrossRef]
47. Ichitani, Y.; Okamura, H.; Matsumoto, Y.; Nagatsu, I.; Ibata, Y. Degeneration of the nigral dopamine neurons after 6-hydroxydopamine injection into the rat striatum. *Brain Res.* **1991**, *549*, 350–353. [CrossRef]
48. Lee, C.S.; Sauer, H.; Björklund, A. Dopaminergic neuronal degeneration and motor impairments following axon terminal lesion by intrastriatal 6-hydroxydopamine in the rat. *Neuroscience* **1996**, *72*, 641–653. [CrossRef]

49. Przedbroski, S.; Leviver, M.; Jiang, H.; Ferreira, M.; Jackson-Lewis, V.; Donaldson, D.; Togasaki, D.M. Dose-dependent lesions of the dopaminergic nigrostriatal pathway induced by instrastriatal injection of 6-hydroxydopamine. *Neuroscience* **1995**, *67*, 631–647. [CrossRef]
50. Sauer, H.; Oertel, W.H. Progressive degeneration of nigrostriatal dopamine neurons following intrastriatal terminal lesions with 6-hydroxydopamine: A combined retrograde tracing and immunocytochemical study in the rat. *Neuroscience* **1994**, *59*, 401–415. [CrossRef]
51. Barata-Antunes, S.; Teixeira, F.G.; Mendes-Pinheiro, B.; Domingues, A.V.; Vilaça-Faria, H.; Marote, A.; Silva, D.; Sousa, R.A.; Salgado, A.J. Impact of aging on the 6-OHDA-induced rat model of Parkinson's disease. *Int. J. Mol. Sci.* **2020**, *21*, 3459. [CrossRef]
52. Ungerstedt, U. 6-hydroxy-dopamine induced degeneration of central monoamine neurons. *Eur. J. Pharmacol.* **1968**, *5*, 107–110. [CrossRef]
53. Johnston, R.E.; Schallert, T.; Becker, J.B. Akinesia and postural abnormality after unilateral dopamine depletion. *Behav. Brain Res.* **1999**, *104*, 189–196. [CrossRef]
54. Blesa, J.; Phani, S.; Jackson-Lewis, V.; Przedborski, S. Classic and new animal models of Parkinson's disease. *J. Biomed. Biotechnol.* **2012**, *2012*, 845618. [CrossRef]
55. Lindner, M.D.; Plone, M.A.; Francis, J.M.; Blaney, T.J.; Salamone, J.D.; Emerich, D.F. Rats with partial striatal dopamine depletions exhibit robust and long-lasting behavioral deficits in a simple fixed-ratio bar-pressing task. *Behav. Brain Res.* **1997**, *86*, 25–40. [CrossRef]
56. Barnéoud, P.; Parmentier, S.; Mazadier, M.; Miquet, J.M.; Boireau, A.; Dubédat, P.; Blanchard, J.C. Effects of complete and partial lesions of the dopaminergic mesotelencephalic system on skilled forelimb use in the rat. *Neuroscience* **1995**, *67*, 837–848. [CrossRef]
57. Winkler, C.; Sauer, H.; Lee, C.S.; Björklund, A. Short-term GDNF treatment provides long-term rescue of lesioned nigral dopaminergic neurons in a rat model of Parkinson's disease. *J. Neurosci.* **1996**, *16*, 7206–7215. [CrossRef]
58. Cousins, M.S.; Salamone, J.D. Involvement of ventrolateral striatal dopamine in movement initiation and execution: A microdialysis and behavioral investigation. *Neuroscience* **1996**, *70*, 849–859. [CrossRef]
59. Olsson, M.; Nikkhah, G.; Bentlage, C.; Björklund, A. Forelimb akinesia in the rat Parkinson model: Differential effects of dopamine agonists and nigral transplants as assessed by a new stepping test. *J. Neurosci.* **1995**, *15*, 3863–3875. [CrossRef] [PubMed]
60. Schallert, T.; Norton, D.; Jones, T.A. A Clinically Relevant Unilateral Rat Model of Parkinsonian Akinesia. *J. Neural Transplant. Plast.* **1992**, *3*, 332–333. [CrossRef]
61. Klein, A.; Metz, G.A.; Papazoglou, A.; Nikkhah, G. Differential effects on forelimb grasping behavior induced by fetal dopaminergic grafts in hemiparkinsonian rats. *Neurobiol. Dis.* **2007**, *27*, 24–35. [CrossRef] [PubMed]
62. Klein, A.; Wessolleck, J.; Papazoglou, A.; Metz, G.A.; Nikkhah, G. Walking pattern analysis after unilateral 6-OHDA lesion and transplantation of foetal dopaminergic progenitor cells in rats. *Behav. Brain Res.* **2009**, *199*, 317–325. [CrossRef]
63. Whishaw, I.Q.; O'connor, W.T.; Dunnett, S.B. The contributions of motor cortex, nigrostriatal dopamine and caudate-putamen to skilled forelimb use in the rat. *Brain* **1986**, *109*, 805–843. [CrossRef] [PubMed]
64. Ciucci, M.R.; Russell, J.A.; Schaser, A.J.; Doll, E.J.; Vinney, L.M.; Connor, N.P. Tongue force and timing deficits in a rat model of Parkinson disease. *Behav. Brain Res.* **2011**, *222*, 315–320. [CrossRef] [PubMed]
65. Plowman, E.K.; Maling, N.; Rivera, B.J.; Larson, K.; Thomas, N.J.; Fowler, S.C.; Manfredsson, F.P.; Shrivastav, R.; Kleim, J.A. Differential sensitivity of cranial and limb motor function to nigrostriatal dopamine depletion. *Behav. Brain Res.* **2013**, *237*, 157–163. [CrossRef] [PubMed]
66. Plowman, E.K.; Maling, N.; Thomas, N.J.; Fowler, S.C.; Kleim, J.A. Targeted motor rehabilitation dissociates corticobulbar versus corticospinal dysfunction in an animal model of parkinson's disease. *Neurorehabil. Neural Repair* **2014**, *28*, 85–95. [CrossRef] [PubMed]
67. Kane, J.R.; Ciucci, M.R.; Jacobs, A.N.; Tews, N.; Russell, J.A.; Ahrens, A.M.; Ma, S.T.; Britt, J.M.; Cormack, L.K.; Schallert, T. Assessing the role of dopamine in limb and cranial-oromotor control in a rat model of Parkinson's disease. *J. Commun. Disord.* **2011**, *44*, 529–537. [CrossRef] [PubMed]
68. Ciucci, M.R.; Ahrens, A.M.; Ma, S.T.; Kane, J.R.; Windham, E.B.; Woodlee, M.T.; Schallert, T. Reduction of Dopamine Synaptic Activity: Degradation of 50-kHz Ultrasonic Vocalization in Rats. *Behav. Neurosci.* **2009**, *123*, 328–336. [CrossRef] [PubMed]
69. Kelm-Nelson, C.A.; Brauer, A.F.L.; Ciucci, M.R. Vocal training, levodopa, and environment effects on ultrasonic vocalizations in a rat neurotoxin model of Parkinson disease. *Behav. Brain Res.* **2016**, *307*, 54–64. [CrossRef] [PubMed]
70. Fleming, S.M.; Schallert, T.; Ciucci, M.R. Cranial and related sensorimotor impairments in rodent models of Parkinson's disease. *Behav. Brain Res.* **2012**, *231*, 317–322. [CrossRef]
71. Grant, L.M.; Barnett, D.G.; Doll, E.J.; Leverson, G.; Ciucci, M. Relationships among rat ultrasonic vocalizations, behavioral measures of striatal dopamine loss, and striatal tyrosine hydroxylase immunoreactivity at acute and chronic time points following unilateral 6-hydroxydopamine-induced dopamine depletion. *Behav. Brain Res.* **2015**, *291*, 361–371. [CrossRef]
72. Ciucci, M.R.; Ma, S.T.; Fox, C.; Kane, J.R.; Ramig, L.O.; Schallert, T. Qualitative changes in ultrasonic vocalization in rats after unilateral dopamine depletion or haloperidol: A preliminary study. *Behav. Brain Res.* **2007**, *182*, 284–289. [CrossRef] [PubMed]
73. Cohen, A.D.; Tillerson, J.L.; Smith, A.D.; Schallert, T.; Zigmond, M.J. Neuroprotective effects of prior limb use in 6-hydroxydopamine-treated rats: Possible role of GDNF. *J. Neurochem.* **2003**, *85*, 299–305. [CrossRef] [PubMed]
74. Tillerson, J.L.; Cohen, A.D.; Caudle, W.M.; Zigmond, M.J.; Schallert, T.; Miller, G.W. Forced Nonuse in Unilateral Parkinsonian Rats Exacerbates Injury. *J. Neurosci.* **2002**, *22*, 6790–6799. [CrossRef]

75. Dutra, M.F.; Jaeger, M.; Ilha, J.; Kalil-Gaspar, P.I.; Marcuzzo, S.; Achaval, M. Exercise improves motor deficits and alters striatal GFAP expression in a 6-OHDA-induced rat model of Parkinson's disease. *Neurol. Sci.* **2012**, *33*, 1137–1144. [CrossRef]
76. Tillerson, J.L.; Caudle, W.M.; Reverón, M.E.; Miller, G.W. Exercise induces behavioral recovery and attenuates neurochemical deficits in rodent models of Parkinson's disease. *Neuroscience* **2003**, *119*, 899–911. [CrossRef]
77. Mabandla, M.; Kellaway, L.; St Clair Gibson, A.; Russell, V.A. Voluntary Running Provides Neuroprotection in Rats After 6-Hydroxydopamine Injection into the Medial Forebrain Bundle. *Metab. Brain Dis.* **2004**, *19*, 43–50. [CrossRef] [PubMed]
78. Ciucci, M.R.; Vinney, L.; Wahoske, E.J.; Connor, N.P. A translational approach to vocalization deficits and neural recovery after behavioral treatment in Parkinson disease. *J. Commun. Disord.* **2010**, *43*, 319–326. [CrossRef]
79. Landers, M.R.; Kinney, J.W.; van Breukelen, F. Forced exercise before or after induction of 6-OHDA-mediated nigrostriatal insult does not mitigate behavioral asymmetry in a hemiparkinsonian rat model. *Brain Res.* **2014**, *1543*, 263–270. [CrossRef] [PubMed]
80. Voon, V.; Potenza, M.N.; Thomsen, T. Medication-related impulse control and repetitive behaviors in Parkinson's disease. *Curr. Opin. Neurol.* **2007**, *20*, 484–492. [CrossRef] [PubMed]
81. Voon, V.; Napier, T.C.; Frank, M.J.; Sgambato-Faure, V.; Grace, A.A.; Rodriguez-Oroz, M.; Obeso, J.; Bezard, E.; Fernagut, P.O. Impulse control disorders and levodopa-induced dyskinesias in Parkinson's disease: An update. *Lancet Neurol.* **2017**, *16*, 238–250. [CrossRef]
82. Simola, N.; Serra, M.; Marongiu, J.; Costa, G.; Morelli, M. Increased emissions of 50-kHz ultrasonic vocalizations in hemiparkinsonian rats repeatedly treated with dopaminomimetic drugs: A potential preclinical model for studying the affective properties of dopamine replacement therapy in Parkinson's disease. *Prog. Neuro-Psychopharmacol. Biol. Psychiatry* **2021**, *108*, 110184. [CrossRef]
83. Edgren, E.; Hedstrand, U.; Nordin, M.; Rydin, E.; Ronquist, G. Prediction of outcome after cardiac arrest. *Crit. Care Med.* **1987**, *15*, 820–825. [CrossRef] [PubMed]
84. Ciucci, M.R.; Grant, L.M.; Rajamanickam, E.S.P.; Hilby, B.L.; Blue, K.V.; Jones, C.A.; Kelm-Nelson, C.A. Early identification and treatment of communication and swallowing deficits in Parkinson disease. *Semin. Speech Lang.* **2013**, *34*, 185–202. [CrossRef]
85. Dowd, E.; Monville, C.; Torres, E.M.; Wong, L.-F.; Azzouz, M.; Mazarakis, N.D.; Dunnett, S.B. Lentivector-mediated delivery of GDNF protects complex motor functions relevant to human Parkinsonism in a rat lesion model. *Eur. J. Neurosci.* **2005**, *22*, 2587–2595. [CrossRef] [PubMed]
86. Azzouz, M.; Ralph, S.; Wong, L.F.; Day, D.; Askham, Z.; Barber, R.D.; Mitrophanous, K.A.; Kingsman, S.M.; Mazarakis, N.D. Neuroprotection in a rat Parkinson model by GDNF gene therapy using EIAV vector. *Neuroreport* **2004**, *15*, 985–990. [CrossRef]
87. Vercammen, L.; Van der Perren, A.; Vaudano, E.; Gijsbers, R.; Debyser, Z.; Van den Haute, C.; Baekelandt, V. Parkin Protects against Neurotoxicity in the 6-Hydroxydopamine Rat Model for Parkinson's Disease. *Mol. Ther.* **2006**, *14*, 716–723. [CrossRef] [PubMed]
88. Field, E.F.; Metz, G.A.; Pellis, S.M.; Whishaw, I.Q. Sexually dimorphic postural adjustments during vertical behaviour are altered in a unilateral 6-OHDA rat model of Parkinson's disease. *Behav. Brain Res.* **2006**, *174*, 39–48. [CrossRef] [PubMed]
89. Tamás, A.; Lubics, A.; Szalontay, L.; Lengvári, I.; Reglodi, D. Age and gender differences in behavioral and morphological outcome after 6-hydroxydopamine-induced lesion of the substantia nigra in rats. *Behav. Brain Res.* **2005**, *158*, 221–229. [CrossRef]
90. Siani, F.; Greco, R.; Levandis, G.; Ghezzi, C.; Daviddi, F.; Demartini, C.; Vegeto, E.; Fuzzati-Armentero, M.T.; Blandini, F. Influence of estrogen modulation on glia activation in a murine model of parkinson's disease. *Front. Neurosci.* **2017**, *11*, 306. [CrossRef]
91. Gillies, G.E.; McArthur, S. Independent influences of sex steroids of systemic and central origin in a rat model of Parkinson's disease: A contribution to sex-specific neuroprotection by estrogens. *Horm. Behav.* **2010**, *57*, 23–34. [CrossRef]
92. Tieu, K. A guide to neurotoxic animal models of Parkinson's disease. *Cold Spring Harb. Perspect. Med.* **2011**, *1*, a009316. [CrossRef] [PubMed]
93. Deumens, R.; Blokland, A.; Prickaerts, J. Modeling Parkinson's Disease in Rats: An Evaluation of 6-OHDA Lesions of the Nigrostriatal Pathway. *Exp. Neurol.* **2002**, *175*, 303–317. [CrossRef]
94. Perlbarg, V.; Lambert, J.; Butler, B.; Felfli, M.; Valabrègue, R.; Privat, A.-L.; Lehéricy, S.; Petiet, A. Alterations of the nigrostriatal pathway in a 6-OHDA rat model of Parkinson's disease evaluated with multimodal MRI. *PLoS ONE* **2018**, *13*, e0202597. [CrossRef] [PubMed]
95. Li-Jessen, N.Y.K.; Ridgway, C. Neuroanatomy of Voice and Swallowing. In *Neurologic and Neurodegenerative Diseases of the Larynx*; Springer International Publishing: Berlin/Heidelberg, Germany, 2020; pp. 21–40.
96. Jean, A. Brain stem control of swallowing: Neuronal network and cellular mechanisms. *Physiol. Rev.* **2001**, *81*, 929–969. [CrossRef] [PubMed]
97. Goedert, M. Alpha-synuclein and neurodegenerative diseases. *Nat. Rev. Neurosci.* **2001**, *2*, 492–501. [CrossRef] [PubMed]
98. Polymeropoulos, M.H.; Lavedan, C.; Leroy, E.; Ide, S.E.; Dehejia, A.; Dutra, A.; Pike, B.; Root, H.; Rubenstein, J.; Boyer, R.; et al. Mutation in the α-synuclein gene identified in families with Parkinson's disease. *Science* **1997**, *276*, 2045–2047. [CrossRef] [PubMed]
99. Tan, E.K.; Chandran, V.R.; Fook-Chong, S.; Shen, H.; Yew, K.; Teoh, M.L.; Yuen, Y.; Zhao, Y. Alpha-synuclein mRNA expression in sporadic Parkinson's disease. *Mov. Disord.* **2005**, *20*, 620–623. [CrossRef]
100. Farrer, M.; Kachergus, J.; Forno, L.; Lincoln, S.; Wang, D.S.; Hulihan, M.; Maraganore, D.; Gwinn-Hardy, K.; Wszolek, Z.; Dickson, D.; et al. Comparison of Kindreds with Parkinsonism and α-Synuclein Genomic Multiplications. *Ann. Neurol.* **2004**, *55*, 174–179. [CrossRef]

101. Singleton, A.B.; Farrer, M.; Johnson, J.; Singleton, A.; Hague, S.; Kachergus, J.; Hulihan, M.; Peuralinna, T.; Dutra, A.; Nussbaum, R.; et al. α-Synuclein Locus Triplication Causes Parkinson's Disease. *Science* **2003**, *302*, 841. [CrossRef]
102. Ibáñez, P.; Bonnet, A.M.; Débarges, B.; Lohmann, E.; Tison, F.; Pollak, P.; Agid, Y.; Dürr, A.; Brice, P.A. Causal relation between α-synuclein gene duplication and familial Parkinson's disease. *Lancet* **2004**, *364*, 1169–1171. [CrossRef]
103. Braak, H.; Del Tredici, K.; Rüb, U.; De Vos, R.A.I.; Jansen Steur, E.N.H.; Braak, E. Staging of brain pathology related to sporadic Parkinson's disease. *Neurobiol. Aging* **2003**, *24*, 197–211. [CrossRef]
104. Holmqvist, S.; Chutna, O.; Bousset, L.; Aldrin-Kirk, P.; Li, W.; Björklund, T.; Wang, Z.Y.; Roybon, L.; Melki, R.; Li, J.Y. Direct evidence of Parkinson pathology spread from the gastrointestinal tract to the brain in rats. *Acta Neuropathol.* **2014**, *128*, 805–820. [CrossRef] [PubMed]
105. Brettschneider, J.; Del Tredici, K.; Lee, V.M.-Y.; Trojanowski, J.Q. Spreading of pathology in neurodegenerative diseases: A focus on human studies. *Nat. Rev. Neurosci.* **2015**, *16*, 109–120. [CrossRef] [PubMed]
106. Krüger, R.; Kuhn, W.; Müller, T.; Woitalla, D.; Graeber, M.; Kösel, S.; Przuntek, H.; Epplen, J.T.; Schöls, L.; Riess, O. Ala30Pro mutation in the gene encoding α-synuclein in Parkinson's disease. *Nat. Genet.* **1998**, *18*, 106–108. [CrossRef] [PubMed]
107. Conway, K.A.; Harper, J.D.; Lansbury, P.T. Accelerated in vitro fibril formation by a mutant α-synuclein linked to early-onset Parkinson disease. *Nat. Med.* **1998**, *4*, 1318–1320. [CrossRef]
108. Yonetani, M.; Nonaka, T.; Masuda, M.; Inukai, Y.; Oikawa, T.; Hisanaga, S.I.; Hasegawa, M. Conversion of wild-type α-synuclein into mutant-type fibrils and its propagation in the presence of A30P mutant. *J. Biol. Chem.* **2009**, *284*, 7940–7950. [CrossRef] [PubMed]
109. Wang, L.; Fleming, S.M.; Chesselet, M.F.; Taché, Y. Abnormal colonic motility in mice overexpressing human wild-type α-synuclein. *Neuroreport* **2008**, *19*, 873–876. [CrossRef]
110. Fleming, S.M.; Tetreault, N.A.; Mulligan, C.K.; Hutson, C.B.; Masliah, E.; Chesselet, M.F. Olfactory deficits in mice overexpressing human wildtype α-synuclein. *Eur. J. Neurosci.* **2008**, *28*, 247–256. [CrossRef]
111. Fleming, S.M.; Salcedo, J.; Fernagut, P.O.; Rockenstein, E.; Masliah, E.; Levine, M.S.; Chesselet, M.F. Early and progressive sensorimotor anomalies in mice overexpressing wild-type human α-synuclein. *J. Neurosci.* **2004**, *24*, 9434–9440. [CrossRef]
112. Kirik, D.; Björklund, A. Modeling CNS neurodegeneration by overexpression of disease-causing proteins using viral vectors. *Trends Neurosci.* **2003**, *26*, 386–392. [CrossRef]
113. Kirik, D.; Rosenblad, C.; Burger, C.; Lundberg, C.; Johansen, T.E.; Muzyczka, N.; Mandel, R.J.; Björklund, A. Parkinson-Like Neurodegeneration Induced by Targeted Overexpression of α-Synuclein in the Nigrostriatal System. *J. Neurosci.* **2002**, *22*, 2780–2791. [CrossRef] [PubMed]
114. Kirik, D.; Annett, L.E.; Burger, C.; Muzyczka, N.; Mandel, R.J.; Björklund, A. Nigrostriatal α-synucleinopathy induced by viral vector-mediated overexpression of human α-synuclein: A new primate model of Parkinson's disease. *Proc. Natl. Acad. Sci. USA* **2003**, *100*, 2884–2889. [CrossRef] [PubMed]
115. Lo Bianco, C.; Ridet, J.L.; Schneider, B.L.; Déglon, N.; Aebischer, P. α-synucleinopathy and selective dopaminergic neuron loss in a rat lentiviral-based model of Parkinson's disease. *Proc. Natl. Acad. Sci. USA* **2002**, *99*, 10813–10818. [CrossRef]
116. Decressac, M.; Mattsson, B.; Lundblad, M.; Weikop, P.; Björklund, A. Progressive neurodegenerative and behavioural changes induced by AAV-mediated overexpression of α-synuclein in midbrain dopamine neurons. *Neurobiol. Dis.* **2012**, *45*, 939–953. [CrossRef] [PubMed]
117. Gombash, S.E.; Manfredsson, F.P.; Kemp, C.J.; Kuhn, N.C.; Fleming, S.M.; Egan, A.E.; Grant, L.M.; Ciucci, M.R.; MacKeigan, J.P.; Sortwell, C.E. Morphological and behavioral impact of AAV2/5-mediated overexpression of human wildtype alpha-synuclein in the rat nigrostriatal system. *PLoS ONE* **2013**, *8*, e81426. [CrossRef]
118. Paumier, K.L.; Luk, K.C.; Manfredsson, F.P.; Kanaan, N.M.; Lipton, J.W.; Collier, T.J.; Steece-Collier, K.; Kemp, C.J.; Celano, S.; Schulz, E.; et al. Intrastriatal injection of pre-formed mouse α-synuclein fibrils into rats triggers α-synuclein pathology and bilateral nigrostriatal degeneration. *Neurobiol. Dis.* **2015**, *82*, 185–199. [CrossRef]
119. Grant, L.M.; Richter, F.; Miller, J.E.; White, S.A.; Fox, C.M.; Zhu, C.; Chesselet, M.F.; Ciucci, M.R. Vocalization deficits in mice over-expressing alpha-synuclein, a model of pre-manifest parkinson's disease. *Behav. Neurosci.* **2014**, *128*, 110–121. [CrossRef]
120. Chesselet, M.F.; Richter, F.; Zhu, C.; Magen, I.; Watson, M.B.; Subramaniam, S.R. A Progressive Mouse Model of Parkinson's Disease: The Thy1-aSyn ("Line 61") Mice. *Neurotherapeutics* **2012**, *9*, 297–314. [CrossRef]
121. Bonifati, V.; Rizzu, P.; Van Baren, M.J.; Schaap, O.; Breedveld, G.J.; Krieger, E.; Dekker, M.C.J.; Squitieri, F.; Ibanez, P.; Joosse, M.; et al. Mutations in the DJ-1 gene associated with autosomal recessive early-onset parkinsonism. *Science* **2003**, *299*, 256–259. [CrossRef]
122. Lev, N.; Roncevich, D.; Ickowicz, D.; Melamed, E.; Offen, D. Role of DJ-1 in Parkinson's disease. *J. Mol. Neurosci.* **2006**, *29*, 215–225. [CrossRef]
123. Taira, T.; Saito, Y.; Niki, T.; Iguchi-Ariga, S.M.M.; Takahashi, K.; Ariga, H. DJ-1 has a role in antioxidative stress to prevent cell death. *EMBO Rep.* **2004**, *5*, 213–218. [CrossRef]
124. Cookson, M.R. Parkinsonism due to mutations in PINK1, Parkin, and DJ-1 and oxidative stress and mitochondrial pathways. *Cold Spring Harb. Perspect. Med.* **2012**, *2*, a009415. [CrossRef]
125. Ryan, B.J.; Hoek, S.; Fon, E.A.; Wade-Martins, R. Mitochondrial dysfunction and mitophagy in Parkinson's: From familial to sporadic disease. *Trends Biochem. Sci.* **2015**, *40*, 200–210. [CrossRef]

126. Espay, A.J.; Lewitt, P.A.; Kaufmann, H. Norepinephrine deficiency in Parkinson's disease: The case for noradrenergic enhancement. *Mov. Disord.* **2014**, *29*, 1710–1719. [CrossRef]
127. Yang, K.M.; Blue, K.V.; Mulholland, H.M.; Kurup, M.P.; Kelm-Nelson, C.A.; Ciucci, M.R. Characterization of oromotor and limb motor dysfunction in the DJ1 -/- model of Parkinson disease. *Behav. Brain Res.* **2018**, *339*, 47–56. [CrossRef]
128. Dave, K.D.; De Silva, S.; Sheth, N.P.; Ramboz, S.; Beck, M.J.; Quang, C.; Switzer, R.C.; Ahmad, S.O.; Sunkin, S.M.; Walker, D.; et al. Phenotypic characterization of recessive gene knockout rat models of Parkinson's disease. *Neurobiol. Dis.* **2014**, *70*, 190–203. [CrossRef]
129. Sun, J.; Kouranova, E.; Cui, X.; Mach, R.H.; Xu, J. Regulation of dopamine presynaptic markers and receptors in the striatum of DJ-1 and pink1 knockout rats. *Neurosci. Lett.* **2013**, *557*, 123–128. [CrossRef] [PubMed]
130. Creed, R.B.; Menalled, L.; Casey, B.; Dave, K.D.; Janssens, H.B.; Veinbergs, I.; van der Hart, M.; Rassoulpour, A.; Goldberg, M.S. Basal and Evoked Neurotransmitter Levels in Parkin, DJ-1, PINK1 and LRRK2 Knockout Rat Striatum. *Neuroscience* **2019**, *409*, 169–179. [CrossRef] [PubMed]
131. Hague, S.; Rogaeva, E.; Hernandez, D.; Gulick, C.; Singleton, A.; Hanson, M.; Johnson, J.; Weiser, R.; Gallardo, M.; Ravina, B.; et al. Early-onset Parkinson's disease caused by a compound heterozygous DJ-1 mutation. *Ann. Neurol.* **2003**, *54*, 271–274. [CrossRef] [PubMed]
132. Zhou, W.; Barkow, J.C.; Freed, C.R. Running wheel exercise reduces α-synuclein aggregation and improves motor and cognitive function in a transgenic mouse model of Parkinson's disease. *PLoS ONE* **2017**, *12*, e0190160. [CrossRef] [PubMed]
133. Valente, E.M.; Brancati, F.; Caputo, V.; Graham, E.A.; Davis, M.B.; Ferraris, A.; Breteler, M.M.B.; Gasser, T.; Bonifati, V.; Bentivoglio, A.R.; et al. PARK6 is a common cause of familial parkinsonism. *Neurol. Sci.* **2002**, *23*, 117–118. [CrossRef]
134. Bentivoglio, A.R.; Cortelli, P.; Valente, E.M.; Ialongo, T.; Ferraris, A.; Elia, A.; Montagna, P.; Albanese, A. Phenotypic characterisation of autosomal recessive PARK6-linked parkinsonism in three unrelated Italian families. *Mov. Disord.* **2001**, *16*, 999–1006. [CrossRef]
135. Thomas, K.J.; Cookson, M.R. The role of PTEN-induced kinase 1 in mitochondrial dysfunction and dynamics. *Int. J. Biochem. Cell Biol.* **2009**, *41*, 2025–2035. [CrossRef] [PubMed]
136. Poole, A.C.; Thomas, R.E.; Andrews, L.A.; McBride, H.M.; Whitworth, A.J.; Pallanck, L.J. The PINK1/Parkin pathway regulates mitochondrial morphology. *Proc. Natl. Acad. Sci. USA* **2008**, *105*, 1638–1643. [CrossRef]
137. Billia, F.; Hauck, L.; Konecny, F.; Rao, V.; Shen, J.; Mak, T.W. PTEN-inducible kinase 1 (PINK1)/Park6 is indispensable for normal heart function. *Proc. Natl. Acad. Sci. USA* **2011**, *108*, 9572–9577. [CrossRef] [PubMed]
138. Kelm-Nelson, C.A.; Lechner, S.A.; Lettenberger, S.E.; Kaldenberg, T.A.R.; Pahapill, N.K.; Regenbaum, A.; Ciucci, M.R. Pink1−/− rats are a useful tool to study early Parkinson disease. *Brain Commun.* **2021**, *3*, fcab077. [CrossRef]
139. Hatano, Y.; Li, Y.; Sato, K.; Asakawa, S.; Yamamura, Y.; Tomiyama, H.; Yoshino, H.; Asahina, M.; Kobayashi, S.; Hassin-Baer, S.; et al. Novel PINK1 mutations in early-onset parkinsonism. *Ann. Neurol.* **2004**, *56*, 424–427. [CrossRef] [PubMed]
140. Li, Y.; Tomiyama, H.; Sato, K.; Hatano, Y.; Yoshino, H.; Atsumi, M.; Kitaguchi, M.; Sasaki, S.; Kawaguchi, S.; Miyajima, H.; et al. Clinicogenetic study of PINK1 mutations in autosomal recessive early-onset parkinsonism. *Neurology* **2005**, *64*, 1955–1957. [CrossRef]
141. Grant, L.M.; Kelm-Nelson, C.A.; Hilby, B.L.; Blue, K.V.; Paul Rajamanickam, E.S.; Pultorak, J.D.; Fleming, S.M.; Ciucci, M.R. Evidence for early and progressive ultrasonic vocalization and oromotor deficits in a PINK1 gene knockout rat model of Parkinson's disease. *J. Neurosci. Res.* **2015**, *93*, 1713–1727. [CrossRef] [PubMed]
142. Marquis, J.M.; Lettenberger, S.E.; Kelm-Nelson, C.A. Early-onset Parkinsonian behaviors in female Pink1-/- rats. *Behav. Brain Res.* **2020**, *377*, 112175. [CrossRef] [PubMed]
143. Darley, F.L.; Aronson, A.E.; Brown, J.R. Clusters of deviant speech dimensions in the dysarthrias. *J. Speech Hear. Res.* **1969**, *12*, 462–496. [CrossRef] [PubMed]
144. Willadsen, M.; Seffer, D.; Schwarting, R.K.W.; Wöhr, M. Rodent ultrasonic communication: Male prosocial 50-khz ultrasonic vocalizations elicit social approach behavior in female rats (rattus norvegicus). *J. Comp. Psychol.* **2014**, *128*, 56–64. [CrossRef]
145. Johnson, R.A.; Kelm-Nelson, C.A.; Ciucci, M.R. Changes to Ventilation, Vocalization, and Thermal Nociception in the Pink1-/-Rat Model of Parkinson's Disease. *J. Parkinsons. Dis.* **2020**, *10*, 489–504. [CrossRef] [PubMed]
146. Pultorak, J.D.; Kelm-Nelson, C.A.; Holt, L.R.; Blue, K.V.; Ciucci, M.R.; Johnson, A.M. Decreased approach behavior and nucleus accumbens immediate early gene expression in response to Parkinsonian ultrasonic vocalizations in rats. *Soc. Neurosci.* **2016**, *11*, 365–379. [CrossRef]
147. Kelm-Nelson, C.A.; Trevino, M.A.; Ciucci, M.R. Quantitative Analysis of Catecholamines in the Pink1 −/− Rat Model of Early-onset Parkinson's Disease. *Neuroscience* **2018**, *379*, 126–141. [CrossRef] [PubMed]
148. Kelm-Nelson, C.A.; Yang, K.M.; Ciucci, M.R. Exercise effects on early vocal ultrasonic communication dysfunction in a PINK1 Knockout Model of Parkinson's Disease. *J. Parkinsons. Dis.* **2015**, *5*, 749–763. [CrossRef]
149. Stevenson, S.A.; Ciucci, M.R.; Kelm-Nelson, C.A. Intervention changes acoustic peak frequency and mesolimbic neurochemistry in the Pink1-/- rat model of Parkinson disease. *PLoS ONE* **2019**, *14*, 1–21. [CrossRef]
150. de Haas, R.; Heltzel, L.C.M.W.; Tax, D.; van den Broek, P.; Steenbreker, H.; Verheij, M.M.M.; Russel, F.G.M.; Orr, A.L.; Nakamura, K.; Smeitink, J.A.M. To be or not to be pink(1): Contradictory findings in an animal model for Parkinson's disease. *Brain Commun.* **2019**, *1*, fcz016. [CrossRef] [PubMed]

151. Ren, X.; Butterfield, D.A. Fidelity of the PINK1 knockout rat to oxidative stress and other characteristics of Parkinson disease. *Free Radic. Biol. Med.* **2021**, *163*, 88–101. [CrossRef]
152. Wang, S.; Zhang, Q.J.; Liu, J.; Wu, Z.H.; Wang, T.; Gui, Z.H.; Chen, L.; Wang, Y. Unilateral lesion of the nigrostriatal pathway induces an increase of neuronal firing of the midbrain raphe nuclei 5-HT neurons and a decrease of their response to 5-HT1A receptor stimulation in the rat. *Neuroscience* **2009**, *159*, 850–861. [CrossRef]
153. Guerra, M.J.; Liste, I.; Labandeira-Garcia, J.L. Effects of lesions of the nigrostriatal pathway and of nigral grafts on striatal serotonergic innervation in adult rats. *Neuroreport* **1997**, *8*, 3485–3488. [CrossRef]
154. Kelm-Nelson, C.A.; Stevenson, S.A.; Ciucci, M.R. Atp13a2 expression in the periaqueductal gray is decreased in the Pink1 -/- rat model of Parkinson disease. *Neurosci. Lett.* **2016**, *621*, 75–82. [CrossRef] [PubMed]
155. Glass, T.J.; Kelm-Nelson, C.A.; Russell, J.A.; Szot, J.C.; Lake, J.M.; Connor, N.P.; Ciucci, M.R. Laryngeal muscle biology in the Pink1/ rat model of Parkinson disease. *J. Appl. Physiol.* **2019**, *126*, 1326–1334. [CrossRef] [PubMed]
156. Kelm-Nelson, C.A.; Gammie, S. Gene expression within the periaqueductal gray is linked to vocal behavior and early-onset parkinsonism in Pink1 knockout rats. *BMC Genom.* **2020**, *21*, 1–13. [CrossRef] [PubMed]
157. Manning-Bog, A.B.; McCormack, A.L.; Li, J.; Uversky, V.N.; Fink, A.L.; Di Monte, D.A. The herbicide paraquat causes up-regulation and aggregation of α-synuclein in mice: Paraquat and α-synuclein. *J. Biol. Chem.* **2002**, *277*, 1641–1644. [CrossRef]
158. Jackson-Lewis, V.; Blesa, J.; Przedborski, S. Animal models of Parkinson's disease. *Park. Relat. Disord.* **2012**, *18*, S183–S185. [CrossRef]
159. Cicchetti, F.; Lapointe, N.; Roberge-Tremblay, A.; Saint-Pierre, M.; Jimenez, L.; Ficke, B.W.; Gross, R.E. Systemic exposure to paraquat and maneb models early Parkinson's disease in young adult rats. *Neurobiol. Dis.* **2005**, *20*, 360–371. [CrossRef]
160. Nandipati, S.; Litvan, I. Environmental exposures and Parkinson's disease. *Int. J. Environ. Res. Public Health* **2016**, *13*, 881. [CrossRef]
161. Greenamyre, J.T.; Cannon, J.R.; Drolet, R.; Mastroberardino, P.G. Lessons from the rotenone model of Parkinson's disease. *Trends Pharmacol. Sci.* **2010**, *31*, 141–142. [CrossRef]
162. Betarbet, R.; Sherer, T.B.; MacKenzie, G.; Garcia-Osuna, M.; Panov, A.V.; Greenamyre, J.T. Chronic systemic pesticide exposure reproduces features of Parkinson's disease. *Nat. Neurosci.* **2000**, *3*, 1301–1306. [CrossRef]
163. Cannon, J.R.; Tapias, V.; Na, H.M.; Honick, A.S.; Drolet, R.E.; Greenamyre, J.T. A highly reproducible rotenone model of Parkinson's disease. *Neurobiol. Dis.* **2009**, *34*, 279–290. [CrossRef] [PubMed]
164. Marey-Semper, I.; Gelman, M.; Lévi-Strauss, M. A selective toxicity toward cultured mesencephalic dopaminergic neurons is induced by the synergistic effects of energetic metabolism impairment and NMDA receptor activation. *J. Neurosci.* **1995**, *15*, 5912–5918. [CrossRef] [PubMed]
165. Höglinger, G.U.; Féger, J.; Prigent, A.; Michel, P.P.; Parain, K.; Champy, P.; Ruberg, M.; Oertel, W.H.; Hirsch, E.C. Chronic systemic complex I inhibition induces a hypokinetic multisystem degeneration in rats. *J. Neurochem.* **2003**, *84*, 491–502. [CrossRef] [PubMed]

brain sciences

MDPI

Article

Chronic Fentanyl Self-Administration Generates a Shift toward Negative Affect in Rats during Drug Use

Angela N. Dao [1], Nicholas J. Beacher [1,2], Vivian Mayr [1], Annalisa Montemarano [1], Sam Hammer [1] and Mark O. West [1,*]

1 Department of Psychology, Rutgers University, Piscataway, NJ 08854, USA; angela.dao@rutgers.edu (A.N.D.); nicholas.beacher@nih.gov (N.J.B.); vvm19@scarletmail.rutgers.edu (V.M.); am2454@scarletmail.rutgers.edu (A.M.); SHammer@epilepsygroup.com (S.H.)

2 Intramural Research Program, National Institute on Drug Abuse, National Institutes of Health, Baltimore, MD 21224, USA

* Correspondence: markwest@rutgers.edu

Abstract: Drug addiction is thought to be driven by negative reinforcement, and it is thought that a shift from positive affect upon initial exposure to negative affect after chronic exposure to a drug is responsible for maintaining self-administration (SA) in addicted individuals. This can be modeled in rats by analyzing ultrasonic vocalizations (USVs), a type of intraspecies communication indicative of affective state based on the frequency of the emission: calls in the 22 kHz range indicate negative affect, whereas calls in the 50 kHz range indicate positive affect. We employed a voluntary chronic, long-access model of fentanyl SA to analyze affective changes in the response to chronic fentanyl exposure. Male Sprague-Dawley rats self-administered either fentanyl (N = 7) or saline (N = 6) for 30 consecutive days and USVs were recorded at four different time points: the day before the first SA session (PRE), the first day of SA (T01), the last day of SA (T30), and the first day of abstinence (ABS). At T01, the ratio of 50 to 22 kHz calls was similar between the fentanyl and saline groups, but at T30, the ratio differed between groups, with the fentanyl group showing significantly fewer 50 kHz calls and more 22 kHz calls relative to saline animals. These results indicate a shift toward a negative affect during drug use after chronic exposure to fentanyl and support negative reinforcement as a main driving factor of opioid addiction.

Keywords: opioids; fentanyl; self-administration; ultrasonic vocalizations; affect

Citation: Dao, A.N.; Beacher, N.J.; Mayr, V.; Montemarano, A.; Hammer, S.; West, M.O. Chronic Fentanyl Self-Administration Generates a Shift toward Negative Affect in Rats during Drug Use. *Brain Sci.* **2021**, *11*, 1064. https://doi.org/10.3390/brainsci11081064

Academic Editor: Fabrizio Schifano

Received: 27 July 2021
Accepted: 11 August 2021
Published: 13 August 2021

1. Introduction

Affective state in response to aversive or appetitive stimuli can be readily modeled in laboratory rats by analyzing a form of intraspecies communication known as ultrasonic vocalizations (USVs). USVs can be categorized into two ranges associated with different meanings, due to their emission during emotionally arousing situations: calls in the 18–33 kHz range are referred to as 22 kHz USVs and calls in the 35–70 kHz range are referred to as 50 kHz USVs [1,2].

Due to their predictable occurrence in response to aversive situations, 22 kHz calls are considered representative of a negative affective state and are generated during common behavioral situations such as exposure to predators [3], unfamiliar touch [4], drug withdrawal [5,6], and exposure to pain such as foot shock [7]. Conversely, 50 kHz calls are reliably emitted during situations involving the anticipation of potential reward [8,9], social contact [10], electrostimulation of mesolimbic sites supporting self-stimulation [11], and the administration of addictive drugs [12,13]. Therefore, 50 kHz calls are representative of an appetitive, hedonic behavioral state associated with positive affect [9,10]. As both 22 and 50 kHz USVs can provide insight into the affective state of animals, they can be recorded and analyzed during periods of drug administration to identify changes in affect associated with chronic opioid exposure.

It is commonly accepted that negative reinforcement is the driving factor behind addiction [14], whereby the escape or avoidance of negative affect during drug use and withdrawal is the main motivating factor maintaining the self-administration of addictive drugs [15]. Furthermore, a shift from positive affect upon initial drug use to the emergence of negative affect after extended drug exposure and development of drug dependence and/or addiction is a salient motivational factor in chronic, long-access animal models [16] and human drug abuse [17]. Analyses of rodent USVs during opioid administration indicated that anticipatory 50 kHz emissions decrease after repeated exposure to morphine, suggesting an aversive influence of morphine exposure [12] and supporting a shift from positive to negative affect in response to repeated opioid administration.

Morphine was shown to dose-dependently suppress both 22 and 50 kHz USVs independent of opioid effects on pain perception, and this response can be attenuated by administration of a mu-opiate receptor antagonist such as naloxone [18], implicating opioid specificity. Additionally, opioid administration suppresses 22 kHz USVs emitted during foot shock, whereas the audible pain-associated squeak is unaffected, further dissociating USV suppression by opioids from their analgesic effects [7]. These results are consistent with claims that opioids influence affective as well as autonomic, somatic, and motor processes, and indicate a role of opioids in modifying the central mechanisms of USVs.

Surprisingly, little emphasis has been placed on understanding the specific affective state of rats during the self-administration of opioids. Therefore, we employed a long-access fentanyl self-administration (SA) model to analyze the changes in affect in response to chronic fentanyl exposure in rats. Intravenous SA is the most translatable animal model for human drug addiction and is thus often referred to as the gold standard for measuring abuse liability [19,20]. The model establishes two key aspects of drug addiction: compulsive drug use and escalation of drug intake over time. The inclusion of USV analysis allows for the identification of the emergence of a negative affective state, which is a third key aspect of drug addiction. Based on previous studies citing a shift from positive to negative affect after repeated drug exposure [12,16,17], we predicted that rats would exhibit increased positive affect upon initial exposure to fentanyl, as indicated by increased 50 kHz call rates during the first session of fentanyl SA, which would then shift toward negative affect after 30 days of fentanyl SA.

2. Materials and Methods

2.1. Animals

Adult, male Sprague-Dawley rats (Charles River, Wilmington, MA, USA) were allowed to self-administer the opiate receptor agonist fentanyl HCL (dose = 2.57 μg/kg per i.v. infusion; fentanyl SA group, N = 7) or saline (saline SA group, N = 6) for 30 consecutive days. Rats were singly housed on a 12:12 h light:dark cycle (lights on at 10:30 a.m.). Prior to surgery, rats were allowed to reach adult weight (350 g) and maintained at this weight thereafter to avoid the addition of fat tissue.

2.2. Catheterization Surgery

Animals were anesthetized with a ketamine/xylazine (K/Xyl) mixture (50 mg/kg, i.p.) and given an injection of atropine (10 mg/kg; i.p.) to decrease fluid buildup in lungs and prevent respiratory arrest. Anesthesia was monitored and maintained throughout surgery by intermittent K/Xyl injections. During surgery, animals were chronically implanted with an intravenous catheter in the right jugular vein. This catheter was threaded subcutaneously and exited at the scalp where it was led through a j-shaped stainless-steel cannula attached to the skull using dental cement and jeweler's screws. The catheter was protected by a metal spring-leash permanently connected to the animal's cannula to prevent damage. Following surgery, the animal was housed in a self-administration operant chamber at all times for the entirety of the SA experiment. Animals were allowed one week to recover from surgery, during which time they received once daily i.v. infusions of antibiotics and NSAID pain reliever (rimadyl and baytril). During all hours other than SA sessions, a

200 μL infusion of saline was delivered every 25 min by a computer-controlled syringe pump to preserve catheter patency. Animals received water ad libitum and received enough food to maintain a weight of 350 g throughout the duration of the experiment.

2.3. Self-Administration Apparatus

The clear Plexiglas chamber in which animals were housed included a corner with a fixed 6-photocell device used to monitor and record head movements [21]. An infusion was administered only when a correct operant response was performed in this corner. A correct operant response consisted of breaking photocells 2 and 3 in succession within 1 s. All rewarded responses (RRs) and unrewarded responses (URs) were recorded. The Plexiglas chamber was housed within a ventilated, sound-attenuating outer shell.

2.4. Self-Administration

SA sessions (6 h/day, 7 days/week) were conducted using the long-access model of Ahmed and Koob [22], which models human addiction, including escalation of intake and persistent increase in the motivation for drug-taking [23,24]. SA sessions ran each day for 30 consecutive days, starting at light onset. Sessions automatically ended upon the completion of 6 h. During the session, drug or saline was available during the entire 6 h on an FR1 reinforcement schedule. A correct response (except during timeout, see below) turned on the pump and automatically dispensed a 0.9 μg/0.075 mL solution of intravenous fentanyl (or an equal volume of saline) through the surgically implanted catheter over 2.5 s, for an average infusion dose of 2.57 μg/kg. This correct response was defined as an RR. An RR immediately triggered a 40 s inter-trial interval timeout as a precautionary measure to prevent overdose, but all responses provided during this time were recorded as URs.

2.5. USV Recording and Scoring

Prior to the commencement of SA recording sessions, a condenser microphone (CM16/CMPA, Avisoft) was suspended 2.5 cm above an arrangement of small holes in the top of the Plexiglas SA chamber. USVs were recorded at a 250 kHz sampling frequency (16 bits) using recording software (Ultrasound Gate, Avisoft, Glienicke/Nordbahn, Germany). Baseline USVs were recorded one week after surgery prior to the start of the first SA session over the same 6 h period as SA sessions were conducted. Subsequent USV recordings were obtained for 6 h during session 1 (T01), session 30 (T30), and the first day of abstinence (ABS; session 31) at the same time of day as a 6 h SA session. As characterization and scoring of USVs are time- and labor-intensive, this limited agenda was designed to capture affective calling during SA for the first time, SA for the 30th consecutive day, and the first time being deprived of the expected drug (18 h withdrawal).

Audio files were run through an automated detector, DeepSqueak [25], to isolate potential calls. These were then manually checked to distinguish between actual calls (which were accepted) and artifacts and background noise (which were rejected). The automatic detector outputs the specific frequency and exact timing of individual calls. Only manually accepted calls were used for analyses. Calls were designated as belonging to the 22 or 50 kHz ranges.

2.6. Data Analyses

Data were analyzed using Prism GraphPad software. Behavioral measures included (i) number of RRs/session, (ii) average drug level (mg/kg) maintained during SA, (iii) slope of escalation of intake, and (iv) 22 and 50 kHz call rates during baseline, T01, T30, and ABS sessions.

2.6.1. Escalation of Intake

The total number of RRs was regressed over sessions 1 through 30 for both fentanyl and saline SA. The analysis was conducted using a simple linear regression, where RR was defined as the dependent variable and session was defined as a continuous independent variable. An additional linear regression was performed in which total fentanyl intake was regressed over the session (1–30), where intake was defined as a dependent variable and session was defined as a continuous independent variable. To incorporate body weight into the calculation of total fentanyl intake, the following equation was used: intake = (#RRs × μg fentanyl per infusion)/body weight.

2.6.2. USV Analysis

One-tailed *t*-tests were conducted to compare the ratio of 50 kHz to 22 kHz USVs for T01 vs. T30. One-tailed t-tests were performed based on expectations derived from the literature, from which we formed the hypothesis that the ratio would shift toward fewer positive and more negative calls at T30 relative to T01.

3. Results

3.1. Acquisition of Fentanyl SA

A total of 28 rats were surgically prepared for SA. Seven rats in the fentanyl SA group, and six rats in the saline SA group completed all phases of SA and USV recording. Animals self-administered fentanyl in a manner consistent with animal models of substance use disorder, in which escalation of intake is a key marker of addiction [26]. The average number of reinforced responses (#RRs) was plotted against SA session (session 1–30). The simple linear regression revealed that the slope of the line for fentanyl SA (0.6242) was significantly different from zero ($F(1,28) = 16.45$, $p = 0.0004$), identifying escalation of intake over time. Accounting for body weight, a separate linear regression similarly identified escalation of fentanyl intake (μg/kg) over time ($F(1,28) = 9.912$, $p = 0.0039$, Figure 1). The average number of RRs in the saline group remained low and did not change across sessions (Figure 1; $F(1,28) = 4.140$, $p > 0.05$).

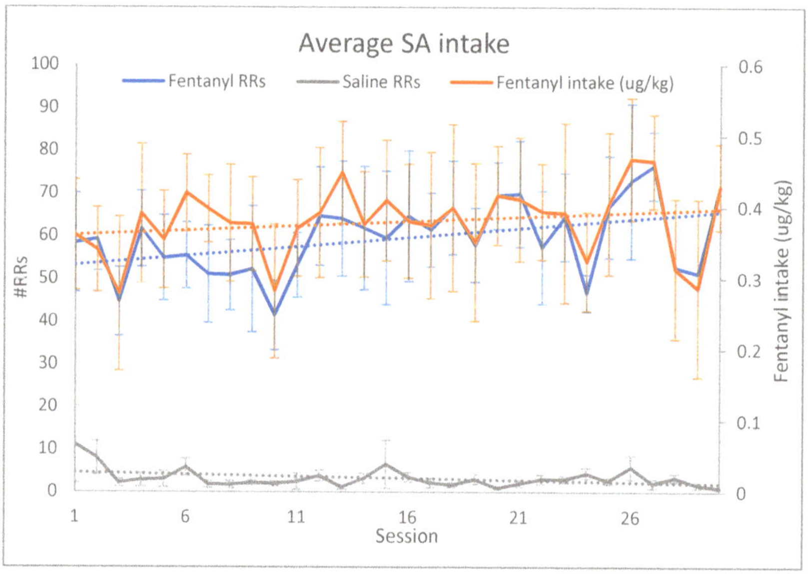

Figure 1. Average SA intake. A simple linear regression revealed that fentanyl animals exhibited escalation of intake over 30 days of SA training as demonstrated by the increased number of fentanyl infusions (RRs; $F(1,28) = 16.45$, $p = 0.0004$) and average daily intake of fentanyl (accounting for body weight; μg/kg; $F(1,28) = 9.912$, $p = 0.0039$). Saline animals did not escalate intake over 30 days of SA ($F(1,28) = 4.140$, $p > 0.05$). Error bars denote the SEM.

3.2. *Shift toward Negative Affect after 30 Days of Fentanyl SA*

Animals exhibited a mix of affective responses to fentanyl SA in session 1 (T01). Relative to baseline (PRE), some rats emitted more 50 kHz than 22 kHz calls during the six hours of T01, whereas others exhibited the opposite pattern (Figure 2). However, by session 30 (T30), no fentanyl SA rats emitted more 50 kHz than 22 kHz calls, with several rats emitting more 22 kHz than 50 kHz calls. Across all sessions, the average 50 kHz call rate per hour peaked at T01 but declined to near zero at T30 (Figure 3). Combined, the seven rats in the fentanyl group emitted a total of one 50 kHz call during the six hours of T30. In contrast, the average 22 kHz call rates peaked at T30.

To test the hypothesis that affective state would become more negative after a month of daily fentanyl SA, a key planned comparison was between the ratio of 50 to 22 kHz call rates at T01 vs. T30. The ratio was computed for each animal in each session using the formula $(B - A)/(B + A)$, where B is the 50 kHz call rate and A is the 22 kHz call rate. Using this formula, positive values indicate a higher prevalence of 50 kHz calls, negative values indicate a higher prevalence of 22 kHz calls, and zero indicates equal rates of each (Figure 2). At T01, the fentanyl group (M = −0.17, SD = 0.65) showed no difference from the saline group (M = 0.06, SD = 0.59) ($t(11) = -0.684$, $p = 0.25$). In contrast, at T30, the fentanyl group (M = −0.37, SD = 0.39) showed a significant shift toward fewer 50 kHz and more 22 kHz calls compared to the saline group (M = 0.10, SD = 0.35) ($t(11) = -2.31$, $p = 0.02$). Thus, the ratio of 50 kHz to 22 kHz calls was not different between groups at T01 but shifted to a significantly more negative value in the fentanyl group at T30 (Figure 2). These results indicate the emergence of a predominately negative affective state during a six-hour session of drug use after a month of chronic fentanyl self-administration.

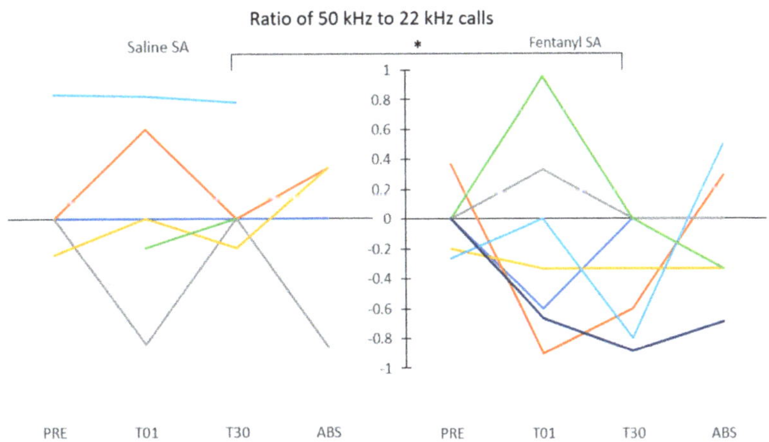

Figure 2. Ratio of 50 to 22 kHz calls. Tracking of all rats' 50 kHz vs. 22 kHz calls across all sessions. Each line represents one rat. Y-axis = $(B - A)/(B + A)$, where B is the rate of 50 kHz USVs and A is the rate of 22 kHz USVs. Thus, numbers above the horizontal line at zero represent a higher ratio of positive to negative calls; numbers below zero indicate a higher ratio of negative to positive calls; zero (horizontal line) indicates equal rates of the two calls. Note the absence of positive USVs and the prevalence of negative USVs in the 30th fentanyl SA session. The ratio did not differ between groups on T01, but the negative ratio of the fentanyl group was significantly different from the ratio of the saline group at T30; * $p = 0.02$.

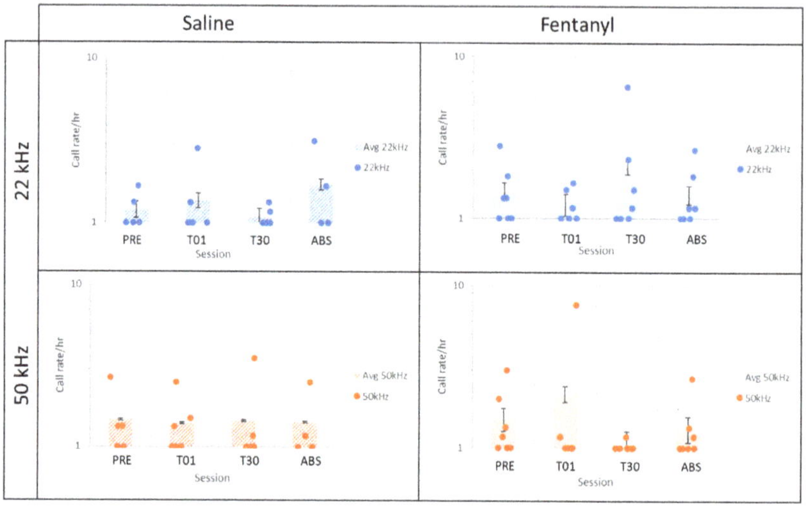

Figure 3. The 22 kHz and 50 kHz call rates per hour. Each bar represents the mean call rate per hour at PRE, T01, T30, and ABS for both saline and fentanyl groups. Each dot represents one animal. The fentanyl group exhibited a shift toward negative affect after 30 days of fentanyl SA as indicated by the increase in 22 kHz calls and decrease in 50 kHz calls at T30 compared to T01. This change was not observed in the saline group. Two data points (one fentanyl T01 and one saline PRE) were outliers and removed from Figure 3, but not from statistical analyses, to avoid distortion of the y-axis, so that the visualization of the data is consistent with the results showing a shift toward negative affect at T30. Note the non-linear scale on the y-axis. Error bars denote the SEM.

4. Discussions

The present results provide evidence of a shift toward negative affect during chronic opioid administration, supporting negative reinforcement as a salient motivating factor driving drug addiction. The decrease in 50 kHz and increase in 22 kHz USVs in session 30 relative to session 1 of fentanyl but not saline SA suggest an aversive response to repeated opioid use, which is consistent with previous reports [12]. The call rate/hour results (Figure 3) are reinforced by the data presented by the ratio of 50 kHz calls to 22 kHz calls (Figure 2), which track animals across sessions and highlight individual differences among animals. This ratio provided a useful measure of general affect, which varied between positive and negative during the long-access sessions in the present study. T30 in the fentanyl group was the only recording session exhibiting an equal or greater ratio of negative to positive calls for all rats, as well as a dearth of 50 kHz calls, indicating not only a lack of positive affect but also a shift toward overall negative affect. Accordingly, these results further corroborate those of previous studies reporting a shift from positive to negative affect after chronic drug abuse [12,16,17].

We observed no significant or uniform increase in positive affect at T01 of fentanyl SA, consistent with previous studies reporting the suppression of 50 kHz calls in response to opioids in drug-naïve rats [27,28]. The initial exposure to opioids does not consistently generate a positive affective response [29,30] (for a review, see Verendeev and Riley [31]) typically observed with other classes of drugs [16,32]. Given the presence of some negative affective responses to fentanyl SA on T01, we cannot rule out the possibility that the increase in 22 kHz calls at T30 could have been associated with a higher drug intake in that session. That increase in negative calls plus the lack of major changes in the call rates of either frequency at ABS highlight the presence of negative affect during drug use after repeated exposure to the drug, not just once the drug has left the body and withdrawal sets in. The mix of both 22 and 50 kHz calls during the 6 h ABS session may reflect fluctuations between a negative affective state associated with opioid withdrawal [6] and a positive affective state associated with anticipation of drug for the 31st consecutive day, since 50 kHz calls can indicate an anticipatory state [8,33,34].

Opioids have been known to suppress USVs in general [12,18,28]. This appears to be the case in the present study, considering the low call rates detected compared to the higher rates of both 22 and 50 kHz calls detected during cocaine SA (e.g., [16,35]). The pattern of USV emission during fentanyl SA appears to be notably different from the SA of other classes of drugs, such as stimulants. Behaviorally, we did not observe an initial "load-up" period in which drug level rises rapidly at the beginning of each SA session as is observed with cocaine SA [36]. Cocaine drug level is a strong predictor of affect, such that the 50 kHz emissions by animals self-administering cocaine coincide with rising drug levels exclusively during initial load-up, coinciding with a decrease in the 22 kHz call rate. Thereafter, 22 kHz calls dominate the maintenance phase, increasing whenever drug level falls [37]. Cocaine generates a positive affect upon first exposure to SA [16], along with motoric activation and emotional arousal, which is the driving force for the emission of USVs [8]. This does not appear to be the case for fentanyl SA, as initial exposure to fentanyl SA generated neither positive nor negative affect in the present study. Opioid SA is not as clearly associated with emotional arousal as that of cocaine, considering environmental preferences with respect to each drug. Badiani et al. [19] reviewed extensive evidence that both rats and humans prefer recreational and/or social use of the psychomotor stimulant cocaine, whereas they prefer taking opioids in private settings. Fentanyl SA may be associated with greater tranquility, less emotional arousal, and less tendency to emit positive vocalizations.

We observed a trend toward elevated 50 kHz call rates during the first SA session, similar to that of Avvisati et al. [38], who also studied opioid SA in rats occupying their home cage. They reported a further increase in 50 kHz calls during the first 30 min of SA sessions after two weeks of heroin SA. That finding, however, differs from the decline we observed in the 50 kHz call rate after a month of daily fentanyl SA. A second difference between the studies is the paucity of 22 kHz calls in their study compared to the present increase in 22 kHz calls during session 30 of fentanyl SA. These differences may relate to the SA of the two different opioids, their use of 3 h sessions for 14 days vs. the 6 h session model of addiction [22] for 30 days used in the present study, or their daily alternation with cocaine SA. Resolving these differences will be important for understanding affective processes in opioid abuse.

Some insight may be gained from our prior studies of cocaine SA. During load-up, the spike in 50 kHz call rates increased [16] or was sustained [37] across 14 sessions of cocaine SA, similar to the increase across sessions observed by Avvisati et al. [38] during the first 30 min of their sessions. Following load-up on cocaine, the 5+ hour maintenance phase was devoid of 50 kHz calls but dominated by 22 kHz calls, indicating that responding was being maintained by negative reinforcement [37]. A similar predominance of 22 kHz calls indicating negative affect and potentially negative reinforcement may have characterized the maintenance of fentanyl SA. However, we observed substantial variability within and between subjects in the timing of fentanyl self-infusions, with no clear separation of load-up from maintenance. Therefore, we chose not to attempt analyses of the initial portion of sessions, but instead analyzed the whole session.

The present decline in positive affect in parallel with an increase in negative affect across chronic opioid SA is consistent with the decrease to below baseline in subjective "liking" of the effects of fentanyl administration in human subjects over time [39,40], in agreement with the incentive sensitization hypothesis of Berridge and Robinson [39]. Our findings in self-administering rats indicate that repeated exposure to fentanyl generates a significant shift toward negative affect even while on the drug. This suggests that chronic fentanyl SA involves aversive effects, despite outward signs that might appear consistent with models of addiction emphasizing positive reinforcement. Multiple ascending pathways originating in the brainstem are responsible for the generation of emotional arousal, including the ventral dopaminergic system originating in the ventral tegmental area (VTA) [8], a pathway strongly implicated in the rewarding effects of both opioids and stimulants. In the case of stimulants such as cocaine, the lateral habenula has been implicated in playing a critical role in the regulation of negatively motivated behaviors by targeting midbrain neuromodulatory systems such as the dopaminergic pathway projecting from the VTA [41,42]. Negative behaviors associated with opioids, however, are predominantly modulated by the paraventricular nucleus of the thalamus, which has significant inputs to the nucleus accumbens, whereby it mediates the negative signs of withdrawal and opioid-related aversive memory [43]. Therefore, it is possible that differences in affective regulation during cocaine versus fentanyl SA may be a result of activity in separate neural circuits during exposure to different classes of drugs.

We previously reported that the only USVs emitted during the maintenance phase of cocaine SA are 22 kHz calls [37]. Those negative calls were associated with falling cocaine levels, suggesting that further study is necessary of the relationship between USVs and fluctuating levels of fentanyl in the blood. Whereas the pattern of fentanyl SA and its associated affect are markedly different from those of cocaine SA, our findings with both fentanyl and cocaine [16] are consistent with the literature, supporting negative reinforcement as a driving factor of addiction and emphasizing that escape/avoidance of negative affect is a salient motivational factor reinforcing drug addiction.

5. Conclusions

Chronic fentanyl SA generates a shift toward negative affect during drug use as indicated by an increase in 22 kHz calls and a decrease in 50 kHz calls at T30 compared to saline control animals. The shift toward a negative affect during drug use is consistent with previous reports of increased negative affect in response to chronic drug exposure [12,16,17] and supports negative reinforcement as a salient motivational factor driving drug addiction.

Author Contributions: Conceptualization, A.N.D., N.J.B. and M.O.W.; methodology, A.N.D., N.J.B. and M.O.W.; validation, A.N.D. and M.O.W.; formal analysis, A.N.D.; investigation, A.N.D., V.M., A.M. and S.H.; resources, M.O.W.; data curation, A.N.D.; writing—original draft preparation, A.N.D.; writing—review and editing, A.N.D., N.J.B. and M.O.W.; visualization, A.N.D. and M.O.W.; supervision, A.N.D. and M.O.W.; project administration, M.O.W.; funding acquisition, M.O.W. All authors have read and agreed to the published version of the manuscript.

Funding: This research was funded by Rutgers' Brain Health Institute, Dorothy and David Cooper Endowment, and the Aresty Research Center.

Institutional Review Board Statement: The study was conducted according to the guidelines of the Declaration of Helsinki and approved by the Institutional Review Board of Rutgers University (protocol TR202000084, approved 24 September 2020).

Data Availability Statement: The data presented in this study are available on request from the corresponding author.

Acknowledgments: We thank Alisa Ray, Srishti Bose, Megha Karnam, and Steven Leichner for their excellent technical assistance.

Conflicts of Interest: The authors declare no conflict of interest. The funders had no role in the design of the study; in the collection, analyses, or interpretation of data; in the writing of the manuscript, or in the decision to publish the results.

References

1. Knutson, B.; Burgdorf, J.; Panksepp, J. Ultrasonic vocalizations as indices of affective states in rats. *Psychol. Bull.* **2002**, *128*, 961. [CrossRef]
2. Portfors, C.V. Types and functions of ultrasonic vocalizations in laboratory rats and mice. *J. Am. Assoc. Lab. Anim. Sci.* **2007**, *46*, 28–34.
3. Blanchard, R.; Blanchard, D.; Agullana, R.; Weiss, S.M. Twenty-two kHz alarm cries to presentation of a predator, by laboratory rats living in visible burrow systems. *Physiol. Behav.* **1991**, *50*, 967–972. [CrossRef]
4. Brudzynski, S.M.; Bihari, F.; Ociepa, D.; Fu, X.-W. Analysis of 22 kHz ultrasonic vocalization in laboratory rats: Long and short calls. *Physiol. Behav.* **1993**, *54*, 215–221. [CrossRef]

5. Covington, H.E.; Miczek, K.A. Vocalizations during withdrawal from opiates and cocaine: Possible expressions of affective distress. *Eur. J. Pharmacol.* **2003**, *467*, 1–13. [CrossRef]
6. Vivian, J.A.; Miczek, K.A. Ultrasounds during morphine withdrawal in rats. *Psychopharmacology* **1991**, *104*, 187–193. [CrossRef] [PubMed]
7. Tonoue, T.; Ashida, Y.; Makino, H.; Hata, H. Inhibition of shock-elicited ultrasonic vocalization by opioid peptides in the rat: A psychotropic effect. *Psychoneuroendocrinology* **1986**, *11*, 177–184. [CrossRef]
8. Brudzynski, S. Biological Functions of Rat Ultrasonic Vocalizations, Arousal Mechanisms, and Call Initiation. *Brain Sci.* **2021**, *11*, 605. [CrossRef]
9. Knutson, B.; Burgdorf, J.; Panksepp, J. High-Frequency Ultrasonic Vocalizations Index Conditioned Pharmacological Reward in Rats. *Physiol. Behav.* **1999**, *66*, 639–643. [CrossRef]
10. Brudzynski, S.M.; Pniak, A. Social contacts and production of 50-kHz short ultrasonic calls in adult rats. *J. Comp. Psychol.* **2002**, *116*, 73–82. [CrossRef] [PubMed]
11. Burgdorf, J.; Wood, P.L.; Kroes, R.A.; Moskal, J.R.; Panksepp, J. Neurobiology of 50-kHz ultrasonic vocalizations in rats: Electrode mapping, lesion, and pharmacology studies. *Behav. Brain Res.* **2007**, *182*, 274–283. [CrossRef]
12. Hamed, A.; Taracha, E.; Szyndler, J.; Krząścik, P.; Lehner, M.; Maciejak, P.; Skórzewska, A.; Plaznik, A. The effects of morphine and morphine conditioned context on 50 kHz ultrasonic vocalisation in rats. *Behav. Brain Res.* **2012**, *229*, 447–450. [CrossRef] [PubMed]
13. Ma, N.; Liu, Y.; Li, N.; Wang, C.-X.; Zhang, H.; Jiang, X.-F.; Xu, H.-S.; Fu, X.-M.; Hu, X.; Zhang, D.-R. Addiction related alteration in resting-state brain connectivity. *NeuroImage* **2010**, *49*, 738–744. [CrossRef] [PubMed]
14. Koob, G.F. Negative reinforcement in drug addiction: The darkness within. *Curr. Opin. Neurobiol.* **2013**, *23*, 559–563. [CrossRef]
15. Baker, T.B.; Piper, M.E.; McCarthy, D.E.; Majeskie, M.R.; Fiore, M.C. Addiction Motivation Reformulated: An Affective Processing Model of Negative Reinforcement. *Psychol. Rev.* **2004**, *111*, 33–51. [CrossRef] [PubMed]
16. Klein, S.; Beacher, N.J.; Kulik, J.M.; Estrin, D.J.; Pawlak, A.P.; West, M.O. Emergence of negative affect as motivation for drug taking in rats chronically self-administering cocaine. *Psychopharmacology* **2020**, *237*, 1407–1420. [CrossRef]
17. Koob, G.F. Drug Addiction: The Yin and Yang of Hedonic Homeostasis. *Neuron* **1996**, *16*, 893–896. [CrossRef]
18. Van Der Poel, A.M.; Noach, E.J.K.; Miczek, K.A. Temporal patterning of ultrasonic distress calls in the adult rat: Effects of morphine and benzodiazepines. *Psychopharmacology* **1989**, *97*, 147–148. [CrossRef]
19. Badiani, A.; Belin, D.; Epstein, D.; Calu, D.; Shaham, Y. Opiate versus psychostimulant addiction: The differences do matter. *Nat. Rev. Neurosci.* **2011**, *12*, 685–700. [CrossRef]
20. Koob, G.F.; Volkow, N.D. Neurocircuitry of Addiction. *Neuropsychopharmacology* **2009**, *35*, 217–238. [CrossRef]
21. Root, D.H.; Barker, D.J.; Ma, S.; Coffey, K.R.; Fabbricatore, A.T.; West, M.O. Evidence for learned skill during cocaine self-administration in rats. *Psychopharmacology* **2011**, *217*, 91–100. [CrossRef]
22. Ahmed, S.H.; Koob, G. Transition from moderate to excessive drug intake: Change in hedonic set point. *Science* **1998**, *282*, 298–300. [CrossRef]
23. Ahmed, S.H.; Walker, J.R.; Koob, G.F. Persistent Increase in the Motivation to Take Heroin in Rats with a History of Drug Escalation. *Neuropsychopharmacology* **2000**, *22*, 413–421. [CrossRef]
24. Lenoir, M.; Guillem, K.; Koob, G.F.; Ahmed, S.H.; Guillem, K. Drug specificity in extended access cocaine and heroin self-administration. *Addict. Biol.* **2011**, *17*, 964–976. [CrossRef] [PubMed]
25. Coffey, K.R.; Marx, R.G.; Neumaier, J.F. DeepSqueak: A deep learning-based system for detection and analysis of ultrasonic vocalizations. *Neuropsychopharmacology* **2019**, *44*, 859–868. [CrossRef] [PubMed]
26. Ahmed, S.H. *Escalation of Drug Use, in Animal Models of Drug Addiction*; Springer: Berlin/Heidelberg, Germany, 2011; pp. 267–292.
27. Best, L.M.; Zhao, L.L.; Scardochio, T.; Clarke, P.B. Effects of repeated morphine on ultrasonic vocalizations in adult rats: Increased 50 kHz call rate and altered subtype profile. *Psychopharmacology* **2017**, *234*, 155–165. [CrossRef]
28. Wright, J.M.; Deng, L.; Clarke, P.B.S. Failure of rewarding and locomotor stimulant doses of morphine to promote adult rat 50-kHz ultrasonic vocalizations. *Psychopharmacology* **2012**, *224*, 477–487. [CrossRef]
29. McAuliffe, W.E. A Second Look at First Effects: The Subjective Effects of Opiates on Nonaddicts. *J. Drug Issues* **1975**, *5*, 369–397. [CrossRef]
30. Verendeev, A.; Riley, A.L. Relationship between the rewarding and aversive effects of morphine and amphetamine in individual subjects. *Learn. Behav.* **2011**, *39*, 399–408. [CrossRef] [PubMed]
31. Verendeev, A.; Riley, A.L. Conditioned taste aversion and drugs of abuse: History and interpretation. *Neurosci. Biobehav. Rev.* **2012**, *36*, 2193–2205. [CrossRef]
32. Ahrens, A.M.; Ma, S.T.; Maier, E.; Duvauchelle, C.; Schallert, T. Repeated intravenous amphetamine exposure: Rapid and persistent sensitization of 50-kHz ultrasonic trill calls in rats. *Behav. Brain Res.* **2009**, *197*, 205–209. [CrossRef] [PubMed]
33. Kuchniak, K.; Wyszogrodzka, E.; Chrapusta, S.J.; Czarna, M.; Michalak, M.; Płaźnik, A.; Krząścik, P.; Mierzejewski, P.; Taracha, E. Using anticipatory and drug-evoked appetitive ultrasonic vocalization for monitoring the rewarding effect of amphetamine in a rat model of drug self-administration. *Behav. Brain Res.* **2019**, *376*, 112187. [CrossRef]
34. Simola, N.; Serra, M.; Marongiu, J.; Costa, G.; Morelli, M. Increased emissions of 50-kHz ultrasonic vocalizations in hemiparkinsonian rats repeatedly treated with dopaminomimetic drugs: A potential preclinical model for studying the affective properties of dopamine replacement therapy in Parkinson's disease. *Prog. Neuro-Psychopharmacol. Biol. Psychiatry* **2021**, *108*, 110184. [CrossRef]

35. Barker, D.J.; Root, D.; Ma, S.; Jha, S.; Megehee, L.; Pawlak, A.P.; West, M.O. Dose-dependent differences in short ultrasonic vocalizations emitted by rats during cocaine self-administration. *Psychopharmacology* **2010**, *211*, 435–442. [CrossRef]
36. Fabbricatore, A.T.; Ghitza, U.E.; Prokopenko, V.F.; West, M.O. Electrophysiological evidence of mediolateral functional dichotomy in the rat accumbens during cocaine self-administration: Tonic firing patterns. *Eur. J. Neurosci.* **2009**, *30*, 2387–2400. [CrossRef]
37. Barker, D.J.; Simmons, S.J.; Servilio, L.C.; Bercovicz, D.; Ma, S.; Root, D.H.; Pawlak, A.P.; West, M.O. Ultrasonic vocalizations: Evidence for an affective opponent process during cocaine self-administration. *Psychopharmacology* **2013**, *231*, 909–918. [CrossRef]
38. Avvisati, R.; Contu, L.; Stendardo, E.; Michetti, C.; Montanari, C.; Scattoni, M.L.; Badiani, A. Ultrasonic vocalization in rats self-administering heroin and cocaine in different settings: Evidence of substance-specific interactions between drug and setting. *Psychopharmacology* **2016**, *233*, 1501–1511. [CrossRef]
39. Berridge, K.C.; Robinson, T.E. Liking, wanting, and the incentive-sensitization theory of addiction. *Am. Psychol.* **2016**, *71*, 670–679. [CrossRef] [PubMed]
40. Zacny, J.P.; Lichtor, J.L.; Flemming, D.; Coalson, D.W.; Thompson, W.K. A dose-response analysis of the subjective, psychomotor and physiological effects of intravenous morphine in healthy volunteers. *J. Pharmacol. Exp. Ther.* **1994**, *268*, 1–9.
41. Hu, H.; Cui, Y.; Yang, Y. Circuits and functions of the lateral habenula in health and in disease. *Nat. Rev. Neurosci.* **2020**, *21*, 277–295. [CrossRef]
42. Clerke, J.A.; Congiu, M.; Mameli, M. Neuronal adaptations in the lateral habenula during drug withdrawal: Preclinical evidence for addiction therapy. *Neuropharmacology* **2021**, *192*, 108617. [CrossRef]
43. Zhu, Y.; Wienecke, C.F.R.; Nachtrab, G.; Chen, Y.Z.C.F.R.W.G.N.X. A thalamic input to the nucleus accumbens mediates opiate dependence. *Nat. Cell Biol.* **2016**, *530*, 219–222. [CrossRef]

brain sciences

MDPI

Review

Social Behavior and Ultrasonic Vocalizations in a Genetic Rat Model Haploinsufficient for the Cross-Disorder Risk Gene *Cacna1c*

Markus Wöhr [1,2,3,4,*], Theresa M. Kisko [3,4] and Rainer K.W. Schwarting [3,4]

1 Social and Affective Neuroscience Research Group, Laboratory of Biological Psychology, Research Unit Brain and Cognition, Faculty of Psychology and Educational Sciences, KU Leuven, B-3000 Leuven, Belgium
2 Leuven Brain Institute, KU Leuven, B-3000 Leuven, Belgium
3 Faculty of Psychology, Experimental and Biological Psychology, Behavioral Neuroscience, Philipps-University of Marburg, D-35032 Marburg, Germany; kiskot@staff.uni-marburg.de (T.M.K.); schwarti@staff.uni-marburg.de (R.K.W.S.)
4 Center for Mind, Brain, and Behavior, Philipps-University of Marburg, D-35032 Marburg, Germany
* Correspondence: markus.wohr@kuleuven.be; Tel.: +32-16-19-45-57

Abstract: The top-ranked cross-disorder risk gene *CACNA1C* is strongly associated with multiple neuropsychiatric dysfunctions. In a recent series of studies, we applied a genomically informed approach and contributed extensively to the behavioral characterization of a genetic rat model haploinsufficient for the cross-disorder risk gene *Cacna1c*. Because deficits in processing social signals are associated with reduced social functioning as commonly seen in neuropsychiatric disorders, we focused on socio-affective communication through 22-kHz and 50-kHz ultrasonic vocalizations (USV). Specifically, we applied a reciprocal approach for studying socio-affective communication in sender and receiver by including rough-and-tumble play and playback of 22-kHz and 50-kHz USV. Here, we review the findings obtained in this recent series of studies and link them to the key features of 50-kHz USV emission during rough-and-tumble play and social approach behavior evoked by playback of 22-kHz and 50-kHz USV. We conclude that *Cacna1c* haploinsufficiency in rats leads to robust deficits in socio-affective communication through 22-kHz and 50-kHz USV and associated alterations in social behavior, such as rough-and-tumble play behavior.

Keywords: Cav1.2; calcium; animal model; rough-and-tumble play; social play; social approach; ultrasonic vocalization; playback; social contact call; alarm call

check for updates

Citation: Wöhr, M.; Kisko, T.M.; Schwarting, R.K.W. Social Behavior and Ultrasonic Vocalizations in a Genetic Rat Model Haploinsufficient for the Cross-Disorder Risk Gene *Cacna1c*. *Brain Sci.* **2021**, *11*, 724. https://doi.org/10.3390/brainsci11060724

Academic Editors: Stefan M. Brudzynski and Jeffrey Burgdorf

Received: 26 April 2021
Accepted: 26 May 2021
Published: 29 May 2021

1. Introduction

The top-ranked cross-disorder risk gene *CACNA1C* is strongly associated with multiple neuropsychiatric dysfunctions. This includes affective disorders, namely depression [1] and bipolar disorder [2], as well as neurodevelopmental disorders, most notably autism spectrum disorder [3] and schizophrenia [4]. *CACNA1C* encodes the pore-forming α1C subunit of the voltage-gated L-type calcium channel (LTCC) Cav1.2. Characterized by a broad tissue expression profile with high expression levels in the central nervous system, Cav1.2 accounts for the majority of all LTCCs in the brain. It is strongly involved in the regulation of depolarization-dependent calcium influx into the cell, triggering intracellular signaling cascades including major pathways involved in neuronal plasticity processes [5].

Single-nucleotide polymorphisms (SNPs) within *CACNA1C* are amongst the best replicated and most robust genetic findings from genome-wide association studies (GWAS) in psychiatry [6]. Up to now, the exact molecular consequences of *CACNA1C* SNPs associated with neuropsychiatric dysfunctions are not fully understood. Often, such SNPs are intronic and it is widely believed that they exert their effects through altering *CACNA1C* gene expression. Albeit not consistently, a substantial number of studies reported such SNPs to result in decreased expression levels [7–12]. For instance, CACNA1C expression levels were found to be decreased in postmortem analyses of brain samples from risk SNP carriers in bipolar disorder [9] and schizophrenia [10], possibly because of inhibitory transcriptional regulation through chromosomal looping [11]. A better understanding of the functional effects of *CACNA1C* dosage and how reductions in *CACNA1C* gene expression alter behavioral phenotypes with relevance to neuropsychiatric dysfunctions is therefore needed.

In a recent series of studies, we applied a genomically informed approach and contributed extensively to the behavioral characterization of a genetic rat model haploinsufficient for the cross-disorder risk gene *Cacna1c* [13–19], flanked by neurobiological analyses [20–23]. In this rat model, Cav1.2 expression in the brain is reduced to about 50% of wildtype littermate controls, both in males [15] and females [16]. In a significant subset of our studies, we focused on social behavior and ultrasonic vocalizations because reduced social functioning was associated with *CACNA1C* SNPs and is commonly seen in neuropsychiatric disorders in humans [15–19]. For example, *CACNA1C* SNPs were associated with reduced socio-affective information processing capacities. This was reflected in altered facial emotion recognition and social outgroup processing [24–26]. Verbal fluency was reported to be reduced in *CACNA1C* SNP risk carriers [27].

2. Social Behavior in Rats

Wild rats live in large social colonies with overlapping generations, often in underground burrow systems with shared tunnels and chambers. Such colonies are typically structured into subgroups, with prominent near-linear dominance hierarchies, particularly in males. In line with the complexity of their social environment, the social life of rats is characterized by a broad variety of social behaviors [28].

In adult rats, the social repertoire includes reproductive behaviors, with both sexes being highly promiscuous. In fact, there is little evidence for mate choice and stable pair bonds. Strong bonds, however, are formed between mother and their infants. Maternal caregiving behavior ranges from nursing to licking and grooming. Another important feature of the social repertoire displayed by adult rats are agonistic behaviors, usually directed against intruders from outside the colony. Males were found to be less socially tolerant than females, with males but not females patrolling and defending territory boarders. In females, aggressive behavior was found to be low except during lactation. Other prominent aspects of the social life of rats are huddling and social grooming [29].

In juvenile rats, a very prominent component of their social repertoire is play fighting, also called rough-and-tumble play [30]. As the first social behavior not directed towards the mother, it is widely believed that rough-and-tumble play helps to prepare the young rats to develop important social skills needed in a complex social environment. In fact, rough-and-tumble play does not only contain sexual and aggressive behavioral components, but it also requires the fast integration of multiple sources of social information during the selection of appropriate behavioral responses. Rough-and-tumble play is initiated by one rat approaching another one and attempting to touch its neck with the snout, called nape contact or pouncing. During rough-and-tumble play, rats wrestle, box, and kick each other with the aim to turn the play partner on the back and to pin it. Regular attempts to escape lead to chasing behavior. A major difference to aggressive interactions in adult rats is that juvenile rats take turns. Rough-and-tumble play thus lacks a clear dominance pattern with one rat dominating the other rat. Typically, males engage in rough-and-tumble play more than females [31]. Rough-and-tumble play is highly rewarding to juvenile rats [32].

While it is notoriously difficult to observe rats in the wild due to the fact that they are nocturnal and have a sub-terrestrial lifestyle, available evidence suggests that laboratory rats display the full behavioral repertoire of wild rats, albeit wild rats were reported to be more aggressive and less playful [28]. Moreover, in line with the daily demands posed by living in a complex social environment, laboratory studies showed that rats can differentiate between individuals. The ability to recognize others is not only relevant for discriminating colony members from intrudes but also within the colony, for instance in the context of mate and food competition [33]. Consistent with studies on social transmission of food preferences in wild rats, laboratory studies demonstrated that rats can learn about positive and negative events through observing others. This includes the transfer of information concerning distant foods [34] and observational learning of fear [35]. Laboratory studies further revealed that rats engage in cooperative actions [36], prefer mutually rewarding options [37], follow rules of general and direct reciprocity [38], and display empathy-driven helping behavior [39].

3. Ultrasonic Vocalizations in Rats

The rich social repertoire of rats includes the emission of vocal signals, so called ultrasonic vocalizations (USV) [40,41]. As opposed to sonic vocalizations, they are not audible to humans because their sound frequency is too high for the human hearing system. Typically, three main types are distinguished, fulfilling multiple biological functions as socio-affective signals [42]. The first type that occurs during early ontogeny are 40-kHz USV emitted by pups in response to separation from mother and littermates within the first two weeks of life [43]. They play a fundamental role in maintaining close mother–pup interactions and 40-kHz USV were shown to induce search and retrieval behavior in the mother, particularly during lactation [44]. Later, in juvenile and adult rats, two different types are prevalent, with their occurrence being strongly dependent on the emotional valence of the situation (Figure 1). In situations characterized by a negative valence, such as predator exposure, 22-kHz USV occur [45]. Such 22-kHz USV induce behavioral inhibition in recipients, suggesting that they serve as alarm calls [46]. In contrast, 50-kHz USV occur in situations characterized by a positive valence, typically appetitive social interactions, such as rough-and-tumble play [47] and mating [48] and were dubbed "rat laughter" [49]. Such 50-kHz USV are characterized by a high level of complexity and many different 50-kHz USV subtypes were described. However, no consensus has been reached on how many different 50-kHz USV subtypes exist and recently applied classification systems, for instance, differentiate between two [50], four [51], or even fourteen [52] subtypes. Commonly distinguished 50-kHz USV subtypes include FLAT, STEP, TRILL, and MIXED calls. It is believed that the frequency-modulated subtypes of 50-kHz USV, i.e., STEP, TRILL, and MIXED calls, are most closely associated with positive affect and reward. There is strong evidence indicating that 50-kHz USV fulfill affiliative functions [53].

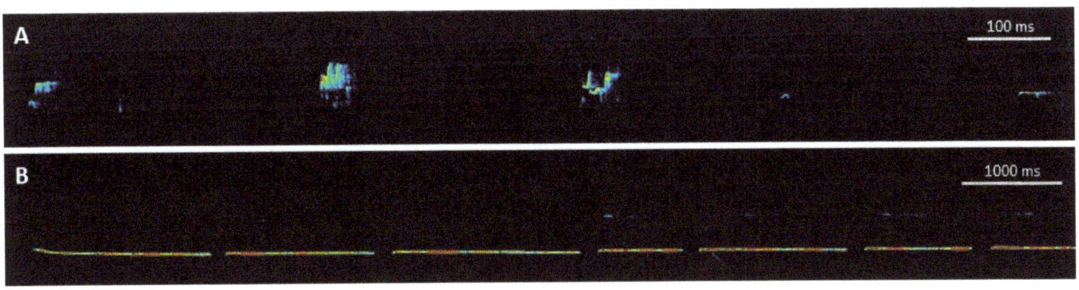

Figure 1. Exemplary spectrograms of ultrasonic vocalizations (USV) emitted by rats. (**A**) Spectrogram of 50-kHz USV serving as social contact calls. Please note the presence of different 50-kHz USV subtypes, including FLAT, STEP, TRILL, and MIXED calls. (**B**) Spectrogram of 22-kHz USV serving as alarm calls. Please note the difference in time scaling.

3.1. Rough-and-Tumble Play and Ultrasonic Vocalizations in Rats

Juvenile rats emit 50-kHz USV in anticipation and during rough-and-tumble play [47]. Anticipatory 50-kHz USV were found to occur at high levels in a context associated with rough-and-tumble play and to increase over days with repeated play sessions. This suggests that the emission of 50-kHz USV is driven by the anticipation of a rewarding social interaction and that it reflects wanting [54]. In fact, while rats without prior play experience vocalized very little during the exposure to a conspecific separated by a screen, high rates of 50-kHz USV occurred under such conditions after a single opportunity to play and rats emitted even more 50-kHz USV than during rough-and-tumble play. It was recently suggested that the reduced emission of anticipatory 50-kHz USV could reflect a depression-related behavioral phenotype [55].

During rough-and-tumble play, the highest numbers of 50-kHz USV occur during actual play phases as opposed to non-playful social interactions [15]. Moreover, it was shown that the emission of 50-kHz USV changes as a function of rough-and-tumble play behavior [56] and 50-kHz USV emission was found to be correlated across individuals with appetitive components of the rough-and-tumble play repertoire, such as dorsal contacts [47]. Later, more detailed spectrographic analyses were performed and revealed that this correlation was essentially driven by the frequency-modulated subtypes of 50-kHz USV, i.e., STEP, TRILL, and MIXED calls, primarily associated with liking [56]. Together, this indicates that 50-kHz USV reflect wanting or liking, depending on the time point of emission [54].

Synchronized recordings with high temporal resolution further allowed insights into how 50-kHz USV are integrated into the rough-and-tumble play repertoire [15]. The synchronized analyses revealed that particularly high 50-kHz USV rates occur during wrestling and chasing. Relatively low numbers are emitted during pinning, with 50-kHz USV emission rates similar to non-play phases. Considering the fast-paced actions displayed during rough-and-tumble play, this indicates that there is a fine temporal control of 50-kHz USV emission and that 50-kHz USV emission and specific components of the rough-and-tumble play repertoire are precisely coordinated. However, not all 50-kHz USV subtypes occur at similar rates. The prevalence is highest for FLAT, TRILL, and MIXED calls, while STEP calls occur les often. This holds true for pinning, wrestling, and chasing, suggesting that distinct 50-kHz USV subtypes are not associated with specific components of the rough-and-tumble play repertoire within individuals [15]. Other studies support a precise temporal control. By focusing on a limited time window before and after a playful interaction, it was shown that 50-kHz USV occur at particularly high rates immediately before a playful attack and that different 50-kHz USV subtypes are associated with specific components of the rough-and-tumble play repertoire, such as a short subtype, which was seen at high rates after evasions but not complete rotations [57].

In line with the more prominent rough-and-tumble play behavior in males [31], emission of 50-kHz USV was repeatedly found to be higher in males than females [16,57]. In particular, FLAT and STEP calls are higher in males than in females. Overlapping calls also occur in higher rates in males than females [17]. The peak frequency of 50-kHz USV emitted by males was found to be lower than in females [17]. Likewise, the close association between 50-kHz USV emission and specific components of the rough-and-tumble play repertoire was shown to be most prominent in males, while not clearly present in females [17]. Moreover, there is also evidence for strain differences. During rough-and-tumble play, more 50-kHz USV were found to be emitted by Sprague–Dawley than Wistar rats, with Wistar rats engaging less in rough-and-tumble play behavior [58,59].

The high prevalence of 50-kHz USV in anticipation and during rough-and-tumble play and their fine temporal integration into the rough-and-tumble play repertoire suggest that 50-kHz USV promote and maintain playful social interactions by functioning as play signals and/or social contact calls. This view is supported by deafening experiments and devocalization studies. For instance, pinning was found to be diminished in deafened rats, while play initiation through dorsal contacts remained unchanged [60]. Moreover, a series of experiments with rats that were unable to vocalize due to surgical devocalization demonstrated that lack of socio-affective communication through 50-kHz USV leads to a disruption of rough-and-tumble play behavior [61–63]. In pairs of devocalized rats, rough-and-tumble play behavior was clearly reduced. Devocalized rats launched fewer playful attacks, displayed an increase in startle responses when contacted by the play partner, and were more likely to defend themselves in response to a playful attack [62].

Moreover, a related study found that rough-and-tumble play behavior is lower in intact pairs that were housed with devocalized cage mates than in intact pairs housed with intact cage mates. This suggests that rough-and-tumble play helps to learn about the social functions of 50-kHz USV. An intact rat exclusively engaging in playful interactions with a devocalized cage mate might lack necessary input associated with hearing 50-kHz USV during the critical time window of the rough-and-tumble play period [63].

In dyads with one devocalized and one intact rat, the effects were milder but alterations in specific components of the rough-and-tumble play repertoire, such as defensive responses, were repeatedly observed [61,62]. Moreover, evidence was provided in support of the notion that the rat that is pinning is emitting 50-kHz USV because 50-kHz USV emission was clearly higher when the intact rat was pinning than when it was pinned by the devocalized partner [63]. Other aspects of the rough-and-tumble play repertoire, however, were not affected by surgical devocalization. As reported before [57], 50-kHz USV were more frequent before playful contact is made than when such contact is terminated, and this pattern remained intact despite surgical devocalization of one of the play partners [62]. Interestingly, pre-contact 50-kHz USV were just as frequent as those of the devocalized play partner initiated the playful interaction as when an intact one was. This suggests that 50-kHz USV are not only emitted by the initiator to signal a playful interaction, but also by the receiving partner, presumably because 50-kHz USV function as enticements by one rat to solicit a playful attack from another [62]. However, this view is challenged by results obtained in a preference test, where rats did not prefer to engage in rough-and-tumble play with an intact play partner over a devocalized play partner. Mute rats were found to be as attractive as rats that were able to vocalize when both were simultaneously available [62]. In fact, it was found that more playful attacks are launched against devocalized than intact play partners [61]. Together, the devocalization studies indicate that 50-kHz USV facilitate rough-and-tumble play, presumably through promoting positive affect in play partners or by the induction of social proximity.

There is compelling evidence indicating that the emission of 50-kHz USV during rough-and-tumble play is under genetic control. A prominent role of genetic factors is highlighted by three independent selective breeding studies targeting distinct behavioral domains. In a first selective breeding study, rats were selected depending on their tendencies to emit low versus high rates of 50-kHz USV in response to rough-and-tumble play mimicked by a human experimenter through tickling [64,65]. Within a few generations, prominent line differences in the emission of 50-kHz USV were detected and rough-and-tumble play was found to be altered. Already in the fourth generation, the emission of 50-kHz USV during rough-and-tumble play was highest in the high line. In the low line, 50-kHz USV were virtually absent. An unselected random control line emitted moderate numbers of 50-kHz USV [65]. Consistent with the idea that 50-kHz USV facilitate rough-and-tumble play, changes in the emission of 50-kHz USV were associated with alterations in rough-and-tumble play and pinning was found to be clearly affected. Mirroring the 50-kHz USV emission pattern, pinning was highest in the high line and lowest in the low line, with the random line displaying an intermediate phenotype [65]. Of note, similar findings were obtained in a replication of the selective breeding study using more detailed spectrographic analyses of 50-kHz USV emission [66–68] and linked to an autism-related behavioral phenotype [69].

A second selective breeding study assessed 50-kHz USV emission during rough-and-tumble play in rats selected for low versus high levels of separation-induced 40-kHz USV as pups [70]. It was found that in both lines rough-and-tumble play behavior and 50-kHz USV emission were lower than in a random control line. This indicates that low social motivation as seen in the low line is associated with reduced 50-kHz USV emission. However, this also indicates that anxiety plays a modulatory role. Rats selected for high rates of separation-induced 40-kHz USV as pups are characterized by high trait anxiety and display enhanced anxiety-related behavior on the elevated plus maze in adulthood [71].

This suggests that high trait anxiety is associated with lower 50-kHz USV emission during rough-and-tumble play [72].

In fact, in a third selective breeding study, rats were selected for low versus high anxiety-related behavior on the elevated plus maze and through this means it was confirmed that 50-kHz USV emission is strongly affected by trait anxiety [73]. Rats selectively bred for high anxiety levels engaged less in rough-and-tumble play and emitted fewer 50-kHz USV than rats selectively bred for low anxiety levels or a random control line. The effects of selective breeding were strong and 50-kHz USV emission was almost completely absent during rough-and-tumble play in rats characterized by high trait anxiety [73]. Together, the selective breeding studies [64–73] demonstrate that the emission of 50-kHz USV is a heritable trait, for which rats can be selected and that is negatively associated with trait anxiety.

Besides genetic factors, the environment has a strong impact on the emission of 50-kHz USV. For instance, play deprivation for about three weeks through individual housing was found to enhance 50-kHz USV emission during rough-and-tumble play, while aversive stimuli, such as bright white light reduced 50-kHz USV emission [47]. Other social experiences, such as social rejection, might also have a negative impact [74].

Various psychoactive agents affect the emission of 50-kHz USV during rough-and-tumble play. This includes morphine [58], amphetamine [58], and the endocannabinoid signaling modulator URB597 [59]. However, their effects were often complex and dependent on other factors, such as strain. For instance, morphine enhanced 50-kHz USV emission during rough-and-tumble play in Sprague–Dawley rats but not Wistar rats and this effect was most prominent during the initial playful encounters [58]. In contrast, amphetamine inhibited 50-kHz USV emission during rough-and-tumble play but enhanced 50-kHz USV during non-social activities, such as cage exploration and self-grooming, in Sprague–Dawley rats but not Wistar rats [58]. Moreover, the anandamide hydrolysis inhibitor URB597 was reported to enhance 50-kHz USV emission during rough-and-tumble play depending on strain and the aversiveness of the situation [59].

A prominent role in modulating the emission of 50-kHz USV is played by the endogenous vasopressin system. While the intracerebroventricular administration of synthetic vasopressin had no effect, blocking the central vasopressin system through injections of a vasopressin 1a receptor antagonist into the brain led to a reduction in rough-and-tumble play and 50-kHz USV [73]. Other studies targeted the glutamatergic system [50,75,76] or the insulin-like growth factor I [77]. Regional brain cholecystokinin levels were found to change as a function of rough-and-tumble play behavior and thus presumably proportional to 50-kHz USV emission [78]. Although ethanol was reported to enhance social behavior, it had no effect on 50-kHz USV during rough-and-tumble play [79].

There is also a significant number of studies assessing the long-term effects of various early life stressors on 50-kHz USV emission during rough-and-tumble play. Such studies focused, for instance, on the effects of prenatal exposure to ethanol [80,81] and valproic acid [82,83]. Recent studies found that prenatal ethanol exposure leads to an enhanced emission of 22-kHz USV at the expense of 50-kHz USV during rough-and-tumble play in males but not females [84]. Likewise, it was shown that rats exposed to the viral mimic poly I:C during prenatal development engage less in rough-and-tumble play and emit fewer 50-kHz USV. Interestingly, these effects were most prominent in males [85,86]. Similar findings were obtained in a study on early life stress, including maternal separation and lipopolysaccharide injections. Early life stress was found to be associated with reduced emission of 50-kHz USV during rough-and-tumble play, an effect that could be prevented by a variety of sensory interventions during neonatal development [87]. Together, this suggests that reduced levels of 50-kHz USV during rough-and-tumble play in the sender reflect atypical development, which is relevant for a wide range of neurodevelopmental disorders in humans, such as autism and schizophrenia.

3.2. Playback of Ultrasonic Vocalizations in Rats

To study the socio-affective communicative functions of the different types of USV in rats, we developed a playback paradigm [88]. Through this means, we showed that 40-kHz USV emitted by isolated pups, but not an artificial 40-kHz sine wave tone, lead to maternal search behavior in the dam, consistent with the notion that pup 40-kHz USV help to maintain a close interaction between pup and dam [89]. We further demonstrated for the first time that 50-kHz USV typically emitted by juvenile and adult rats in appetitive social interactions, such as rough-and-tumble play, evoke social approach behavior, indicating that 50-kHz USV serve as social contact calls [88]. The social approach response is strong and is associated with a prominent increase in locomotor activity. It typically occurs within a few seconds after the playback presentation started. Often, social approach behavior during 50-kHz USV playback is followed by search behavior after the playback presentation ended. Response calls might also occur [88].

Importantly, strong social approach behavior is evoked exclusively in response to the playback of natural 50-kHz USV. Firstly, natural 22-kHz USV lead to behavioral inhibition in line with an alarming function [88], irrespective of the threatening stimulus that caused 22-kHz USV emission [46]. Secondly, background noise and time- and amplitude-matched white noise do not elicit social approach behavior [90]. Thirdly, albeit time- and amplitude-matched 50-kHz USV sine-wave tones do lead to social approach behavior, the response is weaker than during playback of natural 50-kHz USV [88].

Social approach behavior evoked by playback of 50-kHz USV is most prominent in juvenile rats and weaker in adult rats [88]. In adult rats, the social approach response is particularly strong in females [91]. Moreover, rats that emit particularly high levels of 50-kHz USV were found to display stronger social approach behavior than rats that produce few 50-kHz USV [92]. Together, this indicates that developmental aspects, sex-related factors, and personality traits related to sociability play an important modulatory role. It further suggests that rats with a higher level of social motivation, as seen for instance in juvenile rats during the rough-and-tumble play period, display stronger social approach behavior in response to 50-kHz USV playback [88]. This idea is further supported by the fact that social approach behavior was enhanced following a brief period of social isolation of 24 h [93].

Long-term social isolation for four weeks, however, exerted an inhibitory effect and blocked social approach behavior in response to 50-kHz USV playback in juvenile but not adult rats, suggesting that social experiences during the rough-and-tumble play period are important for rats to develop their social behavioral repertoire [93]. Corroborating evidence was obtained in rats exposed to different forms of environmental enrichment [94]. Specifically, long-term exposure to physical environmental enrichment was associated with reduced social approach behavior in response to 50-kHz USV playback. In contrast, social environmental enrichment led to enhanced social approach behavior. In fact, the reduction in social approach behavior following physical environmental enrichment was reverted by additional exposure to social environmental enrichment.

Another important feature of the social approach response is the habituation phenomenon. Strong social approach behavior is typically only seen during the first exposure to playback of 50-kHz USV [90]. Even after a delay of one week, no prominent social approach response is seen during a second exposure. The habituation phenomenon, however, appears to be strain-dependent [95] and can be blocked by administering the amnesia-inducing drug scopolamine immediately following the first exposure [90].

At the neurobiological level, the social approach response is associated with increased neuronal activity in the nucleus accumbens, a key brain area involved in reward processing [96]. Another brain area with increased neuronal activity in response to playback of 50-kHz USV is the prefrontal cortex. While 50-kHz USV activate prefrontal cortex and nucleus accumbens, 22-kHz USV lead to an activation of the perirhinal cortex, the amygdala, and the central gray [96]. In the nucleus accumbens, 50-kHz USV playback leads to phasic release of dopamine and the strength of the dopamine response is positively associated with social approach behavior [97]. Repeated 50-kHz USV playback leads to diminished dopamine release, suggesting that the habituation phenomenon does not only occur at the behavioral but also the neurobiological level. Importantly, dopamine release is seen exclusively in response to 50-kHz USV but not background noise and time- and amplitude-matched white noise, or 22-kHz USV.

The social approach response evoked by playback of 50-kHz USV can be modulated pharmacologically. In line with the observation that increased dopamine release is associated with particularly strong social approach behavior [97], amphetamine treatment was found to result in enhanced social approach behavior [98]. The enhancing effect of amphetamine was most prominent in rats that are characterized by personality traits related to low sociability and that produce relatively few 50-kHz USV [92]. In fact, amphetamine treatment was found to boost social approach behavior in these rats in such a manner that their response was indistinguishable from the one displayed by rats emitting high rates of 50-kHz USV. A prominent modulatory role is also exerted by the opioid system [99]. Specifically, administration of the μ-opioid-receptor antagonist naloxone reduced social approach behavior in response to 50-kHz USV playback, while the μ-opioid-receptor agonist morphine enhanced the social approach response.

Finally, we applied the 50-kHz USV playback paradigm repeatedly in rat models for neuropsychiatric dysfunctions, most notably rat models for neurodevelopmental disorders, such as autism. In a first study, we studied social behavior and ultrasonic communication in a *Shank3* deficient rat model for autism [100]. In humans, *SHANK3* mutations lead to Phelan-McDermid syndrome and are one of the most penetrant causes of autism [101]. During direct reciprocal social interactions, male but not female *Shank3* deficient rats displayed deficits and engaged less in social sniffing and allogrooming behaviors. While social approach behavior in response to 50-kHz USV playback was evident in male but not female *Shank3* deficient rats, male *Shank3* deficient rats did not display search behavior after the playback presentation ended [100]. More prominent deficits were evident in *Ube3a3* deficient rats [102]. In *Ube3a3* deficient rats, the response to 50-kHz USV playback was clearly reduced, as reflected in a less prominent increase in locomotor activity, lower levels of social approach behavior during 50-kHz USV playback, and lack of search behavior after the playback presentation ended. In humans, *UBE3A* deletions cause Angelman syndrome characterized by severe developmental delay and intellectual disability, most notably lack of language acquisition [101]. Finally, we found that environmental risk factors associated with neurodevelopmental disorders, most notably early life exposure to the organophosphorus pesticide chlorpyrifos, reduced social approach behavior in response to 50-kHz USV playback [103]. Specifically, chlorpyrifos exposure led to a dose-dependent inhibition of social approach behavior, with the two highest doses completely blocking the social approach response in females but not males. Together, this indicates that measuring social approach behavior evoked by playback of 50-kHz USV in the receiver can help to reveal social behavior and ultrasonic communication deficits in genetic and environmental models with relevance to human neuropsychiatric dysfunctions.

4. Social Behavior and Ultrasonic Vocalizations in *Cacna1c* Haploinsufficient Rats

As part of a translational effort to better understand major genetic risk factors for neuropsychiatric dysfunctions [104], we applied a longitudinal deep phenotyping approach and contributed to the behavioral characterization of a genetic rat model haploinsufficient for the cross-disorder risk gene *Cacna1c*. Because deficits in processing social signals are associated with reduced social functioning as commonly seen in neuropsychiatric disorders, we focused on socio-affective communication through 22-kHz and 50-kHz USV. In our studies, we compared constitutive heterozygous *Cacna1c*$^{+/-}$ rats to wildtype *Cacna1c*$^{+/+}$ littermate controls [15–19].

In our first study, we focused on juvenile male rats [15]. We applied a reciprocal approach for studying socio-affective communication in sender and receiver by including rough-and-tumble play and playback of 50-kHz USV.

Firstly, we compared rough-and-tumble play and concomitant 50-kHz USV emission between *Cacna1c*$^{+/-}$ rats and *Cacna1c*$^{+/+}$ littermate controls. Rough-and-tumble play was not affected by *Cacna1c* haploinsufficiency in juvenile male rats. Specifically, the total time spent playing did not differ between genotypes. Moreover, the occurrence of specific components of the rough-and-tumble play repertoire, such as pinning, wrestling, and chasing, was not affected.

The emission of 50-kHz USV, however, was strongly affected by *Cacna1c* haploinsufficiency in juvenile male rats (Figure 2A). *Cacna1c*$^{+/-}$ rats emitted fewer 50-kHz USV than *Cacna1c*$^{+/+}$ littermate controls. This genotype difference was evident during actual play phases but also during non-playful social interactions. While 50-kHz USV emission was particularly high during wrestling and chasing in *Cacna1c*$^{+/+}$ littermate controls, this was not the case in *Cacna1c*$^{+/-}$ rats where high 50-kHz USV emission rates were exclusively seen during chasing.

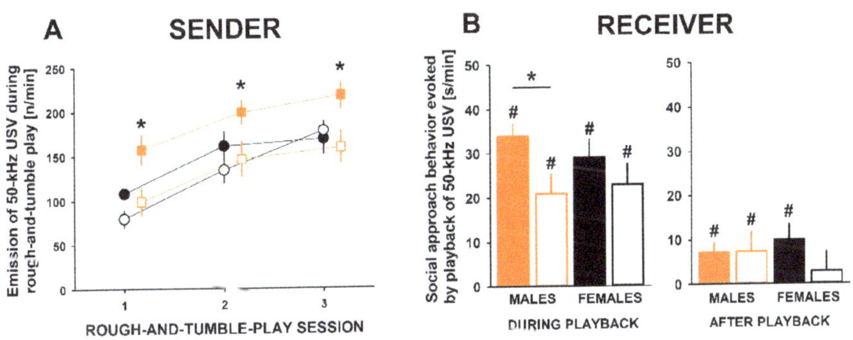

Figure 2. Effects of *Cacna1c* haploinsufficiency on (**A**) the emission of 50-kHz ultrasonic vocalizations (USV) emitted during rough-and-tumble play in the sender (filled orange squares = male *Cacna1c*$^{+/+}$ littermate controls; open orange squares = male *Cacna1c*$^{+/-}$ rats; filled black circles = female *Cacna1c*$^{+/+}$ littermate controls; open white circles = female *Cacna1c*$^{+/-}$ rats; $N = 10$ play pairs each) and (**B**) on social approach behavior evoked by playback of 50-kHz USV in the receiver (filled orange bars = male *Cacna1c*$^{+/+}$ littermate controls; open orange bars = male *Cacna1c*$^{+/-}$ rats; filled black bars = female *Cacna1c*$^{+/+}$ littermate controls; open black bars = female *Cacna1c*$^{+/-}$ rats; $N = 20$ rats each). Please note that social approach behavior is shown as change scores from baseline. Data were previously reported in [15,16]. * $p < 0.05$ vs. male *Cacna1c*$^{+/+}$ littermate controls; # $p < 0.05$ vs. baseline.

Moreover, *Cacna1c* haploinsufficiency affected certain 50-kHz USV subtypes more than others and the overall reduction in 50-kHz USV was primarily driven by reduced numbers of FLAT and MIXED calls. Related to that, there were prominent genotype effects on the prevalence of 50-kHz USV subtypes during specific components of the rough-and-tumble play repertoire. During wrestling and chasing, TRILL calls were increased in *Cacna1c*$^{+/-}$ rats as compared to *Cacna1c*$^{+/+}$ littermate controls, primarily at the cost of MIXED calls. No prominent differences were seen during pinning. Finally, acoustic features of 50-kHz USV differed between genotypes. Most notably, the peak frequency of 50-kHz USV was higher in *Cacna1c*$^{+/-}$ rats than *Cacna1c*$^{+/+}$ littermate controls. Peak amplitude was lower in *Cacna1c*$^{+/-}$ rats, while call duration and frequency modulation were not affected by *Cacna1c* haploinsufficiency.

Secondly, we measured social approach behavior evoked by the playback of 50-kHz USV (Figure 2B). Although both genotypes displayed a social preference, social approach behavior was reduced in *Cacna1c*$^{+/-}$ rats, as compared to *Cacna1c*$^{+/+}$ littermate controls. Moreover, *Cacna1c*$^{+/+}$ littermate controls but not *Cacna1c*$^{+/-}$ rats displayed search behavior after the playback presentation ended. No increase in locomotor activity was evident during and after playback.

Together, the first study indicates that a prominent reduction of Cav1.2 expression in the brain has detrimental effects on socio-affective communication through 50-kHz USV in male rats haploinsufficient for the cross-disorder risk gene *Cacna1c* [15]. Importantly, *Cacna1c* haploinsufficiency affects not only the emission of 50-kHz USV in the sender but also socio-affective information processing in the receiver and leads to an altered response to 50-kHz USV serving an affiliative function as social contact calls. Given that 50-kHz USV emission was associated with positive affect, this indicates that *Cacna1c*$^{+/-}$ rats experience rough-and-tumble play as less rewarding, and it would be interesting to test whether this is reflected in lower levels of social conditioned place preference [105].

In our second study, we focused on juvenile female rats [16,17]. The females in this second study were littermates of the male rats used in the first study and were tested in the exact same behavioral paradigms [15]. This gave us the opportunity to compare males and females and to test whether the effects of *Cacna1c* haploinsufficiency are modulated by sex.

In contrast to males, *Cacna1c* haploinsufficiency exerted strong effects on rough-and-tumble play in females [16]. Specifically, the total time spent playing was clearly higher in female *Cacna1c*$^{+/-}$ rats as compared to female *Cacna1c*$^{+/+}$ littermate controls. In fact, female *Cacna1c*$^{+/-}$ rats engaged even more in rough-and-tumble play than male *Cacna1c*$^{+/+}$ littermate controls. This is particularly remarkable given the typical sex difference in rough-and-tumble play, with male rats playing more than females [31]. The increase in rough-and-tumble play was driven by remarkably high levels of pinning behavior [16]. Pinning behavior in *Cacna1c*$^{+/-}$ rats was not only higher than in female *Cacna1c*$^{+/+}$ littermate controls but also higher than in male *Cacna1c*$^{+/+}$ littermate controls. No prominent genotype effects on wrestling and chasing were found.

The increase in rough-and-tumble play behavior displayed by female *Cacna1c*$^{+/-}$ rats, however, was not paralleled by an increase in the 50-kHz USV emission rate (Figure 2A). In fact, female *Cacna1c*$^{+/-}$ rats emitted a similarly high number of 50-kHz USV as female *Cacna1c*$^{+/+}$ littermate controls and slightly less than male *Cacna1c*$^{+/+}$ littermate controls [16]. Although the 50-kHz USV emission rate was not affected by *Cacna1c* haploinsufficiency, the temporal organization of 50-kHz USV emission was [17]. Similar to male *Cacna1c*$^{+/+}$ littermate controls, female *Cacna1c*$^{+/-}$ rats emitted the highest numbers of 50-kHz USV during actual play phases as opposed to non-playful social interactions. Such a close association with the actual play phase was not seen in female *Cacna1c*$^{+/+}$ littermate controls. Moreover, again similar to male *Cacna1c*$^{+/+}$ littermate controls, the emission of 50-kHz USV was highest during wrestling and chasing in female *Cacna1c*$^{+/-}$ rats. Once more, such a close association with specific components of the rough-and-tumble play repertoire, however, was not seen in female *Cacna1c*$^{+/+}$ littermate controls.

Moreover, *Cacna1c* haploinsufficiency had minor effects on 50-kHz USV subtypes [17]. Female *Cacna1c*$^{+/-}$ rats emitted more STEP calls than female *Cacna1c*$^{+/+}$ littermate controls. Likewise, the number of overlapping calls seen when two rats emitted 50-kHz USV at the same time was higher in female *Cacna1c*$^{+/-}$ rats. In fact, female *Cacna1c*$^{+/-}$ rats but not female *Cacna1c*$^{+/+}$ littermate controls reached levels seen in male *Cacna1c*$^{+/+}$ littermate controls. However, there were no prominent genotype effects on the prevalence of 50-kHz USV subtypes during specific components of the rough-and-tumble play repertoire. Acoustic features of 50-kHz USV did not differ between genotypes.

When measuring social approach behavior evoked by playback of 50-kHz USV in females, both genotypes displayed an increase in locomotor activity with a clear social preference (Figure 2A). No genotype differences were evident during playback [16]. After the playback presentation ended, however, female *Cacna1c*$^{+/+}$ littermate controls but not female *Cacna1c*$^{+/-}$ rats displayed search behavior, similar to the pattern in males. No increase in locomotor activity was evident after playback. Response calls evoked by the playback of 50-kHz USV occurred in *Cacna1c*$^{+/-}$ rats and *Cacna1c*$^{+/+}$ littermate controls of both sexes, although their emission was found to be particularly low in female *Cacna1c*$^{+/-}$ rats [16].

Together, the second study indicates that the effects of *Cacna1c* haploinsufficiency are modulated by sex. Most notably, *Cacna1c* haploinsufficiency in females led to hypermasculinization of rough-and-tumble play, as reflected in a clear increase in pinning behavior, without exerting prominent effects on the emission of 50-kHz USV [16,17]. Because sex differences in rough-and-tumble play were repeatedly associated with differences in testosterone levels [106], it would be interesting to compare *Cacna1c*$^{+/-}$ rats and *Cacna1c*$^{+/+}$ littermate controls of both sexes and to quantify testosterone.

In the third study, we focused on social interactions in adult female rats [18]. In contrast to the studies in juvenile rats, this included not only same-genotype dyads but also mixed-genotype dyads. Through this means, we showed that the emission of 50-kHz USV is highest in dyads consisting of two *Cacna1c*$^{+/+}$ littermate controls, but lowest in dyads with two *Cacna1c*$^{+/-}$ rats. Intermediate levels of 50-kHz USV were seen in mixed-genotype dyads with one *Cacna1c*$^{+/-}$ rat and a *Cacna1c*$^{+/+}$ littermate control.

During social interactions, all major 50-kHz USV subtypes occurred. FLAT and TRILL calls were most prevalent, while STEP calls occurred rarely. *Cacna1c* haploinsufficiency had no prominent effect on their prevalence. Acoustic features did not differ between genotypes, with the exception of peak amplitude, which was lower in dyads including one or two *Cacna1c*$^{+/-}$ rats, particularly for FLAT calls.

In line with the hypermasculinized rough-and-tumble play behavior displayed by female *Cacna1c*$^{+/-}$ rats, *Cacna1c*$^{+/-}$ rats behaved in a more dominant manner in the tube test. During social interactions, however, aggressive behavior was rarely seen and not enhanced in dyads with one or two *Cacna1c*$^{+/-}$ rats. In fact, there were no effects of *Cacna1c* haploinsufficiency on social behavior at the level of dyads, apart from increased physical contact in *Cacna1c*$^{+/-}$ rats. Interestingly, however, non-social behaviors were affected. Rearing and digging were reduced in dyads including *Cacna1c*$^{+/-}$ rats, while self-grooming behavior was strongly enhanced.

Detailed temporal analyses using synchronized high-resolution recordings revealed that the emission of 50-kHz USV was higher during social behaviors than non-social behaviors. Particularly high 50-kHz USV emission rates occurred during following behavior. Moderate levels of 50-kHz USV were emitted during social sniffing and physical contact. Lowest numbers of 50-kHz USV were seen during social grooming. In fact, 50-kHz USV emission during social grooming was not higher than during self-grooming in *Cacna1c*$^{+/+}$ littermate controls.

The reduced emission of 50-kHz USV in dyads including *Cacna1c*$^{+/-}$ rats was seen during social behaviors and non-social behaviors. The genotype effects were most prominent during following behavior, but also seen during rearing, digging, and self-grooming, suggesting that the enhanced level of self-grooming in dyads including *Cacna1c*$^{+/-}$ rats was not associated with positive affect reflected by 50-kHz USV.

We further took advantage of the mixed-genotype dyads and individually analyzed the behavior of *Cacna1c*$^{+/-}$ rats and *Cacna1c*$^{+/+}$ littermate controls while socially interacting with the other genotype. This detailed analysis revealed that *Cacna1c*$^{+/+}$ littermate controls spent more time sniffing the partner than *Cacna1c*$^{+/-}$ rats, indicating that *Cacna1c*$^{+/-}$ rats were extensively sniffed by *Cacna1c*$^{+/+}$ littermate controls but did not reciprocate. When comparing the behavioral profile displayed by *Cacna1c*$^{+/-}$ rats during a social interaction with other *Cacna1c*$^{+/-}$ rats and *Cacna1c*$^{+/+}$ littermate controls, the level of physical contact in *Cacna1c*$^{+/-}$ rats was higher while socially interacting with other *Cacna1c*$^{+/-}$ rats. In *Cacna1c*$^{+/+}$ littermate controls, their social behavior was not affected by the genotype of the partner.

Together, the third study indicates that effects of *Cacna1c* haploinsufficiency on social behavior and socio-affective communication through 50-kHz USV are also evident in adult rats. Most notably, detailed temporal analyses revealed prominent reductions in 50-kHz USV during social but also non-social behaviors, suggesting that their reduced emission in *Cacna1c*$^{+/-}$ rats is not specifically linked to deficits in social behavior. Considering the effects of *Cacna1c* haploinsufficiency on dominance behavior, it would be interesting to test how the presence of *Cacna1c*$^{+/-}$ rats affects aggressive behavior and the social hierarchy in mixed-genotype groups of rats [107].

Finally, in our fourth study, we applied a recently refined 22-kHz USV playback paradigm [46] and tested whether *Cacna1c* haploinsufficiency leads to an altered response to 22-kHz USV serving an alarming function [19]. As expected, male and female *Cacna1c*$^{+/+}$ littermate controls displayed behavioral inhibition in response to playback of 22-kHz USV. Behavioral inhibition evoked by 22-kHz USV was evident in response to 22-kHz USV recorded during both predator urine exposure and a retention test on learned fear. The lack of prominent differences in the potency to elicit behavioral inhibition depending on the threat context used for recording 22-kHz USV supports the generalizability of their alarming effects. However, generalizability was limited in *Cacna1c*$^{+/-}$ rats in a sex-dependent manner and *Cacna1c* haploinsufficiency led to less pronounced and less specific behavioral inhibition in male but not female rats.

Together, the fourth study shows that behavioral inhibition evoked by playback of alarm 22-kHz USV is robust and occurs in response to both sets of 22-kHz USV yet is modulated by *Cacna1c* in a sex-dependent manner. Considering the less pronounced and less specific behavioral inhibition evoked by 22-kHz USV in *Cacna1c*$^{+/-}$ rats, it would be interesting to test whether *Cacna1c* haploinsufficiency affects the emission of 22-kHz USV in response to predator exposure [45] or during fear learning [108].

In summary, the four studies indicate that *Cacna1c* haploinsufficiency in rats leads to robust deficits in socio-affective communication through 22-kHz and 50-kHz USV. *Cacna1c* haploinsufficiency affected the sender in a sex-specific way. In males, *Cacna1c* haploinsufficiency led to reduced 50-kHz USV emission during rough-and-tumble play. In females, *Cacna1c* haploinsufficiency led to hypermasculinization of the rough-and-tumble-play repertoire as juveniles and lower emission of 50-kHz USV and mild alterations in social behavior in adulthood. *Cacna1c* haploinsufficiency also affected the receiver. Social approach behavior evoked by 50-kHz USV was reduced in both male and female *Cacna1c* haploinsufficient rats, although effects were more prominent in males. Moreover, male but not female *Cacna1c* haploinsufficient displayed less pronounced and less specific behavioral inhibition evoked by 22-kHz USV. *Cacna1c* haploinsufficiency is thus associated with a variety of alterations in social behavior, possibly due to lower motivation and/or diminished capability to display appropriate responses to important socio-affective communication signals.

5. Other Behavioral Phenotypes Displayed by *Cacna1c* Haploinsufficient Rats

As compared to the prominent effects of *Cacna1c* haploinsufficiency on socio-affective communication through 22-kHz and 50-kHz USV, reduced expression of Cav1.2 had only moderate effects in our studies on learning and memory. Our studies revealed intact spatial memory and reversal learning capabilities in a radial arm maze using food as reward, with slightly superior memory performance at the cost of reduced cognitive flexibility in *Cacna1c*$^{+/-}$ rats [13]. Such effects were primarily seen in males but not females. Social and physical enrichment had positive effects in *Cacna1c*$^{+/-}$ rats and *Cacna1c*$^{+/+}$ littermate controls and ameliorated slight reversal learning deficits displayed by *Cacna1c*$^{+/-}$ rats [14]. Novel object recognition memory was not affected by *Cacna1c* haploinsufficiency but impaired following post-weaning social isolation in both genotypes.

In an independent series of experiments performed by another laboratory, touchscreens were used for reversal learning with sucrose solution as reward. While learning was intact, *Cacna1c*$^{+/-}$ rats were found to be impaired in reversal learning and made more errors during reversal than *Cacna1c*$^{+/+}$ littermate controls [109]. In a related study, the effects of *Cacna1c* haploinsufficiency on delay and trace auditory fear conditioning was tested [110]. In the delay condition that typically results in strong conditioning to the auditory cue, *Cacna1c*$^{+/-}$ rats were found to display an increased fear response to the context. In the trace condition that typically leads to strong conditioning to the context, however, *Cacna1c*$^{+/-}$ rats displayed the opposite pattern and showed an elevated fear response to the auditory cue. In an unpaired condition, the fear response to both auditory cue and context was enhanced in *Cacna1c*$^{+/-}$ rats and it was suggested that *Cacna1c* haploinsufficiency is associated with inappropriate fear responses.

In the most recent study performed by the other laboratory, a contextual fear conditioning paradigm was applied, and contextual fear memory was found to be unchanged in *Cacna1c*$^{+/-}$ rats [107]. However, following a pre-exposure to the to-be-conditioned context, *Cacna1c*$^{+/-}$ rats displayed a much stronger fear response than *Cacna1c*$^{+/+}$ littermate controls, which was interpreted as evidence for impaired latent inhibition in *Cacna1c*$^{+/-}$ rats.

6. Neurobiological Alterations in *Cacna1c* Haploinsufficient Rats

At the neurobiological level, we obtained evidence for *Cacna1c* playing an important role in mitochondrial integrity and function [22] by showing that *Cacna1c* downregulation promotes resilience against glutamate-induced oxidative stress in neurons [21]. However, *Cacna1c* haploinsufficiency did not affect mitochondrial bioenergetics and reactive oxygen species production in rats and this was independent of whether rats were exposed to post-weaning social isolation or social and physical enrichment [20]. We also did not obtain evidence for the effects of *Cacna1c* haploinsufficiency on brain morphology by comparing volumetric properties of hippocampus and prefrontal cortex [23]. Likewise, adult hippocampal neurogenesis was not affected by *Cacna1c* haploinsufficiency. When quantifying cell proliferation and survival through an immunofluorescent multiple staining approach to ensure neuronal cell type specificity, no genotype differences were seen [23]. In another study, however, reduced cell proliferation in *Cacna1c*$^{+/-}$ rats was reported, albeit in absence of effects of *Cacna1c* haploinsufficiency on immature neurons and the size of the dentate gyrus [111]. *Cacna1c* haploinsufficiency was also reported to be associated with reduced brain-derived neurotrophic factor (BDNF) expression in the prefrontal cortex [109].

In the most recent study on impaired latent inhibition during contextual fear conditioning in *Cacna1c*$^{+/-}$ rats, neurobiological mechanisms implicated in learning and memory were studied by focusing on the dorsal hippocampus [112]. Specifically, associative plasticity at CA1 pyramidal synapses and network synchronization through phase-amplitude coupling between theta and gamma oscillations of the CA1 local field potential were studied. It was shown that synaptic plasticity is affected by *Cacna1c* haploinsufficiency and that the induction of long-term potentiation through theta-burst pairing is impaired in *Cacna1c*$^{+/-}$ rats. It was further found that spine calcium signaling is impaired during postsynaptic spike bursts in CA1 pyramidal neurons form *Cacna1c*$^{+/-}$ rats. Moreover, phase-amplitude coupling during exploration of a novel environment was reduced in *Cacna1c*$^{+/-}$ rats in the absence of behavioral differences in the novelty response. Genotypes did not differ in a familiar environment.

Similar to previous studies [23,111], hippocampal morphology and cellular density did not differ between *Cacna1c*$^{+/-}$ rats and *Cacna1c*$^{+/+}$ littermate controls [112]. However, phosphorylated extracellular-signal regulated kinase (pERK) and cAMP response element-binding protein (pCREB) immunoreactivities were significantly reduced in the hippocampus of *Cacna1c*$^{+/-}$ rats, without genotype effects on total ERK and CREB levels. It was thus suggested that impaired ERK- and CREB-mediated synapse-to-nucleus signaling in *Cacna1c*$^{+/-}$ rats might contribute to hippocampal dysfunctions, eventually translating into impairments in learning and memory. This view is supported by the observation that activation of ERK signaling through the BDNF mimetic TrkB/TrkC neurotrophin receptor co-activator LM22B-10 restored pERK and pCREB levels and led to a reversal of the long-term potentiation deficit in *Cacna1c*$^{+/-}$ rats. Most importantly, intra-hippocampal administration of LM22B-10 treatment also reversed the impairment in latent inhibition during contextual fear conditioning in *Cacna1c*$^{+/-}$ rats. A similar effect was seen following an intra-peritoneal injection of LM22B-10. Together, this suggests that impaired ERK signaling-mediated excitation-transcription coupling underlies the learning and memory deficits seen in *Cacna1c*$^{+/-}$ rats.

7. Conclusions

We conclude that *Cacna1c* haploinsufficiency in rats leads to robust deficits in socio-affective communication through 22-kHz and 50-kHz USV and associated alterations in social behavior, such as rough-and-tumble play behavior (Table 1). These deficits appear to be more severe than related deficits displayed by established rat models for neurodevelopmental disorders, such as the *Shank3* deficient rat model for autism. However, no neurobiological correlate has been identified yet. Comparatively mild effects of *Cacna1c* haploinsufficiency on learning and memory were reported. Such deficits were linked to impaired ERK signaling.

Table 1. Effects of *Cacna1c* haploinsufficiency on ultrasonic communication in rats.

Sender	Receiver	Sex	Age	References
Reduced 50-kHz USV emission rates during rough-and-tumble play	Reduced social approach behavior evoked by playback of pro-social 50-kHz USV	Male	Juvenility	[15]
Unchanged 50-kHz USV emission rates during rough-and-tumble play	Unchanged social approach behavior evoked by playback of pro-social 50-kHz USV [1]	Female	Juvenility	[16,17]
Reduced 50-kHz USV emission rates during social interactions		Female	Adulthood	[18]
	Reduced behavioral inhibition evoked by playback of alarm 22-kHz USV	Male	Adulthood	[19]
	Unchanged behavioral inhibition evoked by playback of alarm 22-kHz USV	Female	Adulthood	[19]

USV = ultrasonic vocalizations; [1] Social approach behavior evoked during 50-kHz USV playback was unchanged, yet search behavior after playback was reduced.

8. Future Perspectives

Because socio-affective communication through 22-kHz and 50-kHz USV is involved in the regulation of many social behaviors, it would be interesting to see whether the effects of *Cacna1c* haploinsufficiency on sender and receiver affect other social behaviors, such as cooperative actions [36], mutual rewarding preferences [37], general and direct reciprocity [38], and empathy-driven helping behavior [39]. For instance, it was suggested that 50-kHz USV emission is involved in cooperative behavior in an instrumental task [113]. Related to that, it would be interesting to test whether the reduction in 50-kHz USV emission in *Cacna1c*$^{+/-}$ rats occurs specifically in a social context or whether this is also seen in response to non-social stimuli. Because *CACNA1C* SNPs are strongly associated with affective disorders, namely depression [1] and bipolar disorder [2], in humans, it would be interesting to measure mania-like elevated mood through amphetamine-induced 50-kHz USV in *Cacna1c*$^{+/-}$ rats. Amphetamine is known to be a very potent elicitor of 50-kHz USV in rats [114]. Finally, given the evidence in support of the notion that impaired ERK signaling-mediated excitation-transcription coupling underlies the learning and memory deficits seen in *Cacna1c*$^{+/-}$ rats, it would be interesting to test whether impaired ERK signaling is associated with the deficits in socio-affective communication through 22-kHz and 50-kHz USV and whether associated alterations in social behavior can be similarly rescued by a BDNF mimetic, such as LM22B-10.

Author Contributions: M.W. wrote the manuscript with the help of T.M.K. and R.K.W.S., M.W. and R.K.W.S. acquired funding. All authors have read and agreed to the published version of the manuscript.

Funding: Our research on the behavioral effects of *Cacna1c* haploinsufficiency in rats, including social behavior and ultrasonic vocalizations as described in this review, was supported by grants from the Deutsche Forschungsgemeinschaft (DFG; German Research Council) to RS (SCHW 559/14-1 and SCHW 559/14-2) and MW (MW 1732/4-1 and MW 1732/4-2).

Acknowledgments: We wish to thank present and former members and collaborators of the laboratory for their important contributions to the research described in this review. The graphical abstract was created with BioRender.com (access date: 24 May 2021).

Conflicts of Interest: M.W. is scientific advisor of Avisoft Bioacoustics. The other authors declare no conflict of interest. The funders had no role in the design of the study; in the collection, analyses, or interpretation of data; in the writing of the manuscript, or in the decision to publish the results.

References

1. Green, E.K.; Grozeva, D.; Jones, I.; Jones, L.; Kirov, G.; Caesar, S.; Gordon-Smith, K.; Fraser, C.; Forty, L.; Russell, E.; et al. The bipolar disorder risk allele at CACNA1C also confers risk of recurrent major depression and of schizophrenia. *Mol. Psychiatry* **2010**, *15*, 1016–1022. [CrossRef] [PubMed]
2. Ferreira, M.A.; O'Donovan, M.C.; Meng, Y.A.; Jones, I.R.; Ruderfer, D.M.; Jones, L.; Fan, J.; Kirov, G.; Perlis, R.H.; Green, E.K.; et al. Collaborative genome-wide association analysis supports a role for ANK3 and CACNA1C in bipolar disorder. *Nat. Genet.* **2008**, *40*, 1056–1058. [CrossRef] [PubMed]
3. Splawski, I.; Timothy, K.W.; Sharpe, L.M.; Decher, N.; Kumar, P.; Bloise, R.; Napolitano, C.; Schwartz, P.J.; Joseph, R.M.; Condouris, K.; et al. Ca(V)1.2 calcium channel dysfunction causes a multisystem disorder including arrhythmia and autism. *Cell* **2004**, *119*, 19–31. [CrossRef] [PubMed]
4. Nyegaard, M.; Demontis, D.; Foldager, L.; Hedemand, A.; Flint, T.J.; Sørensen, K.M.; Andersen, P.S.; Nordentoft, M.; Werge, T.; Pedersen, C.B.; et al. CACNA1C (rs1006737) is associated with schizophrenia. *Mol. Psychiatry* **2010**, *15*, 119–121. [CrossRef]
5. Zamponi, G.W. Targeting voltage-gated calcium channels in neurological and psychiatric diseases. *Nat. Rev. Drug Discov.* **2016**, *15*, 19–34. [CrossRef]
6. Kabir, Z.D.; Lee, A.S.; Rajadhyaksha, A.M. L-type Ca^{2+} channels in mood, cognition and addiction: Integrating human and rodent studies with a focus on behavioural endophenotypes. *J. Physiol.* **2016**, *594*, 5823–5837. [CrossRef]
7. Bigos, K.L.; Mattay, V.S.; Callicott, J.H.; Straub, R.E.; Vakkalanka, R.; Kolachana, B.; Hyde, T.M.; Lipska, B.K.; Kleinman, J.E.; Weinberger, D.R. Genetic variation in CACNA1C affects brain circuitries related to mental illness. *Arch. Gen. Psychiatry* **2010**, *67*, 939–945. [CrossRef]
8. Eckart, N.; Song, Q.; Yang, R.; Wang, R.; Zhu, H.; McCallion, A.S.; Avramopoulos, D. Functional Characterization of Schizophrenia-Associated Variation in CACNA1C. *PLoS ONE* **2016**, *11*, e0157086. [CrossRef]
9. Gershon, E.S.; Grennan, K.; Busnello, J.; Badner, J.A.; Ovsiew, F.; Memon, S.; Alliey-Rodriguez, N.; Cooper, J.; Romanos, B.; Liu, C. A rare mutation of CACNA1C in a patient with bipolar disorder, and decreased gene expression associated with a bipolar-associated common SNP of CACNA1C in brain. *Mol. Psychiatry* **2014**, *19*, 890–894. [CrossRef]
10. Jaffe, A.E.; Hoeppner, D.J.; Saito, T.; Blanpain, L.; Ukaigwe, J.; Burke, E.E.; Collado-Torres, L.; Tao, R.; Tajinda, K.; Maynard, K.R.; et al. Profiling gene expression in the human dentate gyrus granule cell layer reveals insights into schizophrenia and its genetic risk. *Nat. Neurosci.* **2020**, *23*, 510–519. [CrossRef]
11. Roussos, P.; Mitchell, A.C.; Voloudakis, G.; Fullard, J.F.; Pothula, V.M.; Tsang, J.; Stahl, E.A.; Georgakopoulos, A.; Ruderfer, D.M.; Charney, A.; et al. A role for noncoding variation in schizophrenia. *Cell Rep.* **2014**, *9*, 1417–1429. [CrossRef]
12. Yoshimizu, T.; Pan, J.Q.; Mungenast, A.E.; Madison, J.M.; Su, S.; Ketterman, J.; Ongur, D.; McPhie, D.; Cohen, B.; Perlis, R.; et al. Functional implications of a psychiatric risk variant within CACNA1C in induced human neurons. *Mol. Psychiatry* **2015**, *20*, 162–169. [CrossRef]
13. Braun, M.D.; Kisko, T.M.; Vecchia, D.D.; Andreatini, R.; Schwarting, R.K.W.; Wöhr, M. Sex-specific effects of Cacna1c haploinsufficiency on object recognition, spatial memory, and reversal learning capabilities in rats. *Neurobiol. Learn. Mem.* **2018**, *155*, 543–555. [CrossRef]
14. Braun, M.D.; Kisko, T.M.; Witt, S.H.; Rietschel, M.; Schwarting, R.K.W.; Wöhr, M. Long-term environmental impact on object recognition, spatial memory and reversal learning capabilities in Cacna1c-haploinsufficient rats. *Hum. Mol. Genet.* **2019**, *28*, 4113–4131. [CrossRef]
15. Kisko, T.M.; Braun, M.D.; Michels, S.; Witt, S.H.; Rietschel, M.; Culmsee, C.; Schwarting, R.K.W.; Wöhr, M. Cacna1c haploinsufficiency leads to pro-social 50-kHz ultrasonic communication deficits in rats. *Dis. Models Mech.* **2018**, *11*, dmm034116. [CrossRef]
16. Kisko, T.M.; Braun, M.D.; Michels, S.; Witt, S.H.; Rietschel, M.; Culmsee, C.; Schwarting, R.K.W.; Wöhr, M. Sex-dependent effects of Cacna1c haploinsufficiency on juvenile social play behavior and pro-social 50-kHz ultrasonic communication in rats. *Genes Brain Behav.* **2020**, *19*, e12552. [CrossRef]
17. Kisko, T.M.; Schwarting, R.K.W.; Wöhr, M. Sex differences in the acoustic features of social play-induced 50-kHz ultrasonic vocalizations: A detailed spectrographic analysis in wild-type Sprague-Dawley and Cacna1c haploinsufficient rats. *Dev. Psychobiol.* **2021**. [CrossRef]
18. Redecker, T.M.; Kisko, T.M.; Schwarting, R.K.W.; Wöhr, M. Effects of Cacna1c haploinsufficiency on social interaction behavior and 50-kHz ultrasonic vocalizations in adult female rats. *Behav. Brain Res.* **2019**, *367*, 35–52. [CrossRef]

19. Wöhr, M.; Willadsen, M.; Kisko, T.M.; Schwarting, R.K.W.; Fendt, M. Sex-dependent effects of Cacna1c haploinsufficiency on behavioral inhibition evoked by conspecific alarm signals in rats. *Prog. Neuro-Psychopharmacol. Biol. Psychiatry* **2020**, *99*, e109849. [CrossRef]
20. Michels, S.; Dolga, A.M.; Braun, M.D.; Kisko, T.M.; Sungur, A.Ö.; Witt, S.H.; Rietschel, M.; Dempfle, A.; Wöhr, M.; Schwarting, R.K.W.; et al. Interaction of the psychiatric risk gene Cacna1c with post-weaning social isolation or environmental enrichment does not affect brain mitochondrial bioenergetics in rats. *Front. Cell Neurosci.* **2019**, *13*, e483. [CrossRef]
21. Michels, S.; Ganjam, G.K.; Martins, H.; Schratt, G.M.; Wöhr, M.; Schwarting, R.K.W.; Culmsee, C. Downregulation of the psychiatric susceptibility gene Cacna1c promotes mitochondrial resilience to oxidative stress in neuronal cells. *Cell Death Discov.* **2018**, *4*, e54. [CrossRef]
22. Michels, S.; Wöhr, M.; Schwarting, R.K.W.; Culmsee, C. Psychiatric risk gene Cacna1c determines mitochondrial resilience against oxidative stress in neurons. *Cell Death Discov.* **2018**, *9*, e645. [CrossRef]
23. Redecker, T.M.; Kisko, T.M.; Wöhr, M.; Schwarting, R.K.W. Cacna1c haploinsufficiency lacks effects on adult hippocampal neurogenesis and volumetric properties of prefrontal cortex and hippocampus in female rats. *Physiol. Behav.* **2020**, *223*, e112974. [CrossRef]
24. Krautheim, J.T.; Straube, B.; Dannlowski, U.; Pyka, M.; Schneider-Hassloff, H.; Drexler, R.; Krug, A.; Sommer, J.; Rietschel, M.; Witt, S.H.; et al. Outgroup emotion processing in the vACC is modulated by childhood trauma and CACNA1C risk variant. *Soc. Cogn. Affect. Neurosci.* **2018**, *13*, 341–348. [CrossRef]
25. Nieratschker, V.; Brückmann, C.; Plewnia, C. CACNA1C risk variant affects facial emotion recognition in healthy individuals. *Sci. Rep.* **2015**, *5*, e17349. [CrossRef]
26. Soeiro-de-Souza, M.G.; Otaduy, M.C.; Dias, C.Z.; Bio, D.S.; Machado-Vieira, R.; Moreno, R.A. The impact of the CACNA1C risk allele on limbic structures and facial emotions recognition in bipolar disorder subjects and healthy controls. *J. Affect. Disord.* **2012**, *141*, 94–101. [CrossRef]
27. Krug, A.; Nieratschker, V.; Markov, V.; Krach, S.; Jansen, A.; Zerres, K.; Eggermann, T.; Stöcker, T.; Shah, N.J.; Treutlein, J.; et al. Effect of CACNA1C rs1006737 on neural correlates of verbal fluency in healthy individuals. *Neuroimage* **2010**, *49*, 1831–1836. [CrossRef]
28. Schweinfurth, M.K. The social life of Norway rats (*Rattus norvegicus*). *eLife* **2020**, *9*, e54040. [CrossRef]
29. Lukas, M.; de Jong, T.R. Conspecific interactions in adult laboratory rodents: Friends or foes? *Curr. Top. Behav. Neurosci.* **2017**, *30*, 3–24. [CrossRef] [PubMed]
30. Vanderschuren, L.J.; Achterberg, E.J.; Trezza, V. The neurobiology of social play and its rewarding value in rats. *Neurosci. Biobehav. Rev.* **2016**, *70*, 86–105. [CrossRef]
31. Pellis, S.M.; Field, E.F.; Smith, L.K.; Pellis, V.C. Multiple differences in the play fighting of male and female rats. Implications for the causes and functions of play. *Neurosci. Biobehav. Rev.* **1997**, *21*, 105–120. [CrossRef]
32. Trezza, V.; Baarendse, P.J.; Vanderschuren, L.J. The pleasures of play: Pharmacological insights into social reward mechanisms. *Trends Pharmacol. Sci.* **2010**, *31*, 463–469. [CrossRef] [PubMed]
33. Perna, J.C.; Engelmann, M. Recognizing others: Rodent's social memories. *Curr. Top. Behav. Neurosci.* **2017**, *30*, 25–45. [CrossRef]
34. Galef, B.G.; Wigmore, S.W. Transfer of information concerning distant foods: A laboratory investigation of the 'information-centre' hypothesis. *Anim. Behav.* **1983**, *31*, 748–758. [CrossRef]
35. Fendt, M.; Gonzalez-Guerrero, C.P.; Kahl, E. Observational fear learning in rats: Role of trait anxiety and ultrasonic vocalization. *Brain Sci.* **2021**, *11*, 423. [CrossRef]
36. Avital, A.; Aga-Mizrachi, S.; Zubedat, S. Evidence for social cooperation in rodents by automated maze. *Sci. Rep.* **2016**, *6*, e29517. [CrossRef]
37. Hernandez-Lallement, J.; van Wingerden, M.; Marx, C.; Srejic, M.; Kalenscher, T. Rats prefer mutual rewards in a prosocial choice task. *Front. Neurosci.* **2015**, *8*, e443. [CrossRef]
38. Rutte, C.; Taborsky, M. Generalized reciprocity in rats. *PLoS Biol.* **2007**, *5*, e196. [CrossRef]
39. Bartal, I.B.A.; Decety, J.; Mason, P. Empathy and pro-social behavior in rats. *Science* **2011**, *334*, 1427–1430. [CrossRef]
40. Brudzynski, S.M. Ethotransmission: Communication of emotional states through ultrasonic vocalization in rats. *Curr. Opin. Neurobiol.* **2013**, *23*, 310–317. [CrossRef]
41. Wöhr, M.; Schwarting, R.K.W. Affective communication in rodents: Ultrasonic vocalizations as a tool for research on emotion and motivation. *Cell Tissue Res.* **2013**, *354*, 81–97. [CrossRef]
42. Brudzynski, S.M. Biological functions of rat ultrasonic vocalizations, Arousal mechanisms and call initiation. *Brain Sci.* **2021**, *11*, 605. [CrossRef]
43. Hofer, M.A. Multiple regulators of ultrasonic vocalization in the infant rat. *Psychoneuroendocrinology* **1996**, *21*, 203–217. [CrossRef]
44. Allin, J.T.; Banks, E.M. Functional aspects of ultrasound production by infant albino rats (Rattus norvegicus). *Anim. Behav.* **1972**, *20*, 175–185. [CrossRef]
45. Blanchard, R.J.; Blanchard, D.C.; Agullana, R.; Weiss, S.M. Twenty-two kHz alarm cries to presentation of a predator, by laboratory rats living in visible burrow systems. *Physiol. Behav.* **1991**, *50*, 967–972. [CrossRef]
46. Fendt, M.; Brosch, M.; Wernecke, K.E.A.; Willadsen, M.; Wöhr, M. Predator odour but not TMT induces 22-kHz ultrasonic vocalizations in rats that lead to defensive behaviours in conspecifics upon replay. *Sci. Rep.* **2018**, *8*, e11041. [CrossRef]

47. Knutson, B.; Burgdorf, J.; Panksepp, J. Anticipation of play elicits high-frequency ultrasonic vocalizations in young rats. *J. Comp. Psychol.* **1998**, *112*, 65–73. [CrossRef]
48. Sales, G.D. Ultrasound and mating behaviour in rodents with some observations on other behavioural situations. *J. Zool.* **1972**, *168*, 149–164. [CrossRef]
49. Panksepp, J. Psychology. Beyond a joke: From animal laughter to human joy? *Science* **2005**, *308*, 62–63. [CrossRef] [PubMed]
50. Burgdorf, J.; Panksepp, J.; Moskal, J.R. Frequency-modulated 50 kHz ultrasonic vocalizations: A tool for uncovering the molecular substrates of positive affect. *Neurosci. Biobehav. Rev.* **2011**, *35*, 1831–1836. [CrossRef]
51. Pereira, M.; Andreatini, R.; Schwarting, R.K.W.; Brenes, J.C. Amphetamine-induced appetitive 50-kHz calls in rats: A marker of affect in mania? *Psychopharmacology* **2014**, *231*, 2567–2577. [CrossRef] [PubMed]
52. Wright, J.M.; Gourdon, J.C.; Clarke, P.B. Identification of multiple call categories within the rich repertoire of adult rat 50-kHz ultrasonic vocalizations: Effects of amphetamine and social context. *Psychopharmacology* **2010**, *211*, 1–13. [CrossRef] [PubMed]
53. Wöhr, M. Ultrasonic communication in rats: Appetitive 50-kHz ultrasonic vocalizations as social contact calls. *Behav. Ecol. Sociobiol.* **2018**, *72*, e14. [CrossRef]
54. Berridge, K.C.; Robinson, T.E.; Aldridge, J.W. Dissecting components of reward: 'liking', 'wanting', and learning. *Curr. Opin. Pharmacol.* **2009**, *9*, 65–73. [CrossRef]
55. Burke, C.J.; Modlinska, K.; Mauro, M.H.; Aleksandrova, L.R.; Pellis, S.M.; Phillips, A.G.; Euston, D.R. A naturalistic method to test depression: Anticipation of play. *Behav. Brain Res.* **2021**, *398*, e112975. [CrossRef]
56. Burgdorf, J.; Kroes, R.A.; Moskal, J.R.; Pfaus, J.G.; Brudzynski, S.M.; Panksepp, J. Ultrasonic vocalizations of rats (Rattus norvegicus) during mating, play, and aggression: Behavioral concomitants, relationship to reward, and self-administration of playback. *J. Comp. Psychol.* **2008**, *122*, 357–367. [CrossRef]
57. Himmler, B.T.; Kisko, T.M.; Euston, D.R.; Kolb, B.; Pellis, S.M. Are 50-kHz calls used as play signals in the playful interactions of rats? I. Evidence from the timing and context of their use. *Behav. Process.* **2014**, *106*, 60–66. [CrossRef]
58. Manduca, A.; Campolongo, P.; Palmery, M.; Vanderschuren, L.J.; Cuomo, V.; Trezza, V. Social play behavior, ultrasonic vocalizations and their modulation by morphine and amphetamine in Wistar and Sprague-Dawley rats. *Psychopharmacology* **2014**, *231*, 1661–1673. [CrossRef]
59. Manduca, A.; Servadio, M.; Campolongo, P.; Palmery, M.; Trabace, L.; Vanderschuren, L.J.; Cuomo, V.; Trezza, V. Strain-and context-dependent effects of the anandamide hydrolysis inhibitor URB597 on social behavior in rats. *Eur. Neuropsychopharmacol.* **2014**, *24*, 1337–1348. [CrossRef]
60. Siviy, S.M.; Panksepp, J. Sensory modulation of juvenile play in rats. *Dev. Psychobiol.* **1987**, *20*, 39–55. [CrossRef]
61. Kisko, T.M.; Euston, D.R.; Pellis, S.M. Are 50-kHz calls used as play signals in the playful interactions of rats? III. The effects of devocalization on play with unfamiliar partners as juveniles and as adults. *Behav. Process.* **2015**, *113*, 113–121. [CrossRef] [PubMed]
62. Kisko, T.M.; Himmler, B.T.; Himmler, S.M.; Euston, D.R.; Pellis, S.M. Are 50-kHz calls used as play signals in the playful interactions of rats? II. Evidence from the effects of devocalization. *Behav. Process.* **2015**, *111*, 25–33. [CrossRef] [PubMed]
63. Kisko, T.M.; Wöhr, M.; Pellis, V.C.; Pellis, S.M. From Play to Aggression: High-Frequency 50-kHz Ultrasonic Vocalizations as Play and Appeasement Signals in Rats. *Curr. Top. Behav. Neurosci.* **2017**, *30*, 91–108. [CrossRef] [PubMed]
64. Panksepp, J.; Burgdorf, J. 50-kHz chirping (laughter?) in response to conditioned and unconditioned tickle-induced reward in rats: Effects of social housing and genetic variables. *Behav. Brain Res.* **2000**, *115*, 25–38. [CrossRef]
65. Panksepp, J.; Burgdorf, J.; Gordon, N. Towards a genetics of joy: Breeding rats for "laughter". In *Emotions, Qualia, and Consciousness*; Kazniak, A., Ed.; World Scientific: Singapore, 2001; pp. 123–136. [CrossRef]
66. Burgdorf, J.; Panksepp, J.; Brudzynski, S.M.; Beinfeld, M.C.; Cromwell, H.C.; Kroes, R.A.; Moskal, J.R. The effects of selective breeding for differential rates of 50-kHz ultrasonic vocalizations on emotional behavior in rats. *Dev. Psychobiol.* **2009**, *51*, 34–46. [CrossRef]
67. Burgdorf, J.; Panksepp, J.; Brudzynski, S.M.; Kroes, R.; Moskal, J.R. Breeding for 50-kHz positive affective vocalization in rats. *Behav. Genet.* **2005**, *35*, 67–72. [CrossRef]
68. Webber, E.S.; Harmon, K.M.; Beckwith, T.J.; Peña, S.; Burgdorf, J.; Panksepp, J.; Cromwell, H.C. Selective breeding for 50 kHz ultrasonic vocalization emission produces alterations in the ontogeny and regulation of rough-and-tumble play. *Behav. Brain Res.* **2012**, *229*, 138–144. [CrossRef]
69. Burgdorf, J.; Moskal, J.R.; Brudzynski, S.M.; Panksepp, J. Rats selectively bred for low levels of play-induced 50 kHz vocalizations as a model for autism spectrum disorders: A role for NMDA receptors. *Behav. Brain Res.* **2013**, *251*, 18–24. [CrossRef]
70. Brunelli, S.A.; Nie, R.; Whipple, C.; Winiger, V.; Hofer, M.A.; Zimmerberg, B. The effects of selective breeding for infant ultrasonic vocalizations on play behavior in juvenile rats. *Physiol. Behav.* **2006**, *87*, 527–536. [CrossRef]
71. Dichter, G.S.; Brunelli, S.A.; Hofer, M.A. Elevated plus-maze behavior in adult offspring of selectively bred rats. *Physiol. Behav.* **1996**, *60*, 299–304. [CrossRef]
72. Brunelli, S.A. Selective breeding for an infant phenotype: Rat pup ultrasonic vocalization (USV). *Behav. Genet.* **2005**, *35*, 53–65. [CrossRef]
73. Lukas, M.; Wöhr, M. Endogenous vasopressin, innate anxiety, and the emission of pro-social 50-kHz ultrasonic vocalizations during social play behavior in juvenile rats. *Psychoneuroendocrinology* **2015**, *56*, 35–44. [CrossRef]

74. Schneider, P.; Pätz, M.; Spanagel, R.; Schneider, M. Adolescent social rejection alters pain processing in a CB1 receptor dependent manner. *Eur. Neuropsychopharmacol.* **2016**, *26*, 1201–1212. [CrossRef]
75. Burgdorf, J.; Kroes, R.A.; Weiss, C.; Oh, M.M.; Disterhoft, J.F.; Brudzynski, S.M.; Panksepp, J.; Moskal, J.R. Positive emotional learning is regulated in the medial prefrontal cortex by GluN2B-containing NMDA receptors. *Neuroscience* **2011**, *192*, 515–523. [CrossRef]
76. Moskal, J.R.; Burgdorf, J.; Kroes, R.A.; Brudzynski, S.M.; Panksepp, J. A novel NMDA receptor glycine-site partial agonist, GLYX-13, has therapeutic potential for the treatment of autism. *Neurosci. Biobehav. Rev.* **2011**, *35*, 1982–1988. [CrossRef]
77. Burgdorf, J.; Kroes, R.A.; Beinfeld, M.C.; Panksepp, J.; Moskal, J.R. Uncovering the molecular basis of positive affect using rough-and-tumble play in rats: A role for insulin-like growth factor I. *Neuroscience* **2010**, *168*, 769–777. [CrossRef]
78. Burgdorf, J.; Panksepp, J.; Beinfeld, M.C.; Kroes, R.A.; Moskal, J.R. Regional brain cholecystokinin changes as a function of rough-and-tumble play behavior in adolescent rats. *Peptides* **2006**, *27*, 172–177. [CrossRef]
79. Willey, A.R.; Varlinskaya, E.I.; Spear, L.P. Social interactions and 50 kHz ultrasonic vocalizations in adolescent and adult rats. *Behav. Brain Res.* **2009**, *202*, 122–129. [CrossRef]
80. Waddell, J.; Yang, T.; Ho, E.; Wellmann, K.A.; Mooney, S.M. Prenatal ethanol exposure and whisker clipping disrupt ultrasonic vocalizations and play behavior in adolescent rats. *Brain Sci.* **2016**, *6*, 43. [CrossRef]
81. Wellmann, K.A.; George, F.; Brnouti, F.; Mooney, S.M. Docosahexaenoic acid partially ameliorates deficits in social behavior and ultrasonic vocalizations caused by prenatal ethanol exposure. *Behav. Brain Res.* **2015**, *286*, 201–211. [CrossRef]
82. Wellmann, K.A.; Varlinskaya, E.I.; Mooney, S.M. D-Cycloserine ameliorates social alterations that result from prenatal exposure to valproic acid. *Brain Res. Bull.* **2014**, *108*, 1–9. [CrossRef]
83. Raza, S.; Himmler, B.T.; Himmler, S.M.; Harker, A.; Kolb, B.; Pellis, S.M.; Gibb, R. Effects of prenatal exposure to valproic acid on the development of juvenile-typical social play in rats. *Behav. Pharmacol.* **2015**, *26*, 707–719. [CrossRef]
84. Shahrier, M.A.; Wada, H. Effects of prenatal ethanol exposure on acoustic characteristics of play fighting-induced ultrasonic vocalizations in juvenile rats. *Neurotoxicology* **2020**, *79*, 25–39. [CrossRef]
85. Gzielo, K.; Potasiewicz, A.; Hołuj, M.; Litwa, E.; Popik, P.; Nikiforuk, A. Valproic acid exposure impairs ultrasonic communication in infant, adolescent and adult rats. *Eur. Neuropsychopharmacol.* **2020**, *41*, 52–62. [CrossRef]
86. Gzielo, K.; Potasiewicz, A.; Litwa, E.; Piotrowska, D.; Popik, P.; Nikiforuk, A. The effect of maternal immune activation on social play-induced ultrasonic vocalization in rats. *Brain Sci.* **2021**, *11*, 344. [CrossRef]
87. Kentner, A.C.; Scalia, S.; Shin, J.; Migliore, M.M.; Rondón-Ortiz, A.N. Targeted sensory enrichment interventions protect against behavioral and neuroendocrine consequences of early life stress. *Psychoneuroendocrinology* **2018**, *98*, e74–e85. [CrossRef]
88. Wöhr, M.; Schwarting, R.K.W. Ultrasonic communication in rats: Can playback of 50-kHz calls induce approach behavior? *PLoS ONE* **2007**, *2*, e1365. [CrossRef]
89. Wöhr, M.; Schwarting, R.K.W. Maternal care, isolation-induced infant ultrasonic calling, and their relations to adult anxiety-related behavior in the rat. *Behav. Neurosci.* **2008**, *122*, 310–330. [CrossRef]
90. Wöhr, M.; Schwarting, R.K.W. Testing social acoustic memory in rats: Effects of stimulus configuration and long-term memory on the induction of social approach behavior by appetitive 50-kHz ultrasonic vocalizations. *Neurobiol. Learn. Mem.* **2012**, *98*, 154–164. [CrossRef]
91. Willadsen, M.; Seffer, D.; Schwarting, R.K.W.; Wöhr, M. Rodent ultrasonic communication: Male prosocial 50-kHz ultrasonic vocalizations elicit social approach behavior in female rats (Rattus norvegicus). *J. Comp. Psychol.* **2014**, *128*, 56–64. [CrossRef]
92. Engelhardt, K.A.; Schwarting, R.K.W.; Wöhr, M. Mapping trait-like socio-affective phenotypes in rats through 50-kHz ultrasonic vocalizations. *Psychopharmacology* **2018**, *235*, 83–98. [CrossRef] [PubMed]
93. Seffer, D.; Rippberger, H.; Schwarting, R.K.W.; Wöhr, M. Pro-Social 50-kHz ultrasonic communication in rats: Post-weaning but not post-adolescent social isolation leads to social impairments-phenotypic rescue by re-socialization. *Front. Behav. Neurosci.* **2015**, *9*, 102. [CrossRef] [PubMed]
94. Brenes, J.C.; Lackinger, M.; Höglinger, G.U.; Schratt, G.; Schwarting, R.K.W.; Wöhr, M. Differential effects of social and physical environmental enrichment on brain plasticity, cognition, and ultrasonic communication in rats. *J. Comp. Neurol.* **2016**, *524*, 1586–1607. [CrossRef] [PubMed]
95. Berz, A.; de Souza, C.P.; Wöhr, M.; Schwarting, R.K.W. Limited generalizability, pharmacological modulation, and state-dependency of habituation towards pro-social 50-kHz calls in rats. *IScience* **2021**, *24*, 102426. [CrossRef]
96. Sadananda, M.; Wöhr, M.; Schwarting, R.K.W. Playback of 22-kHz and 50-kHz ultrasonic vocalizations induces differential c-fos expression in rat brain. *Neurosci. Lett.* **2008**, *435*, 17–23. [CrossRef]
97. Willuhn, I.; Tose, A.; Wanat, M.J.; Hart, A.S.; Hollon, N.G.; Phillips, P.E.; Schwarting, R.K.W.; Wöhr, M. Phasic dopamine release in the nucleus accumbens in response to pro-social 50 kHz ultrasonic vocalizations in rats. *J. Neurosci.* **2014**, *34*, 10616–10623. [CrossRef]
98. Engelhardt, K.A.; Fuchs, E.; Schwarting, R.K.W.; Wöhr, M. Effects of amphetamine on pro-social ultrasonic communication in juvenile rats: Implications for mania models. *Eur. Neuropsychopharmacol.* **2017**, *27*, 261–273. [CrossRef]
99. Wöhr, M.; Schwarting, R.K.W. Ultrasonic communication in rats: Effects of morphine and naloxone on vocal and behavioral responses to playback of 50-kHz vocalizations. *Pharmacol. Biochem. Behav.* **2009**, *94*, 285–295. [CrossRef]

100. Berg, E.L.; Copping, N.A.; Rivera, J.K.; Pride, M.C.; Careaga, M.; Bauman, M.D.; Berman, R.F.; Lein, P.J.; Harony-Nicolas, H.; Buxbaum, J.D.; et al. Developmental social communication deficits in the Shank3 rat model of Phelan-McDermid syndrome and autism spectrum disorder. *Autism Res.* **2018**, *11*, 587–601. [CrossRef]
101. Bourgeron, T. From the genetic architecture to synaptic plasticity in autism spectrum disorder. *Nat. Rev. Neurosci.* **2015**, *16*, 551–563. [CrossRef]
102. Berg, E.L.; Pride, M.C.; Petkova, S.P.; Lee, R.D.; Copping, N.A.; Shen, Y.; Adhikari, A.; Fenton, T.A.; Pedersen, L.R.; Noakes, L.S.; et al. Translational outcomes in a full gene deletion of ubiquitin protein ligase E3A rat model of Angelman syndrome. *Transl. Psychiatry* **2020**, *10*, 1–16. [CrossRef]
103. Berg, E.L.; Ching, T.M.; Bruun, D.A.; Rivera, J.K.; Careaga, M.; Ellegood, J.; Lerch, J.P.; Wöhr, M.; Lein, P.J.; Silverman, J.L. Translational outcomes relevant to neurodevelopmental disorders following early life exposure of rats to chlorpyrifos. *J. Neurodev. Disord.* **2020**, *12*, 1–15. [CrossRef]
104. Kircher, T.; Wöhr, M.; Nenadic, I.; Schwarting, R.K.W.; Schratt, G.; Alferink, J.; Culmsee, C.; Garn, H.; Hahn, T.; Müller-Myhsok, B.; et al. Neurobiology of the major psychoses: A translational perspective on brain structure and function-the FOR2107 consortium. *Eur. Arch. Psychiatry Clin. Neurosci.* **2019**, *269*, 949–962. [CrossRef]
105. Trezza, V.; Damsteegt, R.; Vanderschuren, L.J. Conditioned place preference induced by social play behavior: Parametrics, extinction, reinstatement and disruption by methylphenidate. *Eur. Neuropsychopharmacol.* **2009**, *19*, 659–669. [CrossRef]
106. Pellis, S.M.; Pellis, V.C.; McKenna, M.M. Feminine dimension in the play fighting of rats (Rattus norvegicus) and its defeminization neonatally by androgens. *J. Comp. Psychol.* **1994**, *108*, 68–73. [CrossRef]
107. Albert, D.J.; Walsh, M.L.; Gorzalka, B.B.; Siemens, Y.; Louie, H. Testosterone removal in rats results in a decrease in social aggression and a loss of social dominance. *Physiol. Behav.* **1986**, *36*, 401–407. [CrossRef]
108. Wöhr, M.; Borta, A.; Schwarting, R.K.W. Overt behavior and ultrasonic vocalization in a fear conditioning paradigm: A dose-response study in the rat. *Neurobiol. Learn. Mem.* **2005**, *84*, 228–240. [CrossRef]
109. Sykes, L.; Haddon, J.; Lancaster, T.M.; Sykes, A.; Azzouni, K.; Ihssen, N.; Moon, A.L.; Lin, T.E.; Linden, D.E.; Owen, M.J.; et al. Genetic variation in the psychiatric risk gene CACNA1C modulates reversal learning across species. *Schizophr. Bull.* **2019**, *45*, 1024–1032. [CrossRef]
110. Moon, A.L.; Brydges, N.M.; Wilkinson, L.S.; Hall, J.; Thomas, K.L. Cacna1c hemizygosity results in aberrant fear conditioning to neutral stimuli. *Schizophr. Bull.* **2020**, *46*, 1231–1238. [CrossRef]
111. Moon, A.L.; Haan, N.; Wilkinson, L.S.; Thomas, K.L.; Hall, J. CACNA1C: Association with psychiatric disorders, behavior, and neurogenesis. *Schizophr. Bull.* **2018**, *44*, 958–965. [CrossRef]
112. Tigaret, C.M.; Lin, T.E.; Morrell, E.R.; Sykes, L.; Moon, A.L.; O'Donovan, M.C.; Owen, M.J.; Wilkinson, L.S.; Jones, M.W.; Thomas, K.L.; et al. Neurotrophin receptor activation rescues cognitive and synaptic abnormalities caused by hemizygosity of the psychiatric risk gene Cacna1c. *Mol. Psychiatry* **2021**. [CrossRef]
113. Łopuch, S.; Popik, P. Cooperative behavior of laboratory rats (Rattus norvegicus) in an instrumental task. *J. Comp. Psychol.* **2011**, *125*, 250–253. [CrossRef]
114. Wöhr, M. Measuring mania-like elevated mood through amphetamine-induced 50-kHz ultrasonic vocalizations in rats. *Br. J. Pharmacol.* **2021**. [CrossRef]

brain sciences

MDPI

Article

Acoustilytix™: A Web-Based Automated Ultrasonic Vocalization Scoring Platform

Catherine B. Ashley [1], Ryan D. Snyder [1], James E. Shepherd [1], Catalina Cervantes [2], Nitish Mittal [3], Sheila Fleming [4], Jaxon Bailey [2], Maisie D. Nievera [2], Sharmin Islam Souleimanova [2], Bill Nyaoga [2], Lauren Lichtenfeld [2], Alicia R. Chen [2], W. Todd Maddox [5] and Christine L. Duvauchelle [2,6,*]

1 Cornerstone Research Group, Miamisburg, OH 45342, USA; ashleycb@crgrp.com (C.B.A.); snyderrd@crgrp.com (R.D.S.); shepherdje@crgrp.com (J.E.S.)
2 Division of Pharmacology and Toxicology, College of Pharmacy, The University of Texas at Austin, 2409 University Avenue, Stop A1915, Austin, TX 78712, USA; catalinacervantes@austin.utexas.edu (C.C.); jaxon.bailey@utexas.edu (J.B.); maisie.nievera@utexas.edu (M.D.N.); sharmin.souleimanova@austin.utexas.edu (S.I.S.); bill.nyaoga@austin.utexas.edu (B.N.); lauren.lichtenfeld@austin.utexas.edu (L.L.); alicia.chen@austin.utexas.edu (A.R.C.)
3 ZS Associates, Empire State Building, 350 Fifth Avenue, Suite 5100, New York, NY 10118, USA; niimits1@gmail.com
4 Department of Pharmaceutical Sciences, Northeast Ohio Medical University, 4209 State Route 44, RGE, Rootstown, OH 44272, USA; sfleming1@neomed.edu
5 Cognitive Design and Statistical Consulting, LLC, Austin, TX 78746, USA; wtoddmaddox@gmail.com
6 Waggoner Center for Alcohol and Addiction Research, The University of Texas at Austin, 2500 Speedway, Stop A4800, Austin, TX 78712, USA
* Correspondence: duvauchelle@mail.utexas.edu

Citation: Ashley, C.B.; Snyder, R.D.; Shepherd, J.E.; Cervantes, C.; Mittal, N.; Fleming, S.; Bailey, J.; Nievera, M.D.; Souleimanova, S.I.; Nyaoga, B.; et al. Acoustilytix™: A Web-Based Automated Ultrasonic Vocalization Scoring Platform. *Brain Sci.* **2021**, *11*, 864. https://doi.org/10.3390/brainsci11070864

Academic Editors: Stefan M. Brudzynski and Jeffrey Burgdorf

Received: 2 June 2021
Accepted: 18 June 2021
Published: 29 June 2021

Abstract: Ultrasonic vocalizations (USVs) are known to reflect emotional processing, brain neurochemistry, and brain function. Collecting and processing USV data is manual, time-intensive, and costly, creating a significant bottleneck by limiting researchers' ability to employ fully effective and nuanced experimental designs and serving as a barrier to entry for other researchers. In this report, we provide a snapshot of the current development and testing of Acoustilytix™, a web-based automated USV scoring tool. Acoustilytix implements machine learning methodology in the USV detection and classification process and is recording environment-agnostic. We summarize the user features identified as desirable by USV researchers and how these were implemented. These include the ability to easily upload USV files, output a list of detected USVs with associated parameters in csv format, and the ability to manually verify or modify an automatically detected call. With no user intervention or tuning, Acoustilytix achieves 93% sensitivity (a measure of how accurately Acoustilytix detects true calls) and 73% precision (a measure of how accurately Acoustilytix avoids false positives) in call detection across four unique recording environments and was superior to the popular DeepSqueak algorithm (sensitivity = 88%; precision = 41%). Future work will include integration and implementation of machine-learning-based call type classification prediction that will recommend a call type to the user for each detected call. Call classification accuracy is currently in the 71–79% accuracy range, which will continue to improve as more USV files are scored by expert scorers, providing more training data for the classification model. We also describe a recently developed feature of Acoustilytix that offers a fast and effective way to train hand-scorers using automated learning principles without the need for an expert hand-scorer to be present and is built upon a foundation of learning science. The key is that trainees are given practice classifying hundreds of calls with immediate corrective feedback based on an expert's USV classification. We showed that this approach is highly effective with inter-rater reliability (i.e., kappa statistics) between trainees and the expert ranging from 0.30–0.75 (average = 0.55) after only 1000–2000 calls of training. We conclude with a brief discussion of future improvements to the Acoustilytix platform.

Keywords: ultrasonic vocalization; automated scoring; dopamine; addiction; mental health; machine learning; drug discovery; drug development

1. Introduction

Identifying neural mechanisms underlying mental health and addiction disorders is necessary to develop successful interventions, but moving from basic science findings to clinical trials is a lengthy and high-risk process. Although not the sole determinant, emotional processing is directly involved in many mental health disorders and addictions. Animal models allow direct measurement and manipulation of the neural circuitry that drives emotional processing, circuitry that often can only be inferred in human studies [1–5].

Rodent ultrasonic vocalizations (USVs) are known to reflect emotional processing, brain neurochemistry, and brain function-key observations in animal model studies. USVs in the 22–28 kHz and 50–55 kHz frequency range are widely recognized forms of social and emotional expression in rats [6]. USVs in the 22–28 kHz range are provoked by external threats (e.g., danger) or unpleasant internal conditions (e.g., sickness, pain) [6–8]. These negative-affect associated 22–28 kHz USVs are initiated through the ascending mesolimbic cholinergic pathway, originating in the laterodorsal tegmental nucleus (LDT) and extending through the midbrain, anterior hypothalamic, preoptic, and septal regions [7,9,10]. Positive-affect associated USVs, commonly referred to as 50–55 kHz FM USVs, are elicited by activation of sites all along the mesolimbic dopaminergic pathway, including the VTA, anteromedial hypothalamus, preoptic area, and the NAc [11,12]. Although the frequency ranges and spectral dynamics of the emitted USVs differ, similar mechanisms also exist in mice [13,14]. Animal models using both mice and rats are utilized in a broad number of research areas, including mental health and addiction [15].

Analysis of rodent USVs is a foundational technique that underlies a diverse cross-section of mental health and addiction research [16–28]. However, this technique is currently underutilized because USV tabulation is time-intensive, manual, and costly—each two minutes of recorded rodent vocalizations can require up to an hour to hand-score depending upon the number of calls present. Hand-scoring requires extensive prior training before a scorer reaches the necessary level of proficiency, and all too often, this training is sub-optimal, reliability is left unchecked, and large individual differences in hand-scoring results. Notably, even with proficient hand-scorers, the intensive manual nature of the analysis limits researchers' ability to employ the effective and nuanced experimental designs (e.g., multi-hour and/or multi-session experiments) needed to fully explore mental health and addiction related topics. In addition, with the increased predictive success of acoustic parameters of USVs over discrete call classification, tools are needed to automate this process [15,29–32]. This represents a bottleneck for current USV researchers and a barrier to entry for other researchers as well.

Recognizing that automation can alleviate the USV analysis bottleneck, several research groups have developed automated USV detection and classification algorithms, including WAAVES [32], XBAT [33], DeepSqueak [34], and MUPET [35]. WAAVES has enabled Duvauchelle and colleagues to rapidly examine emotional profiles in rats across a broad range of experimental settings in a fraction of the time required with manual analyses [30,32,36–39]. However, the main weakness of all of these automated USV scoring algorithms is that each is customized to specific environmental recording conditions and types of animals (e.g., rats vs. mice) and requires end user trial-and-error-driven thresholding and/or extensive training data to ensure accurate results. These algorithms cannot be broadly applied to data collected in other sound environments without tailoring, thus, impeding research progress.

In this manuscript, we present a snapshot of the development and testing of a new web-based automated USV scoring tool called Acoustilytix (see http://acoustilytix.com/; 30 May 2021). Acoustilytix is environment-agnostic and implements machine learning methodology in the USV detection and classification process (We are in the process of applying this approach in mice). The ongoing development of Acoustilytix is funded through the National Institutes of Mental Health Small Business Technology Transfer (STTR) program, whose mission is to support commercialization of products developed through a partnership between research institutions and industry (It is for this reason that many algorithmic details cannot be presented). The ongoing development of Acoustilytix derives from a unique collaboration between the Cornerstone Research Group (CRG), a high-technology research and development company near Dayton Ohio, and Dr. Christine Duvauchelle, an expert in USV animal research in the College of Pharmacy at the University of Texas (UT). The CRG and UT teams were complemented with an international consortium of USV researchers who generously provided their expertise and input, as well as their USV data to be used during software development. The consortium included a dozen USV researchers from around the world who were interested in helping us develop the most effective automated USV scoring solution.

The remainder of the manuscript is organized into two major sections.

The first focuses on the current state of development of the Acoustilytix platform and the rigorous quantitative testing and evaluation of the platform. The current version of Acoustilytix is quite effective at automated call detection, being environment-agnostic and offering a quantitative measure of call detection uncertainty (see Section 3 below). Acoustilytix is less effective when it comes to automated call classification with this still being a work in progress. In this first section, we briefly summarize some of the user features suggested by our consortium members that have been implemented into Acoustilytix, such as rapid automated USV detection, call parsing, detection confidence, call classification, and parameterization from rats in any USV recording environment. For ease of exposition, the main body of this section will offer a high-level summary, with many of the algorithmic and architectural details in Appendices A–C. Next a detailed analysis of the platform output and a comparison of its effectiveness at automated call detection with the popular DeepSqueak algorithm is presented. Briefly, with no user intervention or tuning, Acoustilytix achieves 93% sensitivity (a measure of how accurately Acoustilytix detects true calls) and 73% precision (a measure of how accurately Acoustilytix avoids false positives) in call detection across four unique recording environments and was superior to the popular DeepSqueak algorithm (sensitivity = 88%; precision = 41%). Acoustilytix is also much less prone to detecting false positives than DeepSqueak. We conclude with a discussion of automated call classification and our progress to date. To anticipate, automated call classification has been challenging. We had to balance the desire to classify multiple distinct call types with the need for large numbers of hand-scored calls for model testing. To date, we have developed a machine learning model that automates the classification of a five call scheme suggested by one of our consortium members. Call classification accuracy is in the 71–79% range (well above chance) and will continue to improve as more USV files are scored, which will provide more training data for improving model fit. Perhaps more importantly, discussions with our consortium members suggest that they are less interested in a product that is well calibrated to a single call classification scheme but rather would like a product that allows them to upload a large number of hand-scored calls from a classification scheme of their choice that will then be used to train a machine learning algorithm for subsequent automated scoring. In the General Discussion section, we briefly summarize what a product offering like this might look like.

The second major section focuses on a recently developed feature of Acoustilytix that offers a fast and effective automated method for training hand-scorers without the need for an expert hand-scorer to be present. Building upon the fact that some level of hand-scoring competency is needed in any USV lab, even with Acoustilytix, we describe a new feature that leverages learning science principles to optimize hand-scoring training [40,41]. In short, hand-scorers in training can select the "training" feature in Acoustilytix. They are presented with a list of training files that contain calls that have been hand-scored by an expert in the laboratory whose classification serves as the teaching signal. The trainee can select a file and begin by selecting prototypical calls from a menu that they can visualize using a spectrogram or listen to at a reduced rate. Once the trainee is ready to begin the formal call-by-call training process, they are presented, one by one, with a large number of calls (e.g., 1000), which they may analyze visually using a spectrogram or audibly by listening to the audio clip at a reduced rate. For each call, the trainee selects a call classification and is given immediate corrective feedback (based on the expert scorer's classification). If in error, the trainee is allowed to examine any prototypical calls again and/or make another classification judgement. This process continues until the trainee provides the correct classification for each call and until all calls in the training set have been correctly classified. Immediate corrective feedback of this sort is known to speed learning and enhance proficiency [40,41]. This increases the likelihood that all hand-scorers in the laboratory mimic the scoring behavior of the expert, thus increasing consistency and reliability of USV hand-scoring. We present the results from a reliability study that verifies the effectiveness of the Acoustilytix hand-scorer training approach.

Finally, we conclude the manuscript with a discussion of future Acoustilytix software development plans and how Acoustilytix might be incorporated into and affect the USV research landscape.

2. Acoustilytix Program Structure

In this section, we provide a high-level description of the Acoustilytix program structure. Acoustilytix is a web-based platform that allows USV researchers to easily upload wav files; to trigger automated USV call detection and classification; and to output call classifications, counts, and a number of acoustic characteristics of each call to a csv file for the users' subsequent analysis.

2.1. *Acoustilytix Web-Based User Interface and Features*

An overriding aim of the Acoustilytix web-based interface is to offer a user-friendly experience for USV researchers. Based on extensive discussions with USV consortium members, the website includes separate pages or drop-down menu functions for wav file uploading, sorting, and tracking of wav files; selecting a USV classification scheme (e.g., the 14 call types identified in Wright, et al., 2010) [42]; initiating the automatic detection; manual verification of automated call detection and classification; and download of tabulated results to a csv file. We briefly summarize each of these functions below.

2.1.1. User Interface

Once a research group gains access to Acoustilytix, the principal investigator (PI) in a laboratory creates a username and password, is assigned their own research group, and receives "administrative" privileges. Once logged in to the Acoustilytix platform, a user can select the "upload file" feature and can either drag and drop or use an upload button to select a wav file for upload to the research group. All research group members can access and score files that have been uploaded by any member of the group. Files can be sorted in any number of ways, for example, alphabetically by file name, by file uploader, or by file scorer. In the current version of the platform, users can select one of the default USV classification schemes from a drop-down menu. (As detailed in the General Discussion section, we are currently building the capability of allowing a researcher to create their own call classification scheme. A large number of hand-scored calls from this scheme will be uploaded to Acoustilytix and will be used to train an automated call classification model.) Once a call classification scheme is selected, it will be applied for all subsequent files until a new call classification scheme is selected. A manual verification process has been implemented that allows researchers to review the automated call classification for any calls they choose. Once a wav file is scored by Acoustilytix, it is available for download as a csv. The file will include the start and stop time, user selected call type classification, and a large number of acoustic parameters associated with each call.

2.1.2. Acoustilytix Features

The two primary features of Acoustilytix are automated call detection and call classification. Because different laboratories or individuals within a laboratory might implement a different call classification scheme, users can select unique call classification schemas to be used during the automated call classification and/or manual verification process.

Acoustilytix automatically begins parsing the audio file and detecting calls upon file upload. Automatic call detection requires three distinct processes—initial detection and file parsing, call isolation and detection confidence, and call parameterization.

Initial Detection and File Parsing: Two of the primary guiding tenets of the Acoustilytix automated detection algorithm are: (1) most of the recording time and frequency bins can be attributed to background noise and (2) USVs are connected in time, frequency, acoustic power, or some combination of these. Thus, Acoustilytix begins by estimating an overall measure of the background noise and identifies likely calls as those segments in the file with connected time, frequency, acoustic power, or some combination that is distinct from background noise. Possible calls are expanded by several milliseconds to capture both ends of any potential calls. These expanded segments are organized into a list where any overlapping segments are coalesced together. These segments with potential calls are then passed into the call isolation module. A detailed explanation of this process is included in Appendix A.

Call Isolation and Detection Confidence: To isolate calls and differentiate them from background and noise, Acoustilytix uses a combinatorial approach applying several filters and algorithms. These are combined in various ways to differentiate the call from the background and to assign a detection confidence value (percent certainty). The details can be found in Appendix B.

Parameterization: Once the calls have been isolated, descriptive parameters can be calculated; the current version of Acoustilytix calculates over 200 parameters for each call. These include parameters such as maximum, minimum, and median frequency; maximum acoustic power (and frequency at which this occurs); and median acoustic power along the call maximal ridge. A review of commonly referenced call types for both rat and mice reveals that temporal variations, i.e., slopes, jumps, etc., appear indicative of many call types [14,42]. Using the ridges from the isolation step as an initial guide, many temporal-varying parameters are also calculated, including max, min, median, and standard deviation of slopes for not only the whole call but also the beginning, middle, and end of the call. Jumps and gaps are also tabulated, as well as a few measures of how quickly slopes change over time (slope variability), which relates to "trill" type calls. The current iteration of the software outputs a sub-set of these parameters in the csv file for researchers. We fully expect the use of these parameters in USV research will become more common [29,30,43].

Manual verification of automated call detection is not a mandatory step in the Acoustilytix process. Even so, and especially since software development is ongoing, we urge researchers to incorporate manual verification into their workflow. During software development, the CRG and UT teams worked iteratively to ensure that the manual verification process was easy to implement and streamlined. The result of this development effort is that the platform currently offers three data formats for exploring and interacting with the data in each file-card, tile, and table. Briefly, the "card" format presents a single segment/call combination with the ability for users to play back the audio segment while a line moves through a responsive spectrogram. From this screen, users can modify the call bounds, mark an auto-detected call as a false positive, or provide a USV classification type. The "tile" format provides an infinite scrolling list of the segment and call combinations for a broader overview. The call parameters and any user-selected call classification are shown on the tile. Tiles can be filtered to include all calls or only calls above a threshold detection confidence (i.e., percent certainty) for detection. Finally, the "table" format provides parameterization details for each of the calls, as well as any user added comments or classifications. These data are able to be downloaded as a csv file for each scoring session. Scoring data are organized in "scoring sessions" with data linked to the user, wav file, and call classification scheme. Progress for each scoring session is tracked, and users can easily resume scoring where they left off. In addition, each file can be scored separately by multiple users or using multiple classification schemes. Multiple users can score the same file, and the results can be easily compared to assess inter-rater reliability.

3. Acoustilytix Validation Results

3.1. Acoustilytix: Detection Statistics

To generate detection statistics, the output from Acoustilytix was compared to compiled data from hand-scored USV files and to the output from DeepSqueak. The time stamps were matched between datasets, and then, calls from the Acoustilytix or DeepSqueak dataset were given one of the following three designations:

- Detected: These are defined as calls detected by hand-scorers and by Acoustilytix (or DeepSqueak).
- Missed: These are defined as calls detected by hand-scorers but not detected by Acoustilytix (or DeepSqueak).
- False Positive: These are defined as calls not detected by hand-scorers that were detected by Acoustilytix (or DeepSqueak).

Two metrics were used to quantify the success of Acoustilytix (or DeepSqueak) at matching the hand-scored data.

- Sensitivity = Detected/(Detected + Missed): This is a measure of how many of the hand-scored calls are correctly detected by Acoustilytix (or DeepSqueak).
- Precision = Detected/(Detected + False Positive): This is a measure of how many hand-scored calls Acoustilytix (or DeepSqueak) detects but also the number detected that are not detected by the hand-scorers.

We examined the sensitivity and precision from different experimental conditions across different recording environments. These results are summarized in Table 1.

Table 1. Sensitivity and precision of Acoustilytix and DeepSqueak for four experimental conditions across four recording environments.

			Sensitivity			Precision		
Recording Environment	**Brief Description of Study Manipulation**	**# of Hand-Score Detected USVs**	**Acoustilytix**	**DeepSqueak**	***p*-Value**	**Acoustilytix**	**DeepSqueak**	***p*-Value**
1	Cocaine	8552	92.5	91.5	0.016	74.5	42.0	<0.00001
2	Ethanol	1720	94.0	92.1	0.029	72.0	65.8	0.00008
3	Morphine	5771	90.6	75.8	<0.00001	83.6	33.1	<0.00001
4	Sex	3462	96.4	97.5	0.0078	59.5	46.1	<0.00001

For Acoustilytix, the sensitivity (accurate detection of calls compared to hand-scored) was above 90% in all cases, with an average of 92.7%. This means that Acoustilytix correctly detected over 90% of the calls noted via hand-scoring in a variety of experimental and recording conditions without any user intervention or tuning. For DeepSqueak, the sensitivity was 87.7% on average for 18,770 USV calls. Compared with Acoustilytix, DeepSqueak's sensitivity had a *p*-value less than 0.00001, which showed a significant difference.

Precision is a critical metric for automated detection success, as the number of false positive USVs should be minimized. As the results in Table 1 suggest, precision was lower than sensitivity in all cases, ranging from 59.5–83.6% with a weighted average of 73.2%. DeepSqueak had an average precision of 41.0% with a *p*-value less than 0.00001 when compared with Acoustilytix's 73.2% average precision.

In developing and evaluating the USV detection algorithm, we realized that some metric of call detection certainty (or confidence) could be of value to USV researchers. We developed such an algorithm, and these values are part of the csv output file. The details of the algorithm are being evaluated as potential intellectual property (IP) and, thus, cannot be presented in detail here. At a high level, call detection certainty is a function of acoustic power of the parsed signal relative to the background power, with greater acoustic power for the parsed signal relative to the background being associated with greater call detection certainty.

In the next iteration of Acoustilytix, we plan to incorporate a new feature into the detection workflow that will allow the USV researcher to set a threshold on the call detection certainty so that all calls below the research-selected threshold will be automatically made available for manual verification.

3.2. Acoustilytix: Call Classification Statistics

The goal of this section is to report the progress toward developing an automated USV call classification machine learning algorithm. Our original goal was to develop an automated USV call classification machine learning algorithm that could accurately classify 20 or more unique call types that are used in the literature. These might include the 15 call types identified by Wright et al. [42] but also more complex waveforms, including dysphonation and frequency shifts, which may have particular relevance to translational research [44,45]. After some initial attempts, this was not deemed feasible in a reasonable amount of time. In addition, and based on discussions with our USV consortium members, we were informed that researchers were more interested in having the ability to upload their own hand-scored USV files that could be used to train an automated scoring algorithm that they could use exclusively.

The first step in achieving this aim was to identify which of many machine learning algorithms to implement. Although a number of machine learning algorithms were explored, we settled upon a random forest algorithm whose parameters were estimated from hand-scored USV call classification data. For the initial validation test, we wanted to balance the need to classify a broad range of calls with the need to have the statistical power to train and test the model. We settled on a five-call classification scheme that has been used by one of our consortium members and represents a composite of an existing call classification scheme by Wright et al. [42]. The mapping from the Wright categories to the five-call composite is presented in Table 2. Representative call spectrograms are displayed in Figure 1. It is important to note that we are not arguing that this is the best call classification scheme to be used in translational research. Rather, the goal is to verify that a random forest algorithm can be used to build an automated hand-scoring tool with reasonable accuracy.

Table 2. Call types in the Wright et al. [42] classification scheme and their mapping to the five-call composite with representative spectrograms from Acoustilytix.

Wright et al. [42]	Five-Call Composite
Flat	Fixed Frequency 50
Short	
Upward Ramp	Frequency Modulated 50
Downward Ramp	
Split	
Step Up	
Step Down	
Multi-step	
Inverted-U	
Complex	
Composite	
Trill	Frequency Modulated with Trills 50
Flat-Trill Combo	
Trill with jumps	
22-kHz call	Long 22-kHz
	Short 22-kHz (>300 ms)

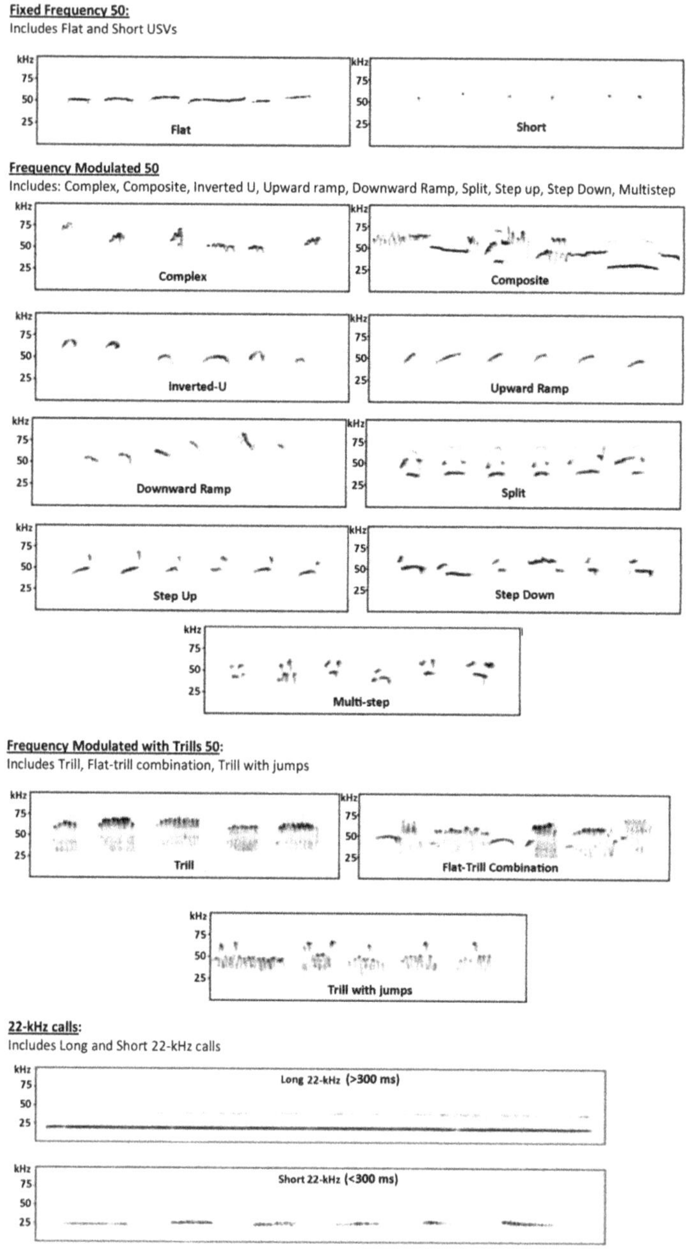

Figure 1. Representative spectrograms for each of the call types in Table 2.

Results

A machine learning algorithm was trained using a supervised random forest classification model where Dr. Duvauchelle's most experienced rater's call classifications were used as the supervised teaching signal for the model. The dataset consisted of 931 rat calls. Table 3 shows the number of each call type in the dataset.

Table 3. Number of each of five call types in the five-call composite dataset classified by an expert scorer.

Call Types	Count
Fixed Frequency 50	202
Frequency Modulated 50	342
Frequency Modulated with Trill 50	132
Long 22-kHz	184
Short 22-kHz	71

The experimental dataset was split into a training dataset consisting of 70% of the calls and a test dataset with the remaining 30% to validate the model's performance and to prevent overfitting. The model features consisted of the parameters that were calculated by the call detection algorithm discussed in the Parameterization section of this paper. Feature selection methods were employed to down-select the number of predictors in the model and to prevent overfitting. A grid-based approach was used to tune the random forest model parameters and to identify the model that yielded the highest test data accuracy.

We began the evaluation process by tallying the number of calls of each type that fall in each of the following four categories:

- True Positive (TP): The hand-scorer selected call type "A", and the classifier selected call type "A". The hand-scorer and classifier agreed that call type "A" was present.
- True Negative (TN): The hand-scorer did not select call type "A", and the classifier also did not select call type "A." The hand-scorer and classifier agreed that call type "A" was absent.
- False Positive (FP): The hand-scorer did not select call type "A", but the classifier selected call type "A." The hand-scorer and classifier disagreed and the hand-scorer's call type selection was considered the true call type.
- False Negative (FN): The hand-scorer selected call type "A", but the classifier did not select call type "A." The hand-scorer and classifier disagreed and the hand-scorer's call type selection was considered the true call type.

These were then used to compute the five model performance metrics displayed in Table 4: These include precision, recall, F1-score, support, and accuracy, which are calculated from the TP, TN, FP, and FN values as follows. The F1-Score is the harmonic mean of precision and recall and gives a better measure of the incorrectly classified cases than the Accuracy metric.

- Precision = TP/(TP + FP)
- Recall = TP/(TP + FN)
- F1-score = 2 * (Precision * Recall)/(Precision + Recall)
- Support = count
- Accuracy = (TP + TN)/Total Population

Table 4. Evaluation of the five-call composite classification model performance on the test dataset.

Five-Call Type	Precision	Recall	F1-Score	Support
Fixed Frequency 50	0.77	0.69	0.73	49
Frequency Modulated 50	0.73	0.87	0.80	12
Frequency Modulated with Trill 50	0.81	0.62	0.70	42
Long 22-kHz	0.92	0.87	0.89	53
Short 22-kHz	0.82	0.75	0.78	24
Overall Accuracy/Support (N-size)			**0.79**	**280**

Precision, recall, F1-score, and accuracy fall between 0 and 1 where values close to 1 indicate strong model performance. Table 4 shows good performance on all of the five call types across all performance metrics with an average accuracy of 79%, which is well above chance level (20%).

For comparison, Table 5 shows how well the model performed on the training data.

Table 5. Evaluation of the five-call composite classification model performance on the training dataset.

Five-Call Type	Precision	Recall	F1-Score	Support
Fixed Frequency 50	0.89	0.80	0.84	153
Frequency Modulated 50	0.82	0.93	0.87	230
Frequency Modulated with Trill 50	0.92	0.81	0.86	90
Long 22-kHz	0.92	0.92	0.92	131
Short 22-kHz	0.86	0.77	0.81	47
Overall Accuracy/Support (N-size)			**0.87**	**651**

Again, model performance was quite good on all performance metrics with an accuracy for the training dataset of 87%. We used a kappa statistic to compare the model to the expert hand-scorer's full data set. Kappa statistics range from 0 to 1 with higher values denoting greater agreement. The model yielded a kappa = 0.79, suggesting strong agreement between the model and the expert hand-scorer. As new data are scored, the model fit will continue to improve and further optimize (We also fit the model with over 30,000 hand-scored calls from a team of six hand-scorers. Seventy percent of the data were used to train the model and 30% to test the model. Test data accuracy was 71% as compared to the 79% with the single expert hand-scorer, and training accuracy was 88% as compared to 87% with the single expert hand-scorer.).

3.3. *Acoustilytix Validation Results General Summary and Future Directions*

In this section we evaluated the effectiveness of Acoustilytix as an automated USV call detection tool and compared it with the currently popular DeepSqueak algorithm. We also examined the ability of a random forest machine learning algorithm to serve as the foundation for automated call classification. We gained a number of important insights from this evaluation.

First, our automated USV detection algorithm was not significantly impacted by background noise. Background noise can obfuscate USVs from both hand-scorers and existing automated call classification solutions. Acoustilytix's proprietary detection algorithm maintained excellent performance even in the presence of background noise. In fact, we showed that USVs missed by hand-scorers during initial scoring were subsequently identified by Acoustilytix and confirmed as true calls by hand-scorers. With no user intervention or tuning, Acoustilytix achieved 93% sensitivity (a measure of how accurately Acoustilytix detects true calls) and 73% precision (a measure of how accurately Acoustilytix avoids false positives) in call detection accuracy across four unique recording environments. Critically, Acoustilytix also generated a call detection certainty metric that reflected the confidence in the model call detection decision. In future iterations of Acoustilytix, we plan to incorporate this into the optional manual call verification process so that USV researchers have the option to manually verify any calls with a detection certainty below some threshold selected by the USV researcher.

Second, Acoustilytix had significantly better sensitivity and precision than DeepSqueak (sensitivity = 88%; precision = 41%) across the four studies analyzed, which included four different environments. DeepSqueak had a high false positive rate across the four studies, which posed a challenge to hand-scoring.

Third, our automated USV call classification algorithm performed quite well on the five-call composite outlined in Table 2. While there is room for continuous improvement, and we have a plan for implementing this (see future directions), Acoustilytix accurately classified 79% of the hand-scored data from nearly 1000 calls classified by an expert into the five-call types. This is well above chance (20%). As additional data are uploaded to the Acoustilytix platform, and as hand-scorers modify or endorse the Acoustilytix classification, model parameters will continue to be optimized in the interest of increased accuracy.

Fourth, and based on conversations with our USV consortium members, it is clear that they prefer an automated call classification platform that allows them to upload a large number of hand-scored USV files that could be used to train an automated scoring algorithm that they could use exclusively. Our progress with the random forest algorithm displayed in Tables 4 and 5 suggest that this feature can be effectively implemented in the next iteration of Acoustilytix.

Fifth, and perhaps most exciting, is the fact that over 200 acoustic parameters associated with each detected USV are being computed and can be outputted in the csv files. As many research groups have shown, including our own [14,30,46–51], the USV acoustic parameter values are highly predictive of a number of addiction and other health related outcomes. It is very likely that the ease of acquisition of these parameters when using Acoustilytix will quicken the pace of research focused on acoustic parameters.

4. Acoustilytix Hand-Scoring Training Feature

As outlined above, the need for highly proficient, fast, and consistent hand-scoring is a clear bottleneck for the use of USVs in research. Few USV researchers are satisfied with their current hand-scoring training procedures, and most report their desire for an automated approach to hand-scorer training.

To accurately hand-score a USV, the scorer must detect the call as distinct from noise in a spectrogram, then, must visually inspect the USV and listen to it to accurately classify the call. This mental representation must then be compared with either visual and auditory mental representations of prototypical calls of each type or actual visual and auditory spectrograms of prototypical calls from each type. The scorer must then assign the call to a USV category.

The best way to train this type of complex visual/auditory matching task is to have a trainee inspect a potential call, classify it, then receive immediate feedback from an expert hand-scorer on the correctness of their classification. This process should be repeated several hundred times to build expertise. Unfortunately, this is time-consuming and costly and, thus, rarely occurs. More commonly, a hand-scorer in training is shown examples along with verbal descriptions of each call type. They are then asked to score a large number of calls (e.g., several hundred), and they only receive corrective feedback after the fact. This is suboptimal, as the best time to strengthen or weaken the neural connection between the stimulus (call to classify) and response (classification) is to provide feedback immediately following the classification [40,41,52].

We recently developed a new function in Acoustilytix to automate the hand-scoring training process that builds upon what is known about learning to optimize the speed, accuracy, and consistency of the resulting hand-scoring capability. This function includes an initial USV call familiarization phase in which a hand-scorer in training can view and listen to spectrograms of numerous calls from any selected call category. Learners can do this at their own pace for as long as they want. This familiarization tool remains available to hand-scorers at all points in time. Once a hand-scorer is ready to practice hand-scoring, they retrieve a USV file from the Acoustilytix platform that has been "tagged" as a training file. This training file has been hand-scored by an expert from the USV laboratory. The expert simply generates a csv file that includes the correct response, uploads it to the platform, and tags it as a training file. Once the hand-scorer in training begins, the first call is presented. The hand-scorer can look at the call and use all of the tools outlined above (e.g., zoom), as well as listen to the call. "Flashcards" of prototypical calls are available for the hand-scorer to view. Once they are ready to classify, they select a call category from a drop-down menu, and they receive immediate feedback from the platform. If their classification is correct, they go on to the next call. If their classification is incorrect, they can either select a new call category, or they can view the prototypical call flashcards and can compare them with the to-be-classified call until they are ready to generate their own classification. This process continues until they classify all calls in the training file. Once complete, an accuracy rate (based only on the first call classification) is displayed. Multiple training files can be scored until the hand-scorer is proficient.

Acoustilytix: Hand-Scoring Trainer Inter-Rater Reliability

A group of five individuals were given traditional (unstructured training), with each subsequently classifying several thousand calls into the five call types described above over the course of several months (ranging from 6000–31,000 calls classified by each hand-scorer). At a later date, all of these individuals were asked to individually score a single test file with approximately 1000 calls that had been hand-scored by an expert (test following unstructured training). They were given one week to complete this task.

We used a kappa statistic to compare each hand-scorer with the expert hand-scorer. An example of the kappa statistic calculation can be found in Appendix C. The kappa statistics for the initial group of five individuals in the test following unstructured training is presented in the top portion of Table 6. Notice that the kappa values are moderate following unstructured training ranging from 0.30–0.55 with a mean of 0.42.

Table 6. Kappa statistics for experienced and novice hand-scorers relative to an expert scorer.

Hand-scorer	Test Following Unstructured Training	Test Following Acoustilytix Training	*p*-Value
	Experienced Scorers		
1	0.42	0.29	<0.00001
2	0.55	0.60	0.031
3	0.30	0.75	<0.00001
4	0.36	0.69	<0.00001
5	0.49	0.64	<0.00001
	Novice Scorers		
6	NA	0.47	NA
7	NA	0.42	NA

Next, we asked two individuals with no prior USV hand-scoring experience to complete Acoustilytix training with approximately 1000 calls, and they were given the same Acoustilytix test. The kappa statistics for the two Acoustilytix-only trained individuals are presented at the bottom of Table 6 (Hand-scorer 6–7). The kappa values were 0.42 and 0.47 with an average of 0.445. This is comparable to the average kappa value for the five extensively trained hand-scorers and, in fact, is slightly higher (0.445 vs. 0.42).

Given the success of Acoustilytix training with the two novice hand-scorers, we asked the five experienced hand-scorers to complete Acoustilytix training with approximately 2000 calls, and they were given the same Acoustilytix test. The kappa statistics for the five Acoustilytix-only trained individuals are presented in the right-most column of Table 6. The kappa values ranged from 0.30–0.75 with an average kappa value of 0.60. Four out of five experienced hand-scorers had significantly better kappa statistics after the Acoustilytix training.

These findings suggest that hand-scoring training in which the learner is given immediate call-by-call feedback, and where the teaching signal is based on the call classification from an expert hand-scorer, leads to more accurate and consistent hand-scoring. Across all seven hand-scorers, and following Acoustilytix training, the average kappa value was 0.55. Further training would likely increase this value. USV researchers could set a minimum threshold on this value if they like and deem hand-scorers "qualified" only when they exceed that threshold. Refresher training could also be incorporated to ensure consistency and accuracy of hand-scoring.

5. General Discussion and Future Directions

Ultrasonic vocalizations in rodents provide a window into emotional and neural processing. These processes are directly involved in many mental health disorders and addictions. Despite the potential of USV analysis in mental health and addiction research, one huge bottleneck exists—namely, the time-intensive, manual and costly nature of USV hand-scoring—specifically, call detection and call classification.

In this report, we provide a snapshot of our current development and testing of Acoustilytix, a web-based automated USV scoring tool that is currently environment-agnostic and implements machine learning methodology in the USV detection and classification process.

Based on discussions with our USV consortium, we identified and implemented a number of features to make the platform easy to use. These include the ability to easily upload USV files, output csv files, and the ability to manually verify or modify an automatically detected call. The initial test of the call detection algorithm in Acoustilytix was promising. With no user intervention or tuning, Acoustilytix achieved 93% sensitivity (a measure of how accurately Acoustilytix detects true calls) and 73% precision (a measure of how accurately Acoustilytix avoids false positives) in call detection accuracy across four unique recording environments and was superior to the popular DeepSqueak algorithm (sensitivity = 88%; precision = 41%). Long calls and short calls are scored separately in DeepSqueak because the functions for each algorithm work differently. While analyzing and comparing the long and short call files, we noticed that DeepSqueak can detect long and short calls that overlap either partially or fully in time. This can create confusion regarding the spectral dynamics of each USV and make analysis more challenging. Because Acoustilytix's call detection algorithm does not require separate analyses of the USV recording files, this problem is avoided. The ability to accurately detect USVs without any user intervention or tuning addresses a major barrier to broad adoption of USV models in addiction and mental health research. Once we incorporate the quantitative metric of call detection certainty into the user facing Acoustilytix dashboard, researchers will be able to set a threshold for manual verification that is fast, accurate, and flexible for users.

We also examined USV call classification accuracy in a five-call composite classification scheme derived from the Wright et al. [42] call categories. Automated call classification accuracy using a random forest machine learning algorithm was 79% for the five-call composite. These accuracy rates are well above chance, but as additional data are uploaded to the Acoustilytix platform, and as hand-scorers modify or endorse the Acoustilytix classification, model parameters will continue to be optimized in the interest of increased USV call classification accuracy. The validation of the random forest model sets the stage for the next iteration of the automated USV call classification algorithm. As requested by our USV consortium members, we plan to incorporate a new feature that allows researchers to upload a large number of hand-scored USV files that could be used to train an automated scoring algorithm that they could use exclusively.

Perhaps most exciting is the fact that over 200 acoustic parameters associated with each detected USV are being computed and can be outputted in the csv files. As our research group and others have shown [31,37,42,53–59], the USV acoustic parameters are highly predictive of a number of addiction and other health related outcomes. We expect that the ease of acquisition of these parameters when using Acoustilytix will quicken the pace of research focused on acoustic parameters.

We also introduced a recently developed feature of Acoustilytix that offers a fast and effective way to train hand-scorers and is built upon a foundation of learning science [40,41]. Critically, trainees are presented with several hundred USVs one at a time, are asked to classify the call, and receive immediate corrective feedback based on the hand-scoring from an expert in the lab. We tested this approach in seven novice hand-scorers and showed that their post training call classifications were highly correlated with the expert (inter-rater reliability: kappa statistic ranged from 0.30–0.75, average = 0.55). This approach increases the likelihood that all hand-scorers in the laboratory mimic the scoring behavior of the expert, thus, increasing consistency and reliability of hand-scoring.

We now briefly discuss some future directions and further refinements planned for Acoustilytix.

Call Detection: With no user intervention or tuning, we were able to achieve high levels of call detection accuracy that were superior to the popular DeepSqueak algorithm. Even so, we are exploring the application of machine learning models in the USV call detection process to further eliminate false positives. As more USV wav files from other research groups are uploaded to the platform and are run through the detection algorithm, this model can be further tuned to provide better separation of valid calls from false positives.

Call Classification: Although automated call classification for the five-call type classification scheme was well above chance, there is still room for improvement. Having explored numerous machine learning types, we are confident that the supervised random forest approach is a good approach, but we need a significantly larger pool of data with hand-scored calls from experts to further improve the predictions from the algorithm. This process is ongoing. As we outline in this manuscript, we began with an expert hand-scorer scoring 1000 calls or so. We used that to train the model. Every time the expert scored a new file, that triggered a model refit so that the fit improved based on the larger set of data. Users can use the model developed from our expert, or they can use their own expert to train the model on a specific call classification scheme. Ultimately, we will use all of these data to train the model using expert scorers from multiple research groups using the same classification scheme. This is beneficial to the whole USV community because it will lead to more uniformity and consistence in scoring within and across research groups, making the research findings more reliable and reproducible.

Acoustic Parameters: Several research groups [10,14,42,46–51,53,54,56,57,59,60], including our own [29–32,36–39,43,61–64], have shown that acoustic parameters are predictive of addiction and mental health related outcomes. For example, our own research group has shown that acoustic parameter values are predictive of whether a call is emitted by a high vs. low alcohol drinking rat or whether that rat is male or female [37,43], as well as a number of other interesting distinctions. Acoustilytix currently computes over 200 acoustic parameters for each call that can be analyzed by USV researchers. Future work should consider applying multi-dimensional similarity-based approaches to create novel call classification schemes that are strictly data driven, in addition to exploring the predictive nature of specific or groups of parameters.

6. Conclusions

This report offers a snapshot of the development of a web-based automated USV detection and classification software solution called Acoustilytix. Acoustilytix accelerates throughput of USV scoring by automating the call detection and classification process. Acoustilytix achieves high detection accuracy across four distinct environments without manual tuning or calibration and is superior to the currently popular DeepSqueak algorithm in sensitivity for three out of four environments and superior in precision in all four environments. Automated call classification was adequate yielding 79% accuracy, but more work is needed. A feature that allows researchers to upload a large number of hand-scored USV files that could be used to train an automated scoring algorithm that they could use exclusively is in demand and under development. The development of an automated hand-scoring feature is innovative and has been well received by the USV community. Taken together, Acoustilytix has the potential to speed the process of USV detection and classification in the interest of developing and testing animal models of mental health and addiction.

Author Contributions: Conceptualization: C.B.A., R.D.S., J.E.S., C.L.D., and W.T.M.; methodology: C.B.A., R.D.S., J.E.S., C.L.D., and W.T.M.; validation: C.B.A., R.D.S., J.E.S., C.C., N.M., S.F., J.B., M.D.N., S.I.S., B.N., L.L., A.R.C., and C.L.D.; formal analysis: C.B.A., R.D.S., and J.E.S.; investigation: C.B.A., R.D.S., J.E.S., C.C., N.M., W.T.M., and C.L.D.; resources: R.D.S. and C.L.D.; data curation: C.B.A., R.D.S., J.E.S., C.C., S.F., W.T.M., and C.L.D.; writing—review and editing: C.B.A., R.D.S., N.M., W.T.M., and C.L.D.; supervision: C.B.A., R.D.S., C.C., W.T.M., and C.L.D.; project Administration: C.B.A., R.D.S., C.C., W.T.M., and C.L.D.; funding acquisition: R.D.S. and C.L.D. All authors have read and agreed to the published version of the manuscript.

Funding: This research was funded by NIMH of the National Institutes of Health under award number R41MH121119.

Institutional Review Board Statement: Not applicable.

Informed Consent Statement: Not applicable.

Data Availability Statement: Not applicable.

Acknowledgments: The content is solely the responsibility of the authors and does not necessarily represent the official views of the National Institutes of Health.

Conflicts of Interest: The authors declare no conflict of interest.

Appendix A. Details of Initial Detection and File Parsing

To determine where a call is most likely to exist, a coarse spectrogram (STFT) of the entire file is calculated. This coarse spectrogram is processed to account for differences due to the recording setup, which can result in varying acoustic power with respect to frequency bins. A maximum acoustic power is then derived from this processed spectrogram at each time interval, resulting in a one-dimensional (1D) signal. To determine where possible calls are located, calls are treated as outliers in this 1D signal.

Appendix B. Call Isolation and Detection Confidence

To isolate calls and differentiate them from background and noise, Acoustilytix uses a combinatorial approach, applying several filters and algorithms. The details of the algorithm are being evaluated as potential IP and, thus, cannot be presented in detail here. These filters are combined in various ways to extract the calls out of the background and provide a detection confidence value (percent certainty)

Appendix C. Example Kappa Statistic Calculation

In calculating the kappa statistic between two scorers, USVs that Acoustilytix missed were not included in the calculation. We were unable to include user-added calls in the analysis because the expert scorer did not add calls when analyzing calls in some recording files. Acoustilytix, at times, detects false USVs. Both scorers can assign the "false positive" category to these USVs. The false positive category was included in the kappa statistic calculation. The method used to calculate the kappa statistic can be found in the University of Nebraska-Lincoln Bivariate Statistics Hand-Computation Cache (2021, February). To begin calculating the kappa statistic, a confusion matrix is generated. The confusion matrix is a 5×5 matrix where each row represents one of the five call types, and each column represents the same five call types. The rows denote the classification for the individual hand-scorer and the columns for the expert hand-scorer. For each call classified, the value in one cell of the 5×5 confusion matrix is incremented. For example, if both the individual and expert classify a call as a Short 22-kHz then the cell in the confusion matrix that represents Short 22-kHz for the individual hand-scorer and Short 22-kHz for the expert is incremented by 1. This process is repeated until all calls are represented in the confusion matrix.

Table A1 shows an example of a confusion matrix for 748 rat calls that were each scored by one individual and by the expert.

Table A1. Example of a confusion matrix comparing call types between two raters.

		Rater 1					
		Fixed Frequency 50	**Frequency Modulated 50**	**Frequency Modulated with Trill 50**	**Long 22**	**Short 22**	**Row Totals**
Rater 2	Fixed Frequency 50	110	75	3	0	1	189
	Frequency Modulated 50	32	305	16	0	0	353
	Frequency Modulated with Trill 50	6	126	60	0	0	192
	Long 22	2	1	0	1	0	4
	Short 22	0	0	0	0	10	10
	Column Totals	**150**	**507**	**79**	**1**	**11**	**748**

Interpreting the confusion matrix is straightforward. On-diagonal elements represent calls where the two raters are in agreement, and off-diagonal elements represent calls where the two raters are not in agreement. For example, Rater 1 and Rater 2 agreed that 110 calls were the "Fixed Frequency 50" call type, but Rater 1 assigned 75 calls to the "Frequency Modulated 50" call type that Rater 2 assigned to "Fixed Frequency 50."

The kappa statistic calculation uses the row sums, column sums, and overall call count. These are shown in the final column and final row in the example in Table A1. The overall total is the same whether you sum the row totals or column totals and is 748 in this example.

Next, the total number of agreements are found by summing the values in the diagonal cells of the confusion matrix. In this example,

$$\Sigma\ agreements = \Sigma\ a = 110 + 305 + 60 + 1 + 10 = 486$$

The expected frequency for the number of agreements is now calculated for each agreement along the diagonal by multiplying the row total by the column total and dividing by the overall total (see Table A2). For example, row #1 multiplied by column #1 and divided by the overall total is:

$$expected\ frequency = ef = 189 * 150/748 = 37.90$$

Table A2. Expected frequencies calculated from confusion matrix in Table A1.

	Expected Frequency
Fixed Frequency 50	37.90
Frequency Modulated 50	239.27
Frequency Modulated with Trill 50	20.28
Long 22	0.0053
Short 22	0.1470

The sum of the expected frequencies of agreement by chance is the sum of the expected frequencies. For example:

$$\Sigma\ expected\ frequencies = \Sigma\ ef = 37.90 + 239.27 + 20.28 + 0.0053 + 0.1470 = 297.60$$

The kappa statistic can now be calculated with the Formula:

$$K = (\Sigma\ a - \Sigma\ ef)/(N - \Sigma\ ef) = (486 - 297.60)/(748 - 297.60) = 0.418$$

References

1. Tóth, M.; Little, P.; Arnberg, F.; Häggkvist, J.; Mulder, J.; Halldin, C.; Gulyás, B.; Holmin, S. Acute neuroinflammation in a clinically relevant focal cortical ischemic stroke model in rat: Longitudinal positron emission tomography and immunofluorescent tracking. *Brain Struct. Funct.* **2016**, *221*, 1279–1290. [CrossRef] [PubMed]
2. Homberg, J.R.; Kyzar, E.J.; Nguyen, M.; Norton, W.H.; Pittman, J.; Poudel, M.K.; Gaikwad, S.; Nakamura, S.; Koshiba, M.; Yamanouchi, H.; et al. Understanding autism and other neurodevelopmental disorders through experimental translational neurobehavioral models. *Neurosci. Biobehav. Rev.* **2016**, *65*, 292–312. [CrossRef] [PubMed]
3. Stewart, A.M.; Kalueff, A.V. Developing better and more valid animal models of brain disorders. *Behav. Brain Res.* **2015**, *276*, 28–31. [CrossRef] [PubMed]
4. Kalueff, A.V.; Stewart, A.M. Modeling neuropsychiatric spectra to empower translational biological psychiatry. *Behav. Brain Res.* **2015**, *276*, 1–7. [CrossRef]
5. Czéh, B.; Fuchs, E.; Wiborg, O.; Simon, M. Animal models of major depression and their clinical implications. *Prog. Neuropsychopharmacol. Biol. Psychiatry* **2016**, *64*, 293–310. [CrossRef]
6. Brudzynski, S.M. Ultrasonic calls of rats as indicator variables of negative or positive states: Acetylcholine-dopamine interaction and acoustic coding. *Behav. Brain Res.* **2007**, *182*, 261–273. [CrossRef]
7. Brudzynski, S.M. Ethotransmission: Communication of emotional states through ultrasonic vocalization in rats. *Curr. Opin. Neurobiol.* **2013**, *23*, 310–317. [CrossRef]
8. Burgdorf, J.; Wood, P.L.; Kroes, R.A.; Moskal, J.R.; Panksepp, J. Neurobiology of 50-kHz ultrasonic vocalizations in rats: Electrode mapping, lesion, and pharmacology studies. *Behav. Brain Res.* **2007**, *182*, 274–283. [CrossRef]
9. Brudzynski, S.M. Communication of adult rats by ultrasonic vocalization: Biological, sociobiological, and neuroscience approaches. *ILAR J.* **2009**, *50*, 43–50. [CrossRef]
10. Bihari, A.; Hrycyshyn, A.W.; Brudzynski, S.M. Role of the mesolimbic cholinergic projection to the septum in the production of 22 kHz alarm calls in rats. *Brain Res. Bull.* **2003**, *60*, 263–274. [CrossRef]
11. Brudzynski, S.M. Pharmacological and behavioral characteristics of 22kHz alarm calls in rats. *Neurosci. Biobehav. Rev.* **2001**, *25*, 611–617. [CrossRef]
12. Thompson, B.; Leonard, K.C.; Brudzynski, S.M. Amphetamine-induced 50 kHz calls from rat nucleus accumbens: A quantitative mapping study and acoustic analysis. *Behav. Brain Res.* **2006**, *168*, 64–73. [CrossRef]
13. Grant, L.M.; Richter, F.; Miller, J.E.; White, S.A.; Fox, C.M.; Zhu, C.; Chesselet, M.-F.; Ciucci, M.R. Vocalization deficits in mice over-expressing alpha-synuclein, a model of pre-manifest Parkinson's disease. *Behav. Neurosci.* **2014**, *128*, 110–121. [CrossRef]
14. Portfors, C.V. Types and functions of ultrasonic vocalizations in laboratory rats and mice. *J. Am. Assoc. Lab. Anim. Sci.* **2007**, *46*, 28–34.
15. Brudzynski, S.M. *Handbook of Ultrasonic Vocalization*; Brudzynski, S.M., Ed.; Academic Press: Cambridge, MA, USA, 2018; ISBN 9780128096000.
16. Burgdorf, J.; Moskal, J.R.; Brudzynski, S.M.; Panksepp, J. Rats selectively bred for low levels of play-induced 50 kHz vocalizations as a model for autism spectrum disorders: A role for NMDA receptors. *Behav. Brain Res.* **2013**, *251*, 18–24. [CrossRef]
17. Clinton, S.M.; Watson, S.J.; Akil, H. High novelty-seeking rats are resilient to negative physiological effects of the early life stress. *Stress* **2014**, *17*, 97–107. [CrossRef]
18. Scattoni, M.L.; Crawley, J.; Ricceri, L. Ultrasonic vocalizations: A tool for behavioural phenotyping of mouse models of neurodevelopmental disorders. *Neurosci. Biobehav. Rev.* **2009**, *33*, 508–515. [CrossRef]
19. Shin, S.; Pribiag, H.; Lilascharoen, V.; Knowland, D.; Wang, X.-Y.; Lim, B.K. Drd3 Signaling in the Lateral Septum Mediates Early Life Stress-Induced Social Dysfunction. *Neuron* **2018**, *97*, 195–208.e6. [CrossRef]
20. Silverman, J.L.; Yang, M.; Lord, C.; Crawley, J.N. Behavioural phenotyping assays for mouse models of autism. *Nat. Rev. Neurosci.* **2010**, *11*, 490–502. [CrossRef]
21. Steele, F.F.; Whitehouse, S.C.; Aday, J.S.; Prus, A.J. Neurotensin NTS1 and NTS2 receptor agonists produce anxiolytic-like effects in the 22-kHz ultrasonic vocalization model in rats. *Brain Res.* **2017**, *1658*, 31–35. [CrossRef]
22. Wöhr, M. Ultrasonic vocalizations in Shank mouse models for autism spectrum disorders: Detailed spectrographic analyses and developmental profiles. *Neurosci. Biobehav. Rev.* **2014**, *43*, 199–212. [CrossRef]
23. Wöhr, M.; Silverman, J.L.; Scattoni, M.L.; Turner, S.M.; Harris, M.J.; Saxena, R.; Crawley, J.N. Developmental delays and reduced pup ultrasonic vocalizations but normal sociability in mice lacking the postsynaptic cell adhesion protein neuroligin2. *Behav. Brain Res.* **2013**, *251*, 50–64. [CrossRef]
24. Yang, M.; Loureiro, D.; Kalikhman, D.; Crawley, J.N. Male mice emit distinct ultrasonic vocalizations when the female leaves the social interaction arena. *Front. Behav. Neurosci.* **2013**, *7*, 159. [CrossRef]
25. Faure, A.; Nosjean, A.; Pittaras, E.; Duchêne, A.; Andrieux, A.; Gory-Fauré, S.; Charvériat, M.; Granon, S. Dissociated features of social cognition altered in mouse models of schizophrenia: Focus on social dominance and acoustic communication. *Neuropharmacology* **2019**, *159*, 107334. [CrossRef]
26. Knutson, B.; Burgdorf, J.; Panksepp, J. High-frequency ultrasonic vocalizations index conditioned pharmacological reward in rats. *Physiol. Behav.* **1999**, *66*, 639–643. [CrossRef]

27. Prus, A.J.; Hillhouse, T.M.; LaCrosse, A.L. Acute, but not repeated, administration of the neurotensin NTS1 receptor agonist PD149163 decreases conditioned footshock-induced ultrasonic vocalizations in rats. *Prog. Neuropsychopharmacol. Biol. Psychiatry* **2014**, *49*, 78–84. [CrossRef]
28. Saunders, J.A.; Tatard-Leitman, V.M.; Suh, J.; Billingslea, E.N.; Roberts, T.P.; Siegel, S.J. Knockout of NMDA receptors in parvalbumin interneurons recreates autism-like phenotypes. *Autism Res.* **2013**, *6*, 69–77. [CrossRef]
29. Duvauchelle, C.L.; Maddox, W.T.; Reno, J.M.; Thakore, N.; Mittal, N.; Cormack, L.K.; Bell, R.L.; Gonzales, R.; Schallert, T. Alcohol-Preferring Rats and 22-kHz Negative-Affect Ultrasonic Vocalizations. In *Handbook of Behavioral Neuroscience*; Brudzynski, S.M., Ed.; Academic Press: Cambridge, MA, USA, 2018; Volume 25, pp. 401–411.
30. Mittal, N.; Todd Maddox, W.; Schallert, T.; Duvauchelle, C.L. Rodent ultrasonic vocalizations as biomarkers of future alcohol use: A predictive analytic approach. *Cogn. Affect. Behav. Neurosci.* **2018**, *18*, 88–98. [CrossRef] [PubMed]
31. Mittal, N.; Maddox, W.T.; Schallert, T.; Duvauchelle, C.L. Spontaneous Ultrasonic Vocalization Transmission in Adult, Male Long-Evans Rats Is Age-Dependent and Sensitive to EtOH Modulation. *Brain Sci.* **2020**, *10*, 890. [CrossRef] [PubMed]
32. Reno, J.M.; Marker, B.; Cormack, L.K.; Schallert, T.; Duvauchelle, C.L. Automating ultrasonic vocalization analyses: The WAAVES program. *J. Neurosci. Methods* **2013**, *219*, 155–161. [CrossRef] [PubMed]
33. Barker, D.J.; Herrera, C.; West, M.O. Automated detection of 50-kHz ultrasonic vocalizations using template matching in XBAT. *J. Neurosci. Methods* **2014**, *236*, 68–75. [CrossRef]
34. Coffey, K.R.; Marx, R.G.; Neumaier, J.F. DeepSqueak: A deep learning-based system for detection and analysis of ultrasonic vocalizations. *Neuropsychopharmacology* **2019**, *44*, 859–868. [CrossRef]
35. Van Segbroeck, M.; Knoll, A.T.; Levitt, P.; Narayanan, S. MUPET-Mouse Ultrasonic Profile ExTraction: A Signal Processing Tool for Rapid and Unsupervised Analysis of Ultrasonic Vocalizations. *Neuron* **2017**, *94*, 465–485.e5. [CrossRef]
36. Ahrens, A.M.; Ma, S.T.; Maier, E.Y.; Duvauchelle, C.L.; Schallert, T. Repeated intravenous amphetamine exposure: Rapid and persistent sensitization of 50-kHz ultrasonic trill calls in rats. *Behav. Brain Res.* **2009**, *197*, 205–209. [CrossRef]
37. Mittal, N.; Thakore, N.; Bell, R.L.; Maddox, W.T.; Schallert, T.; Duvauchelle, C.L. Sex-specific ultrasonic vocalization patterns and alcohol consumption in high alcohol-drinking (HAD-1) rats. *Physiol. Behav.* **2019**, *203*, 81–90. [CrossRef]
38. Mittal, N.; Thakore, N.; Reno, J.M.; Bell, R.L.; Maddox, W.T.; Schallert, T.; Duvauchelle, C.L. Alcohol-Naïve USVs Distinguish Male HAD-1 from LAD-1 Rat Strains. *Alcohol* **2017**, *68*, 9–17. [CrossRef]
39. Mittal, N.; Martinez, A.; Fleming, S.M.; Maddox, W.T.; Schallert, T.; Duvauchelle, C.L. Sex differences in alcohol consumption, object recognition and rearing in male and female high alcohol drinking (had1) rats. In Proceedings of the Society for Neuroscience, Washington, DC, USA, 11–15 November 2017.
40. Maddox, W.T.; Ing, A.D. Delayed feedback disrupts the procedural-learning system but not the hypothesis-testing system in perceptual category learning. *J. Exp. Psychol. Learn. Mem. Cogn.* **2005**, *31*, 100–107. [CrossRef]
41. Maddox, W.T.; Ashby, F.G.; Bohil, C.J. Delayed feedback effects on rule-based and information-integration category learning. *J. Exp. Psychol. Learn. Mem. Cogn.* **2003**, *29*, 650–662. [CrossRef]
42. Wright, J.M.; Gourdon, J.C.; Clarke, P.B.S. Identification of multiple call categories within the rich repertoire of adult rat 50-kHz ultrasonic vocalizations: Effects of amphetamine and social context. *Psychopharmacology* **2010**, *211*, 1–13. [CrossRef]
43. Thakore, N.; Reno, J.M.; Gonzales, R.A.; Schallert, T.; Bell, R.L.; Maddox, W.T.; Duvauchelle, C.L. Alcohol enhances unprovoked 22-28 kHz USVs and suppresses USV mean frequency in High Alcohol Drinking (HAD-1) male rats. *Behav. Brain Res.* **2016**, *302*, 228–236. [CrossRef]
44. Brudzynski, S.M.; Kehoe, P.; Callahan, M. Sonographic structure of isolation-induced ultrasonic calls of rat pups. *Dev. Psychobiol.* **1999**, *34*, 195–204. [CrossRef]
45. Zeskind, P.S.; McMurray, M.S.; Garber, K.A.; Neuspiel, J.M.; Cox, E.T.; Grewen, K.M.; Mayes, L.C.; Johns, J.M. Development of translational methods in spectral analysis of human infant crying and rat pup ultrasonic vocalizations for early neurobehavioral assessment. *Front. Psychiatry* **2011**, *2*, 56. [CrossRef]
46. Brudzynski, S.M.; Silkstone, M.; Komadoski, M.; Scullion, K.; Duffus, S.; Burgdorf, J.; Kroes, R.A.; Moskal, J.R.; Panksepp, J. Effects of intraaccumbens amphetamine on production of 50kHz vocalizations in three lines of selectively bred Long-Evans rats. *Behav. Brain Res.* **2011**, *217*, 32–40. [CrossRef]
47. Brudzynski, S.M. Ultrasonic vocalization induced by intracerebral carbachol in rats: Localization and a dose-response study. *Behav. Brain Res.* **1994**, *63*, 133–143. [CrossRef]
48. Ciucci, M.R.; Ahrens, A.M.; Ma, S.T.; Kane, J.R.; Windham, E.B.; Woodlee, M.T.; Schallert, T. Reduction of dopamine synaptic activity: Degradation of 50-kHz ultrasonic vocalization in rats. *Behav. Neurosci.* **2009**, *123*, 328–336. [CrossRef]
49. Inagaki, H.; Takeuchi, Y.; Mori, Y. Close relationship between the frequency of 22-kHz calls and vocal tract length in male rats. *Physiol. Behav.* **2012**, *106*, 224–228. [CrossRef]
50. Wöhr, M.; Borta, A.; Schwarting, R.K.W. Overt behavior and ultrasonic vocalization in a fear conditioning paradigm: A dose-response study in the rat. *Neurobiol. Learn. Mem.* **2005**, *84*, 228–240. [CrossRef]
51. Simola, N.; Fenu, S.; Costa, G.; Pinna, A.; Plumitallo, A.; Morelli, M. Pharmacological characterization of 50-kHz ultrasonic vocalizations in rats: Comparison of the effects of different psychoactive drugs and relevance in drug-induced reward. *Neuropharmacology* **2012**, *63*, 224–234. [CrossRef]
52. Worthy, D.A.; Markman, A.B.; Todd Maddox, W. Feedback and stimulus-offset timing effects in perceptual category learning. *Brain Cogn.* **2013**, *81*, 283–293. [CrossRef]

53. Basken, J.N.; Connor, N.P.; Ciucci, M.R. Effect of aging on ultrasonic vocalizations and laryngeal sensorimotor neurons in rats. *Exp. Brain Res.* **2012**, *219*, 351–361. [CrossRef]
54. Simola, N.; Ma, S.T.; Schallert, T. Influence of acute caffeine on 50-kHz ultrasonic vocalizations in male adult rats and relevance to caffeine-mediated psychopharmacological effects. *Int. J. Neuropsychopharmacol.* **2010**, *13*, 123–132. [CrossRef] [PubMed]
55. Brudzynski, S.M. Pharmacology of Ultrasonic Vocalizations in Adult Rats: Significance, Call Classification and Neural Substrate. *Curr. Neuropharmacol.* **2015**, *13*, 180–192. [CrossRef] [PubMed]
56. Johnson, A.M.; Grant, L.M.; Schallert, T.; Ciucci, M.R. Changes in rat 50-kHz ultrasonic vocalizations during dopamine denervation and aging: Relevance to neurodegeneration. *Curr. Neuropharmacol.* **2015**, *13*, 211–219. [CrossRef] [PubMed]
57. Simola, N. Rat Ultrasonic Vocalizations and Behavioral Neuropharmacology: From the Screening of Drugs to the Study of Disease. *Curr. Neuropharmacol.* **2015**, *13*, 164–179. [CrossRef]
58. Knutson, B.; Burgdorf, J.; Panksepp, J. Ultrasonic vocalizations as indices of affective states in rats. *Psychol. Bull.* **2002**, *128*, 961–977. [CrossRef]
59. Ciucci, M.R.; Ma, S.T.; Kane, J.R.; Ahrens, A.M.; Schallert, T. Limb use and complex ultrasonic vocalization in a rat model of Parkinson's disease: Deficit-targeted training. *Park. Relat. Disord.* **2008**, *14*, 172–175. [CrossRef]
60. Barker, D.J.; Johnson, A.M. Automated acoustic analysis of 50-kHz ultrasonic vocalizations using template matching and contour analysis. *J. Acoust. Soc. Am.* **2017**, *141*, EL281–EL286. [CrossRef]
61. Reno, J.M.; Duvauchelle, C.L. Response to: Making WAAVES in the vocalization community: How big is the splash? *J. Neurosci. Methods* **2014**, *221*, 230–232. [CrossRef]
62. Reno, J.M.; Thakore, N.; Cormack, L.K.; Schallert, T.; Bell, R.L.; Maddox, W.T.; Duvauchelle, C.L. Negative Affect-Associated USV Acoustic Characteristics Predict Future Excessive Alcohol Drinking and Alcohol Avoidance in Male P and NP Rats. *Alcohol. Clin. Exp. Res.* **2017**, *41*, 786–797. [CrossRef]
63. Maier, E.Y.; Abdalla, M.; Ahrens, A.M.; Schallert, T.; Duvauchelle, C.L. The missing variable: Ultrasonic vocalizations reveal hidden sensitization and tolerance-like effects during long-term cocaine administration. *Psychopharmacology* **2012**, *219*, 1141–1152. [CrossRef]
64. Reno, J.M.; Thakore, N.; Gonzales, R.; Schallert, T.; Bell, R.L.; Maddox, W.T.; Duvauchelle, C.L. Alcohol-Preferring P rats emit spontaneous 22-28 khz ultrasonic vocalizations that are altered by acute and chronic alcohol experience. *Alcohol. Clin. Exp. Res.* **2015**, *39*, 843–852. [CrossRef]

brain sciences

MDPI

Article

Fear Extinction-Based Inter-Individual and Sex Differences in Pain-Related Vocalizations and Anxiety-like Behaviors but Not Nocifensive Reflexes

Peyton Presto [1], Guangchen Ji [1,2], Riley Junell [1], Zach Griffin [1] and Volker Neugebauer [1,2,3,*]

1 Department of Pharmacology and Neuroscience, School of Medicine, Texas Tech University Health Sciences Center, 3601 4th St, Lubbock, TX 79430-6592, USA; peyton.presto@ttuhsc.edu (P.P.); guangchen.ji@ttuhsc.edu (G.J.); Riley.Junell@ttuhsc.edu (R.J.); Zach.Griffin@ttuhsc.edu (Z.G.)

2 Center of Excellence for Translational Neuroscience and Therapeutics, Texas Tech University Health Sciences Center, Lubbock, TX 79430-6592, USA

3 Garrison Institute on Aging, Texas Tech University Health Sciences Center, Lubbock, TX 79430-6592, USA

* Correspondence: volker.neugebauer@ttuhsc.edu; Tel.: +1-806-743-3880; Fax: +1-806-732-2744

Abstract: Inter-individual and sex differences in pain responses are recognized but their mechanisms are not well understood. This study was intended to provide the behavioral framework for analyses of pain mechanisms using fear extinction learning as a predictor of phenotypic and sex differences in sensory (mechanical withdrawal thresholds) and emotional-affective aspects (open field tests for anxiety-like behaviors and audible and ultrasonic components of vocalizations) of acute and chronic pain. In acute arthritis and chronic neuropathic pain models, greater increases in vocalizations were found in females than males and in females with poor fear extinction abilities than females with strong fear extinction, particularly in the neuropathic pain model. Female rats showed higher anxiety-like behavior than males under baseline conditions but no inter-individual or sex differences were seen in the pain models. No inter-individual and sex differences in mechanosensitivity were observed. The data suggest that vocalizations are uniquely suited to detect inter-individual and sex differences in pain models, particularly in chronic neuropathic pain, whereas no such differences were found for mechanosensitivity, and baseline differences in anxiety-like behaviors disappeared in the pain models.

Keywords: vocalizations; fear extinction; pain; sex differences

Citation: Presto, P.; Ji, G.; Junell, R.; Griffin, Z.; Neugebauer, V. Fear Extinction-Based Inter-Individual and Sex Differences in Pain-Related Vocalizations and Anxiety-like Behaviors but Not Nocifensive Reflexes. *Brain Sci.* **2021**, *11*, 1339. https://doi.org/10.3390/brainsci11101339

Academic Editor: Stefan M. Brudzynski

Received: 14 September 2021
Accepted: 6 October 2021
Published: 11 October 2021

1. Introduction

Inter-individual and sex differences have been well documented with regard to anxiety- and depression-like conditions [1–3] and in pain [4–6]. However, neural mechanisms and biomarkers related to pain vulnerability and resilience, including potential sexual dimorphisms, have yet to be fully elucidated. Intricate interactions of sensory, cognitive, and emotional-affective dimensions form the highly complex and intense experience of pain. The strong negative affective component of pain presents a challenge for effective therapeutic strategies, as patients suffering from chronic pain are at increased risk of developing mood and anxiety disorders, and vice versa [7–10]. This suggests that pain may share neurobiological mechanisms, including emotional network neuroplasticity, with negative emotions such as fear [11,12]. Fear learning and extinction networks have been implicated in neuropsychiatric disorders such as anxiety disorders, post-traumatic stress disorder (PTSD), and obsessive compulsive disorder (OCD) [13–15]. Vulnerability to these disorders has been predicted using fear extinction (FE) learning ability as a biomarker for inter-individual differences in the preclinical [16] and clinical [17] setting.

Behavioral studies are a crucial tool for the validation of pain mechanisms and for the assessment of potential pharmacological therapies. A variety of behavioral methods have been developed in preclinical pain models for the evaluation of traits pertaining to sensorimotor function, anxiety- and depressive-like behavior, social interactions, cognitive function, and emotional-affective responses [18]. Higher integrated pain behavior at supraspinal levels has been assessed using vocalizations. Vocalizations are an important method of communication among rodents [19], with frequencies in the audible and ultrasonic ranges. Audible vocalizations of rats in response to a noxious stimulus indicate a nociceptive reaction, whereas ultrasonic vocalizations of the 22 kHz type represent negative emotional-affective responses [20,21]. Ultrasonic vocalizations are considered an effective indicator for measuring negative emotional status and have been used in different experimental models of pain, including arthritis pain [21–24], chronic cancer pain [25,26], and neuropathic pain [27–30]. However, some have called into question the reliability of vocalizations in assessing pain-related behavior [31] and others have found that vocalizations may occur as a response to handling [32]. While a valuable behavioral measure, vocalizations as a pain assessment may be most informative when used in combination with other pain indicators [33]. Inter-individual and sex differences in audible and ultrasonic vocalizations, particularly in the context of pain and fear interactions, have not been determined.

The purpose of this study was to examine the predictive value of fear extinction (FE) learning ability for inter-individual differences in pain-related behavioral responses, particularly emotional-affective pain aspects, with regard to sex. We subjected adult male and female rats to cued fear learning and FE tests and correlated inter-individual differences with pain responses in models of acute arthritis pain and chronic neuropathic pain. We also investigated sex differences in FE phenotypes for measures of sensory (mechanical withdrawal thresholds) and emotional-affective (open field tests for anxiety-like behaviors and audible and ultrasonic components of vocalizations) pain-related behaviors.

2. Materials and Methods

2.1. Animals

A total of 215 male and 190 female Sprague-Dawley rats (150–350 g, 6–12 weeks of age) were group-housed in a temperature-controlled room under a 12 h light/dark cycle with unrestricted access to food and water. On each experimental day, rats were transferred from the animal facility and allowed to acclimate to the laboratory for at least 1 h. Experimental procedures were approved by the Institutional Animal Care and Use Committee (IACUC; protocol #14006) at Texas Tech University Health Sciences Center and conformed to the guidelines of the International Association for the Study of Pain (IASP) and of the National Institutes of Health (NIH).

2.2. Experimental Protocol

Naïve rats were subjected to fear conditioning and FE trials. Rats were then randomly assigned to the arthritis pain model (see Section 2.3) or the neuropathic pain model (see Section 2.4). One week later, the neuropathic pain model was induced or sham surgery was performed. Four weeks after surgery, neuropathic pain-related behavioral changes reach a stable plateau in this model [29]. The arthritis pain model was induced in a separate group of rats at the four-week time point to achieve age-matched experimental groups. Behavioral studies were performed four weeks after surgery or 6 h after arthritis induction when behavioral and neurobiological changes are known to reach a maximum plateau [21]. The experimenter was blinded with regard to the FE phenotype, the neuropathic versus sham condition, and the arthritis versus untreated control condition. The experimental design is illustrated in Figure 1.

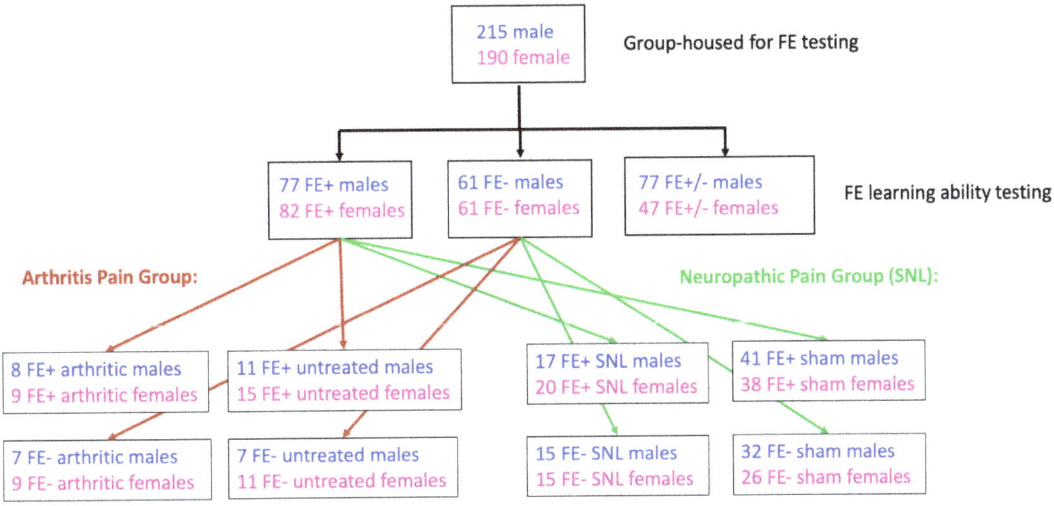

Figure 1. Experimental design. Rats underwent fear conditioning and extinction learning protocols before being separated into FE+ and FE− groups for either the acute arthritis pain (vs. untreated control) groups or the chronic neuropathic pain (vs. sham control) groups. FE: fear extinction.

2.3. Arthritis Pain Model

The well-established mono-arthritis pain model mimics the acute phase of the human osteoarthritis condition and was induced in the left knee joint as described in detail previously [21]. Rats were briefly anesthetized with isoflurane (2–3%; precision vaporizer, Harvard Apparatus, Holliston, MA) and a kaolin suspension (4% in sterile saline, 100 µL) was slowly injected into the joint cavity followed by repetitive flexions and extensions of the leg for 15 min. A carrageenan solution (2% in sterile saline, 100 µL) was then injected into the knee joint cavity and the leg was flexed and extended for another 5 min. This treatment paradigm (the K/C arthritis model) reliably produces a localized inflammation in only one knee joint with damage to the cartilage within 1–3 h. K/C arthritis persists for at least a week and is associated with pain behaviors and neural activity changes in the central and peripheral nervous system. Naïve rats that underwent similar handling but did not receive intraarticular injections were used as a control group, as data from our previous studies demonstrated no differences in the behavior of untreated rats and of those that received intraarticular saline injection [34] or needle insertion [35]. This justified the use of naïve rats as an appropriate control for the K/C pain model, which is well established in our laboratories [36–38].

2.4. Neuropathic Pain Model

The well-established spinal nerve ligation (SNL) model of neuropathic pain [39] was used, which creates stable and long-lasting neuropathic pain behaviors. Rats were anesthetized with isoflurane (2–3%; precision vaporizer, Harvard Apparatus) and underwent sterile surgery where the left L5 spinal nerve was exposed and tightly ligated using 6–0 sterile silk. In the sham-operated control group, the nerve was exposed but not ligated.

2.5. Behaviors

2.5.1. Fear Conditioning and Extinction

Fear conditioning and extinction learning tests were conducted using two chambers of a near infrared Video Fear Conditioning System (Med Associates Inc., Fairfax, VT, USA) as described previously [40–42]. The conditioning chambers were located inside a sound-attenuating isolation cabinet with a metal grid flooring that was connected to a grid stimulator to administer aversive foot shocks. Two distinct chambers with separate visual, olfactory, tactile, dimensional, and lighting environments were used (context A: white light, no fan in chamber, metal grid on chamber floor, lights on in experimental room, rat transported to chamber in transparent box, chamber cleaned with 50% ethanol; context B: near-infrared (NIR) light, fan on in chamber, flat chamber floor, lights off in experimental room, rat transported to chamber in opaque box, chamber cleaned with 70% isopropanol, colored insert with 3 drops of peppermint oil added to alter olfactory environment and physical dimensions). Day 1 consisted of the training phase where rats were habituated to the training chamber (context A) and allowed to explore freely for 5 min, followed by fear conditioning that consisted of a foot shock (0.7 mA, 2 s; the unconditioned stimulus, USA) delivered during the final 2 s of an auditory stimulus (white noise, 80 dB, 4.5 kHz, 30 s; the conditioned stimulus, CS). Two CS-US pairings were used (intertone interval, ITI, 120 s). On day 2, rats were placed in a different chamber (context B) and were habituated for 5 min, followed by extinction training (30 CSs, ITI 5 s). A mounted video camera in the conditioning chambers was used to record the behavior of each rat. Freezing behavior (expressed as a percentage of each 30 s period) was analyzed and quantified using Video Freeze software (Med Associates Inc.) as the conditioned response. Based on their FE learning ability, rats were classified into strong (FE+), "normal" (FE+/−), and weak (FE−) FE groups as determined by evidence of diminishing (below 50%) freezing responses during Phase I (before 600 s), Phase II (600–900 s), or Phase III (after 900 s) of extinction training (see the "Results" section for details). Rats in the two extreme groups (FE+ and FE−) were selected for further behavioral testing and randomly assigned to groups in the arthritis pain model (untreated FE+, untreated FE−, arthritis FE+, and arthritis FE−) or in the neuropathic pain model (sham FE+, sham FE−, SNL FE+, and SNL FE−). Four weeks after SNL or sham surgery, or 6 h after arthritis induction in an age-matched model, behavioral assays (see next paragraphs) were performed.

2.5.2. Mechanosensitivity

Rats were briefly anesthetized with isoflurane (2–3%; precision vaporizer, Harvard Apparatus) and were placed slightly restrained in a customized recording chamber that permitted access to the hindlimbs (U.S. Patent 7,213,538) for stable testing. Hindlimb withdrawal thresholds were evaluated after recovery from anesthesia and after habituation to the recording chamber for 30 min. Hindlimb withdrawal thresholds were evaluated using calibrated forceps with a force transducer whose output was displayed in grams on an LED screen. The calibrated forceps were used to gradually compress the left knee joint (arthritis pain model) or the left hindpaw (neuropathic pain model) with a continuously increasing intensity until a withdrawal reflex was evoked as described in our previous studies [21,35,37,38,43–45]. The withdrawal threshold, defined as the force required to evoke a reflex response, was calculated using the average value from 2 to 3 trials.

2.5.3. Emotional Responses

Components of vocalizations in the audible (20 Hz–16 kHz) and ultrasonic (25 ± 4 kHz) ranges were simultaneously measured after hindlimb withdrawal assays using an automatic computerized vocalization system consisting of a full-spectrum USB ultrasound microphone (max sampling rate: 384 kHz) and UltraVox XT four-channel recording and analysis system (Noldus Information Technology, Leesburg, VA, USA). Rats were briefly anesthetized with isoflurane (2–3%; precision vaporizer, Harvard Apparatus) and placed in the customized recording chamber for stable recordings of vocalizations evoked by natural stimulation. After the rat recovered from anesthesia and habituated to the recording chamber for 30 min, hindlimb withdrawal thresholds were evaluated (see Section 2.5.2) and the calibrated forceps with a force transducer were used for vocalization assays. Vocalizations were evoked by a brief (10 s), continuous noxious stimulus applied to the left knee joint (arthritis pain model; stimulus: 1500 g/30 mm^2) or to the left hindpaw (neuropathic pain model; stimulus: 500 g/6 mm^2) as described in our previous studies [20,28,29,37,38,42].

Vocalizations were automatically detected for 1 min and total durations of audible and ultrasonic components of vocalizations following the onset of mechanical stimulus were analyzed using UltraVox 3.2 software (Noldus Information Technology). For vocalization analyses, audible calls were labeled using frequency ranges of 20 Hz–16 kHz and ultrasonic components of calls were labeled using frequency ranges of 21–29 kHz. The following call descriptions were also specified: minimum amplitude, 50 units; minimum duration, 1 ms; maximum duration, 2000 ms; minimum gap between calls, 1 ms. Calls that fit these criteria were detected for each recording. At the conclusion of each experiment, call statistics for each recording were exported as a text file. The duration (in ms) for each individual call was summed for each 1 min recording period to give the total duration of audible and ultrasonic components of vocalizations for each rat.

2.5.4. Anxiety-Like Behavior

Animal movements within the open field test (OFT) were used to measure anxiety-like behavior. Exploratory behavior in the central or peripheral zones of an arena (70 cm × 70 cm) with acrylic walls (height, 45 cm) was recorded for 15 min using a computerized video tracking and analysis system (EthoVision XT 11 software, Noldus Information Technology) as described previously [42,46]. Time spent in the center of the arena (35 cm × 35 cm) was calculated during the first 5 min. Avoidance of the center of the arena is interpreted to suggest anxiety-like behavior [42,46–48].

2.6. Statistical Analysis

All averaged values are presented as the mean ± SE. Statistical significance was accepted at the level $p < 0.05$. GraphPad Prism 9.0 software was used for all statistical analyses. Statistical analyses were performed on the raw data. For multiple comparisons, a two-way analysis of variance (ANOVA) was used with Bonferroni post hoc tests.

3. Results

3.1. Inter-Individual and Sex Differences in FE Learning Ability of Naïve Male and Female Rats

Fear learning and FE are well-established models of aversive learning that have been used to correlate behavior with neural structure and function, which involve cortico-limbic circuits centered on the amygdala [15]. We previously reported that the identification of distinct behavioral phenotypes based on FE ability in naïve male rats can serve as a predictor for inter-individual differences in pain sensitivity and amygdala neuronal activity in chronic neuropathic pain [42]. Here, we chose to examine whether a similar correlation existed between FE learning ability and acute arthritis pain-related behaviors and if this predictive value could be expanded to include both sexes.

Fear learning and FE were measured in 215 male and 190 female naïve rats (see Sections 2.1 and 2.5.1). During the fear learning session on day 1 of fear conditioning, rats showed minimal freezing behavior during the habituation phase under context A, indicating normal locomotor activity. All rats developed freezing responses after two pairings of CS (white noise, 80 dB, 4.5 kHz, 30 s) and US (0.7 mA foot shock, 2 s) (Figure 2A). During the fear training session on day 2, three groups emerged in both sexes based on differences in the time course and magnitude of declining freezing behavior in the absence of a foot shock (the US) (Figure 2B). For females, 36 rats (35.6%) exhibited a rapid (before 600 s; Phase I) decline in freezing to levels below 50% (per 30 s CS segment), reflecting strong FE learning ability (FE+), while 18 rats (17.8%) maintained freezing levels above 50% past 900 s (Phase III), indicating weak FE learning ability (FE−). The remaining 47 rats (46.5%) showed a decline to below 50% freezing levels between 600 and 900 s (Phase II) of the FE session and were classified as exhibiting "normal" FE learning ability (FE+/−). Males exhibited a different distribution of phenotypes, where 29 rats (19.8%) showed strong FE learning ability (FE+), 47 rats (30.7%) showed weak FE learning ability (FE−), and the remaining 77 rats (50.3%) showed normal FE learning ability (FE+/−) (Figure 2C). Female FE− rats showed a significantly higher percent freezing per 30 s CS segment than those in the female FE+ group ($p < 0.0001$, $F_{1,2080} = 512.8$, two-way ANOVA; Bonferroni post hoc test results are shown in Figure 2B). Similarly, males in the FE− group showed a significantly higher percent freezing per 30 s CS segment than males in the FE+ group ($p < 0.001$, $F_{1,2960} = 1372$, two-way ANOVA; Bonferroni post hoc test results are shown in Figure 2B). Interestingly, FE+ males exhibited significantly lower percent freezing per 30 s CS segment than FE+ females ($p < 0.01$, $F_{12,520} = 12.42$, two-way ANOVA with Bonferroni post hoc tests) while FE− males showed significantly higher percent freezing per 30 s CS segment than FE− females ($p < 0.0001$, $F_{12,520} = 22.75$, two-way repeated-measures ANOVA with Bonferroni post hoc tests). Importantly, no differences in percent freezing were observed between the three groups for either sex during the habituation phases of the fear learning (Figure 2A) or the fear extinction (Figure 2B) sessions.

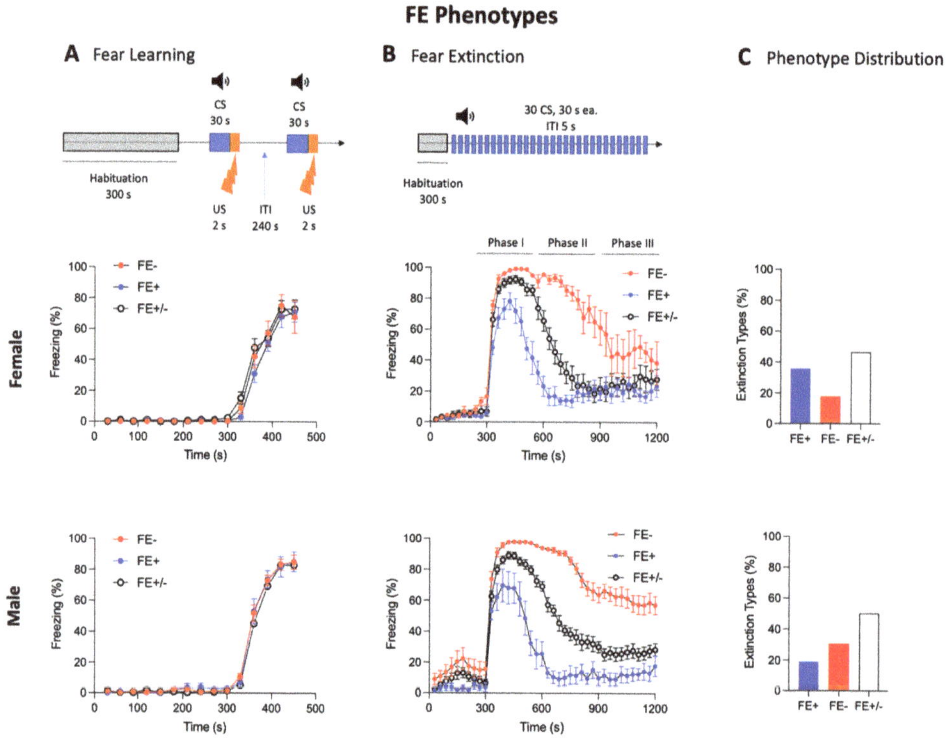

Figure 2. Inter-individual and sex differences in fear extinction learning ability in naïve female and male rats. Fear conditioning on Day 1 (**A**) and extinction (**B**) tests were conducted using two distinct context chambers. (**A**) Fear conditioning on Day 1—rats were habituated to context A followed by fear conditioning (2 CS-US pairs, see Section 2.5.1). The diagram illustrates the experimental protocol. Symbols in the line graph show freezing responses expressed in percent per 30 s segment during fear conditioning with 2 CS-US pairings. (**B**) Fear extinction learning on Day 2—rats were habituated to context B followed by extinction training (30 CSs, no US). The diagram illustrates the experimental protocol. Symbols in the line graph show freezing responses to tone (CS) expressed in percent per 30 s segment. (**C**) Bar histograms show the distribution of rats with strong (FE+), "normal" (FE+/−), and weak (FE−) fear extinction. The population (%) of FE+ was larger in female rats compared to male rats. For details, see the "Methods" and "Results" sections. CS: conditioned stimulus; US: unconditioned stimulus; ITI: intertone interval; FE: fear extinction.

3.2. Inter-Individual and Sex Differences in Arthritis Pain-Related Behaviors of FE+ and FE− Rats

Next, we examined whether inter-individual and sex differences in FE learning ability would correspond with behavioral differences for males and females in an arthritis pain model (K/C arthritis, see Section 2.3) and/or in the untreated control condition. Male and female rats from the FE+ and FE− groups were selected for further behavioral testing and randomly assigned to either the K/C arthritis group or the untreated control group. Five weeks later (corresponding with an age-matched neuropathic pain group), arthritis was induced, and 6 h later, the following behavioral assays were performed: nocifensive reflexes (mechanosensitivity, Figure 3A) and ultrasonic and audible components of vocalizations (emotional responses, Figure 3B,C) evoked by mechanical compression of the knee joint, and the OFT (anxiety-like behavior, Figure 3D).

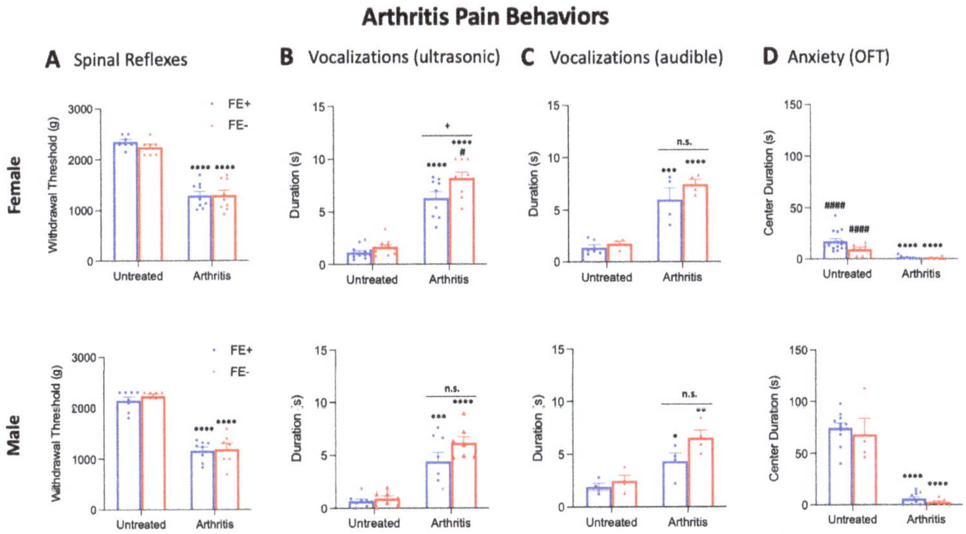

Figure 3. Inter-individual and sex differences in arthritis pain-related behaviors of FE+ and FE− rats. (**A**) Mechanical thresholds tested in untreated control rats and arthritic rats (6 h post-induction) showed no significant differences between FE− (female, $n = 7$; male, $n = 7$) and FE+ (female, $n = 7$; male, $n = 8$) untreated rats or between FE− (female, $n = 9$; male, $n = 7$) and FE+ (female, $n = 9$; male, $n = 8$) arthritic rats, but arthritic FE− and FE+ rats had significantly lower withdrawal thresholds than their untreated controls. **** $p < 0.0001$, ANOVA with Bonferroni post hoc tests (see the "Results" section). (**B**,**C**) Duration (s) of ultrasonic and audible vocalizations, respectively, evoked by a brief (10 s) noxious (1500 g/30 mm^2) mechanical compression of the knee. Significant differences in ultrasonic (but not audible) vocalizations were found between FE− ($n = 9$) and FE+ ($n = 9$) female arthritic rats but not between FE− ($n = 7$) and FE+ ($n = 8$) male arthritic rats or between untreated FE− (female, $n = 11$; male, $n = 7$) and FE+ (female, $n = 13$; male, $n = 8$) rats. For both sexes, arthritic rats had significantly increased vocalizations compared to their untreated controls. n.s.: non-significant; $^+$ $p < 0.05$; $^\#$ $p < 0.05$; * $p < 0.05$; ** $p < 0.01$; *** $p < 0.001$; **** $p < 0.0001$, ANOVA with Bonferroni post hoc tests (see the "Results" section). (**D**) Center duration (s) in the OFT was significantly lower in arthritic FE− (female, $n = 9$; male, $n = 7$) and FE+ (female, $n = 9$; male, $n = 13$) rats compared to the untreated FE− (female, $n = 8$; male, $n = 7$) and FE+ (female, $n = 15$; male, $n = 11$) control rats. No differences were found between FE− and FE+ rats in the untreated control or arthritic groups for either sex. #### $p < 0.0001$; **** $p < 0.0001$, ANOVA with Bonferroni post hoc tests (see the "Results" section). Bar histograms show means ± SEM. FE: fear extinction; OFT: open field test. Asterisk (*) indicates comparison to untreated group; plus sign (+) indicates comparison between phenotypes; pound sign (#) indicates comparison between sexes.

No significant differences in mechanical withdrawal thresholds were found between untreated FE+ rats (female, $n = 7$; male, $n = 8$) or untreated FE− rats (female, $n = 7$; male, $n = 7$) for either sex (Figure 3A). Similarly, no significant differences in mechanosensitivity were found between FE+ rats (female, $n = 9$; male, $n = 8$) or FE− rats (female, $n = 9$; male, $n = 7$) in the arthritis pain model for either sex. However, mechanical withdrawal thresholds were significantly lower for arthritic female FE+ and FE− rats and for arthritic male FE+ and FE− rats compared to their untreated controls ($p < 0.0001$, as shown in Figure 3A), suggesting that both types of rats developed hypersensitivity in the pain model. No significant differences in mechanical withdrawal thresholds were found between female FE+ rats and male FE+ rats or between female FE− rats and male FE− rats for either the arthritis or untreated control groups. For the statistical analyses of mechanical withdrawal thresholds in the four female experimental groups and the four male experimental groups, ANOVA with Bonferroni post hoc tests was used (female, $F_{3,28} = 53.09$; male, $F_{3,26} = 57.02$).

For the ultrasonic and audible components of vocalizations (Figure 3B,C), no significant differences were found between untreated FE+ rats (female, $n = 13$; male, $n = 8$) and untreated FE− rats (female, $n = 11$; male, $n = 7$) for either sex. However, the total duration of vocalizations was significantly higher in female FE− rats ($n = 9$) than female FE+ rats ($n = 9$) in the arthritis pain model ($p < 0.05$, Figure 3B). No significant differences were found in the durations of audible components of vocalizations of these groups or in ultrasonic and audible components of vocalizations of male FE+ rats ($n = 8$) and male FE− rats ($n = 7$) in the arthritis group, though there was a non-significant trend (ultrasonic, $p = 0.1988$; audible, $p = 0.1398$). Total durations of ultrasonic and audible components of vocalizations were significantly increased for arthritic female FE+ and FE− rats and for arthritic male FE+ and FE− rats compared to their untreated controls ($p < 0.05$–0.0001, as shown in Figure 3B,C). Female FE− rats had significantly increased durations of ultrasonic but not audible components of vocalizations compared to male FE− rats ($p < 0.05$, as shown in Figure 3B) in the arthritis model. No differences were seen for durations of ultrasonic and audible components of vocalizations between female FE+ and male FE+ groups (untreated control or arthritis). Together, the data suggest that all groups developed emotional responses to arthritis pain, though it emerged most prominently for female FE− rats. For the statistical analyses of vocalization durations in the four female experimental groups and the four male experimental groups, ANOVA with Bonferroni post hoc tests was used (ultrasonic: female, $F_{3,38} = 80.32$, and male, $F_{3,26} = 23.49$; audible: female, $F_{3,14} = 27.75$, and male, $F_{3,12} = 11.88$).

In the OFT (Figure 3D), no significant difference in arena center duration was found between untreated FE+ rats (female, $n = 15$; male, $n = 11$) and untreated FE− rats (female, $n = 8$; male, $n = 7$) for either sex. Similarly, no significant differences in center duration were found between FE+ rats (female, $n = 9$; male, $n = 13$) and FE− rats (female, $n = 9$; male, $n = 7$) in the arthritis pain model for males or females. In the arthritis pain groups, female FE+ and FE− rats and male FE+ and FE− rats spent significantly less time in the center of the arena compared to their untreated controls ($p < 0.0001$, Figure 3D), suggesting all groups developed increased anxiety-like behavior. However, in the untreated control group, female FE+ and FE− rats spent significantly less time in the arena center compared to male FE+ and FE− rats, respectively ($p < 0.0001$, as shown in Figure 3D), suggesting higher baseline anxiety levels for females of both phenotypes. No significant differences in center duration were seen between females and males in the arthritis pain model for either phenotype. Importantly, no significant differences in locomotor activity were observed between the arthritis pain group and the untreated control group ($p = 0.7327$, Figure 3D), indicating that differences in anxiety-like behavior were not due to a reduction in spontaneous activity following arthritis induction. For the statistical analyses of OFT center duration in the four female experimental groups and the four male experimental groups, ANOVA with Bonferroni post hoc tests was used (female, $F_{3,37} = 16.94$; male, $F_{3,31} = 72.79$).

3.3. Inter-Individual and Sex Differences in Neuropathic Pain-Related Behaviors of FE+ and FE− Rats

As we previously reported that FE learning ability may serve as a predictor for neuropathic pain-related behaviors in male rats [42], we next sought to determine whether inter-individual differences in FE learning ability may also translate into behavioral differences for females in a neuropathic pain model (SNL, see Section 2.4) and/or in the sham control condition. Male and female FE+ and FE− rats were randomly assigned to the neuropathic pain group or sham group, and four weeks after SNL or sham surgery, the same behavioral assays were performed in these animals: nocifensive reflexes (Figure 4A) and ultrasonic and audible components of vocalizations (Figure 4B,C) evoked by mechanical compression of the hindpaw, and the OFT (Figure 4D).

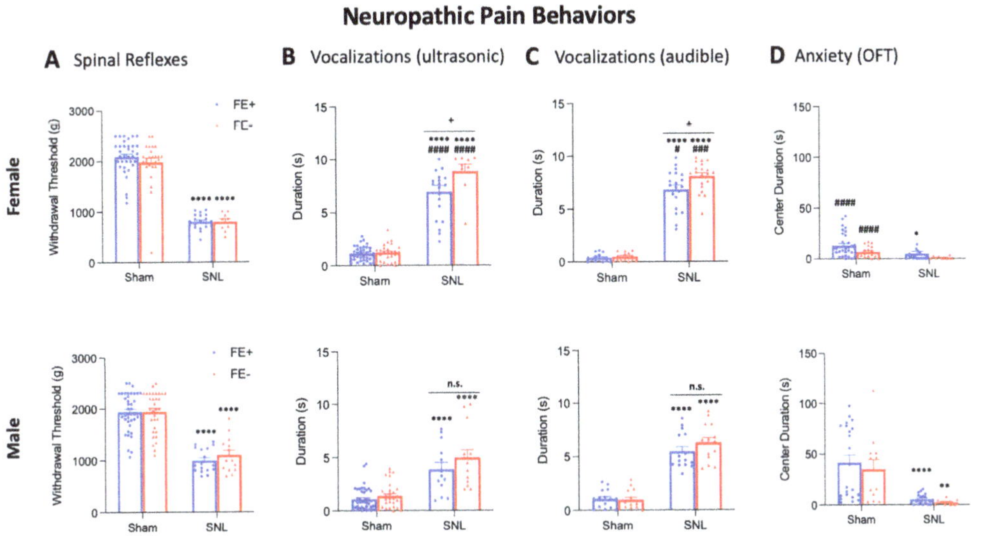

Figure 4. Inter-individual and sex differences in neuropathic pain related behaviors of FE+ and FE− rats. (**A**) Mechanical thresholds tested in sham and chronic neuropathic SNL rats (4 weeks post-induction) showed no significant differences between FE− (female, $n = 26$; male, $n = 32$) and FE+ (female, $n = 35$; male, $n = 40$) sham rats or between FE− (female, $n = 9$; male, $n = 14$) and FE+ (female, $n = 20$, male, $n = 17$) SNL rats, but SNL FE− and FE+ rats had significantly lower withdrawal thresholds than their sham controls. **** $p < 0.0001$, ANOVA with Bonferroni post hoc tests (see the "Results" section). (**B**,**C**) Duration (s) of of ultrasonic and audible vocalizations, respectively, evoked by a brief (10 s) noxious (1500 g/6 mm^2) mechanical compression of the affected hindpaw. Significant differences in ultrasonic and audible vocalizations were found between female FE− ($n = 15$) and FE+ ($n = 20$) SNL rats but not between male FE− ($n = 15$) and FE+ ($n = 15$) SNL rats or between FE− (female, $n = 26$; male, $n = 32$) and FE+ (female, $n = 38$; male, $n = 41$) sham rats. For both sexes, SNL rats had significantly increased vocalizations compared to their sham controls. n.s.: non-significant; $^+ p < 0.05$; $^\# p < 0.05$; $^{\#\#\#} p < 0.001$; $^{\#\#\#\#} p < 0.0001$; **** $p < 0.0001$, ANOVA with Bonferroni post hoc tests (see the "Results" section). (**D**) Center duration (s) in the OFT was significantly lower in FE− (female, $n = 9$; male, $n = 13$) and FE+ (female, $n = 20$; male, $n = 20$) SNL rats compared to their FE− (female, $n = 19$; male, $n = 12$) and FE+ (female, $n = 33$; male, $n = 22$) sham controls. No differences were found between FE− and FE+ rats in the sham or SNL groups for either sex. $^{\#\#\#\#} p < 0.0001$; * $p < 0.05$; ** $p < 0.01$; **** $p < 0.0001$, ANOVA with Bonferroni post hoc tests (see the "Results" section). Bar histograms show means ± SEM. FE: fear extinction; OFT: open field test; SNL: spinal nerve ligation. Asterisk (*) indicates comparison to untreated group; plus sign (+) indicates comparison between phenotypes; pound sign (#) indicates comparison between sexes.

Mechanical withdrawal thresholds showed no significant differences between sham FE+ rats (female, $n = 35$; male, $n = 40$) and sham FE− rats (female, $n = 26$; male, $n = 32$) for either sex (Figure 4A). Similarly, in the neuropathic pain model, there were no significant differences found in withdrawal thresholds between FE+ rats (female, $n = 20$; male, $n = 17$) and FE− rats (female, $n = 9$; male, $n = 14$) for either sex. Both female and male FE+ and FE− rats in the neuropathic pain model showed significantly lower mechanical thresholds compared to their sham controls ($p < 0.0001$, as shown in Figure 4A), suggesting both types of rats developed neuropathic hypersensitivity. No significant differences in mechanical withdrawal thresholds were found between female FE+ and male FE+ rats or between female FE− and male FE− rats for either the neuropathic pain or sham control groups. For the statistical analyses of mechanical withdrawal thresholds between the four female experimental groups and the four male experimental groups, ANOVA with Bonferroni post hoc tests was used (female, $F_{3,86} = 92.25$; male, $F_{3,99} = 46.40$).

For ultrasonic and audible components of vocalizations (Figure 4B,C), no significant differences in duration were found between sham FE+ rats (female, $n = 38$; male, $n = 41$) and sham FE− rats (female, $n = 26$; male, $n = 32$) for either sex. However, the total durations of ultrasonic and audible components of vocalizations were significantly increased in female FE− rats ($n = 15$) compared to female FE+ rats ($n = 20$) in the neuropathic pain model ($p < 0.05$, as shown in Figure 4B,C). No significant differences in the durations of audible or ultrasonic components of vocalizations were found for male FE+ rats ($n = 15$) and male FE− rats ($n = 15$) in neuropathic pain. Both female and male FE+ and FE− rats in the neuropathic model had significantly increased durations of ultrasonic and audible components of vocalization compared to their sham controls ($p < 0.0001$, as shown in Figure 4B,C). In the neuropathic pain group, FE+ female rats had significantly greater durations of ultrasonic and audible components of vocalizations than FE+ male rats (ultrasonic: $p < 0.0001$, as shown in Figure 4B; audible: $p < 0.05$, as shown in Figure 4C) and FE− female rats had significantly greater durations of ultrasonic and audible components of vocalizations than FE− male rats (ultrasonic: $p < 0.0001$, as shown in Figure 4B; audible: $p < 0.001$, as shown in Figure 4C). Together, the data suggest that while all groups developed emotional responses to neuropathic pain, this occurred most prominently for FE− females. Individual examples of real-time waveforms and spectrogram recordings for phenotyped female and male SNL and sham control rats are shown in Figures 5 and 6. Though not reported in this study, our recordings (see Figures 5 and 6) suggest that similar differences between sexes and phenotypes may be observed with regard to the total number of vocalizations. Both the total duration [20,24] and the total number of calls [27,49] have been utilized as effective measures of behavioral responses in the context of pain. For the statistical analyses of ultrasonic and audible components of vocalizations between the four female experimental groups and the four male experimental groups, ANOVA with Bonferroni post hoc tests was used (ultrasonic: female, $F_{3,89} = 125.9$, and male, $F_{3,96} = 25.90$; audible: female, $F_{3,66} = 177.9$, and male, $F_{3,56} = 66.50$).

In the OFT (Figure 4D), no significant differences in arena center duration were found between sham FE+ rats (female, $n = 33$; male, $n = 22$) and sham FE− rats (female, $n = 19$; male, $n = 12$). Similarly, no differences in arena center duration were found between FE+ rats (female, $n = 20$; male, $n = 20$) and FE− rats (female, $n = 9$; male, $n = 13$) in the neuropathic pain model. Female FE+ (but not FE−) rats and male FE+ and FE− rats in the neuropathic pain group spent significantly less time in the center of the arena compared to their sham controls ($p < 0.05$–0.0001, as shown in Figure 4D). In the sham control group, female FE+ and FE− rats spent significantly less time in the center of the arena compared to male FE+ and FE− rats ($p < 0.0001$, as shown in Figure 4D), suggesting higher anxiety levels for females at baseline, as also seen in naïve rats (see Figure 3D). Importantly, there were no significant differences in locomotor activity between the neuropathic pain and sham control groups ($p = 0.8120$, Figure 4D) or between the sham control group and the untreated control group for the arthritis model ($p = 0.4292$, see Figure 3D), indicating that the observed differences in anxiety-like behavior were not due to a reduction in spontaneous activity following surgical procedures. For the statistical analyses of OFT center duration between the four female experimental groups and the four male experimental groups, ANOVA with Bonferroni post hoc tests was used (female, $F_{3,78} = 6.119$; male, $F_{3,63} = 11.53$).

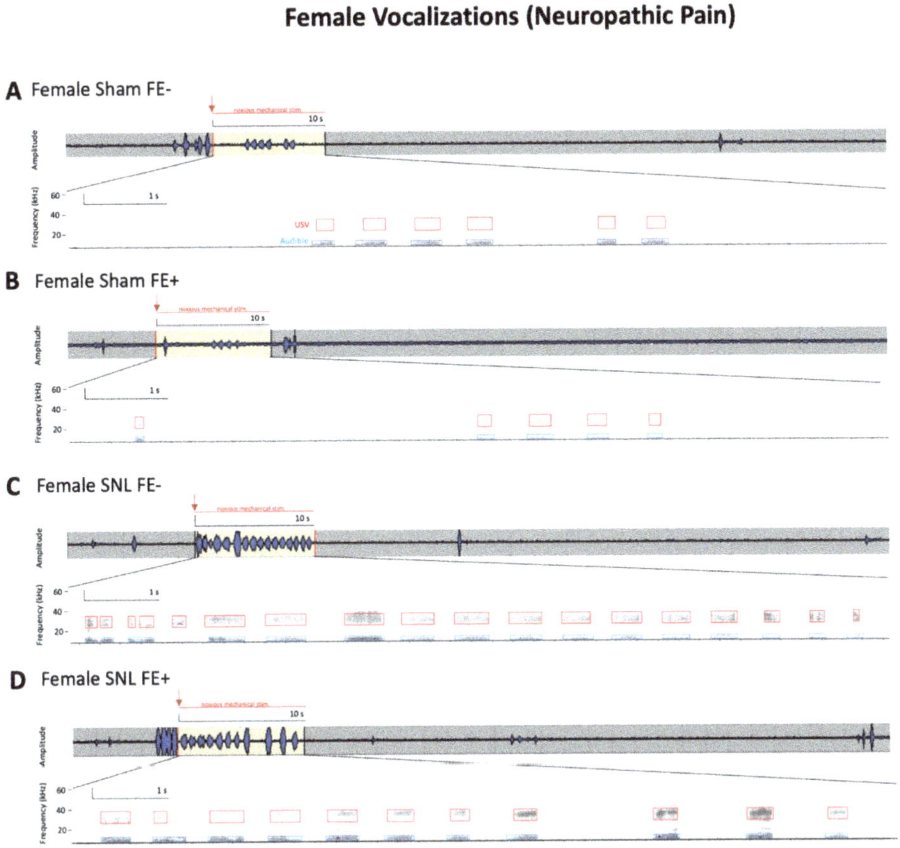

Figure 5. Representative audible and ultrasonic vocalizations from phenotyped female rats in the SNL model of neuropathic pain. Original real-time waveform and spectrogram recordings of vocalizations evoked in response to brief (10 s) noxious (1500 g/6 mm^2) mechanical stimulation of the affected hindpaw 4 weeks after induction of sham (**A**,**B**) or SNL (**C**,**D**) surgery in phenotyped female rats. For details, see Section 2.5.3. Mechanical stimuli were applied to the hindpaw in each recording period, as indicated by the highlighted yellow section of the waveform (upper panel, red arrow indicates initiation of noxious stimulus application); the total duration of the recording is 1 min. Boxes (events) in the spectrogram (lower panel) represent the presence of audible (blue; 20 Hz–16 kHz) and ultrasonic (red; 25 ± 4 kHz) vocalizations during the 10 s application of mechanical stimuli. Female FE− sham rats (**A**) showed more vocalization events in response to noxious stimulus than female FE+ sham rats (**B**). Female FE− SNL rats (**C**) showed more vocalization events in response to noxious stimulus than female FE+ SNL rats (**D**). FE: fear extinction; SNL: spinal nerve ligation; USV: ultrasonic vocalizations.

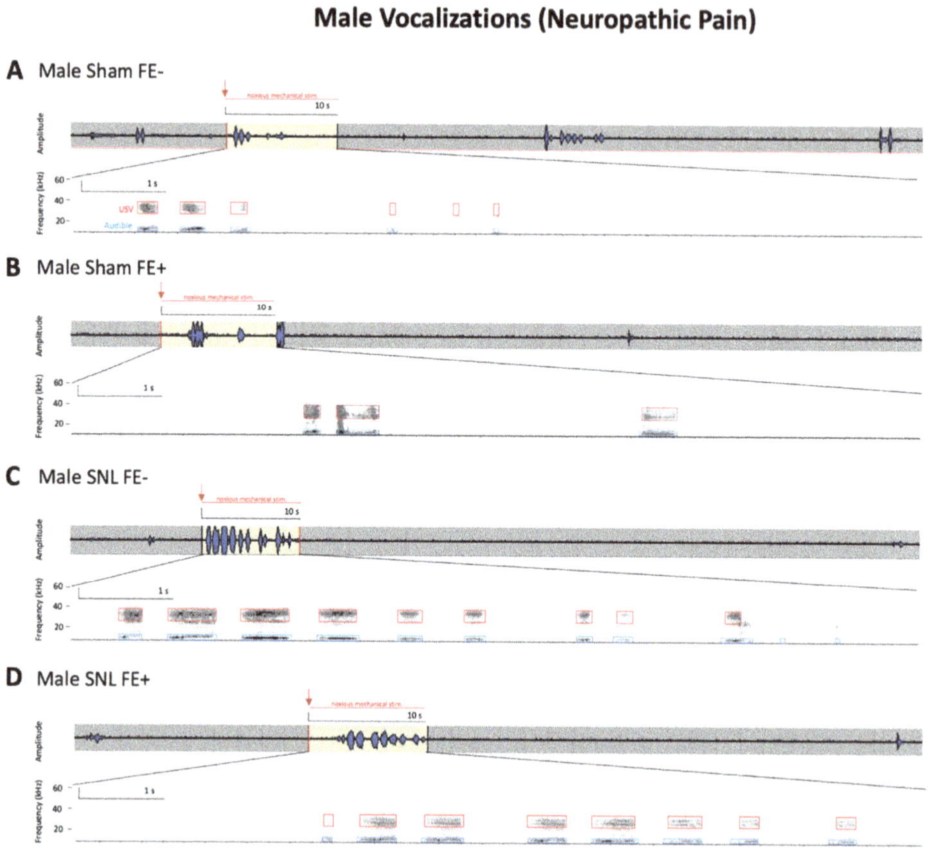

Figure 6. Representative audible and ultrasonic vocalizations from phenotyped male rats in the SNL model of neuropathic pain. Original real-time waveform and spectrogram recordings of vocalizations evoked in response to brief (10 s) noxious (1500 g/6 mm^2) mechanical stimulation of the affected hindpaw 4 weeks after induction of sham (**A**,**B**) or SNL (**C**,**D**) surgery in phenotyped male rats. For details, see Section 2.5.3. Mechanical stimuli were applied to the hindpaw in each recording period, as indicated by the highlighted yellow section of the waveform (upper panel, red arrow indicates initiation of noxious stimulus application); the total duration of the recording is 1 min. Boxes (events) in the spectrogram (lower panel) represent the presence of audible (blue; 20 Hz–16 kHz) and ultrasonic (red; 25 ± 4 kHz) vocalizations during the 10 s application of mechanical stimuli. Male FE− sham rats (**A**) showed more vocalization events in response to noxious stimulus than male FE+ sham rats (**B**). Male FE− SNL rats (**C**) showed more vocalization events in response to noxious stimulus than male FE+ SNL rats (**D**). FE: fear extinction; SNL: spinal nerve ligation; USV: ultrasonic vocalizations.

4. Discussion

This study explored the predictive value of FE learning ability in sensory and affective pain-related behaviors for male and female animals in an acute arthritis and a chronic neuropathic pain model. We previously showed a positive correlation between FE learning ability and neuropathic pain behaviors in adult male rats [42], but it is unclear if these are also found in acute pain conditions and whether female rats exhibit a similar association. The key novelties of this study are the identification of distinct behavioral phenotypes based on FE learning ability for both sexes, with vocalizations being the most effective indicators, and that these behavioral phenotypes show striking differences between male and female rats in both pain models.

Fear learning and extinction assays were selected as approaches to identifying inter-individual differences in pain-related behaviors because these are well-established models for correlating animal behavior with neural structure and function [15]. Previous studies from our lab and others have reported the separation of fast and slow recovery phenotypes based on freezing levels during FE that correlate with differences in anxiety-like behavior [16,42,50–52]. At the clinical level, inter-individual differences in fear response modulation and generalization have been linked to increased vulnerability in the development of anxiety disorders and post-traumatic stress disorder (PTSD) [15,17,53–55], and patients with anxiety disorders, PTSD, and obsessive-compulsive disorder (OCD) have exhibited delayed and/or reduced FE or extinction recall [14,56–59]. A previous epidemiological study reported that most individuals who experience trauma recover, with only a subset going on to develop a psychopathology such as depression or anxiety [60]. Similarly, chronic widespread pain develops in only 10% of the population [61]. However, a major goal of preclinical research is to provide insights into neural processes and behaviors that can predict susceptibility versus resistance to a disorder. This requires the study of neural variability patterns that differ from the central tendency. Thus, we chose to focus on (representatives of) the groups at the two ends of the spectrum (weak FE learning ability (FE−), considered to be "susceptible" rats, versus strong FE learning ability (FE+), considered to be "resistant" rats) within this study instead of including the larger, "normal" FE+/− group.

Little has been studied with regard to sex differences in classic fear conditioning and extinction models. This is an important knowledge gap, as females have twice the lifetime rates of depression and anxiety disorders [62], and human imaging studies revealed structural and functional sex differences in anxiety-relevant brain regions [63]. One preclinical study found that, while fast and slow extinction phenotypes could be identified for both sexes, there were no observable differences between males and females in freezing levels during fear conditioning or extinction [64]. Others have reported impairments in FE recall for female rats when compared to male rats [65,66]. However, several studies have reported that females have greater FE rates when compared to males [67–70]. A recent review suggested that sex hormones may play an important role in conditioned FE, as estrous cycle influences may affect female FE mechanisms [71]. The results of this study are consistent with those from the literature that showed gonadectomized males spent a greater amount of time freezing than gonadectomized females, a pattern that was not affected by estradiol administration [68]. Another study suggested that endogenous estrogen did not affect FE behavior in female rats or naturally cycling women [72]. Therefore, it is unlikely that the estrous cycle significantly affected the freezing levels of females in our study.

Inter-individual differences have been well-documented for pain and pain modulation [5]. Neurobiological mechanisms, including emotional network plasticity, may link pain and fear [11]. This relationship has been explored with regard to the corticolimbic system [73,74], and in particular, the amygdala, a limbic structure that has emerged as a key player in both fear and anxiety networks [75–78] and in the emotional-affective dimensions of pain and pain modulation [8,10,79]. Corticolimbic characteristics involving the amygdala determine the risk of chronic pain and mediate the effects of depression and negative affect on chronic pain [80]. Human studies have investigated the role of the amygdala in pain and fear interactions [73,81], and the amygdala has been implicated in fear-conditioned analgesia in a preclinical setting [82–84]. The amygdala has been implicated in pain-like behaviors [85–87], anxiety-like behaviors [88–90], and in fear learning [12,91]. Pain-related neuroplastic changes lead to hyperexcitability in amygdala output neurons [10], driving pain behaviors in both acute [92,93] and chronic [29,94] pain models. Sex differences with regard to pain conditions have long been recognized, with females greatly outnumbering males as chronic pain patients [6]. However, sex differences in pain-related amygdala neuroplasticity are largely unknown, though one clinical study reported sex differences in resting-state amygdala subnuclei connectivity patterns as a potential explanation for the increased prevalence of conditions of negative affect in women [95]. Even less has been explored about inter-individual and sex differences in fear learning and FE with regard to pain and pain modulation, though a clinical study reported sex differences in pain-related fear conditioning [96]. Ultrasonic vocalizations were previously associated with increased neuronal activity in brain regions regulating fear and anxiety, including the amygdala [97], and have been demonstrated to be an effective indicator of emotional status in pain models [20,21,27]. As ultrasonic vocalizations demonstrated the most striking inter-individual and sex differences in pain-related behaviors for both an acute and a chronic model, insight into potential sexual dimorphisms of pain-related amygdala neuroplasticity is warranted.

The intricate relationship between pain modulation and fear neurocircuitry and mechanisms, particularly in relation to potential discrepancies regarding sex differences, led us to test the hypothesis that FE learning ability can predict pain-related behaviors in both acute (arthritis) and chronic neuropathic (SNL) models of pain, and that these behaviors may differ between males and females. In the present study, distinct behavioral phenotypes differed according to sex in their FE but not fear learning ability. There were no differences at baseline between mechanosensitivity (spinal reflex thresholds) and emotional-affective responses (vocalizations), but females exhibited increased baseline anxiety-like behavior (OFT) compared to males in both the untreated and sham-treated control groups (see Figures 3D and 4D). This confirms findings from the literature that males spent the same or increased time in the center of the OFT compared to females at baseline [3,98], though one study found no sex difference in OFT anxiety-like behavior in a chronic spinal nerve transection pain model [99]. FE+ and FE− phenotypes showed differences in the magnitude of emotional-effective responses not only in the neuropathic pain model, as we previously reported [42], but also in the arthritis pain model (see Figure 3B,C and Figure 4B,C). Additionally, females exhibited significantly increased audible and ultrasonic components of vocalizations compared to males in both of the tested pain models. To the best of our knowledge, sex differences in pain-related vocalizations in the context of FE learning have not been reported. One preclinical study found that male rats vocalized more than female rats despite females exhibiting lower freezing levels during FE, although this effect was strain-specific and did not include any pain models [100]. The novelty here is the identification of sex-specific differences in behavioral phenotypes, which corresponds to sexual dimorphisms in pain-related vocalizations regardless of pain model.

Sonic vocalizations, if emitted with large force and volume, may produce overtones that reach into the ultrasonic frequency range. A note of consideration in the present study is that the ultrasonic components (harmonics) of audible vocalizations presented here cannot be regarded as true ultrasonic aversive vocalizations as rats cannot emit sonic and ultrasonic calls at the same time. However, our results show that harmonic components of vigorous audible vocalizations showed an interesting harmonic spectrum, possibly with additional overtones. Because some of the overtones may depart from the whole multiples of the fundamental frequency, the harmonics and overtones show reinforcement at higher frequencies, creating ultrasonic components of the audible calls that are clearly visible in the spectrograms. Ultrasonic components of vocalizations are of long duration, consistent with the duration of audible calls. Simultaneous audible and ultrasonic vocalization components were demonstrated in response to an acute painful stimulus (tail snip) [101]. Ultrasonic harmonics that were previously reported demonstrated a different duration and lower frequency than presented here [102]. Though the emission of 22 kHz ultrasonic vocalizations has been reported to occur after a significant delay [103–105], in this study, both audible and ultrasonic components were evoked by a continuously present mechanical stimulus for 10 s as opposed to the brief electrical stimuli used in other studies. Repeated vocalizations may be triggered by the continuous noxious stimulus, and thus, latency assessment is not possible with this approach. The use of audible vocalizations in both the audible and ultrasonic ranges, particularly in correlation with other behavioral measures, is a useful measure of pain levels and emotional responses to pain.

The current study provides the rationale for the inter-individual- and sex-specific analysis of synaptic and cellular mechanisms within the amygdala. Future research may address neuroplastic differences between males and females in the context of pain and fear learning, potentially providing insight into the increased prevalence of anxiety, PTSD, and pain in female patients and supporting patient-specific therapeutic strategies for these disorders [15].

5. Conclusions

The data may suggest sexual dimorphisms in FE learning ability that have a predictive value for pain-related behavioral changes, particularly among emotional-affective pain aspects, in both an acute and a chronic pain model. Rats with weak FE learning ability showed an increased magnitude of both arthritic- and neuropathic-pain related affective rather than sensory behaviors, with females demonstrating greater inter-individual differences in affective pain behaviors than males. Vocalizations are strong indicators of inter-individual and sex differences in pain models, particularly in chronic neuropathic pain, whereas no such differences were found for mechanosensitivity, and anxiety-like behaviors showed only baseline differences. The increased correlation between FE learning ability and affective pain-related behaviors in female compared to male rats may be facilitated by amygdala pain mechanisms, though further investigation into sex-specific synaptic and cellular neurobiological mechanisms is warranted.

Author Contributions: Conceptualization, V.N., P.P. and G.J.; methodology, V.N., P.P. and G.J.; validation, P.P., G.J., R.J. and Z.G.; formal analysis, P.P., G.J., R.J., Z.G. and V.N.; investigation, P.P., G.J., R.J. and Z.G.; resources, V.N.; writing—original draft preparation, P.P.; writing—review and editing, V.N. and P.P.; visualization, P.P. and V.N.; supervision, V.N. and G.J.; project administration, V.N. and G.J.; funding acquisition, V.N. and G.J. All authors have read and agreed to the published version of the manuscript.

Funding: This research was funded by the National Institutes of Health, grant number R01 NS038261, R01 NS106902, R01NS118731, R01 NS120395, and R01 NS109255.

Institutional Review Board Statement: Experimental procedures were approved by the Institutional Animal Care and Use Committee (IACUC; protocol #14006, currently approved through 20 June 2022) at Texas Tech University Health Sciences Center.

Informed Consent Statement: Not applicable.

Data Availability Statement: All data generated or analyzed during this study are included in this published article. Data files used for this manuscript are available via a direct and reasonable request to the corresponding author and approval from Texas Tech University Health Sciences Center (TTUHSC).

Conflicts of Interest: The authors declare no conflict of interest.

References

1. Burke, N.N.; Coppinger, J.; Deaver, D.R.; Roche, M.; Finn, D.P.; Kelly, J. Sex differences and similarities in depressive- and anxiety-like behaviour in the Wistar-Kyoto rat. *Physiol. Behav.* **2016**, *167*, 28–34. [CrossRef]
2. Liu, J.; Scira, J.; Donaldson, S.; Kajiji, N.; Dash, G.H.; Donaldson, S.T. Sex and Trait Anxiety Differences in Psychological Stress are Modified by Environment. *Neuroscience* **2018**, *383*, 178–190. [CrossRef] [PubMed]
3. Scholl, J.L.; Afzal, A.; Fox, L.C.; Watt, M.J.; Forster, G.L. Sex differences in anxiety-like behaviors in rats. *Physiol. Behav.* **2019**, *211*, 112670–112678. [CrossRef] [PubMed]
4. Coghill, R.C.; McHaffie, J.G.; Yen, Y.-F. Neural correlates of interindividual differences in the subjective experience of Pain. *Proc. Natl. Acad. Sci. USA* **2003**, *100*, 8538–8542. [CrossRef] [PubMed]
5. Fillingim, R.B. Individual Differences in Pain: Understanding the Mosaic that Makes Pain Personal. *Pain* **2017**, *158* (Suppl. 1), S11–S18. [CrossRef]
6. Sorge, R.E.; Totsch, S.K. Sex Differences in Pain. *J. Neurosci. Res.* **2017**, *95*, 1271–1281. [CrossRef]
7. Moriarty, O.; McGuire, B.E.; Finn, D.P. The effect of pain on cognitive function: A review of clinical and preclinical research. *Prog. Neurobiol.* **2011**, *93*, 385–404. [CrossRef]
8. Veinante, P.; Yalcin, I.; Barrot, M. The amygdala between sensation and affect: A role in Pain. *J. Mol. Psychiatry* **2013**, *1*, 9. [CrossRef]
9. Vachon-Presseau, E.; Centeno, M.V.; Ren, W.; Berger, S.E.; Tétreault, P.; Ghantous, M.; Baria, A.; Farmer, M.; Baliki, M.N.; Schnitzer, T.J.; et al. The Emotional Brain as a Predictor and Amplifier of Chronic Pain. *J. Dent. Res.* **2016**, *95*, 605–612. [CrossRef]

10. Neugebauer, V. Amygdala physiology in Pain. *Handb. Behav. Neurosci.* **2020**, *26*, 101–113.
11. Elman, I.; Borsook, D. Threat Response System: Parallel Brain Processes in Pain vis-à-vis Fear and Anxiety. *Front. Psychiatry* **2018**, *9*, 29. [CrossRef]
12. Palmiter, R.D. The parabrachial nucleus: CGRP neurons function as a general alarm. *Trends Neurosci.* **2018**, *41*, 280–293. [CrossRef] [PubMed]
13. Marin, M.F.; Milad, M.R. Extending the Examination of the Fear Extinction Network Beyond Anxiety and Fear-Based Disorders: Insight into Autism Spectrum Disorder. *Biol. Psychiatry Cogn. Neurosci. Neuroimaging* **2016**, *1*, 302–304. [CrossRef] [PubMed]
14. McLaughlin, N.C.R.; Strong, D.; Abrantes, A.; Garnaat, S.; Cerny, A.; O'Connell, C.; Fadok, R.; Spofford, C.; Rasmussen, S.A.; Milad, M.R.; et al. Extinction retention and fear renewal in a lifetime obsessive–compulsive disorder sample. *Behav. Brain Res.* **2015**, *280*, 72–77. [CrossRef] [PubMed]
15. Milad, M.R.; Quirk, G.J. Fear Extinction as a Model for Translational Neuroscience: Ten Years of Progress. *Annu. Rev. Psychol.* **2012**, *63*, 129–151. [CrossRef] [PubMed]
16. King, G.; Scott, E.; Graham, B.M.; Richardson, R. Individual differences in fear extinction and anxiety-like behavior. *Learn. Mem.* **2017**, *24*, 182–190. [CrossRef] [PubMed]
17. Lonsdorf, T.B.; Merz, C.J. More than just noise: Inter-individual differences in fear acquisition, extinction and return of fear in humans—Biological, experiential, temperamental factors, and methodological pitfalls. *Neurosci. Biobehav. Rev.* **2017**, *80*, 703–728. [CrossRef]
18. Hånell, A.; Marklund, N. Structured evaluation of rodent behavioral tests used in drug discovery research. *Front. Behav. Neurosci.* **2014**, *8*, 252. [CrossRef]
19. Portfors, C.V. Types and functions of ultrasonic vocalizations in laboratory rats and mice. *J. Am. Assoc. Lab. Anim. Sci.* **2007**, *46*, 28–34. [PubMed]
20. Han, J.S.; Bird, G.C.; Li, W.; Jones, J.; Neugebauer, V. Computerized analysis of audible and ultrasonic vocalizations of rats as a standardized measure of pain-related behavior. *J. Neurosci. Methods* **2005**, *141*, 261–269. [CrossRef]
21. Neugebauer, V.; Han, J.S.; Adwanikar, H.; Fu, Y.; Ji, G. Techniques for assessing knee joint pain in arthritis. *Mol. Pain* **2007**, *3*, 8. [CrossRef]
22. Piel, M.J.; Kroin, J.S.; Wijnen AJ van Kc, R.; Im, H.-J. Pain assessment in animal models of osteoarthritis. *Gene* **2014**, *537*, 184–188. [CrossRef]
23. Kim, H.; Thompson, J.; Ji, G.; Ganapathy, V.; Neugebauer, V. Monomethyl fumarate (MMF) inhibits pain behaviors and amygdala activity in a rat arthritis model. *Pain* **2017**, *158*, 2376–2385. [CrossRef]
24. Han, J.S.; Neugebauer, V. mGluR1 and mGluR5 antagonists in the amygdala inhibit different components of audible and ultrasonic vocalizations in a model of arthritic Pain. *Pain* **2005**, *113*, 211–222. [CrossRef]
25. Kurejova, M.; Nattenmüller, U.; Hildebrandt, U.; Selvaraj, D.; Stösser, S.; Kuner, R. An improved behavioural assay demonstrates that ultrasound vocalizations constitute a reliable indicator of chronic cancer pain and neuropathic Pain. *Mol. Pain* **2010**, *6*, 18. [CrossRef] [PubMed]
26. Sevcik, M.A.; Jonas, B.M.; Lindsay, T.H.; Halvorson, K.G.; Ghilardi, J.R.; Kuskowski, M.A.; Mukherjee, P.; Maggio, J.E.; Mantyh, P.W. Endogenous Opioids Inhibit Early Stage Pancreatic Pain in a Mouse Model of Pancreatic Cancer. *Gastroenterology* **2006**, *131*, 900–910. [CrossRef] [PubMed]
27. Burgdorf, J.S.; Ghoreishi-Haack, N.; Cearley, C.N.; Kroes, R.A.; Moskal, J.R. Rat ultrasonic vocalizations as a measure of the emotional component of chronic Pain. *Neuroreport* **2019**, *30*, 863–866. [CrossRef] [PubMed]
28. Thompson, J.M.; Yakhnitsa, V.; Ji, G.; Neugebauer, V. Small conductance calcium activated potassium (SK) channel dependent and independent effects of riluzole on neuropathic pain-related amygdala activity and behaviors in rats. *Neuropharmacology* **2018**, *138*, 219–231. [CrossRef] [PubMed]
29. Ji, G.; Zhang, W.; Mahimainathan, L.; Narasimhan, M.; Kiritoshi, T.; Fan, X.; Wang, J.; Green, T.A.; Neugebauer, V. 5-HT2C Receptor Knockdown in the Amygdala Inhibits Neuropathic-Pain-Related Plasticity and Behaviors. *J. Neurosci.* **2017**, *37*, 1378–1393. [CrossRef]
30. Abraham, A.D.; Leung, E.J.Y.; Wong, B.A.; Rivera, Z.M.G.; Kruse, L.C.; Clark, J.J.; Land, B.B. Orally consumed cannabinoids provide long-lasting relief of allodynia in a mouse model of chronic neuropathic Pain. *Neuropsychopharmacology* **2020**, *45*, 1105–1114. [CrossRef]
31. Jourdan, D.; Ardid, D.; Eschalier, A. Analysis of ultrasonic vocalisation does not allow chronic pain to be evaluated in rats. *Pain* **2002**, *95*, 165–173. [CrossRef]
32. Knutson, B.; Burgdorf, J.; Panksepp, J. Ultrasonic vocalizations as indices of affective states in rats. *Psychol. Bull.* **2002**, *128*, 961–977. [CrossRef] [PubMed]
33. Helmer, L.M.L.; Weijenberg, R.A.F.; de Vries Achterberg, R.W.P.; Lautenbacher, S.; Sampson, E.L.; Lobbezoo, F. Crying out in pain—A systematic review into the validity of vocalization as an indicator for Pain. *Eur. J. Pain* **2020**, *24*, 1703–1715. [CrossRef] [PubMed]
34. Neugebauer, V.; Li, W.; Bird, G.C.; Bhave, G.; Gereau, R.W., IV. Synaptic Plasticity in the Amygdala in a Model of Arthritic Pain: Differential Roles of Metabotropic Glutamate Receptors 1 and 5. *J. Neurosci.* **2003**, *23*, 52–63. [CrossRef]
35. Grégoire, S.; Neugebauer, V. 5-HT2CR blockade in the amygdala conveys analgesic efficacy to SSRIs in a rat model of arthritis Pain. *Mol. Pain* **2013**, *9*, 41. [CrossRef]

36. Kiritoshi, T.; Neugebauer, V. Group II mGluRs modulate baseline and arthritis pain-related synaptic transmission in the rat medial prefrontal cortex. *Neuropharmacology* **2015**, *95*, 388–394. [CrossRef] [PubMed]
37. Mazzitelli, M.; Neugebauer, V. Amygdala group II mGluRs mediate the inhibitory effects of systemic group II mGluR activation on behavior and spinal neurons in a rat model of arthritis Pain. *Neuropharmacology* **2019**, *158*, 107706. [CrossRef] [PubMed]
38. Mazzitelli, M.; Marshall, K.; Pham, A.; Ji, G.; Neugebauer, V. Optogenetic Manipulations of Amygdala Neurons Modulate Spinal Nociceptive Processing and Behavior Under Normal Conditions and in an Arthritis Pain Model. *Front. Pharmacol.* **2021**, *12*, 25. [CrossRef]
39. Bennett, G.J.; Chung, J.M.; Honore, M.; Seltzer, Z. Models of Neuropathic Pain in the Rat. *Curr. Protoc. Neurosci.* **2003**, *22*, 9.14.1–9.14.16. [CrossRef]
40. Dineley, K.T.; Kayed, R.; Neugebauer, V.; Fu, Y.; Zhang, W.; Reese, L.C.; Taglialatela, T. Amyloid-beta oligomers impair fear conditioned memory in a calcineurin-dependent fashion in mice. *J. Neurosci. Res.* **2010**, *88*, 2923–2932.
41. Martin, Z.S.; Neugebauer, V.; Dineley, K.T.; Kayed, R.; Zhang, W.; Reese, L.C.; Taglialatela, T. Alpha synuclein oligomers oppose long-term potentiation and impair memory through a calcineurin-dependent mechanism: Relevance to human synucleopathic diseases. *J. Neurochem.* **2012**, *120*, 440–452. [CrossRef] [PubMed]
42. Ji, G.; Yakhnitsa, V.; Kiritoshi, T.; Presto, P.; Neugebauer, V. Fear extinction learning ability predicts neuropathic pain behaviors and amygdala activity in male rats. *Mol. Pain* **2018**, *14*, 1744806918804441. [CrossRef] [PubMed]
43. Han, J.S.; Li, W.; Neugebauer, V. Critical Role of Calcitonin Gene-Related Peptide 1 Receptors in the Amygdala in Synaptic Plasticity and Pain Behavior. *J. Neurosci.* **2005**, *25*, 10717–10728. [CrossRef] [PubMed]
44. Adwanikar, H.; Ji, G.; Li, W.; Doods, H.; Willis, W.D.; Neugebauer, V. Spinal CGRP1 receptors contribute to supraspinally organized pain behavior and pain-related sensitization of amygdala neurons. *Pain* **2007**, *132*, 53–66. [CrossRef]
45. Palazzo, E.; Fu, Y.; Ji, G.; Maione, S.; Neugebauer, V. Group III mGluR7 and mGluR8 in the amygdala differentially modulate nocifensive and affective pain behaviors. *Neuropharmacology* **2008**, *55*, 537–545. [CrossRef]
46. Hein, M.; Ji, G.; Tidwell, D.; D'Souza, P.; Kiritoshi, T.; Yakhnitsa, V.; Navratilova, E.; Porreca, F.; Neugebauer, V. Kappa opioid receptor activation in the amygdala disinhibits CRF neurons to generate pain-like behaviors. *Neuropharmacology* **2021**, *185*, 108456. [CrossRef]
47. Prut, L.; Belzung, C. The open field as a paradigm to measure the effects of drugs on anxiety-like behaviors: A review. *Eur. J. Pharmacol.* **2003**, *463*, 3–33. [CrossRef]
48. Seibenhener, M.L.; Wooten, M.C. Use of the Open Field Maze to Measure Locomotor and Anxiety-like Behavior in Mice. *J. Vis. Exp.* **2015**, *96*, e52434. [CrossRef]
49. Araya, E.I.; Baggio, D.F.; de Koren, L.O.; Andreatini, R.; Schwarting, R.K.W.; Zamponi, G.W.; Chichorro, J.G. Acute orofacial pain leads to prolonged changes in behavioral and affective pain components. *Pain* **2020**, *161*, 2830–2840. [CrossRef]
50. Bush, D.E.A.; Sotres-Bayon, F.; LeDoux, J.E. Individual differences in fear: Isolating fear reactivity and fear recovery phenotypes. *J. Trauma. Stress* **2007**, *20*, 413–422. [CrossRef]
51. Reznikov, R.; Diwan, M.; Nobrega, J.N.; Hamani, C. Towards a better preclinical model of PTSD: Characterizing animals with weak extinction, maladaptive stress responses and low plasma corticosterone. *J. Psychiatr. Res.* **2015**, *61*, 158–165. [CrossRef]
52. Ueno, M.; Yamada, K.; Ichitani, Y. The relationship between fear extinction and resilience to drug-dependence in rats. *Neurosci. Res.* **2017**, *121*, 37–42. [CrossRef] [PubMed]
53. Duits, P.; Richter, J.; Baas, J.M.P.; Engelhard, I.M.; Limberg-Thiesen, A.; Heitland, I.; Hamm, A.O.; Cath, D.C. Enhancing effects of contingency instructions on fear acquisition and extinction in anxiety disorders. *J. Abnorm. Psychol.* **2017**, *126*, 378–391. [CrossRef] [PubMed]
54. Lissek, S.; Powers, A.S.; McClure, E.B.; Phelps, E.A.; Woldehawariat, G.; Grillon, C.; Pine, D.S. Classical fear conditioning in the anxiety disorders: A meta-analysis. *Behav. Res. Ther.* **2005**, *43*, 1391–1424. [CrossRef] [PubMed]
55. Stegmann, Y.; Schiele, M.A.; Schümann, D.; Lonsdorf, T.B.; Zwanzger, P.; Romanos, M.; Reif, A.; Domschke, K.; Deckert, J.; Gamer, M.; et al. Individual differences in human fear generalization—Pattern identification and implications for anxiety disorders. *Transl. Psychiatry* **2019**, *9*, 307. [CrossRef]
56. Milad, M.R.; Rauch, S.L.; Pitman, R.K.; Quirk, G.J. Fear extinction in rats: Implications for human brain imaging and anxiety disorders. *Biol. Psychol.* **2006**, *73*, 61–71. [CrossRef]
57. Milad, M.R.; Orr, S.P.; Lasko, N.B.; Chang, Y.; Rauch, S.L.; Pitman, R.K. Presence and Acquired Origin of Reduced Recall for Fear Extinction in PTSD: Results of a Twin Study. *J. Psychiatr. Res.* **2008**, *42*, 515–520. [CrossRef] [PubMed]
58. Milad, M.R.; Furtak, S.C.; Greenberg, J.L.; Keshaviah, A.; Im, J.J.; Falkenstein, M.J.; Jenike, M.; Rauch, S.L.; Wilhelm, S. Deficits in conditioned fear extinction in obsessive-compulsive disorder and neurobiological changes in the fear circuit. *JAMA Psychiatry* **2013**, *70*, 608–618. [CrossRef]
59. Duits, P.; Cath, D.C.; Lissek, S.; Hox, J.J.; Hamm, A.O.; Engelhard, I.M.; Van Den Hout, M.A.; Baas, J.M. Updated meta-analysis of classical fear conditioning in the anxiety disorders. *Depress. Anxiety* **2015**, *32*, 239–253. [CrossRef]
60. Galea, S.; Nandi, A.; Vlahov, D. The epidemiology of post-traumatic stress disorder after disasters. *Epidemiol. Rev.* **2005**, *27*, 78–91. [CrossRef]
61. Crofford, L.J. Chronic Pain: Where the Body Meets the Brain. *Trans. Am. Clin. Climatol. Assoc.* **2015**, *126*, 167.
62. Altemus, M.; Sarvaiya, N.; Epperson, C.N. Sex differences in anxiety and depression clinical perspectives. *Front. Neuroendocrinol.* **2014**, *35*, 320–330. [CrossRef]

63. Donner, N.C.; Lowry, C.A. Sex differences in anxiety and emotional behavior. *Pflugers Arch. Eur. J. Psihol.* **2013**, *465*, 601–626. [CrossRef]
64. Gruene, T.M.; Roberts, E.; Thomas, V.; Ronzio, A.; Shansky, R.M. Sex-specific neuroanatomical correlates of fear expression inprefrontal-amygdala circuits. *Biol. Psychiatry* **2015**, *78*, 186–193. [CrossRef] [PubMed]
65. Baran, S.E.; Armstrong, C.E.; Niren, D.C.; Hanna, J.J.; Conrad, C.D. Chronic Stress and Sex Differences on the Recall of Fear Conditioning and Extinction. *Neurobiol. Learn. Mem.* **2009**, *91*, 323–332. [CrossRef]
66. Baran, S.E.; Armstrong, C.E.; Niren, D.C.; Conrad, C.D. Prefrontal cortex lesions and sex differences in fear extinction and perseveration. *Learn. Mem.* **2010**, *17*, 267–278. [CrossRef]
67. Maren, S.; De Oca, B.; Fanselow, M.S. Sex differences in hippocampal long-term potentiation (LTP) and Pavlovian fear conditioning in rats: Positive correlation between LTP and contextual learning. *Brain Res.* **1994**, *661*, 25–34. [CrossRef]
68. Barker, J.M.; Galea, L.A.M. Males show stronger contextual fear conditioning than females after context pre-exposure. *Physiol. Behav.* **2010**, *99*, 82–90. [CrossRef] [PubMed]
69. Daviu, N.; Andero, R.; Armario, A.; Nadal, R. Sex differences in the behavioural and hypothalamic–pituitary–adrenal response to contextual fear conditioning in rats. *Horm. Behav.* **2014**, *66*, 713–723. [CrossRef] [PubMed]
70. Willadsen, M.; Uengoer, M.; Sługocka, A.; Schwarting, R.K.W.; Homberg, J.R.; Wöhr, M. Fear Extinction and Predictive Trait-Like Inter-Individual Differences in Rats Lacking the Serotonin Transporter. *Int. J. Mol. Sci.* **2021**, *22*, 7088. [CrossRef]
71. Glover, E.M.; Jovanovic, T.; Norrholm, S.D. Estrogen and Extinction of Fear Memories: Implications for Posttraumatic Stress Disorder Treatment. *Biol. Psychiatry* **2015**, *78*, 178–185. [CrossRef] [PubMed]
72. Lebron-Milad, K.; Milad, M.R. Sex differences, gonadal hormones and the fear extinction network: Implications for anxiety disorders. *Biol. Mood Anxiety Disord.* **2012**, *2*, 3. [CrossRef] [PubMed]
73. Meier, M.L.; Stämpfli, P.; Humphreys, B.K.; Vrana, A.; Seifritz, E.; Schweinhardt, P. The impact of pain-related fear on neural pathways of pain modulation in chronic low back Pain. *Pain Rep.* **2017**, *2*, e601. [CrossRef] [PubMed]
74. Nees, F.; Becker, S. Psychological Processes in Chronic Pain: Influences of Reward and Fear Learning as Key Mechanisms—Behavioral Evidence, Neural Circuits, and Maladaptive Changes. *Neuroscience* **2018**, *387*, 72–84. [CrossRef] [PubMed]
75. Phelps, E.A.; LeDoux, J.E. Contributions of the amygdala to emotion processing: From animal models to human behavior. *Neuron* **2005**, *48*, 175–187. [CrossRef] [PubMed]
76. Orsini, C.A.; Maren, S. Neural and Cellular Mechanisms of Fear and Extinction Memory Formation. *Neurosci. Biobehav. Rev.* **2012**, *36*, 1773–1802. [CrossRef]
77. Herry, C.; Johansen, J.P. Encoding of fear learning and memory in distributed neuronal circuits. *Nat. Neurosci.* **2014**, *17*, 1644–1654. [CrossRef]
78. Hartley, C.A.; Phelps, E.A. Changing Fear: The Neurocircuitry of Emotion Regulation. *Neuropsychopharmacology* **2010**, *35*, 136–146. [CrossRef]
79. Thompson, J.M.; Neugebauer, V. Cortico-limbic pain mechanisms. *Neurosci. Lett.* **2019**, *702*, 15–23. [CrossRef]
80. Vachon-Presseau, E.; Tétreault, P.; Petre, B.; Huang, L.; Berger, S.E.; Torbey, S.; Baria, A.T.; Mansour, A.R.; Hashmi, J.A.; Griffith, J.W.; et al. Corticolimbic anatomical characteristics predetermine risk for chronic Pain. *Brain* **2016**, *139*, 1958. [CrossRef]
81. Meier, M.L.; Stämpfli, P.; Vrana, A.; Humphreys, B.K.; Seifritz, E.; Hotz-Boendermaker, S. Neural Correlates of Fear of Movement in Patients with Chronic Low Back Pain vs. Pain-Free Individuals. *Front. Hum. Neurosci.* **2016**, *10*, 386. [CrossRef]
82. Butler, R.K.; Ehling, S.; Barbar, M.; Thomas, J.; Hughes, M.A.; Smith, C.E.; Pogorelov, V.M.; Aryal, D.K.; Wetsel, W.C.; Lascelles, B.D.X. Distinct neuronal populations in the basolateral and central amygdala are activated with acute pain, conditioned fear, and fear-conditioned analgesia. *Neurosci. Lett.* **2017**, *661*, 11–17. [CrossRef]
83. Rea, K.; Roche, M.; Finn, D.P. Modulation of Conditioned Fear, Fear-Conditioned Analgesia, and Brain Regional C-Fos Expression Following Administration of Muscimol into the Rat Basolateral Amygdala. *J. Pain* **2011**, *12*, 712–721. [CrossRef]
84. Rea, K.; Olango, W.M.; Harhen, B.; Kerr, D.M.; Galligan, R.; Fitzgerald, S.; Moore, M.; Roche, M.; Finn, D.P. Evidence for a role of GABAergic and glutamatergic signalling in the basolateral amygdala in endocannabinoid-mediated fear-conditioned analgesia in rats. *Pain* **2013**, *154*, 576–585. [CrossRef] [PubMed]
85. Neugebauer, V.; Li, W.; Bird, G.C.; Han, J.S. The amygdala and persistent Pain. *Neuroscientist* **2004**, *10*, 221–234. [CrossRef] [PubMed]
86. Thompson, J.M.; Neugebauer, V. Amygdala Plasticity and Pain. *Pain Res. Manag.* **2017**, *2017*, 1–12. [CrossRef]
87. Liang, S.-H.; Zhao, W.-J.; Yin, J.-B.; Chen, Y.-B.; Li, J.-N.; Feng, B.; Lu, Y.-C.; Wang, J.; Dong, Y.-L.; Li, Y.-Q. A Neural Circuit from Thalamic Paraventricular Nucleus to Central Amygdala for the Facilitation of Neuropathic Pain. *J. Neurosci.* **2020**, *40*, 7837–7854. [CrossRef]
88. Missig, G.A.; Roman, C.W.; Vizzard, M.A.; Braas, K.M.; Hammack, S.E.; May, V. Parabrachial nucleus (PBn) pituitary adenylate cyclase activating polypeptide (PACAP) signaling in the amygdala: Implication for the sensory and behavioral effects of Pain. *Neuropharmacology* **2014**, *86*, 38–48. [CrossRef] [PubMed]
89. Ahrens, S.; Wu, M.V.; Furlan, A.; Hwang, G.-R.; Paik, R.; Li, H.; Penzo, M.A.; Tollkuhn, J.; Li, B. A Central Extended Amygdala Circuit That Modulates Anxiety. *J. Neurosci.* **2018**, *38*, 5567–5583. [CrossRef]
90. Paretkar, T.; Dimitrov, E. The central amygdala corticotropin-releasing hormone (CRH) neurons modulation of anxiety-like behavior and hippocampus-dependent memory in mice. *Neuroscience* **2018**, *390*, 187–197. [CrossRef]

91. Penzo, M.A.; Robert, V.; Tucciarone, J.; De Bundel, D.; Wang, M.; Van Aelst, L.; Darvas, M.; Parada, L.F.; Palmiter, R.D.; He, M.; et al. The paraventricular thalamus controls a central amygdala fear circuit. *Nature* **2015**, *519*, 455–459. [CrossRef] [PubMed]
92. Neugebauer, V.; Li, W. Differential sensitization of amygdala neurons to afferent inputs in a model of arthritic Pain. *J. Neurophysiol.* **2003**, *89*, 716–727. [CrossRef] [PubMed]
93. Kiritoshi, T.; Neugebauer, V. Pathway-specific alterations of cortico-amygdala transmission in an arthritis pain model. *ACS Chem. Neurosci.* **2018**, *9*, 2252–2261. [CrossRef] [PubMed]
94. Jiang, H.; Fang, D.; Kong, L.-Y.; Jin, Z.-R.; Cai, J.; Kang, X.-J.; Wan, Y.; Xing, G.-G. Sensitization of neurons in the central nucleus of the amygdala via the decreased GABAergic inhibition contributes to the development of neuropathic pain-related anxiety-like behaviors in rats. *Mol. Brain* **2014**, *7*, 72. [CrossRef] [PubMed]
95. Engman, J.; Linnman, C.; Van Dijk, K.R.A.; Milad, M.R. Amygdala subnuclei resting-state functional connectivity sex and estrogen differences. *Psychoneuroendocrinology* **2016**, *63*, 34–42. [CrossRef] [PubMed]
96. Labrenz, F.; Knuf-Rtveliashvili, S.; Elsenbruch, S. Sex Differences Linking Pain-Related Fear and Interoceptive Hypervigilance: Attentional Biases to Conditioned Threat and Safety Signals in a Visceral Pain Model. *Front. Psychiatry* **2020**, *11*, 197. [CrossRef]
97. Wöhr, M.; Schwarting, R.K.W. Affective communication in rodents: Ultrasonic vocalizations as a tool for research on emotion and motivation. *Cell Tissue Res.* **2013**, *354*, 81–97. [CrossRef]
98. Knight, P.; Chellian, R.; Wilson, R.; Behnood-Rod, A.; Panunzio, S.; Bruijnzeel, A.W. Sex differences in the elevated plus-maze test and large open field test in adult Wistar rats. *Pharmacol. Biochem. Behav.* **2021**, *204*, 173168. [CrossRef]
99. Won, S.; Park, K.; Lim, H.; Lee, S.J. Sexual dimorphism in cognitive disorders in a murine model of neuropathic Pain. *Behav. Brain Funct.* **2020**, *16*, 1. [CrossRef]
100. Graham, L.K.; Yoon, T.; Lee, H.J.; Kim, J.J. Strain and sex differences in fear conditioning: 22 kHz ultrasonic vocalizations and freezing in rats. *Psychol. Neurosci.* **2009**, *2*, 219–225. [CrossRef]
101. Williams, W.O.; Riskin, D.K.; Mott, K.M. Ultrasonic Sound as an Indicator of Acute Pain in Laboratory Mice. *J. Am. Assoc. Lab. Anim. Sci.* **2008**, *47*, 8. [PubMed]
102. Jourdan, D.; Ardid, D.; Chapuy, E.; Eschalier, A.; Le Bars, D. Audible and ultrasonic vocalization elicited by single electrical nociceptive stimuli to the tail in the rat. *Pain* **1995**, *63*, 237–249. [CrossRef]
103. Tsuzuki, H.; Maekawa, M.; Konno, R.; Hori, Y. Functional roles of endogenous D-serine in pain-induced ultrasonic vocalization. *Neuroreport* **2012**, *23*, 937–941. [CrossRef]
104. Oliveira, A.R.; Barros, H.M.T. Ultrasonic rat vocalizations during the formalin test: A measure of the affective dimension of pain? *Anesth. Analg.* **2006**, *102*, 832–839. [CrossRef] [PubMed]
105. Brudzynski, S.M.; Ociepa, D. Ultrasonic vocalization of laboratory rats in response to handling and touch. *Physiol. Behav.* **1992**, *52*, 655–660. [CrossRef]

MDPI
St. Alban-Anlage 66
4052 Basel
Switzerland
Tel. +41 61 683 77 34
Fax +41 61 302 89 18
www.mdpi.com

Brain Sciences Editorial Office
E-mail: brainsci@mdpi.com
www.mdpi.com/journal/brainsci

www.ingramcontent.com/pod-product-compliance
Lightning Source LLC
Chambersburg PA
CBHW061605220926
51850CB00005B/428